SHIP BUILDING, SALE AND FINANCE

MARITIME AND TRANSPORT LAW LIBRARY

MARITIME AND TRANSPORT LAW LIBRARY

Ship Building, Sale and Finance
edited by Barış Soyer and Andrew Tettenborn
(2016)

International Maritime Conventions
Volume Three
Protection of the Marine Environment
by Francesco Berlingieri
(2015)

International Maritime Conventions
Volume Two
Navigation, Securities, Limitation of Liability and Jurisdiction
by Francesco Berlingieri
(2014)

Maritime Law
3rd edition
edited by Yvonne Baatz
(2014)

Offshore Contracts and Liabilities
by Barış Soyer and Andrew Tettenborn
(2014)

Marine Insurance Fraud
by Barış Soyer
(2014)

International Maritime Conventions
Volume One
The Carriage of Goods and Passengers by Sea
by Francesco Berlingieri
(2014)

International Carriage of Goods by Road: CMR
6th Edition
by Malcolm A. Clarke
(2014)

The Maritime Labour Convention 2006: International Labour Law Redefined
edited by Jennifer Lavelle
(2013)

Modern Maritime Law:
Volume 2: Managing Risks and Liabilities
3rd Edition
by Aleka Mandaraka-Sheppard
(2013)

Modern Maritime Law:
Volume 1: Jurisdiction and Risks
3rd Edition
by Aleka Mandaraka-Sheppard
(2013)

Carriage of Goods by Sea, Land and Air: Uni-Modal and Multi-Modal Transport in The 21st Century
edited by Barış Soyer and Andrew Tettenborn
(2013)

The Law of Yachts and Yachting
by Filippo Lorenzon and Richard Coles
(2012)

Freight Forwarding and Multimodal Transport Contracts
2nd edition
by David A. Glass
(2012)

Marine Insurance Clauses
5th Edition
by N. Geoffrey Hudson, Tim Madge and Keith Sturges (2012)

Pollution at Sea: Law and Liability
edited by Barış Soyer and Andrew Tettenborn
(2012)

Contracts of Carriage by Air
2nd Edition
by Malcolm A. Clarke
(2012)

Places of Refuge: International Law and the CMI Draft Convention
by Eric Van Hooydonk
(2010)

Maritime Fraud and Piracy
2nd Edition
by Paul Todd
(2010)

The Carriage of Goods by Sea under the Rotterdam Rules
edited by D. Rhidian Thomas
(2010)

The International Law of the Shipmaster
by John A. C. Cartner, Richard P. Fiske and Tara L. Leiter
(2009)

The Modern Law of Marine Insurance: Volume 3
edited by D. Rhidian Thomas
(2009)

The Rotterdam Rules: A Practical Annotation
by Yvonne Baatz, Charles Debattista, Filippo Lorenzon, Andrew Serdy, Hilton Staniland and Michael Tsimplis
(2009)

The Evolving Law and Practice of Voyage Charters
Edited by D. Rhidian Thomas
(2009)

Risk and Liability in Air Law
by George Leloudas
(2009)

Legal Issues Relating to Time Charterparties
edited by D. Rhidian Thomas
(2008)

Contracts of Carriage by Land and Air
2nd Edition
by Malcolm A. Clarke and David Yates
(2008)

Bills of Lading and Bankers' Documentary Credits
4th Edition
by Paul Todd
(2007)

Liability Regimes in Contemporary Maritime Law
edited by D. Rhidian Thomas
(2007)

Marine Insurance: The Law in Transition
edited by D. Rhidian Thomas
(2006)

Commencement of Laytime
4th Edition
by Donald Davies
(2006)

General Average: Law and Practice
2nd Edition
by F. D. Rose
(2005)

War, Terror and Carriage by Sea
by Keith Michel
(2004)

Port State Control
2nd Edition
by Oya Ozcayir
(2004)

Modern Law of Marine Insurance: Volume Two
edited by D. Rhidian Thomas
(2002)

Commercial and Maritime Statutes
edited by Peter MacDonald Eggers and Simon Picken
(2002)

Bills of Lading: Law and Contracts
by Nicholas Gaskell, Regina Asariotis and Yvonne Baatz
(2000)

Shipbrokers and the Law
by Andrew Jamieson
(1997)

SHIP BUILDING, SALE AND FINANCE

EDITED BY

BARIŞ SOYER

Professor in Law

Director, Institute of International Shipping and Trade Law, Swansea University

ANDREW TETTENBORN

Professor in Law,

Institute of International Shipping and Trade Law, Swansea University

informa law
from Routledge

First published 2016
by Informa Law from Routledge
2 Park Square, Milton Park, Abingdon, Oxon OX14 4RN

And by Informa Law from Routledge
711 Third Avenue, New York, NY 10017

Informa Law from Routledge is an imprint of the Taylor & Francis Group, an Informa business

British Library Cataloguing in Publication Data
A catalogue record for this book is available from the British Library

Library of Congress Cataloguing in Publication Data
Ship building, sale, and finance / edited by Barıs Soyer and Andrew Tettenborn.
pages cm
Includes bibliographical references and index.
ISBN 978-1-138-84111-6 (hardback : alk. paper) – ISBN 978-1-315-68964-7 (ebook : alk. paper)
1. Shipbuilding contracts–England. 2. Ship mortgages–England. 3. Shipbuilding industry–Law and legislation–England. I. Soyer, Barıs, editor. II. Tettenborn, A. M., editor. III. Title: Shipbuilding, sale, and finance.
KD1642.S55S55 2016
343.4207'862382–dc23
2015006424

ISBN: 978-1-138-84111-6
eISBN: 978-1-315-68964-7

Typeset in Plantin by
Servis Filmsetting Ltd, Stockport, Cheshire

MIX
Paper from responsible sources
FSC www.fsc.org FSC® C013056

Printed and bound in Great Britain by
TJ International Ltd, Padstow, Cornwall

TABLE OF CONTENTS

AUTHORS' BIOGRAPHIES

EDITORS AND CONTRIBUTORS

Professor Barış Soyer, BA (Ankara), LLM, PhD (Southampton)
Director
Institute of International Shipping and Trade Law
Swansea University

Professor Soyer is the Director of the Institute of International Shipping and Trade Law at Swansea University. He is a member of the British Maritime Law Association and British Insurance Law Association. He is the author of *Warranties in Marine Insurance* published by Cavendish Publishing (2001) and *Marine Insurance Fraud* published by Informa Routledge in 2014, and of an extensive list of journal articles published in elite journals such as *Lloyd's Maritime and Commercial Law Quarterly*, *Journal of Business Law*, *Torts Law Journal* and *Journal of Contract Law*. His book on *Warranties in Marine Insurance* was the joint winner of the Cavendish Book Prize 2001 and was awarded the British Insurance Law Association Charitable Trust Book Prize in 2002 for its contribution to insurance literature. A new edition of this book was published in 2006 and a third edition is due in 2015. He is the also the general editor of the *Reforming Marine and Commercial Insurance Law* published by Informa in 2008 and joint editor of *Pollution at Sea: Law and Liability* published by Informa Law in 2012; *Carriage of Goods by Sea, Land and Air* published by Informa Routledge in 2013 and *Offshore Contracts and Liabilities* published by Informa Routledge in 2014. He has also contributed to various books including *Modern Law of Modern Insurance* volumes 2 and 3, *Marine Insurance: The Law in Transition*, *Liability Regimes in Contemporary Maritime Law* and *Legal Issues relating to Time Charterparties*. He is one of the editors of the *Journal of International Maritime Law* and is also on the editorial board of *Shipping and Trade Law* and *Baltic Maritime Law Quarterly*. He currently teaches Admiralty Law, Oil and Gas Law and Marine Insurance on the LLM Programme and also is the Head of the Postgraduate Legal Studies Department.

Professor Andrew Tettenborn, LLB (Cantab), MA
Professor in Law
Institute of International Shipping and Trade Law
Swansea University

Professor Andrew Tettenborn has been attached to the IISTL at Swansea Law School since 2010.

He has also taught at the universities of Cambridge, Exeter and Geneva, has written extensively on widespread aspects of private, shipping and commercial law, and sits on the editorial board of *Lloyd's Maritime and Commercial Law Quarterly*. He has held visiting positions at Melbourne University, the University of Connecticut and at Case Law School, Cleveland, Ohio.

He is author and co-author of books on torts, damages and maritime law, and of numerous articles and chapters on aspects of common law, commercial law and restitution.

CONTRIBUTORS

Professor Simon Baughen, MA (Oxon)
Professor of Shipping Law
Institute of International Shipping and Trade Law
Swansea University

Simon Baughen joined Swansea in September 2013 as Professor of Shipping Law. He had previously been a Reader at Bristol Law School, and yet earlier worked as a shipping lawyer in a London P&I Club and then as a solicitor with the former Horrocks & Co. Professor Baughen's principal research area is in shipping law; his book on the subject, called simply *Shipping Law*, has run to five editions and is well known to academics and students alike as by far the most learned and approachable work on the subject. A sixth edition is due in 2015. He is also the author of the very well-established practitioner's work *Summerskill on Laytime* (2013). His other main area of research is the regulation of the activities of multi-national corporations in the developing world. This is the theme of his forthcoming book *Holding Corporations to Account: Closing the Governance Gap* which is due to be published in 2015. He has previously touched on this theme in his book *International Trade and the Protection of the Environment* (2007) which deals with the relationship between trade and investment treaties and environmental protection. He has also written and taught extensively about the law of trusts.

Charles Buss
Partner
Watson, Farley & Williams LLP, London

Charles Buss joined Watson, Farley & Williams in 1995 and specialises in maritime litigation and arbitration. He regularly acts for lenders in relation to foreclosures, and in disputes of all kinds with borrowers, syndicates and third parties. He also acts for owners and charterers in operational disputes, with particular emphasis on charterparties, newbuildings contracts, MOAs, pool and management agreements and JVs, brokerage disputes and marine insurance claims.

Professor Dr Olivier Cachard
Emeritus Dean of the Faculty of Law, Université de Nancy, France
Head of the International and European Business Law programme
Head of the IDIC-Institute François Gény

Professor Cachard graduated in 1997 at the Université Paris Panthéon-Assas, where in 2001 he obtained his PhD *cum laude* for his doctoral thesis. After his success at the *agrégation de droit privé*, he was appointed as a tenured Professor of Law at the Faculty of Law in Nancy in 2003. He lectures in International Arbitration Law, Admiralty and Transportation Law and Conflict of Laws.

Professor Cachard has held visiting appointments in a number of universities, including Heidelberg, Case Western Reserve (Cleveland, Ohio), and Rikkyo (Tokyo). He was invited as a Director of Studies for the 2009 session at the Hague Academy of International Law and in addition has lectured in Aircraft Passenger Transportation at its 2014 session.

He is an active arbitrator in international commercial disputes and in admiralty law disputes, taking regular appointments as an arbitrator under the ICC Rules and under those of the *Chambre Arbitrale Maritime de Paris*. He is a Member of the board of the *Association française de droit maritime* and a Member of the Maritime Cluster of Luxembourg.

His research interests lie in shipping and transportation law, commercial law, and conflict of laws. In the English language, he has recently published *Luxembourg Maritime Code* (2014). His French language publications include *Droit international privé* (2014) and *Droit du commerce International* (2011).

Simon Curtis, MA (Oxon), BCL, FCIArb
Senior Partner
Curtis Davis Garrard LLP, London

Simon Curtis is regarded internationally as one of London's most experienced maritime lawyers. He started his own City firm, which he still heads, in 1996, and has an enviable reputation in shipbuilding and offshore procurement matters.

As the author of *The Law of Shipbuilding Contracts* published by Informa Law Library, now in its fourth (2012) edition, he is the leading expert under English law on marine construction. He has received several nominations as a leading expert in his field. His expertise has been noted in every edition of *The World's Leading Shipping and Maritime Lawyers* since 1998, and has been highly recommended in the Projects, Energy & Natural Resources (Oil & Gas) and the Transport (Shipping) sections of the UK *Legal 500*.

Simon lectures regularly on a range of shipbuilding and offshore issues, including contracts for the construction and employment of FSOs and FPSOs, and is also a contributing author to *Force Majeure and Frustration of Contract*, also published by Informa Law Library.

John Forrester
Partner
Holman Fenwick Willan LLP, London

John Forrester is a partner in the London office of Holman Fenwick Willan LLP. He is head of the firm's international shipping and offshore finance group, and has over 25 years' experience in the ship finance field.

He has acted for a wide range of banks and export credit agencies on their shipping and offshore loans, including many of the industry's leading banks. He has also represented shipowners, operators, shipbuilders, private equity houses, shipping funds and other investors on various shipping and offshore related transactions, including financing, construction, charters/leasing, sale and purchase, debt restructuring/work-outs, ship registration and joint ventures.

He is the author of the chapter on ship finance in *Secured Finance Transactions* (2007), and has written and presented papers on a wide range of subjects relating to the construction, financing and sale of ships.

Matt Hannaford
Partner
Curtis Davis Garrard LLP, London

Matt Hannaford is a partner at Curtis Davis Garrard LLP in the City, specialising in commercial shipping law with particular expertise in ship finance and ship sale and purchase (including newbuilding projects). His clients include ship owners and operators, major international banks, shipbuilders and private equity companies. In addition, he regularly advises international maritime clients on mergers and acquisitions, joint ventures and pooling arrangements. He lectures extensively on shipping law aspects, and is co-author of *Goldrein's Ship Sale and Purchase* (6th edition). He was a member of the BIMCO drafting committee for the Norwegian Saleform 2012.

Chris Kidd
Partner
Head of Shipbuilding, Offshore Construction and Renewables
Ince & Co LLP, London

Chris Kidd is dual-qualified as a solicitor in England and Hong Kong. Between 1995 and 2001 he practised in Ince's Hong Kong office, during which time he was the Secretary of the Hong Kong Maritime Law Association, and still retains strong industry links in Hong Kong, China and Korea. He is a Fellow of the Chartered Institute of Arbitrators and a member of the LMAA Supporting Members Liaison Committee.

His practice encompasses shipping and offshore matters. He heads the firm's Shipbuilding, Offshore Construction and Renewables teams and frequently advises buyers and shipyards in relation to building, conversion and repair contracts, as well as dispute resolution, for conventional vessels and specialist offshore vessels and floating platforms. He worked with BIMCO to develop NEWBUILDCON, a standard form newbuilding contract, and is a regular speaker and chairman of seminars on shipbuilding, conversion and repair.

He also advises on the drafting and negotiation of contracts, as well as dispute resolution, in the renewables sector including offshore wind farm installation contracts, charterparties and contracts for the construction of wind farm installation vessels. He has been involved in (among others) the offshore projects at Greater Gabbard, Gunfleet Sands, London Array, Borkum Riffgat, Kentish Flats, Westermost Rough, and Gwent-Y-Mor. He has also worked with BIMCO to develop WINDTIME, a standard offshore wind farm personnel transfer and support vessel charterparty.

Mr Kidd has also advised in connection with accidents to offshore vessels, fixed and floating platforms, and acted in a significant number of high-profile cases including BP Thunderhorse, Bombay High North and Piper Alpha.

Simon Kverndal QC
Quadrant Chambers, London

Simon Kverndal QC is a leading maritime barrister and a member of Quadrant Chambers, specialising in charterparty, collision, salvage, offshore and shipbuilding/sale/finance disputes. He comes from a family, originally from Norway, which has been involved for many generations in the shipping industry and, in particular, in maritime claims. As well as his principal practice as an advocate, he is frequently appointed as arbitrator (including Lloyds Open Form salvage, where he is one of the five panel arbitrators); he is also a popular choice as mediator. He took silk in 2002 and is ranked as a 'Leading Silk' in the current editions of both *Chambers UK* and the *Legal 500* directories.

Mr Kverndal is a supporting member of the London Maritime Arbitrators' Association, a member of COMBAR and of the LCLCBA, on the Council of the London Shipping Law Centre and a senior Member of the Court of Assistants of the Worshipful Company of Shipwrights (the shipping industry's livery company).

Peter MacDonald Eggers QC
7 King's Bench Walk, London

Peter MacDonald Eggers QC, a Visiting Fellow at the Institute of International Shipping and Trade Law, is a barrister at 7 King's Bench Walk, specialising in all aspects of commercial law, including insurance and reinsurance, shipping, energy, commodities, international trade and international arbitration. He practised as a solicitor before being called to the Bar by the Middle Temple in 1999 and taking silk in 2011. He is author of *Deceit: The Lie of the Law*; co-author of *Good Faith and Insurance Contracts*; a Contributing Editor of *Chitty on Contracts*; and a contributor to a number of other important works, including *Marine Insurance: The Law in Transition*.

Peter has appeared in numerous commercial cases, including *Rathbone Brothers plc v Novae Corporate Underwriting*, *The Princess of the Stars*, and *Habas Sinai v VSC Steel Co.*

William Maclachlan
Senior Associate
Holman Fenwick Willan LLP, London

William trained and qualified as a solicitor at Norton Rose, before joining Holman Fenwick Willan in 2008, where he is an associate. His practice encompasses all aspects of the financing, operation, construction and management of ships; and in the course of it he advises a wide variety of companies and financial institutions on a range of transactional shipping matters in both the commercial shipping and yachting industries. He has particular expertise in ship sale and purchase, shipbuilding and ship repair contracts and spent some eight months seconded to a leading European shipbuilder. He also represents a large number of private maritime security companies, logistics providers and shipowners in matters concerning maritime security.

Mr Maclachlan has given a number of seminars on ship sale and purchase for the Lloyds' Maritime Academy, and contributed pieces on piracy and armed guards to several industry publications.

Prior to university, William worked as a supernumerary officer cadet with P&O Nedlloyd.

Associate Professor Theodora Nikaki, LLB (Aristotle), LLM (Aristotle, Tulane), PhD (Soton)
Associate Professor
Institute of International Shipping and Trade Law
Swansea University

Dr Nikaki joined the School of Law in 2005 and was promoted to Associate Professor in 2013. Having worked in private practice for several years, she obtained an LLM in Admiralty and Maritime Law from Tulane in 2001 with distinction. She then went to work in a maritime law firm in the US before undertaking a PhD in the UK on the subject of carriage by sea. Her principal research interest lies in the field of carriage of goods by sea, but her interests extend widely and include private international law and multimodal transport. She has published in journals including the *Journal of Business Law*, the *Journal of Maritime Law and Commerce*, the *Berkeley Journal of International Law*, and the *Tulane Maritime Law Journal*.

Subjects she has taught include carriage of goods by sea, land and air; charterparties; e-commerce; international trade; oil and gas law and private international law. She also helps deliver the professional courses offered by the Institute of International Shipping and Trade Law.

David Osborne
Partner
Watson, Farley & Williams LLP, London

David Osborne joined Watson, Farley & Williams on its establishment in 1982 shortly after qualifying with a large City firm; he was made a partner in 1986 and subsequently spent six years in the firm's New York office. His practice concentrates on asset and project finance, with a particular emphasis on the shipping and offshore oil and gas industries. His clients include banks, lessors, shipping companies and companies in the offshore services industry. He also advises on asset financing more generally (where he is listed as a leading practitioner by *Chambers*), together with acquisition finance, balance sheet lending and securitisation. In addition, his practice encompasses more general commercial contract advice. He has spoken and written on a wide range of legal topics, including occasional notes in the LMCLQ and the *Cambridge Law Journal*.

Simon Rainey QC
Barrister
Quadrant Chambers, London

Simon Rainey QC is a barrister at, and former joint head of, Quadrant Chambers, London, one of the leading shipping and commercial sets. In addition he been a visiting research fellow at the Institute of International Shipping and Trade Law, University of Swansea since 2010 and was appointed Honorary Professor of Law at Swansea University in 2013.

Simon has been in practice at the Bar since 1983; he took silk in 2000. In shipping work he is regarded by both *Chambers & Partners* and the *Legal 500* as one of the leaders at the Bar, specialising in all aspects of dry shipping and charterparty disputes. He has been notable in recent years in heavy shipbuilding disputes (he has advised on everything from oil rigs to FPSOs, new generation dry cargo vessels, tankers and super-yachts), and in complex cases involving large casualties arising out of dangerous cargo incidents. He is also highly regarded in the areas of commodities, international arbitration, energy and natural resources, and insurance. He was counsel for the successful buyer on the arbitration appeal in *The Union Power* in 2013 on the application of the Sale of Goods Act 1979 to the Norwegian Saleform; and is the author of *Tug and Tow and Offshore Contracts*, the leading text on offshore industry contracts (the third edition was published by Informa in December 2011).

Appointed a recorder in 2000, Mr Rainey was authorised to sit as a deputy High Court judge in 2008. He also sits as an arbitrator under LMAA, ICC, LCIA, UNCITRAL and other appointments and is a Fellow of the Chartered Institute of Arbitrators.

FOREWORD

I am ashamed to say that until I went to the opening of the colloquium at the Institute of International Shipping and Trade Law (IISTL) at Swansea University in September 2014 I knew nothing about either IISTL or the colloquium. I now appreciate it has a well-deserved reputation of which I was ignorant. It is always a pleasure to follow Sir David Steel, against whom I fought many cases in years gone by. Almost everything he said with regard to the book containing the papers from last year's event is applicable to this year's book. So it would probably be sufficient for me to say (as they do in the Court of Appeal), that I agree with Sir David and there is nothing I can usefully add. However, that invites the response: I agree.

As last year, IISTL has collated a serious of works by experts in many different areas. Their success is thanks, not only to the contributors, but also (as the Preface shows) to the vision and enthusiasm of the editors, Barış Soyer and Andrew Tettenborn, whose powerful intellect is as vibrant as ever.

One of the great disappointments of the Supreme Court is that we almost never hear an appeal in a maritime case. So almost everything in this valuable book came as a surprise to me. I can see that if I ever put up my flag as an arbitrator there will be a great deal to learn.

I cannot pretend that I have read every word, but the book is a tour de force. Six chapters on legal issues relating to shipbuilding followed by ten chapters on ship sale contracts and practice add up to a most impressive body of work on a whole area of the law. The book is a must for students and practitioners. I wish both it and the IISTL every success.

I particularly enjoyed Simon Rainey QC's analysis of the expression 'as is' in the context of the decision in the case of *The Union Power*, although it would not be appropriate for me to express my own view on the points he discusses. Perhaps I will have the opportunity of doing so some time in the future. Finally, as evidence that I have studied the whole book, I record my admiration that it has no less than 1,165 footnotes!

Anthony Clarke,
Lord Clarke of Stone cum Ebony,
Justice of the Supreme Court of the United Kingdom.
April 2015

PREFACE

For most readers the IISTL, otherwise known as the Institute of International Shipping and Trade Law, will need no introduction. For the few who have yet to encounter this organisation, it was founded in 2000 in Swansea Law School to promote the academic study and development of maritime and commercial law, and to organise regular colloquia on selected topics of importance. Since that year it has seldom, if ever, looked back; today the first week in September, when these meetings are invariably held, is pre-emptively blanked out at an early stage in commercial law offices up and down the kingdom. The present volume marks the tenth IISTL Colloquium, held in Swansea in 2014 and dedicated to shipbuilding, ship sales and marine finance. As ever, contributions came from an eclectic mixture of top academics and practitioners, from the City, and from Swansea and other universities.

Like Gaul of old, this particular work is divided into three parts. Its divisions reflect the themes of the gathering from which it grew. The first section, on shipbuilding, covers standard form contracts, the application to them of general contractual principles, and specialist topics such as builders' risks insurance and refund guarantees (the latter topic having in recent years kept a goodly number of lawyers in claret). On sale contracts, something much changed in legal practice since the 2008 crash knocked the stuffing out of freight rates, there is more on standard forms; a vade mecum of what can go wrong at the sharp end and what to do about it; and a sharp analysis of the *cause célèbre* of *The Union Power*. As regards financing, coverage varies widely, much in the same way as the disparate financial schemes available to support shipowners. Enforcement of mortgages and the frustrating effects of international insolvency law, demise charters (now almost overwhelmingly a financing device), derivatives, and mortgagees' interest insurance go to make up a varied collation with something for everyone.

It remains for us as editors to express our heartfelt thanks to our contributors and chairmen, without whose generosity there would have been nothing to put in this book; and even more importantly, our publishers Informa Group plc, without whom it could never have come out. Informa have faithfully supported our colloquia for some years now, and put up with editorial and other quirks with their usual good-humoured but unfailing patience. For any mistakes, of course, it is we who must accept full responsibility.

Barış Soyer
Andrew Tettenborn
February 2015
Swansea, Wales

TABLE OF CASES

TABLE OF STATUTES

PART 1

LEGAL ISSUES RELATING TO SHIP BUILDING

impressive precision, the dimensions, weight, speed, fuel consumption and other readily calculable details of the vessel to be built (or at least provide boxes where that information can be filled in), they are curiously vague and variable on the actual *quality* of design and construction that the customer is entitled to expect. Thus the buyer under the AWES form gets a reassuring but somewhat fuzzy promise that his ship will be built 'in accordance with normal shipbuilding practices [in the place of building] for new vessels of the type and general characteristics of the vessel'.[11] Under the Norwegian form, by contrast, the vessel must be constructed 'in accordance with first class shipbuilding practice in Western Europe'.[12] Newbuildcon similarly uninformatively mandates work 'in accordance with good international shipbuilding and marine engineering practice'. As for the SAJ and CMAC forms, these coyly (and rather curiously) say nothing whatsoever about quality. This creates, to say the least, potential for uncertainty and disagreement: even if there is such a thing as 'normal shipbuilding practice', it may well take two or more experts hired at huge expense to decide what it is. The matter may be especially relevant in the relatively short time after sea-trials and before the contractual delivery date, where there is most scope for argument (and consequent arbitral expense and delay) over what amounts to a shortcoming which the yard is legally bound to put right before handing over the vessel.[13] Is there, for example, room for a contention that there is a difference in standard between first class (Norwegian form) and normal (AWES) shipbuilding practice (with the intriguing implication that all ordinary shipbuilding is by definition somehow second-rate)? More to the point, diffuse provisions of this type, and *a fortiori* the complete non-existence of any standard laid down in some contracts, may leave the argument open that the default standards of s 14 of the Sale of Goods Act 1979[14] have a part to play, especially in the light of Flaux J's recent decision in the related context of ship sale in *Dalmare SpA v Union Maritime Ltd*[15] that only pretty clear words can oust them.[16] It is true that the problem can be, and often is, overcome by the use of more certain and arbitrable standards such

11 AWES, cl. 1(c). Throughout this chapter, when referring to standard forms capitalisation and other common editorial practices will be ignored.

12 Norwegian form, cl. II.1.

13 After delivery has taken place, it seems clear that all the forms are generally sufficient to exclude the Sale of Goods Act duties and indeed almost any implicit duty on the builder: see in particular *China Shipbuilding Corp v Nippon Yusen Kabukishi Kaisha* [2000] 1 Lloyd's Rep 367.

14 On which there is some authority in the shipbuilding context, but much of it elderly and of limited help in the case of modern cargo vessel construction. Representative examples are *McDougall v Aeromarine of Emsworth Ltd* [1958] 1 WLR 1126; *Dixon Kerly Ltd v Robinson* [1965] 2 Lloyd's Rep 404; and *Britain SS Co Ltd v Lithgows Ltd*, 1975 SC 110. See, too, the ship sale case of *Dalmare SpA v Union Maritime Ltd* [2012] EWHC 3537 (Comm); [2013] 1 Lloyd's Rep 509, referred to below.

15 [2012] EWHC 3537 (Comm); [2013] 1 Lloyd's Rep 509. The case concerned a ship sale under Saleform 1993, but the reasoning is wider. Its result was disconcerting; significantly, those drafting cl. 18 of the present Saleform 2012 hastily took the opportunity to put the exclusion of statutory implied terms beyond question for the future. For a comprehensive analysis of this judgment see Chapter 8 by Simon Rainey QC.

16 Such words appear, it is suggested, only in Newbuildcon, cl. 37(d), and possibly in CMAC, cl. XIX.4. In the Newbuildcon version, the wording of the guarantee explicitly 'replaces and excludes any other liability, guarantee, warranty and/or condition and/or innominate term imposed or implied by the law, customary, statutory or otherwise, by reason of the construction and sale of the Vessel by the Builder for and to the Buyer'. Presumably this extends to pre-delivery defects.

as those promoted by the International Standards Organisation;[17] but the point remains that the forms as given are inadequate and can amount to a trap for unwary practitioners.

1.1.2 The issue of insurance

In any shipbuilding contract there is invariably a provision for property and risk to remain in the yard until delivery at which point they are transferred to the buyer.[18] Back-to-back with this is a duty in the yard pending delivery to insure at its own charge the vessel and any buyer's supplies, a provision aimed partly at protecting the buyer's right to get his money back in the event of total destruction,[19] and (no doubt) partly at making sure that the yard is in a position to continue with the work in the event of lesser damage.[20] Nevertheless, there are nagging uncertainties here as to two matters: first, precisely what has to be insured against, and second, what happens if the term is broken? As regards the former, the forms vary from the specific (Newbuildcon, as might be expected, cuts straight to the chase and mandates the 1988 Institute Clauses for Builder's Risk terms including war and strikes),[21] through to the not-entirely-precise, which clearly leaves worrying scope for argument ('customary "all-risk" terms' in the Norwegian form)[22] to the maddeningly vague (the CMAC requirement that any policy 'shall cover the damages or losses of the vessel's materials, hull and equipments which incurred [sic] by various marine perils, inland perils or the builder's errors and omissions').[23] The difficulty here is simply that, except for the Newbuildcon form, it is often going to be difficult or impossible to say whether or not the yard is in breach of its obligation. Such indeterminacy ought to worry negotiators: might the yard be able to satisfy its obligation to the letter and yet leave the buyer unprotected against some significant risk? But that leads on to another point: assuming we can get over any uncertainty as to the *content* of the obligation to insure, what happens where the yard is in breach of it? Any immediate harm to the buyer from non-insurance is likely to be nil as long as the risk has not eventuated; if so, any damages from such failure are apt to be nominal. Hence the only live issue is whether failure by the yard might give rise to some other remedy. Specific performance is one possibility, though perhaps not a very practical

17 Which has a large number of specific shipbuilding standards for those who care to use them: for those interested, they can be found in Part 47 of the ISO catalogue.

18 See e.g. Newbuildcon, cl. 31. But note a curiously insidious provision in the SAJ form, which any well-advised buyer must insist on rewriting in a sensible way. 'Title and risk of loss of the vessel and her equipment', it is said, 'shall be in the builder, excepting risks of earthquake, war and tidal waves' (SAJ form, VII.5). If this Delphic provision means that the buyer has to insure his building vessel right from the outset against these risks alone, or take the risk of having to pay for the ship in the event (a not insignificant chance in Japan) of damage to it caused by tidal waves or similar occurrences, it is not attractive.

19 Hence its universal limitation to the total amount paid by the buyer from time to time, this being the amount at stake as regards that party. Although refunds are largely taken care of today by the custom of requiring the buyer to furnish a third-party refund guarantee, it has to be remembered that some of the older forms, such as the SAJ, contain no such provision.

20 For a comprehensive analysis on standard contracts used in insuring new built ships see Chapter 6 below.

21 Newbuildcon, cl. 38.

22 Norwegian form, cl. XI.2.

23 CMAC, cl. XXVIII.1.

one.[24] More importantly, does failure to insure allow suspension or cancellation by the buyer? Even here, the prospects for those seeking certainty do not look good. None of the forms gives any such right expressly and the prospects of demonstrating that a failure to insure, especially against the background of the requirement of a refund guarantee to protect many of the buyer's interests, do not seem good. Indeed, in *Wuhan Ocean Economic Cooperation Co Ltd v Schiffahrts-Gesellschaft Hansa Murcia mbH*,[25] Cooke J held that where there was no immediate threat to the buyer's security even failure to maintain a refund guarantee, a rather more important obligation, was not repudiatory. In the light of this decision anyone arguing that a different rule should apply to failure to insure faces a somewhat uphill task. This is a matter that must be addressed in any properly drafted contract.

1.1.3 Payment and refund guarantees

Payment and refund guarantees issued by a bank or financier are a universal feature of shipbuilding contracts.[26] Payment guarantees cover the buyer's obligation to pay instalments as and when due at various stages of construction; refund guarantees cover the converse case of the yard's obligation in the event of rightful cancellation to reimburse sums paid by the buyer. Both effectively provide the parties with a vital element of credit insurance. However, with the possible exception of the Newbuildcon form, none of the standard forms deals clearly or satisfactorily with the vital point of precisely what must be done in this respect.[27]

To begin with, what happens if a required guarantee is not forthcoming, or ceases to be effective (for example, because of governmental action, the insolvency of the guarantor, or for that matter, simple expiry)? It is not hard to see why this matters. As regards the buyer, without an effective refund guarantee in place there is a big risk that his very large investment (in which, it will be remembered, he has no ownership rights before delivery) will vanish into thin air if the yard becomes insolvent. Conversely, when it comes to the payment guarantee, the yard's cash-flow and the security available to its financier (who will doubtless have lent against the yard's claim to future instalments) are seriously imperilled without assurance of the ability not only to sue for, but actually to collect, instalments as and when due. The position at common law is, unfortunately, not entirely clear here. In the one case where the issue arose as to whether the failure to preserve a guarantee (there a refund guarantee) was repudiatory so as to allow cancellation,[28] it was held that it was not; but this was partly on the special ground that there was no prejudice to the buyer

---

24 There seems no reason why there should not under English law be specific performance of a contract to insure, in the same way as there can be specific performance of an obligation to repair (e.g. *Rainbow Estates Ltd v Tokenhold Ltd* [1999] Ch 64) or to service property (*Posner v Scott-Lewis* [1987] Ch 25).

25 [2012] EWHC 3104 (Comm); [2013] 1 Lloyd's Rep 273.

26 See Newbuildcon, cl. 14(b),(c) and CMAC, cl. V.7, where the term appears in the boilerplate. The SAJ and AWES forms require a buyer's payment, but not a seller's refund, guarantee: see SAJ, Appendix, and AWES, cl. 7(b).

27 For a legal analysis of refund guarantees in shipbuilding context see Chapter 5 below.

28 See *Wuhan Ocean Economic & Technical Cooperation Co Ltd v Schiffahrts-Gesellschaft Hansa Murcia mbH & Co KG* [2012] EWHC 3104 (Comm); [2013] 1 Lloyd's Rep 273.

on the facts,[29] hence the decision is not necessarily the last word on the point. Yet, in such a case, any properly advised party will want a clear option to escape: the seller to avoid having to continue construction without assurance of payment,[30] and the buyer in order (a) to have an option to cancel and get back his money from the yard while the going is good, and (b) to avoid having to pay further instalments which may well turn out to be irrecoverable from a bankrupt builder. Nevertheless, the majority of the standard forms give no such indication. Even where they contain a provision for a guarantee,[31] only Newbuildcon and CMAC state clearly what is to happen if it is not there: namely, by making provision for withdrawal if no guarantee is provided at the beginning, and cancellation if the guarantor becomes insolvent and the guarantee is not replaced by another within 30 days.[32] With the others the issue has to be settled by way of agreed amendment or fought out, unsatisfactorily, at common law.

Secondly, it is all very well for a shipbuilding contract to require a guarantee of a given liability: but there are guarantees and guarantees. For instance, their wording, especially in the case of refund guarantees, may give rise to important doubts as to just what obligations are being secured.[33] Again, both a payment and a refund guarantee may take the form either of an old-fashioned 'see-to-it' guarantee, conditioned on the guaranteed sum actually being presently due, or a purely documentary obligation in the nature of a performance bond or demand guarantee. Differentiating between the two can be difficult, depending as it does on whether the bank's promise to pay sums unpaid by the contractual counterparty can plausibly be construed as dependent on those sums actually being outstanding or only on presentation of particular documents by the beneficiary.[34] The difference matters as

29 This was a case where the guarantee expired by effluxion of time. The reason for the lack of prejudice to the buyer was that under the guarantee's own terms it had only to seek arbitration, in which case it would have stood automatically reinstated.

30 An option to suspend, rather than cancel, may suffice here in the event of short-term problems: but for any long-term difficulties it is vital for the yard to avoid having a half-built ship on its hands subject to possible commitments to a bankrupt buyer.

31 Which all save the Norwegian form do, though the details vary (as will appear below). The provision is patchy. The SAJ form appends a suggested buyer's guarantee in a somewhat outdated form, but does not demand it. It says nothing about a refund guarantee. AWES brusquely demands a buyer's guarantee (see – with the exception of the Newbuildcon form – cl. 14) but not satisfactorily.

32 Newbuildcon, cl. 14(b), (c). See too cl. 39(a)(i) (while no effective refund guarantee in place, buyer is protected from having to pay any instalments even if he cannot cancel). CMAC does something similar, though without the suspension right. Clause XXVII provides for a right of cancellation if the guarantee is rendered ineffective, while cl. XXXII makes the entire effectiveness of the contract depend on the provision of the guarantees in the first place. The difference between this and Newbuildcon is that apparently under CMAC there is no actual duty to provide the guarantee: hence a person who signs a shipbuilding contract but fails to do so is not liable in damages, the only 'remedy' being his right to plead the lack of any effective contract at all.

33 An obvious example is *Rainy Sky SA v Kookmin Bank* [2011] UKSC 50; [2011] 1 WLR 2900, a case which has passed into the canon of authorities on contractual construction. The issue there was whether a refund guarantee drafted with more haste than accuracy covered the case where the yard failed to repay instalments, not because of cancellation by the buyer, but because of its own insolvency and what were in essence Chapter 11 proceedings against it in its home state of Korea. An earlier instance, also ensconced in the contractual canon – this time on the nature of the right of cancellation for breach – was the payment guarantee case of *Hyundai Heavy Industries Co Ltd v Papadopoulos* [1980] 1 WLR 1129.

34 The difference will not be gone into here, save to say that where a document is issued by a bank in a solidly commercial context, and especially where it contains wording such as 'on first demand', the courts lean fairly strongly in favour of a demand guarantee: see *Caja de Ahorros del Mediterraneo v Gold*

a demand bond proper (a) gives the beneficiary an enormous cash-flow and inertial advantage, forcing the other party to take out separate proceedings to dispute any liability and recover his money once the bank has disbursed payment; and (b) more importantly, may bypass the co-contractor's ability to delay payment by seeking arbitration.[35] When it comes to the requirements of the various forms, however, there is (save possibly in the case of Newbuildcon, referred to below) no consistent answer as to what kind of guarantee is required. This is obviously so in the case of the Norwegian form, which says nothing at all about the subject and thus leaves it entirely up to the parties to stipulate what they want: but the same problem appears elsewhere. The SAJ form appends a suggested primitive form of 'see-to-it' payment guarantee (though, oddly enough, the text of the agreement itself does not explicitly require the buyer to provide it): refund guarantees are left unmentioned. The AWES form, another builder-drafted document, says laconically that, as regards payment, '[b]ank guarantees for the different instalments have to be provided by the purchaser before the effective date of the contract . . . to the satisfaction of the contractor',[36] leaving plenty of room, in the absence of a properly drafted provision, for lawyers to earn handsome remuneration by arguing about what kind of guarantee a reasonable builder should be satisfied with.[37] Once again, there is nothing at all on refunds. CMAC appends a form of both payment and refund guarantee, which the respective parties are bound to provide.[38] But neither makes it explicit whether it is a demand guarantee: the former would probably be so construed,[39] while the proper interpretation of the latter is anyone's guess.[40] The least unsatisfactory in

Coast Ltd [2001] EWCA Civ 1806; [2002] CLC 397 at [15]–[27] (Tuckey, LJ), itself a shipbuilding case concerning a refund guarantee. Other shipbuilding instances include *WS Tankship II BV v Kwangju Bank Ltd, Seoul Guarantee Insurance Co* [2011] EWHC 3103 (Comm); *Wuhan Guoyu Logistics Group Co Ltd v Emporiki Bank of Greece SA* [2012] EWCA Civ 1629; [2014] 1 Lloyd's Rep 266; and *Sea-Cargo Skips A/S v State Bank of India* [2013] EWHC 177 (Comm). See generally on the point G Andrews & R Millett, *Law of Guarantees* 6th edn, (Sweet & Maxwell, 2012) Ch 16.

35 This was the point at issue in *Caja de Ahorros del Mediterraneo v Gold Coast Ltd* [2001] EWCA Civ 1806; [2002] CLC 397 and *Wuhan Guoyu Logistics Group Co Ltd v Emporiki Bank of Greece SA* [2012] EWCA Civ 1629; [2014] 1 Lloyd's Rep 266, above. The same point arose in *Rainy Sky SA v Kookmin Bank* [2009] EWHC 2624 (Comm); [2010] 1 All ER (Comm) 823, where a similar result was reached. This part of Simon J's judgment was not appealed when the case went to the Supreme Court at [2011] UKSC 50; [2011] 1 WLR 2900 on another point.

36 See cl. 7(b).

37 There is little doubt that some term in the nature of reasonableness falls to be implied here.

38 See cls V.6 (payment) and V.7 (refund). The forms are in Appendix A (refund) and B (payment). Certainty is not helped by the fact that cl. V.7 in its wording provides for a refund guarantee to be payable in the event of cancellation by the *builder*. No doubt this is a typo and should read 'buyer'.

39 The relevant phrase is in Paragraph 4: 'Should the buyer fails [sic] to punctually pay any installment or interest and such failure is last for fifteen (15) days, we will, upon receipt by us from you of the first written demand for the same, pay to you or to your order the amount of the second, third and fourth installments and relevant interest.' Despite the conditionality of the first part, the reference to a first written demand is probably sufficient here.

40 The phrase runs 'Should the seller fails [sic] to repay to you such any or all [sic] installments due as provided by the articles under the contract, and you suspend to terminate the contract [sic] due to the extension of the delivery date, we will make such payment to you without interest. In the event that the delivery date is delayed for [ ] days and you terminate the contract in accordance with the clause 3 of Article 8 or clause 1 (3), 2 (3), 3 (3), or 4 (3) of Article 3 of the contract, we shall pay to you the aforesaid amount of installments together with interest at the rate of [ ] percent (%) per annum, or [ ] percent (%) per annum in other circumstances. Within thirty (30) running days upon receipt by us from you of a repayment demand, we shall pay to you the sun [sic] as follows.' Whether the reference to a 30-day period

this respect is Newbuildcon, which again appends forms and requires them to be provided: but at least in this case it seems fairly clear the respective forms are indeed demand guarantees.[41] Nevertheless, this is a matter which could do with careful scrutiny from anyone negotiating a contract and, *faute de mieux*, insistence on a bond which in terms states whether it is a demand bond, and precisely what obligations it covers, together with back-to-back terms in the contract itself.

1.2 Variations between standard forms

As mentioned above, most contracts to build substantial cargo vessels follow essentially the same structure. (True, there may be a need for substantial alterations to them as regards particular specialist vessels, such as those working on offshore installations,[42] but this is by-the-by.) There is almost invariably provision for payment in instalments at various stated stages of construction: for at least some form of guarantee or performance bonds to secure payment or repayment, or both, against an insolvent counterparty: for assignment: for standards of build: for supervision by the buyer during the build: for later changes to specifications: for subcontracting: for ultimate performance of the vessel and class compliance: for sea-trials and acceptance: for the builder's guarantee of quality once the vessel has been completed and handed over: and to a greater or lesser extent for the exclusion of other liabilities. However, the fine details of many of these provisions vary considerably between the different forms, either owing to a greater or lesser degree of care taken in drafting them,[43] or alternatively because some of these forms are naturally more pro-builder than others (the SAJ template, for example, was drafted largely by builders' interests, whereas Newbuildcon and the NSC were explicitly aimed at holding the ring between builders and buyers). It is some of these small differences that provide a lesson in the dangers and difficulties of standard forms.

1.2.1 Subcontracting and its effects

Whenever work is subcontracted, the yard invariably remains liable for the work of the subcontractor, independently of whether it is itself at fault.[44] Nevertheless, subcontracting remains a point of major importance for both parties. The yard obviously wants to retain as free a hand as possible.[45] The buyer, by contrast, unless

for payment, or a subsequent provision in Paragraph 2 for delay of payment in the event of a dispute going to arbitration, is sufficient to turn this into a documentary demand guarantee is an open question.

41 See Appendices A(i) to (iii). By referring to the precise documentation to be provided (notably a demand and a copy of any demand served on the other party), it seems clear that the obligation here must be a purely documentary one.

42 Where, oddly enough, there seem to be few, if any, specialised forms; as a rule the ordinary forms are simply adapted in a belt-and-braces operation.

43 Without wishing to offend, it is obvious from a brief reading that (for instance) a great deal more meticulous effort went into drafting the Newbuildcon than the SAJ form.

44 This is true of contractual liability at common law (e.g. *Photo Production Ltd v Securicor Transport Ltd* [1980] AC 827, at 848 (per Lord Diplock)). In fact, however, this is explicitly stated under every form (e.g. SAJ, cl. IX.1).

45 To some extent this may also be supported by the buyer: a wide discretion as to choice of subcontractors may reduce costs and accelerate delivery.

he is simply using a well-known yard as a front-end for cheapskate construction elsewhere,[46] is likely to be concerned that the yard he has appointed will actually do the work: he has, after all, presumably chosen this yard for its reputation and personal competence as well as other things.[47] What is significant is that the standard forms vary spectacularly here. Under the SAJ contract, as might be expected with a builder-produced document, a laconic provision simply licenses the builder to subcontract 'any portion' of the construction work,[48] which with the possible exception of laying off the work completely to someone else (since one can plausibly argue that the whole of a project is not a 'portion' of it), gives virtual *carte blanche* to the yard. All the other forms, by contrast, have some limits, all based to a greater or lesser extent on the concept of an approved 'maker's list' of approved subcontractors appended to the agreement. But even here buyers need to be aware of the differences. Under the Newbuildcon form, there remains a general right to subcontract, albeit limited to firms on the 'makers' list', plus a provision that while the buyer's consent is required to go outside the list, in the case of minor work this consent is not to be unreasonably refused.[49] In the Norwegian form, the right to subcontract at all is excluded in the case of the 'hull and major sections', with everything else being subject to the maker's list principle.[50] Under CMAC, there is a somewhat obscure provision which starts out by allowing anything to be delegated, but then provides that 'delivery and final assembly into the vessel' has to be at the builder's own yard.[51] The result of leaving this clause in place, which should be borne in mind by buyers' representatives, is that large sections even of the hull, provided they are capable of being transported, may still turn out to have been fabricated in the back of beyond, and somewhere completely different from the yard originally signed up.

All this may cause problems for a buyer, for a number of reasons. For one thing, it is all very well to say that the original yard guarantees the quality of the work; but the sensible buyer wants a good ship, not a lawsuit, a right to refuse acceptance until defects are corrected, or a promise of free repairs during a guarantee period.[52] For another, while it is true that the builder invariably assumes an obligation to produce a finished product satisfactory to class,[53] the potential for difficulty and delay is much reduced in so far as all subcontractors have been pre-approved by the classification society involved, which will not necessarily be the case where the

46 Which happens. Giveaway bespoke clauses on the following lines are not unknown: 'The construction of the hull and major sections are accepted by the buyer to be subcontracted in a low-cost country. The builder shall remain fully liable for the due performance of such work as if done by the builder at the builder's yard.'

47 Hence at common law a contractor cannot normally subcontract skilled work without the customer's agreement: see J Beatson, *Anson's Law of Contract* 29th edn, (OUP, 2010) at 448.

48 See cl. I.4 SAJ.

49 Newbuildcon, para. 19: 'The Builder shall employ the sub-contractors as set out in the Specification or Maker's List. Except for minor work, the Builder shall not employ other sub-contractors without the Buyer's approval, which shall not be unreasonably withheld.'

50 Norwegian form, cl. II.4.

51 See cl. X CMAC.

52 Especially as the terms of the guarantee, when coupled with the invariable exclusion of other liabilities, may leave substantial defects uncovered, as shown by *China Shipbuilding Corp v Nippon Yusen Kabukishi Kaisha* [2000] 1 Lloyd's Rep 367. They also often exclude other expenses such as downtime while defects are being put right. The effect of builders' guarantees is discussed below.

53 E.g. Newbuildcon, cl. 3; Norwegian form, cl. II.3.

builder has an entirely free hand in deciding whom to give work to. And yet again, assume that, as soon as the ink on the contract is dry, the buyer has – as is extremely common – agreed to time-charter the vessel when complete. Although it is often assumed that would-be charterers of newbuilds are not concerned with the details of building,[54] this cannot be guaranteed to be the case,[55] and under some forms in the case of substantial subcontracting there could be considerable difficulties.[56] As a result, it is not surprising that the variety of different provisions on subcontracting is matched by the tendency to replace or supplement them with bespoke provisions of the parties' own making: for example, requiring subcontractors to be on a classification society's approved list,[57] or demanding consultation before appointing subcontractors for certain purposes.[58]

1.2.2 Supervision and modifications

Having a ship constructed is not like ordering a new Ford Focus. Shipbuilding contracts are invariably a highly cooperative long-term exercise between the buyer, the yard and others involved in the transaction. Since there is invariably a central guarantee that the finished vessel will be acceptable to the buyer's nominated classification society,[59] class is closely involved throughout the process (so much so, indeed, that none of the standard forms even bothers to provide specifically for its inclusion).[60] Similarly, all construction contracts take place under the watchful eye of a buyer's representative, whose right to be present on the spot is stipulated in some little detail.[61] Apart from pointing out possible defects (of which more below), one of his functions is to negotiate over post-contract alterations in the specifications, which are an inevitable feature of any large-scale construction project taking place over an extended period. Such alterations may arise from a number of causes.

---

54 The facts in the well-known tanker case of *Reardon Smith Line Ltd v Yngvar Hansen-Tangen* [1976] 1 WLR 989 are an obvious illustration. Faced with the certainty of ruinous losses due to the 1974 oil crisis, the Shelltime charterer of a new build tanker tried to throw up the deal merely because actual construction had been (as was entirely usual) subcontracted to another Japanese yard; hence the builder was not the one mentioned in the charter. He lost. A virtual carbon copy, also arising under a Shelltime form, was *Sanko Steamship Co Ltd v Kano Trading Ltd* [1978] 1 Lloyd's Rep 156.

55 In *Reardon Smith v Yngvar Hansen-Tangen*, above, the charterers were opportunistic and entirely undeserving. They were invoking a technical arrangement between yard and subcontractor, both equally reputable, which was completely standard (since the yard itself could not produce ships of the requisite size). It is not difficult to imagine situations where this would not be the case, and where a build discrepancy might allow cancellation: see below.

56 For instance, under SUPPLYTIME 2005, unlike the Shelltime charter in *Reardon Smith v Yngvar Hansen-Tangen*, many of the details of the vessel are expressly warranted by the owner to be correct: see para. 3(a) ('The Owners undertake that at the date of delivery under this Charter Party the Vessel shall be of the description and Class as specified in ANNEX "A", attached hereto . . .'). This clearly creates a potential liability in damages in the owner: and it seems not unlikely that, at least in some instances, it would justify rejection by the charterer as well.

57 See, e.g., a provision on the lines of 'the Builder may subcontract the fabrication of [components] to European Subcontractors who are certified by the Classification Society as meeting DNV MPQA standards'.

58 Indeed, this is explicitly required by one of the standard forms, namely the Norwegian: see cl. II.4.

59 See, e.g., Newbuildcon, cl. 3; Norwegian form, cl. II.3.

60 The Norwegian and Newbuildcon forms, however, provide directly for their presence at sea trials: see respectively cl. VII.3 and cl. 27(c).

61 One such representative is allowed under SAJ (SAJ, cl. IV.2): a reasonable number elsewhere.

They include requests by the buyer to make changes; the builder's desire to take advantage of new developments or construction techniques, or his need to provide a substitute for particular components or methods of construction now impracticable; and (vitally) regulatory changes and the updating of detailed class requirements, which are matters that can happen quite suddenly in the course of any new building. Obviously the ideal position here is swift and amicable agreement on such matters; indeed this is often forthcoming, since neither party wishes any delay or deadlock during construction while things are sorted out. But this can only take place against a clear background position on how far such alterations can be insisted on, by whom, what their effect is on the timetable of construction and, most importantly, who pays. Here there are significant variations.

We can begin with alterations requested by the buyer. It is, of course, always possible for buyer and builder to agree *ad hoc* on such changes, with any consequent price and delivery adjustments. But this point is trivial: the important question is how far the contract goes further and actually gives any kind of right to the buyer to demand them. On this the SAJ and CMAC forms are spectacularly unimpressive, merely reiterating what we all know – that is, that the parties can agree, provided the buyer is happy to pay the extra.[62] The AWES form is little better, giving the buyer the right to ask for modifications and obliging the seller to carry them out, but then emptying these provisions of all content by saying that any obligation lapses unless the parties within ten days 'fully agree expressly and in writing' on the modifications, any extra payment, and so on.[63] As regards the Norwegian form and Newbuildcon, the former makes any duty dependent on agreement,[64] but an accompanying provision for an objective means of assessing any price increase suggests that there must be at least some element of obligation,[65] thus creating the possibility of liability in a builder who refused even to negotiate. The latter, drawing doubtless on dissatisfaction with the other forms, makes things largely clear, by not only giving the buyer the right to ask for reasonable alterations which do not in the builder's reasonable judgment interfere with the latter's planning or programmes in respect of other commitments, but also saying that in the absence of agreement the builder is to do as requested, with matters of extra payment referred to arbitration.[66] In practice, of

62 SAJ, cl. V: 'The Specifications may be modified and/or changed by written agreement of the parties hereto, provided that such modifications and/or changes or an accumulation thereof will not, in the Builder's judgment, adversely affect the Builder's other commitments, and provided, further, that the Buyer shall first agree, before such modifications and/or changes are carried out, to alteration in the Contract Price, the Delivery Date and other terms and conditions of this Contract and Specification occasioned by or resulting from such modification and/or changes . . .'. (See too CMAC, cl. XII.1, which is much the same, only longer). Even then one is left wondering why or how the parties' ability to vary a contract by mutual agreement could be excluded where inconvenient to one of them. To be fair, however, it might just be arguable that if (as is entirely possible) the governing law was that of a civil law jurisdiction there might be some sort of good faith obligation on the builder to consider the buyer's request.

63 AWES, cl. 3. The statement that any requirement lapses in the absence of quick express agreement in writing would seem to exclude any possibility of a reference of the matter to arbitration under the dispute resolution clause in cl. 15.

64 See cl. VI (' . . . provided further that the parties shall first agree to possible adjustment in contract price, . . . ').

65 Since otherwise the words would be idle. Compare *Sudbrook Trading Estate Ltd v Eggleton* [1983] 1 AC 444 (obligation to pay reasonable price where means of ascertainment of price fail).

66 See cl. 24(a), (b) and (e).

course, parties may prefer a bespoke provision; one along the lines of that contained in Newbuildcon, even if that form is not used, is ideal. Alternatively it would be possible to stipulate for an obligation in the builder to negotiate in good faith on a request for modification and then act accordingly.[67] This is problematical, at least where the governing law is a common law system,[68] since in such jurisdictions there remains a nagging doubt over whether an obligation to negotiate in good faith is one that can be enforced, and if so, how far.[69]

Builders, for their part, stipulate for the right to make changes for the entirely legitimate reason of allowing maximum flexibility in the face of inevitable alterations in circumstances during the currency of the build. As a result, all shipbuilding contracts contain terms allowing minor builder's alterations to accommodate changed circumstances and also substitution of materials or components where this is necessary to keep construction on track. Nevertheless, this still leaves problems. While the builder needs flexibility, the buyer will want to have at least some control, if only because what is minor is often not self-evident and is open to genuine disagreement. Of course, even when such changes are made, there is always the vexed question of who pays for any overshoot (or, almost as important, who gets the credit for any cost savings).[70]

In all these there is again noticeable variation, which those negotiating such contracts are apt to miss but in fact would do well to draft round. Thus minor changes to suit the builder's working methods are generally allowed, though not universally (the CMAC form, puzzlingly, has no provision at all). Where changes are permitted, most forms are sensitive to the important fact that there may be room for argument over whether a change is indeed minor or whether it is something which, however apparently inconsequential to a builder in a hurry, is something that a buyer might legitimately feel strongly about. They thus require consent from the buyer, not to be unreasonably withheld.[71] By contrast, and with potential disadvantage to buyers,

67 For instance, a possible clause reads: 'This contract, the plans and the specification may be modified from time to time by agreement of the parties. The builder shall act in good faith and on an open book basis to implement modifications requested by the buyer, . . . subject to the buyer agreeing to necessary modifications to the contract price, the delivery date and any other relevant provisions of this contract. The builder agrees to act in good faith and on an open book basis to implement any such modifications (i) at the lowest cost reasonably possible; (ii) within the shortest period of time reasonably possible; and (iii) without any loss in the relative priority of the building work for the ship compared to other construction work in the shipyard . . . '.

68 If it is a civil law system there is less difficulty, since such systems almost invariably accept the possibility of an obligation to negotiate in good faith.

69 Because of suggestions by Lord Denning MR in *Courtney & Fairbairn Ltd v Tolaini Brothers (Hotels) Ltd* [1975] 1 WLR 297, at 301–302 and Lord Ackner in *Walford v Miles* [1992] 2 AC 128, at 138, and also the decision in *Barbudev v Eurocom Cable Management Bulgaria EOOD* [2012] EWCA Civ 548; [2012] 2 All ER (Comm) 963. It is true that the matter is still in flux: see, e.g., *Petromec Inc v Petroleo Brasileiro SA Petrobras (No 3)* [2005] EWCA Civ 891; [2006] 1 Lloyd's Rep 121 at [115]–[121] (Longmore LJ) and *Butters v BBC Worldwide Ltd* [2009] EWHC 1954; [2009] BPIR 1315 at [151] (Peter Smith J). But, perhaps in contrast to an enthusiastic law professor, no sane buyer wants a clause incorporated that could encourage such arguments.

70 A matter of some significance; indeed, to an increasing extent, contracts actually contain terms requiring one or both parties to seek and promote possible cost savings.

71 See SAJ, cl. V.1; Norwegian form, cl. VI.1; Newbuildcon, cl. 25. Presumably if consent were unreasonably withheld there would be liability in damages represented by the lost opportunity to engage in any cost savings: cf. the recent charter case of *Falkonera Shipping Co v Arcadia Energy Pte Ltd* [2014] EWCA Civ 713; [2014] 2 Lloyd's Rep 406.

one form (AWES) does not require consent, instead allowing changes willy-nilly in so far as they count as 'not affecting the vessel's performance characteristics' (whatever that means).[72] Again, in all such affairs, cost matters. While in most cases changes of this sort do not affect the overall price, one form is different from the others. The carefully drafted Newbuildcon template, while protecting the buyer from any surcharge, specifically requires any cost savings obtained thereby to be passed on to him:[73] a difference which could involve sizeable sums.

Similarly, while all the standard forms, without exception, allow the use of replacement materials (for entirely understandable reasons), how they go about it varies. The SAJ, rather primitively, regards a claim to use this in the same way as a buyer's modification, subject to agreement between the parties.[74]The CMAC form provides for the use of replacement components, but starkly requires the permission of the buyer before this can be done at all, without giving further details of what happens if that consent is not forthcoming.[75] Going to the other extreme, the AWES form dispenses with the need for consent altogether.[76] In practice this may frequently be satisfactory, since the buyer will probably be consulted and may well be only too willing to agree; but there may equally well be substitutions with which he is not happy, and there is room for putting in some sort of provision for consent, at least if this is not to be unreasonably withheld, and for dispute resolution if differences remain irrevocable – which is the solution chosen by the Newbuildcon drafters.[77]

Turning to a slightly different matter, compliance with class and other regimes (such as SOLAS and MARPOL) is absolutely central to all shipbuilding contracts. Unfortunately this has to be reconciled with the inconvenient fact that such regimes are apt to change frequently and unpredictably between contract and delivery. As a result, one of the few areas of near uniformity arises from the fact that all the standard forms contain a regime under which the builder is bound to build in compliance with the elements of such regimes as are compulsory[78] at the time of delivery, even if this involves a change of specifications in mid-construction, with associated adjustments in the price and, where necessary, in the delivery date as well. The ideal and most straightforward scheme is that in Newbuildcon, under which the builder is

72 AWES form, cl. 3(b).

73 See cl. 25.

74 See cl. V.3. It is not said what happens if the parties fail to agree, or what the builder can do if faced by an intransigent buyer demanding (say) a reduction in price. In order to avoid the possibility of deadlock, any person using the SAJ form is well advised to replace this article with a properly drafted one, containing a provision for the buyer's consent not to be unreasonably withheld, for some form of arbitration in the event of irreconcilable disagreement, and for a formula for dealing with any extra costs (or savings).

75 See cl. XII.3. The same comment applies as above. Query whether this is one of the cases where the English courts would imply a ban on entirely arbitrary refusal. Compare Leggatt LJ in *Abu Dhabi National Tanker Co v Product Star Shipping Co Ltd* [1993] 1 Lloyd's Rep 397, at 404 ('where A and B contract with each other to confer a discretion on A, that does not render B subject to A's uninhibited whim . . . [T]he authorities show that not only must the discretion be exercised honestly and in good faith but . . . it must not be exercised arbitrarily, capriciously or unreasonably').

76 See cl. 3(b).

77 A result achieved by a combination of cls 25 and 42(b). Unlike the other forms, which put the risks of substitution on, and conversely allow any savings to enure to, the builder, Newbuildcon also gives the buyer credit for any costs saved as a result of such substitution.

78 Changes merely advisory are invariably treated as buyers' modifications. See e.g. SAJ, cl. V.2(b); Newbuildcon, cl. 26(c).

bound to carry out any necessary alterations announced post-contract, with adjustments in so far as unagreed established by the dispute resolution procedure.[79] The CMAC form proceeds on the same lines, but suffers from some obscure drafting in the relevant clauses, which need careful adjustment and rewriting.[80] Similarly, the SAJ scheme requires work provided the buyer agrees to any extra costs and other changes proposed by the builder,[81] though significantly it lacks any provision for reference to arbitration in the event of dispute.[82]

Normally the criterion of whether the new regime applies is simply whether the changes come into effect on or after the signing of the contract, even though they may have been formally announced earlier.[83]

1.2.3 Termination of the contract

All the standard forms allow, as might be expected, for termination by the yard in respect of non-payment of instalments, and do so fairly similarly.[84] The right of the buyer to cancel for the builder's shortcomings, by contrast, is more unevenly covered, despite the fact that it, and the clarity with which it is expressed, can matter a great deal. To take a straightforward example, suppose contractual difficulties arise shortly before an instalment becomes payable. A swift and incontrovertible cancellation by the buyer will immeasurably strengthen his hand in preventing a call by the builder on the guarantor – and hence inevitably on the buyer – for the instalment concerned.[85] The absence of a rightful cancellation, on the other hand, will

---

79 Newbuildcon, cl. 26(b). The AWES scheme is similar, save that it makes provision for a specific cut-off date to be stated in the contract (see arts 1(c) and 3(c)). This can be a useful provision where at the time of signing the contract it is known that changes will take effect in the near future, and the quoted price has already taken account of them.

80 It says in cl. XII.2(2) that in the absence of agreement on price adjustment, extension of delivery, increase or decrease of speed, deadweight or the need for additional security, the builder may ignore the rule changes and proceed as before. It then says in cl. XII.2(3) that in the absence of agreement on price adjustment or extension of delivery, the builder 'shall' proceed with such modifications as required by the rules with any consequential matters being referred to arbitration. Unless 'shall' is read as 'may', these clauses flatly contradict each other. With respect, the drafting here can be politely described as a dog's breakfast and, for anyone seeking to use this form, a completely new and comprehensible *ad hoc* provision is essential.

81 SAJ, cl. V(2)(a). Presumably it would be implied that any proposals must be reasonable.

82 The duty to proceed being simply expressed to be conditional on agreement. A proper bespoke clause dealing with failure to reach that agreement is vital here.

83 Note, however, the Norwegian form, which under cl. VI.2 excludes from any adjustment of the contract terms changes publicised but not effectual at the time of contracting, on the basis that the builder ought already to have taken account of these in setting the price. It does not really matter which solution is adopted, provided those negotiating shipbuilding contracts are clear over the prospective rule changes which they do or do not want to be in account at the time of contracting.

84 Two of the more modern forms (Newbuildcon and the Norwegian form) go further and say, rightly, that where the buyer is clearly insolvent the seller need not wait for non-payment but can cancel straight away: see Newbuildcon, cl. 39(d) and Norwegian form, cl. 12.3. With the other forms, this obvious piece of boilerpate has to be affixed by hand.

85 True, if the guarantee is a demand guarantee and there is a genuine dispute it will not invalidate a good faith demand: see e.g. *Turkiye Is Bankari v Bank of China* [1998] 1 Lloyd's Rep 250 and the recent decision in *Wuhan Guoyu Logistics Group Co Ltd v Emporiki Bank of Greece SA* [2012] EWCA Civ 1629; [2014] 1 Lloyd's Rep 266 (the latter involving the exact scenario in the text). But even there it still strengthens the buyer's hand considerably in negotiations to be able to put the beneficiary (or the beneficiary's financier) on clear notice of a valid termination.

leave him helpless against such a demand, with the attendant effects on cash-flow.[86] Furthermore, uncertainty in cancellation rights provides a further trap for the buyer. If he does not cancel, he faces the consequences just mentioned. Assuming he does, he also takes a big risk. If he is right, at least he can activate the repayment guarantee to recover sums already paid. If he is wrong, he is deprived even of this, on the basis that his cancellation will itself be regarded as an unlawful repudiation, which is not covered by the guarantee.[87]

By way of background, it seems clear that even if the contract does not say so expressly, where there is a genuine repudiation by the builder, the buyer retains the right to accept it and cancel.[88] But such cases are rare and, in any case, are of limited use to the buyer because of the uncertainty of what amounts to a repudiatory breach and the enormous risks to the buyer if he gets it wrong. It follows that normally the buyer will have to rely on specific cancellation rights under the contract. Unfortunately, these can vary subtly in at least two cases.

First, take late delivery. Shipbuilding contracts invariably use a sliding scale model for late completion, with short delays triggering a price rebate and delays beyond a certain date allowing cancellation.[89] But there is a catch. What are referred to as permissible delays (in effect, delays due to *force majeure*, default of the buyer or agreed changes in specification) are outside this scheme.[90] The result can be surprising. In the SAJ and AWES forms, for example, there is theoretically no end to the buyer's duty to accept in so far as the only delay is due to *force majeure* or otherwise permissible. Unless steps are taken to rewrite the relevant provisions, he must apparently continue cooling his heels (and keeping his payment guarantee active) until the sounding of the last trump, or at least until he has a chance of persuading a sceptical arbitrator that the delay has been such as to frustrate the contract. There is an obvious need for an absolute 'walk-away' date even where the delay is no-one's fault. To be fair, the other forms go some way to dealing with this. The Norwegian form gives a long-stop of 270 days including *force majeure*;[91] Newbuildcon and CMAC, 270 days including all forms of delay.[92]

---

86 It is true that if the demand is indeed unjustified the buyer may in the fullness of time have a claim on the builder's refund guarantee to repay him the amount reimbursed to his bank. But no sane businessman wishes to be without many millions of dollars for months or years on end, however cast-iron any assurance that he will get them back eventually.

87 Since, however much the buyer may be in good faith, a cancellation without the right to do so is a repudiation; invariably refund guarantees apply only where the buyer *rightfully* cancels under the terms of the contract. Nor, for that matter, is it likely that the buyer can recover from the builder direct: cf *Hyundai Heavy Industries Co Ltd v Papadopoulos* [1980] 1 WLR 1129.

88 Such a rule formed a clear assumption lying behind *Wuhan Ocean Economic & Technical Cooperation Co Ltd v Schiffahrts-Gesellschaft Hansa Murcia mbH & Co KG* [2012] EWHC 3104 (Comm); [2013] 1 Lloyd's Rep 273 (though in that case there was held to be no repudiatory breach).

89 For example, 210 days under the SAJ form (cl. III); 180 days under the Norwegian form (cl. IV.1(b)). On the operation of such clauses, where contractual provision is made for arbitration of a claim to terminate, see *Nanjing Tianshun Shipbuilding Co Ltd v Orchard Tankers Pte Ltd* [2011] EWHC 164 (Comm); [2011] 2 All ER 789.

90 See, e.g. SAJ, cl. III.1(e); Norwegian form, cl. VIII.1.

91 See cl. IV.1(c).

92 Newbuildcon, cl. 39(a)(iii); CMAC, cl. XXVII.3.

A second point of difficulty is failure of the yard to progress the building, before this has developed into an actual postponement of the delivery date. Suppose it becomes clear to the buyer that work on the vessel has stopped or virtually ground to a halt – an event that often, though not invariably, indicates insolvency or financial difficulties. The buyer has an important interest in being able to escape from the contract at that point, without necessarily having to wait for an actual delay in delivery that would justify cancelling the contract some months hence.[93] At common law, any right to do so is limited. An express or implied representation that work has not only stopped but has been terminated, or that the vessel will not be delivered before the right to cancel is triggered, will do.[94] If it is clear that there will inevitably be a failure to perform entitling the other side to escape, there is the authority (though not the certainty) that this can be treated by the other party as a repudiation.[95] Short of this, the other party must, it seems, bide his time, even if he reasonably thinks it rather unlikely that he will, in fact, receive performance.[96]

Those negotiating on behalf of buyers need to note that the treatment this subject gets from the different forms is various. Under SAJ and AWES, nothing is said about stoppage. It follows that without an amendment of the form the buyer remains bound when work grinds to a halt, unless the yard itself makes it clear by words or conduct that it will not perform the contract. The Norwegian form does deal with the matter, but cumbersomely and badly. Having started promisingly by giving the buyer a right to demand security from a builder who is insolvent or unlikely to be able to perform it would, one might have thought, have followed this through and allowed cancellation if no such security was forthcoming. But no: the only right given is to pay any sums owing into an escrow account instead of to the yard and then deduct them from any final payment.[97] This is better than nothing, but hardly satisfactory to the buyer. It is left to Newbuildcon (and to CMAC which copies Newbuildcon word-for-word on the point) to produce a satisfactory solution. Under clause 39 of Newbuildcon, where work stops for a stated number of days, the buyer is given the right to demand that it restarts within a given period; if it does not the

93 In particular, he will want to avoid the risk that by cancelling the contract he will himself be regarded as being in repudiatory breach, thus imperilling his right to a refund. There are other reasons too: for example, the buyer may have an entirely understandable wish not to have to maintain a payment guarantee in force for a vessel he is highly unlikely to get.

94 Implicit in *Primera Maritime (Hellas) Ltd v Jiangsu Eastern Heavy Industry Co Ltd* [2013] EWHC 3066 (Comm); [2014] 1 Lloyd's Rep 255 (where, however, the contract had been affirmed after the renunciation). That repudiation remains as an independent ground for cancellation in a shipbuilding contract was confirmed in *Wuhan Ocean Economic & Technical Cooperation Co Ltd v Schiffahrts-Gesellschaft Hansa Murcia mbH & Co KG* [2012] EWHC 3104 (Comm); [2013] 1 Lloyd's Rep 273.

95 *Universal Cargo Carriers Corporation v Citati* [1957] 2 QB 401, at 441 (Devlin J); *Shepherd (FC) & Co Ltd v Jerrom* [1987] QB 301, at 323 (Mustill LJ); see too *Geden Operations Ltd v Dry Bulk Handy Holdings Inc* [2014] EWHC 885 (Comm); [2014] 2 Lloyd's Rep 66. Sometimes apparent incapacity coupled with silence or prevarication may be taken as renunciation of any intent to perform (see, e.g. the recent charter case of *SK Shipping (S) Pte Ltd v Petroexport Ltd* [2009] EWHC 2974 (Comm); [2010] 2 Lloyd's Rep 158).

96 So held in *Universal Cargo Carriers Corporation v Citati* [1957] 2 QB 401. See too *The Madeleine* [1967] 2 Lloyd's Rep 224, and *The Mihalis Angelos* [1971] 1 QB 164, at 200–201, 207–208 (Edmund Davies and Megaw LJJ).

97 See cl. III.3.

buyer can cancel the contract.[98] In addition there is an automatic right to cancel forthwith in the event of insolvency.[99]

1.2.4 Rejection

Buyers do not lightly refuse to accept a brand new vessel tendered to them which is already freighted with a big investment of time and dollars, particularly if, as frequently happens, they have pre-chartered her and inability to perform that charter will cost them a lot of money in profits foregone or damages payable to an irate charterer, as the case may be. Nevertheless, the right to decline delivery remains very significant. There may be matters of very serious complaint, over build quality or otherwise; furthermore, it is worth remembering that with refusal to accept delivery the underlying issue may be a claim not to outright refusal, but rather to a demand to delay acceptance pending correction of real or imagined defects.[100]

In a few cases the matter is straightforward; indeed is tied in with cancellation. All the standard forms, for instance, say that deficiencies in speed, fuel consumption or deadweight tonnage lead to a sliding scale of price reductions up to a limit, and beyond that allow cancellation;[101] not infrequently this scheme is extended further by bespoke agreement to cover other readily measurable matters.[102] The difficulties come in a different area: namely, where at or after sea-trials the vessel is alleged to have some other defect in construction or design which means she does not comply with the contractual specification.[103] The SAJ and CMAC forms, perhaps surprisingly, are curiously pro-buyer here: taken literally, the buyer only has to take the vessel if she wholly conforms to all specifications; they can thus insist on the correction of any defect, however minor, as a condition of acceptance.[104] The other three forms have a more up-to-date regime, incorporating a 'minor defects' exception: but even here there are variations which can mean a lot to the buyer. AWES, for example, having referred to conformity with the contract and class and regulatory control,

---

98 Newbuildcon, cl. 39(a)(ii). The periods themselves are for negotiation: boxes are supplied to fill them in. The equivalent in CMAC is cl. XXVII.1(2).

99 Newbuildcon, cl. 39(d). The equivalent in CMAC is cl. XXVII.4.

100 Nor should the buyer's bargaining position be ignored in this respect. Where negotiations over previous contractual problems are ongoing, a yard's mind may be concentrated wonderfully by the prospect of having a completed but undelivered vessel on its hands taking up room, obstructing further work, and tying up cash-flow.

101 See, e.g., SAJ form, cl. III; Norwegian form, cl. IV.

102 For example, it may extend to volumetric capacity (cubic or bale, as the case may be), as in the Norwegian form, cl. IV.5 or AWES, cl. 5(c). The Newbuildcon form encourages this further, not only dealing with capacity in cl.11 but suggestively leaving a space blank for other matters to be similarly dealt with; see cl.12. Obvious candidates would be TEU container capacity, or vehicle numbers on a ro-ro ferry (on the latter of which, cf *Cenargo Ltd v Empresa Nacional Bazán* [2002] EWCA Civ 524; [2002] CLC 1151).

103 By way of background explanation, the universal scheme of all shipbuilding contracts is that delivery is due on a particular date, but that at some time before that date sea-trials are to be carried out. At a given (short) period after these sea-trials, the buyer must indicate whether or not he will accept delivery of the vessel. Delivery itself is a formal act consisting of the handover of the ship and a sheaf of papers.

104 See the unqualified words of SAJ, cl. VI.4: ' . . . should the results of the trial run indicate that the Vessel, or any part or equipment thereof, does not conform to the requirements of this Contract or the Specifications, . . . then the Builder shall take necessary steps to correct such non-conformity. . . .' CMAC, cl. XIII.4, is almost word-for-word the same.

goes on to say[105] that the buyer must accept the vessel despite non-conformities that are 'of minor importance and do not prevent safe operation of the vessel' against a commitment to deal with them within the guarantee period.[106] This could leave the buyer in an awkward situation since, taken literally, it could make the buyer accept a vessel even though she had (minor) defects going to class and hence to tradeability as such;[107] furthermore, although the yard agrees to deal with the defects, any downtime to the vessel is, it would seem, for the buyer's account. The latter point is better dealt with by the Norwegian and Newbuildcon forms, which explicitly make the yard pay for downtime[108] and, in the case of Newbuildcon, require it to give third-party security as well.[109] Only Newbuildcon is entirely satisfactory on the first point, with a specific limitation of minor defects to those that 'do not affect Class or the operation of the vessel'.[110]

1.2.5 Extent of liability for defects

In connection with defects found after delivery, all forms follow the same broad 'clean slate' model – that is, providing a carefully modulated guarantee, essentially involving an unconditional promise by the yard to repair defects within a certain time wherever the vessel is,[111] and then using a broad brush to exclude all other liability that might otherwise arise.[112] There is invariably an exclusion of responsibility for consequential losses such as downtime, and any costs of bringing the vessel in for repair are always for the buyer's account. Nevertheless, elsewhere, parties may discover all too late that the extent of the buyer's rights can still vary considerably. Most importantly, there is the issue of the extent of the free work to be done where parts are defective. Obviously the cost of removing and replacing the defective part is covered. But what of repairing damage done to neighbouring components (for example, a shaft tunnel cracked because of vibration caused by a defective propeller-shaft) or to the vessel as a whole (as with damage to shell plating caused by the

105 See cl. 4(d).

106 A period that varies from newbuild to newbuild, but is typically set at one year from delivery.

107 True, in practice, this might be unlikely, and one suspects that class might often be open to negotiation. But the potential for uncompensated trouble and disruption to the buyer remains highly noticeable.

108 Norwegian form, cl. VII.4(d)(ii); Newbuildcon, cl. 27(d)(iv)(2).

109 Newbuildcon, cl. 27(d)(iv)(3).

110 Newbuildcon, cl. 27(d)(iv).

111 More precisely, repairs are either done at the original yard, or arranged by the buyer elsewhere in consultation with the builder. E.g. SAJ, cl. IX.1–3; Newbuildcon, cl. 35.

112 E.g. SAJ, cl. IX.4 ('The Builder shall have no responsibility or liability for any other defect whatsoever in the Vessel than the Defects specified in Paragraph 1 of this Article. Nor the Builder shall [sic] in any circumstance be responsible or liable for any consequential or special losses, damage or expenses including, but not limited to, loss of time, loss of profit or earning or demurrage directly or indirectly occasioned to the Buyer by reason of the Defects specified in Paragraph 1 of this Article or due to repairs or other works done to the Vessel to remedy such Defects'); and Norwegian form, cl. X.1 ('. . . the Builder shall have no responsibility for defects or the consequences thereof (including loss of profit and loss of time) discovered after Delivery and Acceptance of the Vessel'). See, too, Newbuildcon, cl. 37(b), (d). Consistent with their reputation for upholding freedom of contract, the English courts are apt to take such broad exclusions entirely seriously: see, e.g., *China Shipbuilding Corp v Nippon Yusen Kabukishi Kaisha* [2000] 1 Lloyd's Rep 367 (in substance concerning the SAJ clause above, holding that it exonerated the yard even for breaches of express terms of the contract).

disintegration of a badly made screw? CMAC is highly generous here, imposing responsibility for 'defects and damages caused by any of the [guarantee] defects';[113] which seems to create a pretty open-ended liability for any physical damage, allowing both above examples. Newbuildcon is not quite as munificent, allowing claims for consequential damage only if 'caused as a *direct and immediate* consequence of such guarantee defects',[114] but still probably has the same effect. The Norwegian form restricts any duty to repair other damaged parts that 'can be considered to form part of the same equipment or system', and even then suggesting a ceiling on recovery:[115] the result would presumably be that the first, but not the second, instance above would be covered. As for the SAJ and AWES forms, by limiting the guarantee liability to repair of the defect concerned and excluding all other liability, they effectively provide the buyer with a new shaft or propeller and then tell him he is on his own. There are other differences too. For instance, where repairs are done other than at the original yard, the SAJ, CMAC and Newbuildcon give the buyer full recovery of all reasonable costs incurred;[116] by contrast, AWES and the Norwegian form limit reimbursement to the yard's own would-be charges,[117] thus heaping up serious potential problems for owners trading in high-cost areas and unwilling to send their vessels thousands of miles away for repair at their own charge as to lost time.

1.2.6 Assignment and security

Financing is crucial in shipbuilding contracts. Third-party payment and refund guarantees aside, particularly important here is financing through assignment of the benefit of the contract to a lender (whether by yard or buyer), and the creation of security interests in the vessel in the course of construction.

We can begin with assignment. When it comes to assignment of the benefit of the contract to a potential lender, both parties need assurance that this is permissible. Furthermore, since English law, in common with a number of other systems, provides that the validity of any transfer of rights by assignment depends on the terms of the contract giving rise to the right assigned,[118] the financier has to be assured that any assignment will be effective to create a valid security.

In view of its significance, not surprisingly assignment is touched on in the boilerplate of all the forms. Nevertheless the treatment varies sufficiently between them, both in type and completeness, as possibly to trip up an unwary negotiator. The SAJ form, drafted in the early days when financing was not as developed as it is now, contains a bald ban on assignment without the consent of the other party.[119] Unless this

113 CMAC, cl. XIX.4.

114 Newbuildcon, cl. 35(b).

115 Norwegian form, cl. X.3.

116 SAJ, cl. 12(a); CMAC, cl. XIX.3; Newbuildcon, cl. 35(d)(i).

117 AWES, cl.12(a) (reimbursement 'shall not exceed the estimated costs of carrying out the guarantee work at the CONTRACTOR'S yard(s)'); Norwegian form, cl. X.3(b).

118 A point finally established by *Linden Gardens Trust Ltd v Lenesta Sludge Disposal Ltd* [1994] 1 AC 85. Within the EU this will, it seems, go further and preclude any other law being applied that might otherwise recognise the assignment: Rome I Regulation, Article 14.2.

119 SAJ, cl. XIV. For good measure, there is also an entirely obsolete requirement of government

is re-written – and this must be a top priority for any competent adviser faced with the SAJ form – any party needing finance (or re-finance) may find himself in difficulties, either because he can only do so on the terms dictated by the other party, or because lenders may well shy away from contracts of this sort. The other forms, it must be admitted, are more satisfactory here, either providing for consent coupled with a vital 'not to be unreasonably withheld',[120] or (more radically) giving an absolute right to assign to a financier, and limiting any right of veto to other assignments.[121] The latter seems more satisfactory, in that it is not entirely clear what advantage is gained by giving the other party any kind of control over financing matters.[122]

As with assignment, so with mortgages and hypothecs. To be sure, in contrast to assignment, the *validity* of any such mortgage in the hands of the financier depends not on the contract but on the provisions of the law of the place of building.[123] Contractual controls remain important, if only because the interests of the yard, the buyer and their respective creditors must be reconciled. To see why, take the provisions of the SAJ form. This, like the Newbuildcon version, puts no limits on the power of the yard to hypothecate the half-built vessel: a feature it shares with the CMAC and Norwegian forms, which indeed go further and expressly provide for such an unqualified right.[124] The problem here is that unless discharged before delivery, such a mortgage on principle binds the vessel in the hands of the buyer (indeed, if it did not there would be little point in it). Hence the buyer not only has no guarantee of receiving an unencumbered vessel at delivery, but may also face further financing difficulties. He will almost invariably have agreed to create a mortgage over the vessel on completion and registration: as his prospective financier will be aware, this may prove difficult if she is already subject to an existing mortgage created by a now-insolvent yard in favour of another bank. The answer, of course, is to give the buyer at least some right to consent to such mortgages. The AWES form does just this,[125] simply prohibiting their creation without the buyer's written approval: but however convenient for the buyer, this may go too far. It is quite possible to protect the buyer's interests even if a mortgage is given: all that is necessary is for any mortgage to provide that it lapses on delivery of the vessel, with the lender's interests being protected by an assignment to it of the yard's right to payment of the relevant instalments of the price. Rather better in this respect, it is suggested, would

consent for any assignment by the buyer, which any competent negotiator will immediately demand struck.

120 Norwegian form, cl. XIII; AWES, Article 18.

121 See Newbuildcon, cl. 45; CMAC, cl. XXIV. Such other assignments normally concern cases where the buyer decides he does not want personally to take delivery but to re-sell, or with matters such as corporate re-organisation. The yard may justifiably want to retain some control over who it is dealing with here.

122 Indeed, where the assignment is of a right to payment, no more and no less, some legal systems simply override any contractual restrictions on assignment for precisely this reason. See, e.g. the German Commercial Code (HGB), § 354a, following the precedent of the UCC, § 9-408 (less relevant in these circumstances).

123 On the basis that the validity of the creation of any proprietary interest in a chattel depends on the *lex rei sitae*: L. Collins *et al, Dicey & Morris on the Conflict of Laws* 14th edn, (Sweet & Maxwell, 2008) ch. 24. How far a mortgagee can hold a mortgage known to have been given in breach of contract will no doubt vary between legal systems.

124 CMAC, cl. XXV.2; Norwegian form, cl. XI.

125 AWES, cl. 8(b).

be a right of veto in the buyer, but subject to the 'consent not to be unreasonably withheld' proviso, it being understood that the criterion of reasonableness would be whether the buyer's interests were properly protected.

1.3 Conclusion

The conclusion to be drawn from this chapter will be fairly obvious from what has gone before. The savings of time and trouble that come from using standard forms can only be fully utilised by those who know the strengths and weaknesses of each of them, and which bits of the boilerplate need to be discarded. A close look at the variations between them and the details of the problems they raise, such as is provided in this chapter, is as essential as it ever was.

CHAPTER 2

The BIMCO NEWBUILDCON Standard Form Shipbuilding Contract: salient features and pitfalls

*Chris Kidd**

2.1 Introduction

BIMCO's development of a shipbuilding contract fulfils its longstanding ambition to produce a standard document for every sector of the shipping industry representing the entire life cycle of a ship from construction through trading to demolition.

The objective was to provide builders and buyers with a modern, clearly worded, balanced and comprehensive shipbuilding contract which would provide a solid contractual platform from which the parties can reach agreement improving upon the standard forms currently in existence. While all of these forms deal with the essential elements of a shipbuilding contract they contain omissions, inadequacies and unclear provisions.[1] They are inadequate in many ways to the point that it is difficult to recommend their use without amendment and updating them to reflect modern practices.

Many will be familiar with the SAJ form published in 1974 by the Shipbuilders' Association of Japan, which forms the basis of various standard forms used in South Korea, China, Singapore and Taiwan. Unfortunately, the language used is, in some respects, not ideal, and many provisions often have to be amended. It is also weighted in favour of the builder: see for example the provisions dealing with sub-contracting (art. 1.4), modifications (art. V), *force majeure* (art. VII) and the exclusion of any liability of the builder for damages in the event of the buyer's rescission (art. X). There are also a number of gaps in the wording, for example, design liabilities are not addressed, the builder's liquidation or other financial default is not included in a builder's default clause, and there is no provision for a refund guarantee or other security for repayment of pre-delivery instalments in the event of the buyer's rescission following the builder's default.

The more recent Association of West European Shipbuilder's Form[2] adopted in 2004 is widely used in Europe but it has many unsatisfactory and unclear provisions. The Standard Form Norwegian Shipbuilding Contract 2000 is an 'agreed document',

* Partner, Head of Shipbuilding, Offshore Construction and Renewables, Ince & Co LLP, London. The author was a member of the BIMCO Drafting Group that produced NEWBUILDCON and is heavily indebted to William Buckley of Ince & Co LLP for his substantial assistance in researching and writing this chapter.

1 For a comprehensive analysis highlighting main differences between standard shipbuilding contracts see Chapter 1 of the present book written by Professor A Tettenborn.

2 Also known as the AWES form.

negotiated between the Norwegian Shipowners' Association and the Norwegian Shipbuilders' Association, making it the only standard contract which has been negotiated, but it still has some shortcomings. The US Maritime Administration – MARAD contract is the primary US shipbuilding contract. It has influenced contracts in North and South America, but following the decline of new building activity in that part of the world, it is seldom used except by the US Navy and the coastguard.

In China, the China Maritime Arbitration Commission published in October 2011 the 'CMAC Standard Newbuilding Contract' (Shanghai form) for use in the Chinese export market. This contains a number of features which are unfavourable to the buyer, for example relating to design responsibility, approval of plans and drawings, currency fluctuation, price escalation, Chinese-language notices and Chinese arbitration clauses. The form suffers in places from rather loose drafting, for example relating to the builder's right to adjust the contract price following modifications, its duty to procure insurance and its duty to repair defects at delivery. There are also several poorly drafted or unclear provisions, for example the duties of the buyer's supervisor, its rights of rejection after sea trials, and the parties' rights of assignment. Finally, the form contains a number of apparent drafting errors, for example in respect of the Buyer's right to resist suggested modifications to the vessel and the effective date of the contract.

All of these standard forms have various shortcomings and, in many respects, were outdated – not reflecting modern shipbuilding practices; there was a need for a new, modern and more balanced form containing the essential elements of a modern shipbuilding contract in clear language and which overcomes the omission, inadequacies and unclear provisions of the previously existing standard forms.

Although BIMCO's earlier form REPAIRCON provided BIMCO with useful guidance on managing the contractual relationship between owner and yard, it was plainly inadequate for building a new vessel. NEWBUILDCON was therefore written entirely from scratch, adopting a novel structure intended to reflect the chronological stages of a shipbuilding project. Drafted and fine-tuned over three years, commencing in 2005, with the objective of producing a modern, comprehensive, clearly worded and balanced form, its creation was guided by BIMCO's Steering Committee and featured input from the International Association of Classification Societies (IACS), ship-owner and insurance representatives, and Asian and European shipyards. A consultation draft was produced and there was an exhaustive consultation process which produced some significant changes in the finalised version.

This chapter will discuss NEWBUILDCON's salient features as well as the particular strengths and weaknesses of the form that have been identified since its release, with particular reference to the SAJ form and the CMAC form.

2.2 General layout

NEWBUILDCON uses BIMCO's standard contractual layout. Part I consists of three pages of boxes in which the main details of the contract are entered, providing a useful 'checklist' for the parties' negotiations and ensuring that the most important details are easily accessible. Part II then contains the standard terms and conditions that govern the parties' rights and obligations, subject to the parties' amendments; the two parts feature extensive cross-referencing for ease of use and interpretation. Annex A features a number of standard-form guarantees to be given by both parties, which provide the underlying financial securities that underpin the viability of any substantial shipbuilding project and which were vetted by specialist banking lawyers before the form's release. Annexes B and C contain the vessel's specification and a maker's list; these are both left blank to accommodate the particular technical needs of the project.

2.3 Part I – The 'box layout'

This part provides boxes for the parties to fill in the most important details of the contract: the names and details of the parties, the vessel's type, the main dimensions, features and details of the vessel (Boxes 4–5), the classification society, the parties' guarantors, the contractual date of delivery, and the agreed price and instalment amounts. The optional Box 4E provides space for the parties to state particular technical requirements for the vessel. Part I also allows the parties to provide for contractual details such as its effective date, conditions that must be satisfied before it becomes effective and the number of days within which these conditions must be performed (Boxes 25–26). There is also space for the parties to conveniently agree options for further vessels to be built on the same terms and conditions.

In particular, Part I allows the parties to provide for the builder's failure to meet exact contractual specifications; the parties can address key points early in negotiations without needing to confront the detailed underlying contractual wording. Boxes 13–16 allow the parties to specify the amount of liquidated damages for the builder's account in the event of the vessel not meeting the main specifications (speed, fuel consumption, deadweight and cubic capacity) and the maximum amount of such damages permitted before the buyer gains the right to terminate the contract. Additionally, Box 17 allows the parties to agree the details of any other particular deficiencies that will give rise to liquidated damages or the right to terminate the contract. Finally, Box 18 contains the builder's liquidated damages payable per day for late delivery of the vessel and a maximum cap on such damages.

The box format has been criticised for providing insufficient space for more complex projects such as LNG carriers.[3] However, it was not intended for such vessels, its primary purpose being for use in building conventional vessels, for example tankers, bulk carriers, and container vessels. The form has nevertheless been used for a variety of other specialist vessels including specialist wind farm installation vessels and OSVs. Whatever type of vessel is being constructed, the format provides a useful 'checklist' of the most salient contract terms and a convenient executive summary of the main commercial details which senior management and bankers find useful.

3 K. Krut, 'The BIMCO "NEWBUILDCON" Shipbuilding Form: Welcome Clarity but Poor Timing' p. 1.

Briefing note available on Curtis Davis Garrard's website at: https://www.cdg.co.uk/knowledge-centre/briefings/welcome-clarity-poor-timing-newbuildcon-shipbuilding-form (last accessed 31 December 2014) (hereinafter referred to as Krut).

In contrast to the SAJ and CMAC forms, NEWBUILDCON does not contain a preamble setting out the parties' general obligations and the scope of the project; instead, it has a short 'preamble text'[4] setting out the order of precedence between the constituent parts of the Contract (in general, the contract takes precedence over other documents), between Parts I and II (Part I, as the 'agreed' clauses, takes precedence over the 'printed' clauses of Part II) and different iterations of the plans and/or drawings ('*the later in time shall prevail*'). Particular clauses of Part II alter this order of precedence where appropriate by prioritising the specification.

2.4 Part II – Terms and conditions

Part II of NEWBUILDCON comprises a Definitions/Interpretation clause and Sections 1–6 each deal with a specific aspect of the shipbuilding project.

2.4.1 Section 1 – Vessel

(a) General obligations

Section 1 governs the parties' primary obligations, in particular those of the builder. Clause 1 establishes the standard to which the builder is expected to construct the vessel: '*in accordance with good international shipbuilding and marine engineering practice*'. Although this has been criticised as possibly lacking in '*any real meaning*',[5] it is submitted that this wording provides a realistic and commercially achievable standard of performance where the specification may not be clear in describing the standard of performance expected of the builder. Other formulations of the standard of performance can impose an unrealistically high standard (e.g. 'first class shipbuilding standards') or a potentially unacceptably low standard (e.g. 'builder's usual standard'), while providing a reasonable level of quality assurance for the buyer. Such protection is absent from the SAJ form.

Section 1 also clearly puts responsibility for the design of the vessel onto the builder,[6] which duty is less clearly established under the SAJ and CMAC forms.

(b) Description

Clause 2 sets out default provisions (i.e. where not provided in the specification) for defining, measuring and calculating the vessel's speed, fuel consumption, deadweight, propulsion, and cargo capacity as agreed in Box 4. Curtis regards the provisions on ascertaining the vessel's speed as '*admirably detailed*' and '*descriptive of the conditions under which the vessel is required to reach the guaranteed speed*', although it would appear that where one or more of the vessel's parameters in Box 4 differs from those of the actual vessel being tested, '*the measured speed must be adjusted mathematically for the purposes of comparison with the guaranteed speed*'.[7]

4 S. Curtis, *The Law of Shipbuilding Contracts* 4th edn, (Informa 2012) at p. 17 (hereinafter referred to as Curtis).

5 *Ibid*, at pp. 33–34.

6 See cl. 1(a), 4(a) and 20(f).

7 See Curtis at p. 75.

(c) Classification, rules and regulations
Clause 3 requires the builder to design, construct, survey, test and deliver the vessel in compliance with the requirements of its classification society and the relevant regulatory authorities (defined as including the flag state and other bodies stated in the specification). The clause makes the builder responsible for ensuring that the vessel complies with both rules in force at the date of signing the form *and* those which enter into effect after execution of the contract, if they were '*ratified and promulgated*' on or before the date of the contract. This is particularly relevant in the modern market, where years may elapse between contract execution and delivery. Under the SAJ form, the buyer retained the potentially unacceptable risk of the vessel falling foul of requirements entering into force between execution and delivery; NEWBUILDCON's '*more elaborate and perhaps more modern approach*'[8] shifts some of the compliance risk to the builder. Clause 26, with which this clause interlinks, obliges the parties to endeavour to agree on how to adjust the contract terms in consequence of such compulsory changes. If this cannot be done, disputes are to be resolved by the relevant classification society or regulatory authority under the multi-tiered dispute resolution Clause 42. This represents a much clearer position than under the CMAC form, which reserves the builder's right to adjust the contract price in this situation without any guidance on the exercise of its discretion.[9]

(d) Protective coatings and source of origin
Clauses 4 and 5 are novel clauses implementing a number of modern environmental IMO standards, emphasising the modern requirement for parties to comply with IMO regulations and developments in accordance with the IACS Common Structural Rules.[10] These include the IMO 'hazardous materials inventory' clause, which requires the builder to '*take due account of the Vessel's ultimate disposal when designing and constructing the Vessel*'; a 'green passport' clause, under which the builder must provide the buyer with a list of materials used in the vessel's construction which are known to be hazardous; and a 'protective coatings' clause, implementing the minimum requirements of the IMO Performance Standard for Protective Coatings subject to any higher standard in the specification. Clause 4 has been criticised as imposing an obligation on yards that they would be ordinarily unwilling to accept.[11] However, Curtis regards the CMAC form as, '*whether or not intentionally*', imposing an even stronger obligation on the builder to use environmentally friendly materials in the construction of the vessel generally, rather than only with regard to its disposal.[12]

2.4.2 Section 2 – Financial

NEWBUILDCON allows the parties to choose the currency for payment rather than imposing yen (as per the SAJ form) or US dollars (as in the CMAC form).

8 *Ibid*, at p. 92.
9 *Ibid*, at p. 93.
10 A. Mandaraka-Sheppard, *Modern Maritime Law Volume 2: Managing Risks and Liabilities* 3rd edn, (Informa 2013), at para. 7.2 (hereinafter referred to as Mandaraka-Sheppard).
11 See Krut p. 2.
12 See Curtis at pp. 282–283.

However, reflecting BIMCO's aim of producing a clear contract which reflects current market practice, NEWBUILDCON does not include provision for adjustment of the contract price via a price escalation clause (e.g. where the cost of materials increases), a feature which is built into the CMAC form, but is not accepted by most buyers.

(a) Instalments

Clause 15(a) of the form gives substantial freedom to the parties to decide exactly when each instalment will be due and the amount, using the boxes in Part I; the unamended form stipulates only that the first instalment will be due within five days of the builder obtaining a refund guarantee. This approach contrasts with the SAJ and CMAC approach of tying each instalment to a particular stage in the construction and delivery of the vessel. Clause 16 of the form also effectively divides responsibility for taxes by providing a mutual indemnity and timeframe for reimbursal of those paid by the wrong party.

(b) Liquidated damages and set-off

Clauses 8–13 and Clause 15(c) in this section provide the form's contractual mechanisms for calculating reductions in the contract price, either for deficiencies in the vessel (as discussed above) or late delivery. The form also includes an optional clause 12, under which the parties can include a mechanism for determining the level of liquidated damages payable for additional types of technical deficiency (e.g. where the vessel contains specialist equipment or has specific requirements), as specified in Box 4E of Part I. The clauses link with the appropriate boxes in Part I to provide a 'maximum' level of deficiency, at which point the buyer has the option of terminating the contract. Conversely, clause 15(b) provides that additional costs referable to modifications to the vessel and responsibilities of the buyer may be added to the final price.

In contrast to the SAJ and CMAC forms, NEWBUILDCON excludes the buyer's rights of set-off except against the final instalment;[13] the correct final amount is determined by adjusting the final instalment. Surprisingly, despite this exclusion, Curtis submits[14] that a retrospective claim for defects in the vessel could nonetheless be brought under the NEWBUILDCON form even if the buyer failed to make such a claim at delivery; this argument succeeded in the *Cenargo* case despite the existence of a more prescriptive contractual regime governing the deduction of liquidated damages.[15] Curtis notes that, while section 27(1) of the Unfair Contract Terms Act 1977 means that the Act will not apply to the bulk of shipbuilding contracts, it may well prevent the enforcement of this exclusion of set-off in those limited cases where the Act does apply,[16] chiefly where the vessel is built in the UK and bought by a UK entity.

13 See cl. 17.

14 See Curtis at p. 77.

15 *Cenargo Ltd v Empresa Nacional Bazan de Construcciones Navales Militares SA* [2002] EWCA Civ 524; [2002] CLC 1151.

16 See Curtis at p. 188.

(c) Refund guarantee
NEWBUILDCON differs from the SAJ form in expressly requiring the builder to obtain a refund guarantee at clause 14(b) in the *'form and substance'* of the guarantee supplied in Annex A(iii). Given that these guarantees are usually regarded as the 'financial cornerstone' of newbuilding projects, its inclusion is *'broadly in line with modern practice'*[17] and provides the buyer with essential protection from defaults of the builder. The buyer is also required to furnish an instalment or performance guarantee in the form supplied in Annex (i) or (ii). These clauses provide the builder with more security than it would have under the SAJ form,[18] which requires only that the buyer procure an irrevocable letter of credit in favour of the builder from a bank acceptable to it at least seven days prior to payment of the fourth instalment[19] and attaches a performance guarantee but contains no provision requiring the buyer to provide one.

2.4.3 Section 3 – Production

It is this section that has attracted some of the most detailed commentary from those supporting and criticising NEWBUILDCON. While the clause somewhat alters the rights and obligations of both builder and buyer from the SAJ form, several of these shifts favour the builder in contrast with the usual perception of the form.

(a) Sub-contracting
To begin with, Clause 19 restricts the builder's right to sub-contract from its general freedom to do so under the SAJ form; only sub-contractors specified in the specification/maker's list or those authorised by the buyer may be used (except for 'minor work', which is not defined). Though this is *prima facie* more favourable to the buyer and has been described as a *'major restriction on the Builder's conduct of the design and construction of the Vessel'*,[20] in practice the SAJ form's provision is often amended[21] and it is submitted that NEWBUILDCON's restriction reflects a more reasonable position for the buyer, who will very often be keen to have some control over who performs significant work on the vessel. It was difficult in drafting a standard form to provide a clause that would suit all projects, but having provided a wording, this is one area where there may be a need for the parties to consider and negotiate a regime which is suitable for each particular project. If for some reason they fail to do so, the form does at least have a fall-back default wording.

(b) Approvals
Clause 20 sets out a comprehensive default timetable for the issuing and approval of plans; in keeping with the form's objective of flexibility, however, the parties can override this in the specification. The builder has a number of obligations under this clause which are absent from the SAJ form, including providing the buyer with

17 *Ibid*, at p.293.
18 See Mandaraka-Sheppard, para. 7.3.1
19 SAJ form, Art. 2(4)(d).
20 See Krut at p.2.
21 See Curtis at p.36.

proposed building and testing schedules within 60 days after the effective date of the contract and the obligation to allow the buyer extra time to consider plans and drawings which have been provided. One part of the clause that has drawn criticism is the builder's obligation to provide plans and drawings at least 30 days before the construction works commence; some have suggested that this may cause problems if many drawings are submitted at the same time. Others criticise it as potentially leading to delays. However, once again, it is difficult to provide a wording that will suit all projects and the parties are free to amend the provisions to suit the project; it is not unusual for the specification to have more detailed provisions concerning the timing for production and approval of drawings. Also, modern shipbuilding practice often resembles an 'assembly line', with many vessels being produced to a standard design; this significantly reduces the number of drawings that will need to be completed from scratch in the majority of cases. Furthermore, the parties can very easily provide more bespoke provisions for specialist vessels or new designs in the specification according to their circumstances.

(c) Buyer's supplies

Clause 21, governing materials and equipment supplied by the buyer, is generally regarded as better balanced than the SAJ and CMAC forms.[22] The builder has both the right to reject materials or equipment supplied by the buyer if these are not in good condition or unsuitable for installation, and the right to an extension of time to the delivery date where this has caused actual delay in delivery. On the other hand, the builder's risk insurance[23] must cover the buyer's supplies, an addition which rather extends the builder's common law duty as bailee of the goods,[24] but which is an obligation that buyers usually require.

(d) Buyer's representatives, inspections, test and trials

Clauses 22 and 23 govern the buyer's right to appoint a representative and their rights of inspection, providing much more detailed provisions than the SAJ form. The buyer has the right to appoint both a representative and a reasonable number of assistants, officers and crew accompanying him, as opposed to the one representative allowed under the SAJ form. It is submitted that this is a more realistic position for the buyer given the sheer scale of modern yards and projects, not to mention the potential for prejudice to the buyer where lone representatives are ill or otherwise unavailable. Additionally, the representative may carry out his own inspections of the vessel, attend all tests wherever these take place (rather than merely attend tests at the builder's yard) and may communicate with class, so long as this does not unreasonably interfere with the builder's own communications. This encourages a constructive dialogue that can assist both parties during the building of the vessel.

On the other side, the builder gains the right to request that the representative be replaced if they can show that the representative has carried out his duties in an unreasonable manner, with which the buyer must comply. Under the SAJ form, the

22 See, e.g., Curtis at p.259; Mandaraka-Sheppard at para. 7.4.3.

23 See cl. 38.

24 See Curtis at p.258; this is a duty of reasonable care (*Houghland v R R Low (Luxury Coaches) Ltd* [1962] 1 QB 694).

buyer is only required to make such a replacement following his own investigations; this addition therefore provides added protection to the builder.

Finally, clause 23(d) makes it clear that the buyer's attendance at tests, trials and inspections as well as any failure to point out non-conformities in the vessel does not relieve the builder of its obligations under the contract. This position is not dramatically different from that under the SAJ form or indeed that at common law, wherein very clear evidence would be required of the buyer's knowledge and subsequent concealment of the defect for it to assume responsibility.[25] It is worth noting that the CMAC form in this regard is more pro-builder, as the buyer gives an undertaking that its representative shall minimise any increase in building costs.

(e) Modifications and changes

Clause 24 sets out the procedure for the buyer to request modifications and changes to the vessel. It interlinks with clauses 20(c) and (d), which differentiate between changes and mere comments, amendments or reservations to plans and drawings which are not of such a nature or extent as to constitute a modification or change to the specification. It also interlinks with clause 42, which provides a dispute resolution procedure where the parties cannot agree whether the suggested changes should constitute such modifications or what the effects of the modification should be. This clause has attracted much commentary to the effect that it strongly favours buyers. In brief, the buyer has the right to request reasonable modifications to the vessel's specification. The builder then gives the buyer a written proposal of the consequences of the modification (e.g. on the contract price, delivery date, characteristics of the vessel, etc.) or gives notice that it will not carry out the modifications as doing so would affect its other commitments. If the buyer disagrees with the builder's proposal or with the builder's notice of refusal, it may require the builder to proceed with the work; the consequences of the modification are then decided under the dispute resolution procedure.

Curtis comments that while the parties' rights concerning modifications are much more clearly defined under NEWBUILDCON than the SAJ form,[26] the right of the buyer to require the builder to carry out the work makes it a difficult clause for yards to accept: '*Given the "production line" approach to shipbuilding adopted by the majority of the successful north Asian shipyards, their ability to refuse buyer's modifications which will materially impact upon the performance of other projects is likely to remain a crucial protection and is a significant potential obstacle to the acceptance of the unamended NEWBUILDCON language*'.[27] Dr Stefan Schrandt-Zimmer of Dabelstein & Passehl also sees this as a particularly buyer-friendly clause.[28]

While there is some force in these comments, not all vessels are produced on a 'production line' and in many cases the buyer will want to have some degree of flexibility to request or require changes. The SAJ form, on the other hand, is pro-builder and enables it to refuse to implement a change if agreement is not reached on the time and cost consequences. The buyer cannot insist on changes being made and

25 *Ibid*, at p. 86.
26 *Ibid*, at p. 83.
27 *Ibid*, at p. 91.
28 Quoted in *Shipping World & Shipbuilder*, (211) 2011 at p. 19.

has no recourse if the builder demands an unreasonable price adjustment or extension of time. Moreover, the criticism that under NEWBUILDCON the builder is being deprived of a crucial protection, namely the ability to refuse the buyer's modifications, which will materially impact upon the performance of other projects, is ill-founded because Clause 24(b) of NEWBUILDCON expressly provides that: '*If in the Builder's reasonable judgement, such modifications and/or changes will adversely affect the Builder's planning or programme in relation to the Builder's other commitments, the Builder shall notify the Buyer that it declines to give such a proposal for the requested modifications and/or changes.*'

Of course, if a standard design vessel is being produced on a production line at a price which reflects the fact the builder will not have the flexibility to entertain modifications or changes, the standard NEWBUILDCON wording can be amended to something more in line with the SAJ wording.

(f) Sea trials and delivery defects

Clause 27 deals with the vessel's sea trials, notice of delivery where the vessel meets the specification, and procedures for where the buyer informs the builder of non-compliance with the contract ('delivery defects') and/or rejects the vessel. The clause's provisions for handling disputes arising from any such alleged defects are a particularly distinguishing feature of the form. Curtis regards the provisions concerning the sea trials as administratively more straightforward than the SAJ, dividing responsibility clearly between buyer and builder as to who provides what for the trials.[29]

However, Clause 27(d) has attracted criticism from some quarters as overly buyer-friendly. The clause provides the builder with the right to compel the buyer to take delivery of a vessel with delivery defects where these are of minor importance, do not affect class or the operation of the vessel in its intended trade, and cannot be rectified by the builder within a reasonable time and in any event before the accrual of the buyer's right to terminate. Curtis regards the requirement that delivery defects meet all three criteria as making it difficult to use in practice, with only very basic and limited defects allowing use of the procedure. In particular, he criticises the third criterion as possibly requiring calculation of the future date on which the buyer's termination rights will accrue, which date will depend on a number of matters including future permissible delays; he submits that the clause is only workable if such potential delaying events are left out of the calculation. Additionally, the builder is required to undertake to remedy the faults as soon as possible and provide an indemnity and third-party guarantee to the buyer covering any consequential losses from its accepting the vessel in this state. The model guarantee in Annex A(iv) is unlimited in time, which is unlikely to be acceptable to most financial institutions. He therefore regards the procedure as unlikely to be used by shipyards.[30]

Despite these criticisms, the clause has been accepted by major shipyards, largely because it does help to resolve in advance the potential impasse which frequently arises in the run-up to delivery when a buyer seeks to refuse delivery because of a

29 See Curtis at p. 103.
30 *Ibid*, at pp.113–114.

long punch list of minor defects. Very often the parties end up negotiating a solution along these lines, albeit after tense negotiations which delay delivery to the detriment of both parties.[31] One should also not lose sight of the fact that use of the procedure is entirely at the option of the builder, who gets paid in full notwithstanding its own failure to comply with its obligations under the contract to meet the contractual specification. Furthermore, it might be thought preferable that the builder be able to effect delivery of the vessel in circumstances where a great deal of delay has already occurred and the buyer is fast approaching the point at which it will be able to rescind the contract for excessive delay, especially where such delays are not the builder's fault and the buyer is playing the game of coming up with a long list of minor defects which the builder cannot remedy before the cancellation date so that it can take advantage of the situation and cancel. While it may be conceded that the builder takes on a great deal of risk in indemnifying the buyer for consequential losses, it is unlikely that such losses will be significant where the defects are so minor in the first place to have passed all three criteria in clause 27(d).

2.4.4 Section 4 – Delivery

Clause 28 provides the procedure for delivery of the vessel, the point at which the '*interests of the buyer and builder are likely. . . at their most divergent*'. NEWBUILDCON provides that delivery must take place at least 15 days after completion of the sea trials. This brings some temporal certainty to the most important stage of the project, in contrast to the SAJ and CMAC forms, which do not prescribe a timeframe for delivery. However, the fact that the builder risks being in breach of contract if it cannot complete the trials earlier than 15 days before the date of delivery may be perceived as hard on it.[32] Curtis also asserts that the requirement that the vessel be '*ready for service*' is a potential area of dispute, being difficult to define and possibly conflicting with the right of the builder to force the buyer to take delivery of the vessel notwithstanding minor defects. However, in the author's view, it is unlikely that the parties would understand the phrase to impose anything more onerous than the vessel and its equipment being available and set up ready for use for the ordinary purposes of a vessel of its type.

Clause 29 provides a convenient checklist of the documents for the builder to provide the buyer with at delivery. NEWBUILDCON adds the IMO Hazardous Material Inventory Statement of Compliance and Certificate of Non-Registration to the usual list. It includes a procedure in sub-clause (e) for the provision of interim certificates if final ones are unavailable, including a warranty that these will allow the vessel to be registered and trade and operate without restriction. The builder also undertakes to compensate the buyer for any loss suffered by consequence of its failure to procure such certificates. Curtis considers the builder's potential liability under this clause to be very substantial and potentially beyond the builder's control, for example where subsequent restrictions on the vessel are imposed by class. He

31 An alternative approach in the event of a dispute concerning defects on delivery would be for the buyer to retain a sum from the final instalment (and provide a bank guarantee or other security) until the defects are corrected.

32 See Curtis at p. 128.

therefore questions whether yards will be prepared to accept the exposure.[33] While this is certainly a more buyer-friendly position than that of the SAJ form, it should be noted that the builder is not under an obligation to provide interim certificates and can wait until final certificates are available if it does not wish to take the risk.

Additionally, the right of the buyer to request '*any other documents reasonably required by the Buyer*'[34] is a wide one and in Curtis' view '*somewhat exceptional in light of usual shipbuilding practice*'.[35] It is submitted that, interpreted *ejusdem generis*, this should not be seen as a particularly onerous obligation on the builder given the conventional list of certificates preceding it, except perhaps for unusual projects where the buyer's reasonably requested documentation may be more various than usual.

Clause 30 provides a procedure for payment of the final instalment; it is similar to CMAC in reflecting the modern practice of having the bank hold the deposit with instructions to send it to the builder when the protocol of delivery and acceptance is received, but adds provisions whereby the deposit must be returned if the vessel is not delivered within seven days of the delivery date.[36] The clause also provides for the situation where the parties disagree as to the amount of the final instalment, allowing delivery to be made on the appointed day with the dispute later resolved under clause 42. However, the language of the clause has been criticised as failing to oblige the buyer to take delivery in such circumstances, in common with most yard contracts.[37]

2.4.5 Section 5 – Legal

(a) Permissible delays – *force majeure*

Clause 34 governs the contract's calculation of the delivery date of the vessel and the level of any resulting liquidated damages for late delivery. The clause sets out two types of permissible delay: (i) *force majeure* events and (ii) other events either caused by the buyer or under the responsibility of the buyer. The *force majeure* provisions are based on the International Chamber of Commerce's recommended standard clause, and it contains a similar list of events to the SAJ form, with some additions to reflect modern practice such as terrorism.[38] Additionally, sub-clause (a)(i)(10) provides a 'sweep-up' clause which allows the builder to claim a *force majeure* event for '*any other cause of a similar nature to the above beyond the control of the Builder or its sub-contractors*', which potentially allows a wider range of events to be considered permissible delays than would be the case under the SAJ form. Provided that the builder did not cause or reasonably foresee the event and made all reasonable efforts to minimise the effects of the event on delivery of the vessel, the clause then provides a clear procedure by which the builder can provide notice to the buyer of the delaying event's start and end.[39]

---

33 *Ibid*, at p. 133.

34 Clause 29(l).

35 See Curtis at p. 135.

36 *Ibid*, at p. 53.

37 K. E. Rygh, 'Shipbuilding – legal and practical problems when taking delivery', 20 November 2013. Available at: www.sjorettsforeningen.no/. . ./KER-Young-CMI-Sjørettsforeningen-Klaveness-20-11-2013.pps (last accessed 31 December 2014).

38 See Mandaraka-Sheppard, at para. 10.3.

39 *Ibid*, at para. 7.4.4.

(b) Builder's guarantee

Clause 35, a critical clause for the builder, sets out the scope of the builder's guarantee; in keeping with the form's objective of flexibility, the parties set their own guarantee period in Box 20 (no default being provided). The clause is more buyer-friendly than the SAJ form in expressly including design defects, extending the guarantee to cover sub-contracted work, which is not, in the author's experience, contrary to common practice as some have suggested.[40] The guarantee also extends to damage caused as a result of defects, bringing consequential damage within the guarantee,[41] although this is no different to the position under CMAC.[42] Sub-clause (e), interlinking with Box 21, allows the parties to agree an additional guarantee period, which will apply separately to repairs or replacements made under the builder's guarantee so long as this does not end before the main guarantee period. This is more favourable to the buyer than the unamended SAJ form, but reflects common practice where the builder will often agree a separate six-month warranty from the date of installation.[43] The buyer may also extend the guarantee period if the vessel lays idle for more than 30 days as a result of repair work, which appears to apply even if the works are delayed by the acts or omissions of the builder's sub-contractor. While Curtis finds this surprising,[44] the builder's assumption of responsibility for his sub-contractor's defaults in carrying out guarantee work is consistent with its general responsibility in clause 19 for the acts of sub-contractors.

(c) Responsibilities and exclusions from liabilities

Clause 37 draws together both the builders' and buyers' responsibilities and exclusions of liability; to achieve balance, they are expressed mutually as far as this is logically possible. Sub-clause (b) is the builder's main exclusion clause for liability for defects in the vessel; it is much more widely drafted than many yard contracts, expressly excluding tortious liability (including negligence), with the apparent aim of overcoming the usual English law presumption that exclusion clauses are not intended to cover negligence.[45] The rest of the clause follows the standard shipbuilding contract provisions of expressly limiting the buyer's remedy for defects discovered before or at delivery to liquidated damages, exclusion of liability for defects repaired by third parties, the replacement of any implied warranty or guarantee with the guarantee set out in clause 35 and the exclusion of post-termination liability except to the extent provided in clause 39.

Finally, sub-clauses (f) and (g) provide 'knock-for-knock' clauses whereby each party accepts responsibility and liability for the personal injury and damage to property of its own employees, regardless of negligence (absent wilful misconduct).

It should be noted that the builder's warranty does not replace the buyer's right

---

40 See Curtis at p. 177.

41 S. Zimmer, *Shipping World & Shipbuilder, (211)* 2011 at p. 19; and see *China Shipbuilding Corp v Nippon Yusen Kabukishi Kaisha and Galaxy Shipping PTE Ltd (The Seta Maru)* [2000] 1 Lloyd's Rep 367 on the treatment of consequential damage under the SAJ form.

42 See Curtis at p. 178.

43 *Ibid*, at p. 176.

44 *Ibid* at p. 172.

45 See, e.g., *Canada Steamship Lines v The King* [1952] 1 Lloyd's Rep 1; *HIH Casualty and General Insurance Ltd v Chase Manhattan Bank* [2003] UKHL 6; [2003] 2 Lloyd's Rep 61.

to claim liquidated damages for defects in the vessel which were not claimed at the time of delivery.[46] Additionally, reflecting the position at common law,[47] sub-clause (e) provides that the parties' liability to each other post-termination is not limited or excluded where one party has repudiated the contract. Given the wave of contractual repudiations following the financial crash of 2008, there are good policy reasons for excluding this.

(d) Insurance

Clause 38 sets out the builder's obligations in respect of procuring insurance for the project, which must cover both the total of instalments paid by the buyer to builder and the value of the buyer's supplies at the builder's yard.

Additionally, the clause requires the builder to make good any damage caused by an insured cause (whether or not it receives the insurance proceeds) and in the event of total loss requires the builder to repay the instalments paid by the buyer with interest. Although more buyer-friendly than the CMAC or SAJ forms,[48] it is submitted that this more fairly reflects the position of the buyer. Unlike the builder, the buyer has no control over the effect of an insured cause on the project; it should fairly receive a return on its money in the event of actual or constructive total loss for its having been out of pocket for some time without any vessel to show for it.

The clause terminates the parties' future rights or obligations upon the builder returning the instalments following a total loss, but not their existing accrued rights under the contract. This could, for example, assist a buyer who has been forced to accept the return of its instalments because of a total loss caused by the builder's negligence.[49]

(e) Suspension and termination

Clause 39 sets out a clear and inclusive regime for the parties' termination rights,[50] a 'checklist' of the most common scenarios that might occur. In this regard the form improves upon the SAJ in several respects. Under (a)(i) and (b)(i), both parties have the right to terminate if the other party's main guarantor is deemed insolvent; this provides both sides with important protection as they would otherwise be forced to continue with the contract in extremely risky circumstances with little chance of its being deemed frustrated. Under sub-clause (d), both parties have the right to terminate for the insolvency of the other, which the buyer lacks under the SAJ form; this reflects modern practice[51] and provides the buyer with vital protection.

Turning to the buyer's specific rights of termination, sub-clause (a)(ii) gives the buyer the right to terminate if the builder fails to carry out work for a period of time decided by the parties in Box 22 of Part I. This is intended to provide the buyer with protection in situations where the yard's insolvency is clearly impending, as failure

46 *Cenargo Ltd v Empresa Nacional Bazan de Construcciones Navales Militares SA* [2002] EWCA Civ 524; [2002] CLC 1151.

47 *Stocznia Gdynia SA v Gearbulk Holdings Ltd* [2009] EWCA Civ 75; [2009] 1 Lloyd's Rep 461.

48 See Curtis at pp. 198 and 231

49 *Ibid*, at p. 232

50 See Mandaraka-Sheppard, para. 9.1.

51 See Curtis at p. 196.

to perform contractual obligations is often an early warning sign of insolvency. Additionally, formal insolvency may be difficult to ascertain in certain jurisdictions. A criticism of this provision is that it can reduce the builder's flexibility in terms of scheduling,[52] but the ability of the parties to negotiate a suitable length of time should ensure that the buyer only requires this protection in situations where the yard has encountered drastic unforeseen difficulties.

The buyer also has the standard right of termination after 180 days' delay by *force majeure* events, 180 days of non-permissible delays or an aggregate of 270 days of both. This provides a useful 'long-stop' date for the parties. To prevent the buyer from creating delays in an attempt to get out of its contractual obligations, the clause does not count the 'other events' listed in clause 34(a)(ii) (discussed above) towards this aggregate. Additionally, the timing of the clause intersects with the length of the model refund guarantee in Annex A(iii) (contractual date of delivery plus 300 days) by providing a 'buffer period' during which the parties can try to reach agreement on whether to persevere with the contract or terminate it, consult with financiers, seek board approvals, etc. As discussed in Section 2, the buyer also has the right to terminate the contract under sub-clauses (a)(iv) to (viii) if the sea trials reveal that the vessel has not met the contract's requirements.

The builder's rights of termination mostly reflect those under the SAJ, providing a more comprehensive list of events of default with slightly more generous time limits for the repair of defaults by the buyer.[53] For example, the builder's rights of termination accrue after the buyer has failed to pay an instalment for 21 banking days, after which it must give five banking days' notice of default – in practice, over a month, compared to the SAJ form's 15-day period before default. However, in common with the CMAC form, the builder gains an additional right, not possessed at common law,[54] to suspend work if the buyer has failed to pay for 15 banking days. This counts as an event of permissible delay under clause 34(b), although the builder must give notice to be able to claim for an extension of time rather than having the contractual date of delivery automatically extended as occurs under the SAJ form. Curtis identifies the very specific possibility of a buyer taking advantage of this 15-day period by refusing to take delivery of the newbuilding with the intention of seeking to terminate for delay shortly afterwards;[55] however, this may be a risky course for a buyer given that the builder has the express right to terminate where the buyer fails to take delivery.[56] The clause therefore provides the builder with a useful right where it believes that the buyer will repair its default at some point in the future, and if there are concerns that the clause is too favourable to the buyer, they can easily be remedied by amending the time periods and notice provisions or providing for an automatic extension of time without the need for notice albeit the latter would be unusual without some prior notice. Provision might also be made for the builder to recover the additional costs incurred during any period of suspension.

Following the buyer's termination, the builder must return all sums paid plus

52 See Krut at p.3.
53 See Mandaraka-Sheppard, at para. 12.1.
54 See Curtis at p.281.
55 *Ibid*, at p.212.
56 See cl. 39(b)(iii).

interest thereon and return the buyer's supplies or, if they cannot be returned, pay the buyer's cost of obtaining them.[57] The builder's rights in the event of the buyer's default are, broadly, to complete construction of the vessel and sell it to another party, or to sell the unfinished vessel to another party; any profit on the vessel over what would have been made under the contract is to be paid to the buyer (but without interest as the sale is due to the buyer's default).[58] Importantly, the form provides protection to the buyer by requiring the builder to sell at the '*best price reasonably obtainable*', in contrast to the SAJ's '*rather autocratic*' terms of sale.[59] The builder can also recover any shortfall by first selling the buyer's supplies and then claiming the remainder from the defaulting buyer. NEWBUILDCON's requirement that in both cases the vessel be sold by the builder, rather than providing for the buyer to take delivery of the unfinished vessel, avoids any potential complications surrounding passing of title and comprehensively accounts for the application of the buyer's funds.

Curtis regards the clause as arguably more favourable to the buyer than its equivalent in the SAJ or CMAC forms.[60] In particular, he submits that it is rather arbitrary that the builder has no contractual right to claim additional costs incurred as a result of the buyer's non-payment prior to delivery, but may recover these if the contract is terminated and the vessel sold. However, he notes that this does not preclude the builder's right to claim against the buyer for such costs as a breach of the contract without actually terminating it.[61] It is not easy to see how such costs might arise, and it should be remembered that the builder has both a general right to interest under clause 18 at a level determined by the parties to recompense for any such late payment and its right to stop working during the period of default (see above).

(f) Intellectual property

Clause 40 deals with intellectual property rights. NEWBUILDCON's provisions are far more comprehensive here than the SAJ form in applying to all IP rights, rather than merely patents.[62] Additionally, the form provides a very practical exception to the restriction on disclosure where this is required for an on-purchaser of the vessel.

(g) Choice of law

The form allows the parties to choose whichever law they want (with English law in default of choice), although it should be remembered that whether this choice applies to collateral contracts and to the main contract will depend on the circumstances.[63]

(h) Dispute resolution

Clause 42 is NEWBUILDCON's dispute resolution clause. Compared with those in the SAJ or CMAC forms, it is a greatly modernised, multi-tiered clause, based

57 See cl. 39(e).
58 See cl. 39(f)
59 See Mandaraka-Sheppard, at para. 12.5.1.
60 See Curtis at p.210.
61 *Ibid*, at p.211.
62 *Ibid*, at p.255.
63 *Lurssen Werft GmbH & Co KG v Halle* [2010] EWCA Civ 587; [2011] 1 Lloyd's Rep 265.

on the BIMCO Standard Dispute Resolution Clause and adapted for use in the shipbuilding context. It provides four different methods of resolving disputes:

- classification society or regulatory authority reference;
- expert determination;
- mediation; and
- arbitration.

The first of these is provided for specific disputes relating to the relevant body, on the basis that this refers determination of the dispute to those best placed to determine it. It should be noted that while such decisions are '*final and binding*',[64] this is intended to refer to decisions of the body's head office rather than local officials.[65]

The inclusion of expert determination is a fairly novel concept for shipbuilding contracts despite its widespread use in other areas of construction. The clause includes a timetable for the submission of written submissions within seven days of the expert's appointment, with the expert to report within 28 days of appointment. It provides a quicker and potentially more practically focused method of dispute resolution than arbitration or mediation, although as it is not mandatory, its use depends on the parties agreeing on an appropriate expert.

The inclusion of arbitration is standard for shipbuilding contracts; as with the choice of law, the parties are free to choose the seat of the arbitration (although Curtis notes that, somewhat vaguely, the form does not provide for their choosing the procedural rules)[66] with London Arbitration to apply in default of choice. Unlike CMAC, the clause does not prevent the parties appealing an award to a court, providing the parties with a measure of protection from dubious decisions at the expense of finality.

NEWBUILDCON's mediation clause is intended to operate alongside the arbitration procedure rather than in place of it. The parties may agree to mediate any dispute before arbitration proceedings commence; if they have commenced, either party can call on the other party to agree to mediation without the arbitration proceedings halting. Refusal to agree to mediation can be subsequently taken into account by the tribunal when allocating the costs of the proceedings.[67] The use of mediation in NEWBUILDCON has been criticised as '*vulnerable to delays and imprecision in timing*',[68] although the fact that the tribunal may take the mediation timetable into account when setting the arbitration proceedings' timetable[69] should significantly ameliorate this drawback.

64 See cl. 42(a).
65 BIMCO NEWBUILDCON Explanatory Notes at p. 12.
66 See Curtis p. 235.
67 See cl. 42(e)(iii).
68 F.X. Nolan, '*Shipbuilding Disputes: Selecting and Tailoring Arbitration to Suit*' Available at: http://www.vedderprice.com/files/Publication/46a1676d-14b0-4f0c-a187-f736deda8034/Presentation/PublicationAttachment/16d91d9e-fc8d-4b0d-b1dc-bc188f07891f/Shipbuilding%20Disputes%20-%20Selecting%20and%20Tailoring%20Arbitration%20to%20Suit.pdf (last accessed 31 December 2014).
69 See cl. 42(e)(v).

2.4.6 Section 6 – Sundry

(a) Notices

This section concludes Part II by dealing with various matters not fitting elsewhere. In particular, clause 43 (Notices) provides a much more up-to-date list of procedures than the slightly dated SAJ form,[70] eschewing the '*cable*' in favour of broader '*electronic*' methods of communication.

(b) Effective date

NEWBUILDCON allows the parties to choose their own conditions for the contract becoming effective (discussed above in Part I), in contrast with the use of conditions precedent and subsequent in the CMAC and SAJ forms.[71]

(c) Assignments

It is also somewhat freer in allowing the parties to assign the contract to their financiers without the consent of the other, and unlike the SAJ form does not require that the original buyer remain liable under the contract.[72]

(d) Options

Clause 46 provides a simple options clause, allowing the parties to agree for additional vessels to be built on the same terms as the contract with logical amendments. Though this is an 'unusual feature',[73] which could have been extended with the inclusion of a draft notice of exercise of the option or provisions for adjustment of the purchase price or other terms,[74] the clause allows parties who are happy to use the same terms for optional vessels to do so without the added time and expense of a new agreement. Additionally, in circumstances where the parties have agreed that optional vessels will be at the same price and under the same terms, it is difficult to see why a pro forma notice should be necessary.

(e) Entire agreement

Finally, clause 47 includes an 'entire agreement' clause, preventing further disputes over whether terms have been incorporated in the contract from extra-contractual discussions or correspondence. Curtis' view is that the wording employed is sufficiently broad to exclude any allegation of non-fraudulent pre-contractual misrepresentation. This provides greater protection for the builder than the SAJ form, which fails to include express non-reliance language.[75]

70 See Curtis at p.262.
71 *Ibid*, at p.265.
72 *Ibid*, at p.252.
73 *Ibid*, at p.308.
74 See Krut at p.4.
75 See Curtis at p.278.

2.4.7 Annexes

(a) Guarantees
NEWBUILDCON requires that both parties obtain financial guarantees which are '*substantially in the form and substance*' of those in Annex A.[76] The guarantees supplied were each vetted by banking lawyers with a particular focus on their being suitable for projects in China. For the buyer, the form includes a choice of letter of guarantee (given by the buyer's bank) or buyer's performance guarantee (given by a parent company). The form also provides a model refund guarantee for the builder and model letter of guarantee for the builder's obligations where it has forced the buyer to take delivery of a vessel with defects (see clause 27 above).

The guarantees contain vital 'anti-discharge' wording, designed to protect against the 'rule in *Holme v Brunskill*'[77] which discharges the guarantor's obligations if the underlying contract becomes materially different without his consent, though they do not exclude the beneficiary's initial duty of disclosure to its guarantor.[78] Additionally, the guarantees expressly make the guarantor liable as 'primary obligor', which provides important protection for the parties in the event of the other's insolvency, allowing recovery of sums due without having to deal with insolvency proceedings. Finally, each guarantee's effectiveness is subject to any disputes being determined by '*final unappealable judgment*', providing each party's guarantor with a measure of protection from illegitimate calls on the security.

(b) Specification and maker's list
Annexes B and C are left blank, providing spaces for the parties to add a specification for the vessel and a maker's list respectively.

2.5 Conclusion

Although there was much initial scepticism that the form would be accepted by shipyards, many of whom apparently perceived the form as somewhat buyer-friendly in what was then a seller's market,[79] NEWBUILDCON has steadily grown in popularity. As at June 2014, over 200 new projects are estimated to have used the form since its January 2008 release, with an average of 42 projects a year known to have used the form between 2010 and 2013 inclusive. The versatility of the form is evidenced by its use in a variety of newbuilding projects, including bulkers and tankers, chemical carriers, passenger ships and heavy lift vessels. Additionally, BIMCO has received many requests from Chinese yards for the form notwithstanding its buyer-friendly reputation, indicating that it will be used as a basis for future projects for many years to come. The reasons for this are readily apparent.

The first thing to notice is that NEWBUILDCON has a newly developed structure

76 See cl. 14.
77 *Holme v Brunskill* [1878] 3 QBD 495.
78 See Curtis at p. 299.
79 See, e.g., Krut p. 1, writing in September 2007.

which guides parties through the sequencing of events involved in a newbuilding. BIMCO's distinctive box layout provides a convenient summary of the main commercial terms which are a useful checklist for negotiations. As with BIMCO's other widely used documents, NEWBUILDCON is also available in an electronic format using BIMCO's online pay-as-you-go editing system, Idea 2, which provides an electronic document which can be edited and is easy to navigate with cross-reference hyperlinks.

The clear structure and language improves the readability of the contract. But there is more to it than a change of appearance. Considerable efforts have been made to provide a degree of clarity often missing from existing forms so that parties know where they stand from the outset. Particular features worthy of note are:

- The form is readily adaptable to all types of vessels in all jurisdictions, and includes features which will be needed for projects in China.
- There are more comprehensive provisions detailing obligations during production, particularly sub-contracting, approvals and inspection, tests and trials, modifications, and buyer's suppliers.
- The buyer's relationship with class is more clearly defined.
- Mechanisms are included to avoid delays in delivery in the event of minor defects on delivery.
- Clear provisions relating to provision of refund and performance guarantees.
- Inclusion of wordings for refund and performance guarantees, vetted by banking lawyers, and which are suitable for use in projects in China.
- Harmonised refund provisions in the event of termination.
- Clearer legal provisions dealing with permissible delays, builder's guarantees, responsibilities and exclusions from liabilities, insurance and termination. Responsibility clauses are expressed mutually where possible, producing a more balanced contract.
- A comprehensible *force majeure* clause with an easy-to-check list of *force majeure* events.
- A checklist of documentation required on delivery.
- New IMO Hazardous Materials Inventory and Protective Coatings clauses.
- Modernised dispute resolution provisions clause incorporating reference to the latest dispute resolution methods, including class and expert determination, mediation and arbitration.

NEWBUILDCON has the potential to deliver huge improvements and time savings in the drafting and negotiation process. The improved clarity should also enable parties to easily identify their rights and obligations and minimise the risk of disputes relating to the interpretation of the contract. Indeed, it is notable that despite the torrent of shipbuilding disputes making their way into arbitration and the courts since the dramatic collapse in shipbuilding prices following the financial crisis in 2008, NEWBUILDCON has not to the author's knowledge so far been the subject of an award or judgment which in itself is testimony to its clarity.

Alongside its growing use in shipbuilding projects, parts of the wording of NEWBUILDCON have also been used in negotiations between buyers and builders and as a precedent for drafting particular clauses of bespoke contracts. The process of having shipyards adopt new forms for contracts is a long one, but NEWBUILDCON has taken its place as one of the main standard forms for major projects and will continue to play an important role as a shipbuilding form of global application.

CHAPTER 3

Remedies for breach of shipbuilding contracts – is English law 'fit for purpose'?

*Simon Curtis**

3.1 Introduction

The post-war era has seen numerous periods in which global shipbuilding has experienced either 'boom' or 'bust' market conditions. During such periods demand for new ships has either increased or declined rapidly in response to changes in global transportation needs. The most pronounced of recent global market movements occurred in the early part of this century which, in the period between 2005 and 2008, saw huge increases in world demand for new vessels and the prices of newbuilding tonnage, followed by a cataclysmic market collapse as a consequence of the 2008 'credit crunch' and subsequent global economic recession. In 2010, the Organisation for Economic Co-operation and Development (OECD) reported that global shipbuilding orders had fallen from a total of 22.2 million cgt (compensated gross tons) in the third quarter of 2007 to just over 1 million cgt in the last quarter of 2008 and the first quarter of 2009, a fall of some 90 per cent in under two years.[1] Much of the new shipbuilding capacity which had developed in Asia, particularly in China, in 2005–2008 proved in consequence surplus to requirements; world order books remain very 'thin' and newbuilding prices in many vessel sectors continue to decline.

Such marked and rapid changes in market conditions obviously occur in many industrial sectors, but their impact in a global shipbuilding context is particularly powerful. Shipbuilding projects are almost invariably undertaken on a fixed-price, 'lump sum' basis, with the contract price for the vessel payable by the purchaser in various instalments before and upon her delivery. These factors, coupled with the extensive time frame required for the construction of a commercial newbuilding – sometimes as much as four years from the date of contract signing to delivery – generate significant risks for both parties. These exposures include in particular the risk, normally borne exclusively by her purchaser, that the completed vessel will at the time of delivery be worth substantially more, or substantially less, than the price at which she was originally contracted.

As is well known, during periods of market contraction there have frequently occurred instances in which shipowners, facing the prospect of taking delivery of newbuildings at prices which are substantially less than their market value, have

* Senior Partner, Curtis Davis Garrard LLP, London. The author gratefully acknowledges the assistance of Helen Conybeare Williams in the preparation of this chapter.

1 See www.oecd.org/industry/ind/fallinshipbuildingsettocontinueforsometimesaysoecdcouncilworkingpartyonshipbuilding.htm (last accessed 31 January 2015).

either refused delivery on the grounds that these do not meet the quality or performance standards they were expecting, or sought to delay delivery in the hope that they will, in due course, become entitled to cancel the contract and recover their pre-paid instalments of the price. It should, of course, be recognised in this context that the decision to 'walk away' may not necessarily be a matter for the purchaser alone – if his financing banks have refused to maintain their funding commitment in light of a serious decline in newbuilding values, the purchaser may have no means of paying for the vessel even if he wishes to do so.

Similarly, the less frequent boom periods in world shipbuilding in the last 20 years have given rise to demands by shipbuilders that their customers should pay an increased contract price, or provide other financial inducements, as a condition of agreeing to complete and deliver newbuildings which have increased in market value since they were first contracted. In the frenetic period between 2003 and the global 'crash' in Autumn 2008, the market witnessed numerous instances of supposedly reputable shipbuilders demanding additional payments under shipbuilding contracts for vessels which had substantially increased in market value.

A significant number of the shipbuilding contracts concluded each year for export vessels are subject to English law and provide for the resolution of disputes in London, usually by arbitration. It is therefore relevant to consider which remedies are available under English law to assist both shipbuilders and purchasers in dealing with contractual counterparts who are not behaving entirely honourably. Analysis suggests that English law is unsatisfactory in this respect, lacking certain remedies which, in the interests of justice and commercial efficiency in this sector, should readily be available to shipbuilders and shipowners.

3.2 Enforcing delivery in a depressed market

The first area of concern arises where the newbuilding market is depressed and the purchaser is seeking either to renegotiate the contract price down to current market levels or to avoid taking delivery altogether. The purchaser may, in these circumstances, refuse to accept delivery of the vessel when tendered to him and justify his actions on the grounds that she is not sufficiently compliant with the requirements of the contract. Whilst there will be a proportion of such cases in which the purchaser's conduct can be justified on technical or other grounds, it is remarkable how these disputes tend to emerge in difficult market conditions and to evaporate when economic conditions have improved.

From the shipbuilder's perspective, the purchaser's conduct in seeking to renegotiate the contract price, or alternatively to avoid payment altogether, can obviously have serious financial consequences. The shipbuilder will undoubtedly have committed substantial resources to the construction of the vessel and, particularly where the contract price instalments are 'back loaded', so that a significant proportion of the price is payable only upon delivery, may be substantially exposed if the purchaser is unable or unwilling to take delivery of the vessel and to pay the final instalment.

At this point, the shipbuilder may well seek legal advice as to his options and receive an unwelcome surprise in the form of the United Kingdom Sale of Goods Act 1979. English sale of goods legislation, introduced in the UK in the 1890s and now largely embodied in the 1979 Act, applies in principle to an international contract to build a ship as much as it does to a consumer purchase; this is simply because in English law, ships are 'goods'[2] and shipbuilding contracts are agreements for the sale of 'future goods' by description.[3]

This categorisation has important implications, in particular that, unless the parties have clearly agreed otherwise, there will automatically be imported into the shipbuilding contract so-called 'statutory implied terms' relating to the vessel's 'compliance with description', 'satisfactory quality' and 'reasonable fitness for purpose' as defined in sections 12–14 of the 1979 Act.

But a further issue, concerning the scope of the legal remedies available to the shipbuilder, also arises in consequence of the application of the 1979 Act to shipbuilding contracts governed by English law.

Sections 49 and 50 of the 1979 Act, which comprise a chapter of the legislation entitled 'Seller's Remedies', provide in this respect that:

> 49(1) Where, under a contract of sale, the property in the goods has passed to the buyer and he wrongfully neglects or refuses to pay for the goods according to the terms of the contract, the seller may maintain an action against him for the price of the goods...
>
> 50(1) Where the buyer wrongfully neglects or refuses to accept and pay for the goods, the seller may maintain an action against him for damages for non-acceptance.

As will be clear, whether the different remedies provided by sections 49 and 50 are available to the seller depends entirely upon the issue of title (i.e. whether he continues to own the goods at the time at which he seeks to enforce his contractual rights).

Where title to the goods has already passed to the buyer, but he has '*wrongfully neglect[ed] or refuse[d]*' to pay for them, the seller is entitled under section 49 to sue for the entirety of the outstanding purchase price – his claim is in debt, rather than damages and he is not obliged to show, in terms of causation or foreseeability, that he has suffered recoverable loss; nor is he affected by English common law rules relating to mitigation of that loss. Where, however, title to the goods has *not* passed because the buyer has wrongfully refused to accept and pay for them, the seller's remedy under section 50 of the 1979 Act is an action in damages for non-acceptance, in relation to which the usual principles of causation and foreseeability apply.

It will be noted that the language of section 49(1) is expressed in permissive, rather than mandatory terms. In stating that the '*seller may maintain an action*' for the price, the statute provides no guidance as to whether he must do so. For many years uncertainty existed as to whether, in the absence of an express or implied term in the contract, a right to claim the price (as opposed to damages) for 'undelivered' goods existed either under the 1979 Act or in English common law.[4]

2 *Behnke v Bede Shipping Company* [1927] 1 KB 649.

3 *Reid v Macbeth and Gray* [1904] AC 223 and *McDougall v Aeromarine of Emsworth Ltd* [1958] 2 Lloyd's Rep 345. On this subject generally, see S. Curtis, *The Law of Shipbuilding Contracts* (4th edn), (Informa 2012), at pp. 1–4.

4 See the conflicting views of the Court of Appeal in *Harry & Garry v Jariwalla* [1988] WL 1608652 and *Otis Vehicle Rentals Ltd v Cicely Commercials Ltd* [2002] EWCA Civ 1064.

This debate appears, however, recently to have been settled by the 2013 Court of Appeal decision in *Caterpillar (NI) Ltd v John Holt & Company (Liverpool) Ltd*[5] in which, after considering earlier authorities on the point, Longmore LJ held that:

> .... if there is a potential claim for damages for non-acceptance, then, if property has not passed to the buyer, the seller should be confined to that claim rather than a claim for the price ...[6]

In a shipbuilding context, this is very significant. Although in the eighteenth and nineteenth centuries, many shipbuilding contracts were concluded on the basis of a 'continuous transfer of title' (under which the purchaser progressively acquired title to the vessel as she was constructed by the shipbuilder), the overwhelming proportion of twentieth century and contemporary contracts for the construction of ships provide expressly for title in the vessel and her components to remain in the shipbuilder until she has been accepted and paid for in full by the buyer.[7] Following successful completion of the sea trials, the newbuilding is 'tendered' for delivery by the seller. At a mutually convenient date shortly thereafter, a closing meeting takes place at which, in exchange for a certificate or bill of sale permitting her flag registration by the purchaser and other documents, the purchaser pays the delivery instalment, the parties each sign a so-called Protocol of Delivery and Acceptance and the purchaser's crew assume both factual and legal possession of the vessel prior to sailing her from the shipyard.

Since, in most modern shipbuilding contracts, title does not pass prior to the acceptance of the vessel by the purchaser, but there will almost invariably exist an entitlement to damages for 'non-acceptance', there can accordingly be no claim by the shipbuilder for payment of the delivery instalment if the purchaser wrongfully refuses to accept delivery. The shipbuilder's remedies in these circumstances are either (i) to exercise the contractual default mechanism in the shipbuilding contract, terminating the agreement for the buyer's breach and selling the completed vessel elsewhere; or (ii) to treat the purchaser's unlawful refusal to accept delivery as a repudiation of the contract which the shipbuilder then accepts, thereby replacing the parties' primary obligations under the contract with a secondary obligation on the part of the purchaser to pay damages.[8] In neither case, however, can the shipbuilder maintain a claim in debt for the unpaid balance of the contract price of the vessel represented by the delivery instalment.

In a limited category of cases, this may not represent a serious problem for the shipbuilder – if demand for newbuildings is strong, the shipbuilder will almost

---

5 [2013] EWCA Civ 1232; [2014] 1 Lloyd's Rep 180.

6 *Ibid* at [52] of the judgment. Longmore LJ also noted that s 49(2) of the 1979 Act provided for the seller of goods to maintain an action for the price where property in the goods had not passed if that price was payable on a 'day certain' irrespective of delivery. He considered at [41] that this indicated that s 49(1) was intended to specify the exclusive circumstances in which the seller might maintain an action for the price – s 49 would otherwise in his view be '*largely otiose*'.

7 See, e.g., Art. VII.5 of the Shipbuilders' Association of Japan (SAJ) 1974 standard form and Art. 31 of the BIMCO NEWBUILDCON form published in S. Curtis, *The Law of Shipbuilding Contracts* (4th edn), (Informa, 2012) and also at www.bimco.org (last accessed 31 January 2015).

8 See on this subject Curtis, *ibid*, at pp. 123–125 and 217–221. See also H. Beale *et al*, *Chitty on Contracts* (31st edn), (Sweet & Maxwell, 2012) Chap. 24 on the circumstances in which one party is entitled to treat the other as in breach.

certainly be able to sell the vessel quickly and recover his losses from the defaulting purchaser.

In practice, however, this is rarely the position, and the shipbuilder will often regard the remedy of terminating the shipbuilding contract and selling the vessel elsewhere as extremely unattractive. For many Asian shipbuilders in particular, there are likely to be significant reputational issues associated with terminating a substantial shipbuilding contract, involving a public 'loss of face' even in circumstances in which the termination has resulted squarely from the buyer's default. Furthermore, the delay and uncertainty generated by the need for the shipbuilder to go into the market to resell the vessel may well cause the shipbuilder losses – for example, in respect of his own financing or currency swap arrangements – which will not be capable of recovery in English law by reason of their remoteness from the original breach. To add to the problem, where the purchaser is a single purpose entity and the shipbuilder holds a parent company guarantee of the purchaser's payment obligations under the contract, such guarantee may not extend to the shipbuilder's claims in damages consequent upon his termination of the contract and sale of the vessel elsewhere – the shipbuilder is therefore left unsecured to the extent that its losses on a resale exceed the instalments he has previously been paid under the shipbuilding contract.

All of these problems are likely to be particularly acute in circumstances in which the newbuilding in question is *sui generis* (i.e. the vessel has been built to the buyer's particular design or the requirements of a particular trade or project and as a consequence there are limited resale opportunities available to the shipbuilder).

Whilst it may be perfectly reasonable to require the seller of a commodity, such as oil, coal or grain, to go 'into the market' and sell such goods elsewhere in circumstances in which the buyer has wrongfully refused to accept delivery, this can, accordingly, be very inequitable where the seller has spent significant time and resources in completing a complex manufactured, and possibly illiquid, good such as a newly constructed ship.

In these circumstances, the vessel having been completed in accordance with the shipbuilding contract, there appears to be no sound commercial reason why the purchaser should not be 'held to his bargain' and required – effectively by way of specific performance of the contract – to pay the price he originally agreed to pay. While it may often be impractical for an English court or arbitration tribunal to compel a foreign shipowner to sign the Protocol of Delivery and Acceptance,[9] or to put a crew on board the vessel to remove her from the shipyard, these considerations obviously do not affect the principle involved.[10]

It should be noted in this context that, in the converse situation in which the seller of goods is refusing to deliver them to the buyer, there is a 'specific' remedy available under the 1979 Act. Section 52 provides in this respect that:

9 Although there are powers exercisable by the High Court to authorise the execution of documentation which one party to an English law contract is wrongfully refusing to sign – *The Messiniaki Tolmi* [1983] 2 AC 787 (HL).

10 A shipowner required to pay in full the contract price for the vessel he has contracted to purchase would, in any event, undoubtedly take delivery in order to realise the value for which he has paid.

> In any action for breach of contract to deliver specific or ascertained goods the court may, if it thinks fit, on the plaintiff's application, by its judgment or decree direct that the contract shall be performed specifically, without giving the defendant the option of retaining the goods on payment of damages.

A newly built vessel is plainly 'specific or ascertained goods' within the meaning of the section; upon the buyer's application, there is accordingly a discretionary power granted to the court to order that the shipbuilder should deliver the vessel to the purchaser in circumstances in which the shipbuilder is refusing to do so.[11] It is both surprising and inequitable in such circumstances that the shipbuilder (or indeed any other manufacturer of complex industrial machinery and equipment) should not be entitled to the equivalent specific remedy of a claim for the delivery instalment of the contract price where the buyer is – in breach of contract – refusing to accept delivery. This is particularly so in circumstances in which an award of damages consequent upon the sale of the goods to a third party would not represent an adequate remedy to compensate the seller for the buyer's default.

In a newbuilding context, this problem could – at least in theory – be addressed by the parties themselves in the shipbuilding contract between them. The pre-delivery instalments of the contract price under the overwhelming majority of international shipbuilding contracts are agreed to be payable against specific events (e.g. keel-laying) and/or the provision of third-party documentation, often to be provided by the vessel's classification society, evidencing such events; it is clear from numerous authorities[12] that the builder can in such circumstances sue for the instalment in question when the 'triggering' event has occurred and/or the relevant evidential documentation has been supplied. If, therefore, the shipbuilder and the purchaser choose to agree that the shipbuilder's entitlement to the final instalment should arise irrespective of delivery (against, for example, the issuance of a certificate from the classification society that the vessel has been completed), they could circumvent the limitations of the 1979 Act and thereby entitle the shipbuilder to claim the final instalment regardless of any prior transfer of title in the vessel.

In practice, however, these are – for good reason – not the 'terms of trade' on which global shipbuilding is conducted. Shipowners, and more importantly their financiers, will almost never agree that the shipbuilder's entitlement to payment of the full contract price should accrue ahead of delivery of, and transfer of title in, the completed vessel as the purchaser's ability to draw down the financing required to pay the final instalment of the price is invariably conditional upon his acquiring title to the vessel. Without such title, he cannot grant to his financiers the mortgage which represents the latter's principal security for the funds being advanced.

Furthermore, that the parties to a shipbuilding contract could, at least in theory, create a contractual remedy where one does not apparently exist at law is, in any event, not an argument against reform. There are countless instances in English law in which Parliament or the courts have created the contractual remedies needed

11 Although see *infra* in relation to the problems of securing an order for specific performance of a shipbuilding contract.

12 See, in particular, *Hyundai Heavy Industries Co v Papadopoulos* [1980] 2 Lloyd's Rep 1 and *Wuhan Guoyu Logistics Group Co Ltd v Emporiki Bank of Greece SA* [2012] EWCA Civ 1629; [2014] 1 Lloyd's Rep 266.

to ensure fairness or business efficiency in circumstances in which such remedies could, if they had so chosen, have been agreed by contracting parties themselves. Indeed, if one were to accept the argument that legal reform is inappropriate where a contractual 'self-help' remedy is available to contracting parties, most of the provisions of the Sale of Goods Act 1893 and its various statutory descendants, including the 1979 Act, would never have been enacted into legislation.

Against this background, there are various statutory and judicial solutions to the 'problem' created by section 49(1). One approach would be for Parliament to pass legislation effectively reversing the judgment of the High Court in *Behnke v Bede Shipping Co Ltd*[13] and establishing that a commercial cargo or passenger vessel does not constitute 'goods' within the meaning of the 1970 Act – this would have significant benefits in other respects,[14] although it is perhaps unrealistic to believe that Parliamentary time could be found for a specialised reform of this type.[15]

A simpler solution would be for a senior English court[16] to reverse the decision in *Caterpillar* and to hold that section 49 of the 1979 Act is permissive rather than mandatory in its application. Provided that the vessel was in a condition complying with the requirements of the contract and he had otherwise met the conditions for delivery (he had, for example, obtained, all necessary class and regulatory approvals and certificates) the shipbuilder could then formally tender the vessel for delivery and enforce his claim for payment of the delivery instalment of the contract price.

An alternative judicial solution to the problem would be to re-categorise shipbuilding contracts in English law – a process which may already be underway. Although, as already indicated, these have historically been classified as contracts for the sale of goods, rather than for the provision of work and materials, there have been at least two important decisions, those of the House of Lords in *Hyundai Heavy Industries Co v Papadopoulos* in 1980[17] and *Stocznia Gdanska SA* v *Latvian Reefers* in 1998,[18] in which it has been recognised that this legal categorisation is problematic. Although these cases concerned, *inter alia*, the rather different issue of whether the buyer's guarantor was liable to the shipbuilder to pay a pre-delivery instalment of the contract price following cancellation, this issue was directly influenced by the legal categorisation of the contract, in relation to which Lord Fraser of Tullybelton[19] said in *Hyundai*:

> Much of the plausibility of the argument on behalf of the guarantor seemed to me to be derived from the assumption that the contract price was simply a purchase price. That is

13 [1927] 1 KB 649.

14 It would, in particular, avoid the requirement, rendered necessary by the surprising decision of the High Court in *Dalmare SpA v Union Maritime Limited and Valor Shipping Limited (The Union Power)* [2012] EWHC 3537 (Comm); [2013] 1 Lloyd's Rep 509, for buyers and sellers of a second-hand ship to decide whether or not to exclude the statutory implied terms of 'satisfactory quality' and 'reasonable fitness for purpose' in an 'as is' vessel sale.

15 It is, of course, possible that the amendment could be made within the framework of a more wide-ranging reform of sale of goods legislation – but this is not currently in prospect.

16 Such decision would need to be taken by the Supreme Court or, applying the principles in *Young v Bristol Aeroplane Co Ltd* [1944] KB 718 (CA), by the Court of Appeal in light of the conflicting decisions in *Caterpillar* and *Otis Vehicle Rentals Ltd v Cicely Commercials Ltd* discussed above.

17 [1980] 2 Lloyd's Rep 1.

18 [1998] 1 Lloyd's Rep 609.

19 [1980] 2 Lloyd's Rep 1, at p. 13.

> not so, and once that misconception has been removed I think it is clear that the shipbuilding contract has little similarity with a contract of sale and much more similarity, so far as the present issues are concerned, with contracts in which the party entitled to be paid had either performed work or provided services for which payment is due by the date of cancellation. In contracts of the latter class, which of course includes building and construction contracts, accrued rights to payment are not (in the absence of express provisions) destroyed by cancellation of the contract.

In stating that shipbuilding contracts exhibit '*little similarity with a contract of sale...*', and in refusing to apply sale contract principles to the resolution of the dispute, his Lordship displayed both courage and perspicacity.[20] If, when a suitable opportunity arises in the future, the senior courts of England and Wales are willing to confirm the commercial reality that large shipbuilding contracts are indeed construction agreements, rather than simple contracts of sale and purchase, the problem generated by section 49(1) of the 1979 Act could effectively be consigned to history – global shipbuilders contracting under English law, having completed their side of the shipbuilding bargain, would then have at least a reasonable prospect of successfully claiming the agreed price for the work they had undertaken.

3.3 Enforcing delivery in an 'overheated' market

In the converse and – in a shipbuilding context – less usual situation in which the market is 'overheating' and prices are rising rapidly, the seller of goods (i.e. the shipbuilder) may seek to gain further profit by refusing to deliver them to the purchaser because they are worth much more on the open market and he wishes to capitalise on that additional value. Where – as occurred on numerous occasions during the shipbuilding boom between 2003 and 2008 – the shipbuilder is breaching the shipbuilding contract by refusing to complete and deliver the vessel at the original contract price, what can the newbuilding purchaser do to enforce his English law contract?

One answer to this question lies, yet again, in the Sale of Goods Act 1979, which provides at section 51(1) that:

> Where the seller wrongfully neglects or refuses to deliver the goods to the buyer, the buyer may maintain an action against the seller for damages for non-delivery.

The damages recoverable under section 51(1) of the 1979 Act are represented by the estimated loss directly and naturally resulting in the ordinary course of events from the seller's breach of contract. Where there is an 'available market' for the goods in question, the measure of damages is *prima facie* to be ascertained by the difference between the contract price and the market price of the goods at the time they ought to have been delivered.

However, an award of damages may not be the most appropriate or adequate remedy for the purchaser. What if he has a long term relationship with a charterer which will be ruined if he fails to deliver the vessel on time under the charter? Or his bank will not agree to extend his financing to cover the acquisition in the market place of the replacement vessel at a much higher price? A further possibility is that

20 A similar approach was taken by Viscount Dilhorne.

the shipbuilder is in financial difficulties and may not be 'good' for the damages which might ultimately be awarded against him.

In these circumstances, the newbuilding purchaser contracting under English law has, in practice, just two options.

Firstly, he can bow to the shipbuilder's demand for more money and pay this under protest with the intention of seeking the recovery of his overpayment after delivery of the newbuilding. Second, and alternatively, he can seek – although with considerable difficulty in light of current English law – to force the shipbuilder to complete and deliver the vessel to him at the original contract price.

The first course of action, even assuming that it is available to the purchaser (i.e. that he can afford to pay the 'extra' amount demanded by the shipbuilder) is obviously problematic. Although it is, in principle, open to the purchaser to recover the amount of the overpayment on the grounds that this was paid under economic duress, the success of such a claim will very much depend upon the precise factual circumstances in which the additional payment was demanded, including the state of mind of both the shipbuilder and the shipowner at that time, and their subsequent conduct in the performance of the contract.

In the 1979 case of *North Ocean Shipping Co v Hyundai Construction Co (The Atlantic Baron)*,[21] for example, a Korean shipbuilder had demanded a 10 per cent increase in the contract price of a newbuilding, allegedly to compensate him for the devaluation of the US dollar against the won, and threatened that he would otherwise not perform the contract. The purchaser, who was at that time negotiating a lucrative charter for the newbuilding, agreed to pay the extra 10 per cent demanded but 'without prejudice' to his rights; however, various further instalments of the price were then paid at an increased rate without a further formal protest being made. The High Court held in these circumstances that, although the shipbuilder's threat amounted to economic duress (such that the agreement to pay more was, in principle, voidable at the shipowner's option and the overpayments recoverable); the latter's payment of various instalments of the price without reserving his rights had waived his right of recovery.

On the reasonable assumption that most shipowners would prefer to avoid expensive and uncertain litigation or arbitration proceedings in London to recover monies they were not obliged to pay out in the first place, the second and alternative course of action open to the purchaser is to seek to compel the shipbuilder to complete and deliver the vessel in accordance with his original contractual promise.

Here again, however, the English law remedies available – this time to the purchaser – are arguably inadequate. Although, as previously mentioned, section 52 of the 1979 Act expressly grants an English court or arbitration tribunal the power, by way of an order for specific performance, to order the delivery up of '*specific or ascertained goods . . . without giving the defendant [i.e. the seller] the option of retaining the goods on payment of damages*', such power is in practice almost never used in a shipbuilding context.

This is not because section 52 does not apply to newbuildings – plainly a vessel

21 [1979]1 Lloyd's Rep 89.

under construction is, at least from the point in time at which her keel is laid, '*ascertained goods*'[22] – nor because an award of damages in favour of the buyer will always represent an '*adequate remedy*'.[23] It is instead because the English courts have taken the view that it is not open to them or arbitration tribunals to supervise the shipbuilder's performance of the shipbuilding contract and that this in itself prevents an order for specific performance being made.

The classic exposition of this view was provided in the 1989 High Court decision in *Gyllenhammar & Partners International Ltd v Sour Brodogradevna Split*,[24] in which a Croatian shipbuilder had refused to perform a shipbuilding contract unilaterally created by the purchaser's exercise of an option. In response to the buyer's claim for specific performance, Hirst J stated:[25]

> . . . this is a very complex contract requiring extensive co-operation between the parties on a number of matters, in particular modifications, optional variations, and, perhaps most important of all, matters of detail (some by no means unimportant) left undefined in the specification. In my judgment, these factors, coupled with the consideration that the work would take place in a foreign yard outside the court's jurisdiction, would tell strongly against an order for specific performance being in principle appropriate in the present case.

The Court's refusal to grant specific performance was thus based on the need for supervision of any order, which it felt could not be adequately undertaken.[26]

However, in its 1998 decision in *Co-operative Insurance Society Ltd v Argyll Stores*,[27] which concerned an obligation to operate a supermarket, the House of Lords drew a distinction between orders for specific performance which require a defendant to carry on an activity (e.g. the continuing operation of a business) and those which require him to achieve a specific result (e.g. presumably, the construction and delivery of a ship). Lord Hoffmann held that, in the latter case, the fact that an order for specific performance might necessarily be somewhat imprecise in its terms, or might involve difficulties of supervision, did not represent an objection to its making, even in relation to complex projects. His Lordship observed that:

> Even if the achievement of the result is a complicated matter which will take some time, the court, if called upon to rule, only has to examine the finished work and say whether it complies with the order . . . This distinction between orders to carry on activities and orders to achieve results explains why the courts have in appropriate circumstances ordered specific performance of building contracts and repairing covenants . . .[28]

---

22 *See* H. Beale *et al*, *Chitty on Contracts* (31st edn), (Sweet and Maxwell, 2012), at para. 27-016 on the meaning of '*specific or ascertained*' goods.

23 *Ibid*, at para. 27-005 noting that, although the historical foundation of the equitable remedy of specific performance was the absence of an adequate remedy in damages, the modern judicial approach is to ask whether specific performance represents the most *appropriate* remedy in all the circumstances of the case.

24 [1989] 2 Lloyd's Rep 403.

25 *Ibid*, at 422.

26 Note that *Gyllenhammar* concerned a situation in which construction of the vessel had never commenced (the shipyard having refused to 'honour' the option in light of the apparent refusal on the part of its financing banks to support the project). It was as such clear that such supervision would have extended over a substantial period of time and involved a significant commitment of resources, which considerations obviously do not apply to a substantially completed newbuilding.

27 [1998] AC 1.

28 *Ibid*, at [13].

A similar approach has also been taken in other non-marine cases dealing with agreements for the provision of services and the enforcement of a tenant's repair obligations.[29] In consequence, *Chitty on Contracts* (31st edn) has expressed the view that:

> The difficulty of supervision is . . . one ground that has been given for the refusal of the courts in some cases specifically to enforce contracts to do building work or to keep buildings in repair. But in such cases specific performance is sometimes ordered and no practical difficulty seems to have arisen in enforcing such orders. This suggests that the "difficulty" of supervision has been somewhat exaggerated; and various devices at the court's disposal can be used to overcome it.[30]

There is clear authority, again derived from *Behnke v Bede Shipping Co Ltd*,[31] that, if damages would not represent an adequate remedy, specific performance can be granted in respect of a contract for the sale and purchase of a second-hand vessel. Against this background, it is submitted that, particularly in circumstances in which construction of the newbuilding is substantially advanced, the discretionary remedy of an order for specific performance should in principle be available to her purchaser. Plainly, in light of the views expressed by Lord Hoffmann in *Co-operative Insurance Society Ltd*, the need for supervision should represent only a limited factor in the exercise by the court of its power to order the completion and delivery of the vessel in circumstances in which the shipbuilder is wrongfully refusing to progress the project and transfer title in the vessel to her purchaser.

It is therefore hoped that our courts will, at an appropriate early opportunity, renounce *Gyllenhammar* and hold that discretionary relief by way of an order for specific performance is available to enforce the performance of an English law shipbuilding contract – this should be *readily* available at the later stages of construction, when the need for supervision will clearly be less than at earlier stages of the newbuilding process.

3.4 Conclusion

If England and Wales are to maintain their position as the leading providers of legal services to the global maritime community, it is vital that their unified legal system should continue to meet the business needs of that community, at least insofar as these needs can be clearly established and to do so would not be contrary to our national interests. Analysis suggests that the legal remedies available to both shipowners and shipbuilders to enforce the performance of international shipbuilding contracts require early Parliamentary or judicial attention and reform.

---

29 See *Posner v Scott-Lewis* [1987] Ch 25 and *Rainbow Estates Ltd v Tokenhold* [1999] Ch 64.

30 See para. 27-026.

31 [1927] 1 KB 649.

CHAPTER 4

The application of the 'prevention principle'[1] in the shipbuilding context

*Simon Kverndal QC**

4.1 Introduction

All synallagmatic contracts which involve delivery against instalment payments are subject to a tension between the respective rights of buyer and seller arising out of the seller's obligation to deliver by the contract date. Where time is or is made of the essence a failure to deliver on time may give rise to catastrophic consequences for the seller; equally, however, if the seller is relieved of his obligation to deliver by a particular date the consequences may be similarly drastic for the buyer: for this reason there is no presumption that time is of the essence in sale of goods cases unless the parties so agree. The gulf between the two extremes no doubt explains why so many shipbuilding cases are fought to a conclusion, especially since the severity of the consequences is so often exacerbated by the impact that time extensions may have on the validity of refund guarantees.

The importation of the so-called 'prevention principle' into the law of shipbuilding contracts has, in our view, increased rather than diminished this tension and introduced unnecessary uncertainty. It is a principle (if we can call it that) which has a sensible and secure place in mitigating unfairness in the operation of land-based construction contracts, where the construction work almost invariably takes place on the buyer's property. However, shipbuilding contracts are generically and practically different and to apply an all-embracing principle of such grave potential consequences is in our submission unprincipled and unhelpful. It increases the potential for opportunistic plays by either builder or seller based not on what the parties intended when they entered into the contract, but on market forces and the chance to make substantial short-term gains. It fails to recognise, we suggest, the need for a flexible approach which arises from the international character and the range and breadth of ship construction projects

* Quadrant Chambers. The author wishes to thank Ms Emily McCrea-Theaker for her invaluable contribution to this chapter.

1 Jackson J in *Multiplex v Honeywell* [2007] EWHC 447, at [47–48], described the principle in the following fashion:

> The essence of the principle is that the promisee cannot insist upon the performance of an obligation which he has prevented the promisor from performing. In the field of construction law one consequence of the prevention principle is that the employer cannot hold the contractor to a specified completion date, if the employer has by an act or omission prevented the contractor from completing by that date. Instead, time becomes at large and the obligation to complete by the specified date is replaced by an implied obligation to complete within a reasonable time.

and the differences between one contract and another – who designs, whether off-the-peg or bespoke, the parties' respective bargaining positions, the nature of the refund guarantees on offer, financing pressures, national and cultural differences between the parties, and so on. Sometimes, we might add, one of the parties – usually the buyer – may be remarkably naïve with regard to the complexities of the shipbuilding contract and likely to be unaware of much of what he is signing himself up to. Can a 'one size fits all' approach as harsh in its consequences as the prevention principle fairly apply to this lumpy and amorphous species? Our view is that, despite recent judicial application in the *Adyard* case[2], we do not have to apply it and we should not.

This chapter considers whether the prevention principle should be imported into this field and concludes that it should not; that, at most, it should apply only to deny the buyer his right to liquidated damages but should not deprive him of his right to cancel. It contains:

1. A look at the way in which shipbuilding contracts have been and are now categorised, and the trend towards regarding them as akin to construction contracts, which has allowed importation of the prevention principle.
2. A brief look at the prevention principle itself.
3. A review of *Adyard.*
4. Our views.

4.2 The traditional categorisation of shipbuilding contracts

Shipbuilding contracts, by their very nature, consist of an agreement not only to manufacture, but also to sell, completed vessels. The obligations assumed by the parties on entering into such a contract are therefore an amalgamation of those found in both the sale of goods context (the obligation to transfer ownership in the vessel) and non-marine construction (the obligation to build in accordance with the contractual specification).

Despite this, shipbuilding contracts were treated historically as contracts for the sale of goods. In *McDougall v Aeromarine of Emsworth Ltd*,[3] Diplock J (as he then was) observed that: 'it seems well settled by authority that, although a ship building contract is, in form, a contract for the construction of the vessel, *it is in law a contract for the sale of goods*'.[4]

This categorisation had important practical consequences. In *McDougall* it meant that the plaintiff was entitled to a full refund of the money which he had paid to the defendants in respect of the construction of a yacht which he had rejected because of its non-compliance with the terms of the agreement. Despite the considerable amount of work expended by the defendants in building the yacht, there had been a total failure of consideration.[5]

---

2 *Adyard Abu Dhabi v SD Marine Services* [2011] EWHC 848 (Comm); (2011) 27 Const LJ 594.

3 [1958] 2 Lloyd's Rep 345.

4 See pp. 355–356 (emphasis added); also *Re Blyth Shipbuilding & Dry Docks Co Ltd (No 3)* [1926] 1 Ch 494, 499, where Romer J said that a shipbuilding contract was 'unquestionably a contract for the sale of future goods'.

5 See *ibid.* at p.359.

4.2.1 The analogy with non-marine construction contracts

The traditional approach was, therefore, to draw a clear divide between marine and non-marine construction. Unlike a shipbuilding contract, other construction contracts were treated as '*entire contract[s] for the sale of goods and work and labour*'.[6]

Unlike shipbuilding contracts, in a typical construction contract the work will be carried out on the land or in the buildings of the buyer. The fundamental characteristic of such agreements, which distinguishes them from other major commercial contracts (such as those for sale or services), is therefore that:

> as work proceeds and becomes fixed or attached to the land of the owner it progressively and irretrievably becomes the property of the Owner, whatever the financial rights or obligations of the parties may be at the time.[7]

By contrast, in the shipbuilding context, the yard generally retains ownership of the ship until a late stage. In this respect, there is a much closer analogy between shipbuilding contracts and contracts for the sale of future goods, than with other non-marine construction. Rescission by the buyer arising out of serious breach by the builder has extremely serious consequences: recovery of all instalments in addition to a common law right to claim damages. Conversely, of course, shipyards also retain the option to sell their work on to another buyer, an option that is not open to the majority of land-based constructors.

In many other respects, there are, however, similarities. Like many construction projects, shipbuilding involves the regulation of a substantial and complex project, in which each party undertakes long-term obligations towards the other and accepts the accompanying significant commercial risks.[8] As a result, the view of many is that the similarities outweigh the differences:

> Although the ultimate purpose of such a contract is to transfer legal title to goods in return for payment of an agreed price, the nature and extent of the commitments assumed by both parties in order to achieve this objective are more akin to non-marine construction than a mere agreement of sale and purchase.[9]

4.2.2 The modern approach to the categorisation of shipbuilding contracts

As a result, in recent years there has been an increasing judicial willingness to apply non-marine construction principles in the shipbuilding context. Following the decisions of the House of Lords in *Hyundai Heavy Industries Co v Papadopoulos*[10] and *Stocznia Gdanska SA v Latvian Shipping Co*,[11] it is now clearly recognised that these contracts exhibit characteristics of both sale of goods and construction law.

In both cases, the question for the House of Lords was whether unpaid instalments of the contract price which had accrued prior to the builder's termination of a

6 See *Gilbert-Ash (Northern) Ltd v Modern Engineering (Bristol) Ltd* [1974] AC 689, at 717 (Lord Diplock).

7 *Hudson's Building and Engineering Contracts* (12th edn, 2010), 1–001.

8 See S. Curtis, *The Law of Shipbuilding Contracts* (4th edn, Informa Law, 2012), at p. 1.

9 *Ibid.*

10 [1980] 2 Lloyd's Rep 1.

11 [1998] 1 Lloyd's Rep 609.

shipbuilding contract remained payable. The conclusion was that they did; the work already completed by the builders in partially constructing the vessels was good consideration. As Viscount Dilhorne explained in *Hyundai*:

> . . . it was a contract to 'build, launch, equip and complete' a vessel and 'to deliver and sell' her. The contract price included 'all costs and expenses for designing and supplying all necessary drawings for the Vessel . . .' It was a contract which was not simply one of sale but which so far as the construction of the vessel was concerned, resembled a building contract.[12]

Lord Fraser similarly distinguished the cases cited to the court concerning contracts of sale:

> in both these cases the contracts were simply contracts of sale which did not require the vendor to perform any work or incur any expense on the subjects of sale. But the contract in the instant case is not of that comparatively simple character. The obligations of the builder were not confined to selling the vessel but they included designing and building it and there were special provisions . . . that the contract price . . . shall include payment for services in the inspection, tests, survey and classification of the vessel [– and also –] all costs and expenses for designing and supplying all necessary drawings for the vessel in accordance with the specifications.
>
> Accordingly the builder was obliged to carry out work and to incur expense, starting from the moment that the contract was signed, including the wages of designers and workmen, fees for inspection and for cost (sic) of purchasing materials. . . . [T]he builder was bound to incur considerable expense in carrying out his part of the contract long before the actual sale could take place.[13]

Lord Fraser went on to explain that the fallacy in the appellants' argument was the assumption that the *contract* price was simply a *purchase* price and, once that misconception had been removed, it was clear that 'the shipbuilding contract has little similarity with a contract of sale and much more similarity . . . with contracts in which the party entitled to be paid had either performed work or performed services for which payment is due by the date of cancellation', which, of course, included building and construction contracts.[14]

The same reasoning was applied in *Stocznia*; the design and construction of the vessel was part of the yard's contractual duties and formed part of the consideration for which the price was to be paid.[15] They were obliged to carry out work and incur expense from the moment that the contract was signed, despite the fact that the transfer of ownership would not occur until a late stage.[16]

On the facts of these two cases, the willingness of the House of Lords to ensure that the builders were properly remunerated is understandable, but neither decision addresses the underlying question of the proper legal categorisation of such contracts.[17] As such, it is arguable that they do little to change the categorisation of shipbuilding contracts as a whole.[18] In contrast, the more recent decision of the

12 See [1980] 2 Lloyd's Rep 1, at 5.
13 *Ibid.* at 13–14.
14 *Ibid.* at 14.
15 See [1998] 1 Lloyd's Rep 609, at 619–620 (Lord Goff).
16 *Ibid.* at 627(2) (Lord Lloyd).
17 See Curtis, *The Law of Shipbuilding Contracts* (4th edn, Informa Law, 2012), at p.2.
18 *Ibid.*

High Court in *Adyard* has been interpreted by some as having more wide-ranging effect. One way or another the door is ajar – if not open – to the migration and infusion of non-marine construction law.

4.3 The prevention principle: its origins and recent developments

There can be much debate as to whether the prevention principle is a rule of law or an implied term of the contract. It is submitted that the latter view is to be preferred: the proposition that 'a promisee cannot insist upon the performance of an obligation which he has prevented the promisor from performing' is merely an extension of the principle that a party cannot rely on his own wrong or default to establish his claim. The question of whether developments and variations of that principle are to be applied to a particular contract must, in our view, depend on implication not on some rule of law.

A long line of English authority starting with *Holme v Guppy*[19] extended the basic proposition and held that, unless the contract expressly or impliedly provided for the extension of time in such circumstances, the buyer was precluded from claiming any liquidated damages whatsoever for late delivery where his own conduct (irrespective of whether or not it amounted to a breach) had in fact delayed completion of the contract works. So far, so good, in the right context: historically, the courts have treated liquidated damages provisions with suspicion, particularly in cases where damages could be assessed with a reasonable degree of accuracy.[20] The general approach was to view such clauses as oppressive and penal in effect.[21] As a result, there was a general judicial willingness to avoid their application. The prevention principle was one way of doing so.

Liquidated damages operate by stipulating for some definite contractual completion date (whether specifically named in the contract or determined by reference to some specific event), after which a set daily or weekly rate becomes payable in respect of the delay in completing the work. If that specified contractual completion date ceases to apply owing to an act of prevention, and there is no provision for the extension of that date, there is no date for the liquidated damages provision to attach to and the right to such damages is lost. But where there is an extension of time clause the prevention principle does not apply because the ability to extend means that the builder still ought to be able to complete in time.

As a result, whilst extension of time clauses may appear to be in the interests of the builder, in fact they are often of more benefit to the buyer (or employer), in that they serve to keep alive a liquidated damages clause which would otherwise be lost. This, as we shall see, complicates and confuses application of the *contra proferentem* principle.

Because of the antipathy to liquidated damages clauses there has been a

19 (1838) 3 M & W 387.

20 *Hudson's Building and Engineering Contracts* (12th edn, 2010), 6-022.

21 Cf. the more modern approach that commercially experienced parties should be free (or indeed encouraged) to agree liquidated damages clauses in the interests of certainty and the avoidance of protracted arguments about the level of losses actually suffered (*Hudson's Building and Engineering Contracts*, above, 6-022).

tendency towards strict construction of extension of time clauses which might otherwise provide an escape from the application of the prevention principle. Thus Salmon LJ:

> [T]he liquidated damages and extension of time clauses in printed forms of contract must be construed strictly *contra proferentem*. If the employer wishes to recover liquidated damages for failure by the contractor to complete on time in spite of the fact that some of the delay is due to the employer's own fault or in breach of contract, then the extension of time clause should provide, expressly or by necessary inference, for an extension on account of such fault or breach on the part of the employer.[22]

Thus generalised grounds for an extension of time (e.g. 'any matter beyond the control of the contractor' or 'any matter such as fairly to entitle the contractor to an extension of time') have been held not to cover acts of prevention or breach by the employer.[23]

However, the prevention principle in construction law appears to have moved on from its original application to the right to claim liquidated damages; it is here that the real controversy begins. It was not always so: a mere 40 or so years ago in *Trollope & Colls Ltd v North West Metropolitan Regional Hospital Board*,[24] the House of Lords held that an agreed delivery date remained in place in spite of a delay which might otherwise have given rise to the application of the prevention principle. The appellants in that case had contracted to construct a district hospital and clinical research centre for the respondents. The works were to be carried out in three phases. The specified date for completion of phase I was 30 April 1969, but the conditions provided for the possibility of an extension of time for completing the works in the event of delays caused by certain specified events (including exceptionally inclement weather, fire damage, variations to the work and delay on the part of nominated sub-contractors amongst others).

In relation to phase III, the conditions provided that the contractor was to begin work six months after the date of issue of the certificate of practical completion of phase I, and set a date for completion of 30 April 1972. The conditions provided for a similar extension as applied to the phase I works, but only where there were delays to the phase III work. Crucially, there was no express provision in the contract by which the completion date for phase III could be extended where delays to phase I had a knock-on effect.

The combined effect of the terms was that if the certificate of practical completion for phase I was issued on 30 April 1969 (as planned), the time for commencing phase III would be six months later (31 October 1969), leaving a period of 30 months to complete the final phase. If, however, there was for any reason a delay in reaching practical completion of phase I, the period for completing phase III would be curtailed.

In the event, there was a delay of 59 weeks in completing phase I and extensions were granted for a total of 47 weeks. As a result, the certification of practical completion for phase I was not provided until 22 June 1970. The effect of these delays,

22 *Peak Construction (Liverpool) v McKinney Foundations Ltd* (1970) 1 BLR 114, at 121.
23 See *Hudson's Building and Engineering Contracts* (12th edn, 2010), 6-026.
24 [1973] 1 WLR 601.

if the express provisions of the contract were construed literally, was that phase III would commence on 22 December 1970, leaving just 16 months to complete the work before the contractual deadline of 30 April 1972.

At first instance, Donaldson J decided in favour of the builders, holding that no term extending the time for completing phase III following delays to phase I could be implied; the meaning of the contract was plain.[25] The majority of the Court of Appeal disagreed, and found for the employer as a matter of construction of the contract, and possibly also on the basis of the prevention principle.[26] Lord Denning relied on *Dodd v Churton*[27] as establishing the following:

> (1) It is well settled that in building contracts – and in other contracts too – when there is a stipulation for work to be done in a limited time, if one party by his conduct – it may be quite legitimate conduct, such as ordering extra work – renders it impossible or impracticable for the other party to do his work within the stipulated time, then the one whose conduct caused the trouble can no longer insist upon strict adherence to the time stated. He cannot claim any penalties or liquidated damages for non-completion in that time.
>
> (2) The time becomes at large. The work must be done within a reasonable time – that is, as a result, the stipulated time plus a reasonable extension for the delay caused by his conduct.[28]

In the House of Lords, Lord Pearson (with whose reasoning Lord Guest, Lord Diplock and Lord Cross agreed) did not apply the prevention principle, but instead held that the express terms of the contract were entirely free and clear from ambiguity; the date for completion of phase III was set as 30 April 1972.[29] As to Lord Denning's interpretation of *Dodd v Churton*, Lord Pearson observed that it did establish the first part of the passage, but did not establish or afford any support to the second part.[30] The decision merely established that where:

> the building owner has ordered extra work beyond that specified by the original contract which has necessarily increased the time requisite for finishing the work, he is thereby disentitled to claim *the penalties for non-completion provided for by the contract.*[31]

Apparently, therefore, the law was both clarified and settled by the House of Lords.

By 2007, however, and the decision of Jackson J, in *Multiplex Constructions (UK) Ltd v Honeywell Control Systems Ltd (No. 2)*,[32] the principle had moved on significantly:

> In the field of construction law, one consequence of the prevention principle is that the employer cannot hold the contractor to a specified completion date, if the employer has by act or omission prevented the contractor from completing by that date. Instead, time becomes at large and the obligation to complete by the specified date is replaced by an implied obligation to complete within a reasonable time.[33]

---

25 See [1973] 1 WLR 601, at 606D–G.
26 See [1973] 1 WLR 601, at 607E–608H.
27 [1897] 1 QB 562.
28 See [1973] 1 WLR 601, at 606F–G.
29 *Ibid.* at 607E–608D, 610E, 612F, 614E–F.
30 See [1973] 1 WLR 601, at 607H.
31 *Ibid.* at 608C–D (emphasis supplied).
32 [2007] EWHC 447 (TCC); (2007) 111 Con LR 78.
33 *Ibid.* at [47]–[48].

Hand-in-hand with this extended approach was a direction by Jackson J (in sharp contrast to Salmon LJ, in the *Peak* case referred to below) to be more flexible towards the construction of the extension of time provision which might be relied upon by the buyer to disapply application of the principle:

> It seems to me that, in so far as an extension of time clause is ambiguous, the court should lean in favour of a construction which permits the contractor to recover appropriate extensions of time in respect of events causing delay.[34]

This appears to suggest that the prevention principle applies to cancellation dates, as happened (as we shall see) in the *Adyard* case. Furthermore, whilst *Trollope & Colls* was referred to in passing in *Multiplex*, and it was observed that the first part of Lord Denning's interpretation of *Dodd v Churton* was accepted by Lord Pearson, Jackson J did not go on to consider the points on which their Lordships disagreed – namely the crucial question of whether time became generally 'at large'.[35] No authority at all was cited in support of Jackson J's summary of the principle quoted above. At a stroke – and without citation of authority – Jackson J applied the prevention principle to cancellation clauses.

Further, the absence of any clear authority on this point prior to *Multiplex* and *Adyard* is itself suggestive of a general understanding that the prevention principle did not and should not apply to cancellation clauses. Cases such as *Holme v Guppy*,[36] *Dodd v Churton*[37] and *Peak Construction v McKinney*,[38] from which the principle is derived, are all concerned with liquidated damages, as opposed to rights of cancellation.[39] In the context of the shipbuilding contract, where the right to cancel is an essential and carefully carved-out buyer's right, it may fairly be asked: if you are going to agree such a right, why subject it to so precarious an existence?

4.4 The decision in *Adyard*

The claimant in *Adyard Abu Dhabi* v *SD Marine Services*,[40] a small-to-medium shipyard, had agreed to construct two 50-metre moorings and special operations support vessels for the defendant SDMS. The contracts were design and build contracts. The basic design of the vessels was set out in specifications appended to the

34 *Ibid.* at [57].

35 *Ibid.* at [54]–[55].

36 (1838) 3 M & W 387.

37 [1987] 1 QB 566.

38 [1970] 1 BLR 111.

39 It may be suggested that this is explained by the context in which the prevention principle developed; in the nineteenth century, when the prevention principle was first established, the rule as to entire obligations held sway. The effect of that rule was that the performing party was not entitled to any of the contract price unless the entire contract was performed. The classic example is the famous case of *Cutter v Powell* (1795) 6 TR 320 in which a seaman was to be paid a lump sum for his services when he completed a voyage, but died before he did so. His executor could not recover *pro tanto* wages because it was an 'entire contract'. Such a rule was capable of having startling and unfair consequences, particularly in the construction context where the employer would not only be free from any obligation to pay, but would also have become the owner of whatever the contract had managed to construct before the contract came to an end. Against this background, the courts were understandably reluctant to hold that time was of the essence.

40 [2011] EWHC 848 (Comm); (2011) 27 Const LJ 594.

contracts, but in so far as that design had to be developed in order to meet class or other regulatory requirements (which were not a result of any subsequent change thereto), that was the risk and responsibility of Adyard.[41] The contracts did, however, provide a mechanism for dealing with changes to the regulatory regime. Where such changes occurred during the period of construction, unless SDMS instructed Adyard otherwise, any such alterations were to be incorporated into the construction of the vessel, provided that SDMS agreed to 'reasonable adjustments ... in the Contract Price, Delivery Date and other terms and conditions ... and the Specifications'. The contracts required the two vessels to be ready for sea trials by 30 September and 30 November 2009 respectively, and conferred a right on SDMS to rescind in the event that they were not ready by those dates.

In the event, it was common ground that the vessels were not ready in time. The dispute concerned whether or not this entitled SDMS to rescind the contracts. At trial, Adyard claimed that SDMS could not rescind since the delays in question had been caused by their own failure to decide promptly whether or not to agree reasonable adjustments to the contractual terms so as to incorporate changes imposed by the UK Maritime Coastguard Agency (the relevant intended flag authority). As a result, Adyard argued, the prevention principle applied and time was at large.

Having considered the evidence, Hamblen J concluded that Adyard's factual case on variation failed. The applicable regulatory requirements had not changed during the relevant period. He nonetheless went on to consider whether Adyard could have relied on the prevention principle. It does not appear that the buyers (represented by construction bar counsel) argued – as it is submitted that they should have done – that it had no application at all in a shipbuilding contract; Hamblen J therefore adopted without comment Jackson J's explanation of the principle in *Multiplex Constructions (UK) Ltd v Honeywell Control Systems Ltd (No. 2)*:[42]

> ... The essence of the prevention principle is that the promisee cannot insist upon the performance of an obligation which he has prevented the promisor from performing.
>
> In the field of construction law, one consequence of the prevention principle is that the employer cannot hold the contractor to a specified completion date, if the employer has by act or omission prevented the contractor from completing by that date. Instead, time becomes at large and the obligation to complete by the specified date is replaced by an implied obligation to complete within a reasonable time. The same principle applies as between main contractor and sub-contractor.
>
> It is in order to avoid the operation of the prevention principle that many construction contracts and sub-contracts include provisions for extension of time. Thus, it can be seen that extension of time clauses exist for the protection of both parties to a construction contract or sub-contract.

Hamblen J then went on to summarise the effect of the relevant authorities in construction law as follows:

> (a) In a basic shipbuilding contract, which simply provides for a Builder to complete the construction of a vessel and to reach certain milestones within specific periods of time, the Builder is entitled to the whole of that period of time to complete the contract work.

41 *Ibid.* at [71]–[72].
42 [2007] EWHC 447; (2007) 111 Con LR 78 at [47]–[49].

> (b) In the event that the Buyer interferes with the work so as to delay its completion in accordance with the agreed timetable, this amounts to an act of prevention and the Builder is no longer bound by the strict requirements of the contract as to time.
> (c) The instruction of variations to the work can amount to an act of prevention.
> ...
> [T]he prevention principle does not apply if the contract provides for an extension of time in respect of the relevant events.[43]

On its true construction, Hamblen J concluded that the contract provided a mechanism for an extension of time where there was no agreement to an adjustment. In doing so, Hamblen J relied on Article VIII of the contract:

> DELAYS AND EXTENSIONS OF TIME FOR DELIVERY
> (FORCE MAJEURE)
> ...
> 3.1 Delays on account of such causes as specified in Clause 1 of this Article and any other delays of a nature which under the terms of this Contract permits postponement of the Delivery Date shall be understood to be permissible delays and are to be distinguished from unauthorised delays on account of which the Contract Price is subject to adjustment . . .
> ...
> 3.3 If the Builder fails to complete either of the stages contained in 3.1 (c) or (e) [sea trials] by the dates specified therein, then the Buyer may, at its option, rescind this Contract in accordance with the provisions of Article X hereof, provided always that, to the extent that any delays are caused by the Buyer's default or any Permissible Delay, that period shall be extended to the same extent.

Clause 1 of Article VIII related to usual *force majeure* events, such as delay or default caused by fire, flood, war, strikes and lockouts. Despite this, Hamblen J noted that Article VIII clause 3.1 covered those causes 'and any other delays of a nature which under the terms of this Contract permits postponement of delivery'. As a result, that article was intended to be 'a broad sweep up clause encompassing the numerous different bases on which the Delivery Date can be extended under these contracts' and the failure to agree an adjustment to the contractual terms was therefore a permissible delay.[44]

In reaching this conclusion, one of the factors which Hamblen J took into consideration was:

> the potentially extreme circumstances of the application of the prevention principle, as explained by Colman J in *Balfour Beatty v Chestermount Properties* (1993) 62 BLR 1, 27. A trivial variation may lead to the loss of the right to liquidated damages for a long period of culpable delay and, in this case, loss of the right to rescind as well.[45]

Both the reasoning and the result are surprising and have come in for some criticism.[46] Having held – without any analysis whatsoever – that the *Multiplex* version of the prevention principle applied generally to shipbuilding contracts, Hamblen J then bent over backwards to *disapply* it on the construction of the *Adyard* contract itself – muddling as he did so delays caused by modifications and caused by *force*

43 See [2011] EWHC 848 (Comm); (2011) 27 Const LJ 594 at [242]–[243].
44 *Ibid.* at [250].
45 See [2011] EWHC 848 (Comm); (2011) 27 Const LJ 594 at [255(4)].
46 See, e.g., an article by Simon Curtis and Louise Elmes in *Lloyd's List* for 11 May 2011.

majeure and accepting an extension of time clause with regard to the *latter* as capable of ousting the operation of the prevention principle. In disapplying the prevention principle he referred to the potentially extreme results of holding it applicable, quoting Colman J's statement in *Balfour Beatty Building Ltd v Chestermount Properties Ltd*[47] that 'a trivial variation may lead to the loss of the right to liquidated damages for a long period of culpable delay and, in this case, loss of the right to rescind as well'.[48] But if he was so concerned about the effects of the wholesale importation of the prevention principle into English shipbuilding contract law, why import it at all?

4.5 The prevention principle and shipbuilding contracts

Our antipathy for the prevention principle will already be apparent. Though not expressed as strongly, it is worth noting that it has also engendered a good deal of judicial unease. Colman J's statement in *Balfour Beatty* has already been cited – a shipping judge hearing a construction case. Lloyd LJ in *Rapid Building Group Ltd* v *Ealing Family Housing Ltd*[49] confessed to being 'startled' to be told that a liquidated damages clause became inoperative as the result of any delay, no matter how slight, on the part of the employer;[50] in doing this he was moreover echoing similar remarks made by Phillimore LJ, in the earlier *Peak Construction* case.[51] Recently, moreover, a leading Technology and Construction Court judge has noted, more than a little disparagingly, that 'the principle is of some antiquity and has a surprising effect on the contractual obligations as to the time for completion', though he then said that since the extension of time mechanism in that contract did cover acts of prevention this was 'not the opportunity to consider the underlying basis for the principle'.[52] Perhaps, even among construction lawyers and because of the increasing proportion of design and build contracts in the construction industry, the 'principle' may be up for fundamental review. Perhaps there is a growing recognition that the 'principle' is a very blunt instrument for knocking out in their entirety carefully carved out and negotiated buyer's rights to liquidated damages and to rescission refunds.

The editor of the latest edition of *Hudson* refers to those '*rare cases*' where there is an act of prevention for which there is no power under the contract to grant an extension of time.[53] Rare, perhaps, in the non-marine field; but it is far from clear that this is the case with shipbuilding contracts, where (notwithstanding the commercial sophistication of the parties) drafting is not always of the first order and where a canny builder may well try to slip into his contract a deliberately narrowly drafted extension of time provision so as to give himself the opportunity to invoke the 'principle'. Should this niche point of English law, 'at first sight somewhat

47 (1993) 62 BLR 1.

48 *Ibid.* at 27.

49 (1984) 29 BLR 1.

50 *Ibid.* at 19.

51 *Peak Construction (Liverpool) v McKinney Foundations Ltd* (1970) 1 BLR 114, at 121.

52 See *Bluewater Energy Services BV v Mercon Steel Structures BV* [2014] EWHC 2132 (TCC); (2014) 155 Con LR 85 at [518]. The judge was Ramsey J.

53 *Hudson's Building and Engineering Contracts* (12th edn, 2010), 6–028.

implausible' even to a great English judge,[54] have any place in the multinational context of an English law shipbuilding contract?

First, we do not need to apply it. There are three points worth making here. First, it is strongly arguable that it is not so much a principle of law as simply a term sometimes implied in a contract (e.g. look for references to it in *Chitty on Contracts*[55]). That being so, tribunals should be slow to apply it in the shipbuilding context where the contractor/builder is likely to be savvy and the contract carefully negotiated. We would, of course, encourage tribunals to be bold and reject the implication. Secondly, there is no binding authority which states that it should apply to shipbuilding contracts. *The Cape Hatteras*,[56] in which the principle was found 'at first sight implausible', was a ship repair case. As for *Adyard* itself, this is, strictly speaking, *obiter*: the yard had lost the argument long before its invocation of the prevention principle came to be considered and the buyer simply failed to argue against its implication. Third, what is noteworthy is the careful way in which Jackson J expressed himself in the *Multiplex* case when introducing the refined statement of the 'principle': 'in the field of construction law . . .'. There is some recognition here of, as well as good reason (in my view) for, restricting its application – or at the very least its refined application – to that field alone.

Second, on any view, its consequences may be 'startling' or 'surprising' and the principle 'at first sight somewhat implausible'. At worst, the consequences may be disastrous, for example where a buyer misguidedly cancels on the basis that the 'drop dead date' has passed, only to be met with a prevention principle argument which, if made good, will put him in repudiatory breach. On any view, commercial certainty is thrown out of the window: does the builder rely on the 'principle' to try to get rid of the contractual delivery date (and liquidated damages provisions connected with it) and the 'drop dead date' (and rescission refunds plus damages) altogether, or does he say that the extension of time clause *does* cover the acts of prevention, abandon reliance on the 'principle' and take his chances instead on the extension of time regime?[57] How can the buyer assess whether the extension of time regime covers the acts of prevention, given that Salmon LJ says that you should construe that regime *contra proferentem* and Jackson J says that you should not? How, in those circumstances, is he to know whether or not he has the right to cancel for missing the 'drop dead date'? For both, the additional impact of any notification regime – has proper notice been given of time extending events? – increases the uncertainty. As Mary Poppins's employer Mr Banks once so memorably said – 'In short we have a ghastly mess'.

Third, the sophistication of shipbuilding contracts and of the parties to them provides sufficient material for negotiating terms which fully protect the builder from the consequences of his being prevented from completing in time because of buyer-related delays. The uncertain grey areas which we have just been considering disappear if the prevention principle is discarded in its entirety and the parties

54 I.e. Staughton J: see *The Cape Hatteras* [1982] 1 Lloyd's Rep 518, 525.

55 See *Chitty on Contracts* (31st edn, 2012), 12-082, 13-013.

56 [1982] 1 Lloyd's Rep 518.

57 In *Adyard*, the yard in fact started out on the latter route and later changed to run the former: see [2011] EWHC 848 (Comm); (2011) 27 Const LJ 594 at [297].

use the express terms of the contract to create a fair and balanced regime in which both the contractual delivery date and the 'drop dead date' may be extended, along with the rights to liquidated damages and rescission refunds and damages which go with failure to meet those dates. Better and fairer by far that these issues are decided on the basis of ordinary principles of construction. Better and fairer by far that the builder recovers as damages for breach of the buyer's obligations (e.g. timely provision of owners' furnished equipment) the liquidated damages referable to the delay caused by that breach than that the buyer's entire right to such damages be discarded.

Fourth, when construction lawyers blithely speak of time being set at large and the contract being completed within a reasonable time their focus is on the right to liquidated damages. For shipbuilding contracts the Exocet missile represented by the prevention principle hits two quite different targets, the liquidated damages *and* the right to rescind. At the very least the right to cancel for passing the 'drop dead date' should not be lost, given the huge impact that that can have on refund guarantees. The date set as a longstop can take into account all foreseeable eventualities and set a fair balance between what the builder and the buyer can live with. If delay is caused by buyer's breaches and if the builder has not negotiated for extensions to the 'drop dead date' arising from those breaches or cannot be protected in damages, that is an outcome which cannot come as an unfair surprise to a savvy builder. There is no need to apply and invoke the blunt instrument of the prevention principle to save him in such circumstances.

Simon Curtis suggests that this split approach is wrong in principle. We are not so sure. Indeed, it seems to us that it is both justifiable (see the House of Lords' decision in the *Trollope & Colls* case referred to above) and workable. By far the better approach is to abandon the prevention principle altogether. In our submission we don't need it, it doesn't work in the shipbuilding context and we don't have to have it.

CHAPTER 5

Drafting and interpreting payment refund guarantees in the shipbuilding context

*John Forrester**

5.1 Introduction

This chapter is about drafting and interpreting refund guarantees. I shall qualify that immediately by saying that it is concerned only with refund guarantees which are governed by English law.

To my mind, drafting a refund guarantee should be easy and, if the author does his or her job properly, there will be no need to worry about the interpretation side of things. As I will explain, there is generally only one fundamental legal issue to address when drafting a refund guarantee (being whether the refund guarantee gives rise to a primary or secondary payment obligation). The rest is mostly about being attentive enough to ensure that the guarantee is fully consistent with the underlying shipbuilding contract and incorporating a few 'boilerplate' provisions.

In terms of legal content, much of this talk will focus on the issue of primary versus secondary payment obligations. However, I will conclude by running briefly through some of the other key points to be considered when drafting or negotiating a refund guarantee.

5.2 Why do problems arise?

Having said that refund guarantees should be easy to draft, it is apparent from the case law and my own experience as a practitioner that problems often arise.

One of the reasons for this is that there is no accepted market standard wording for refund guarantees in the same way as for, say, vessel sale contracts where standard forms such as the Norwegian Sale Form and the Nippon Sale Form are widely used. A recommended form of refund guarantee is attached to BIMCO's NEWBUILDCON form of building contract but that form of contract is not widely used in practice. The SAJ form, which remains the most common template for building contracts in the international market, does not attach any recommended form of refund guarantee. This means that each builder which uses the SAJ Form has had to develop its own refund guarantee wording. At first glance, these different forms tend to look broadly similar, but there are subtle differences.

Another factor, I suspect, is that refund guarantees look quite innocent to those without knowledge of the nuances of English guarantee law and English law's

* Partner, Holman Fenwick Willan LLP.

approach to the interpretation of contracts. They are rarely more than three or four pages in length and, being ancillary to the main shipbuilding contract, the parties will sometimes focus only on the commercial content and ignore the advantages of taking specialist advice on English law.

In other cases, the buyer does involve an English lawyer, but the refund guarantor and builder are unwilling to agree changes to their standard wording (even if that is imperfect) and the buyer makes a commercial decision to proceed notwithstanding those imperfections. That was certainly the case between 2003 and 2008 when owners were falling over themselves to get building slots during the shipping boom. I recall being asked on several occasions during that period to review refund guarantees on behalf of owners or their banks. The sub-text when being sent these documents was usually 'Don't screw up my deal, Mr Lawyer, by getting all fussy over the wording'. These deals often involved Chinese yards who, at that time, were far less experienced than today and whose documents were then sometimes far from perfect. However, unless the deficiencies were fundamental, the owners and their banks (who were themselves keen to support their owners) were often willing to accept the imperfections so as to avoid losing valuable building slots.

None of this beats the story of a Norwegian owner who signed one of the first international newbuilding contracts in China sometime around 2000. I was asked by his bank to review the proposed refund guarantee, which I pronounced to the bank as worthless. I heard nothing more until a couple of years later when the owner's lawyer contacted me to say that the owner had gone ahead with the contract, ignored my advice (if it was ever passed to him by the bank) and paid various instalments against the self-same guarantee wording that I had reviewed. His right to cancel for excessive delay was about to arise and he wanted to check his position under the guarantee if and when he terminated the contract. You can imagine the consternation when I advised that he had no right to call the guarantee because the guarantee had a stated expiry date falling before the contractual termination date. I later discovered that the owner was notoriously reluctant to pay lawyers' fees and presumably thought he could manage on his own. I will tell more of his story later . . .

5.3 The scope of refund guarantees

Before proceeding any further, we should establish exactly what we mean by a refund guarantee. The background to this is that most building contracts provide for instalments of a vessel's contract price to be paid by the buyer at various milestones during construction, but for title to the vessel to pass to the buyer only upon contractual delivery taking place. This means that the buyer is taking a performance risk on the builder because, if the builder fails to construct and deliver the vessel in accordance with the building contract, the buyer will have paid out substantial sums of money in advance payments without receiving the vessel for which it has bargained. In such cases, the buyer's principal contractual remedy will be to rescind the contract and claim a refund of the pre-delivery instalments already paid by it (together with interest accrued at an agreed rate). However, the buyer is then taking a credit risk on the builder who may be financially incapable, or simply unwilling, to pay the required refund amount. It is in respect of this credit risk that refund guarantees are issued.

As we will see later, some instruments described as refund guarantees are not strictly guarantees at all. Nevertheless, I shall continue to use the expression 'refund guarantee' to refer generically to the class of instruments used to secure a builder's refund obligation.

Given that refund guarantees are in the nature of financial guarantees (i.e. they only guarantee payment of a specific sum rather than contractual performance as a whole), they are typically issued by a financial institution such as the builder's bank or an export credit agency in the builder's jurisdiction. In the late 1990s, during the then Asian financial crisis, several Korean yards even turned to US insurers and pension funds to issue refund guarantees on their behalf. Occasionally guarantees are given by a company in the builder's group, although that is rare and is unlikely to equate in financial strength to a bank guarantee. I have even seen one case where the builder issued the refund guarantee itself. No amount of emphasis on the worthlessness of that made any difference to the parties.

At this juncture, it is important to reiterate that refund guarantees only secure the payment of refunds (plus interest) due under the contract; they do not cover other amounts payable by the builder. In particular, if the buyer accepts a repudiatory breach of the building contract by the builder, its claim for damages will not be covered by the refund guarantee. Going back to my Norwegian owner, the problem that he encountered was that three months before the contractual cancellation date for excessively late delivery the yard had not even built the dry dock in which the vessel was to be constructed. On any analysis, the vessel was clearly not going to be completed within that three-month period, but the contract contained no express right of rescission in those circumstances. The only available contractual remedy was for the owner to wait until the cancellation date arrived. However, as mentioned earlier, the refund guarantee expired before that date. The owner then wanted to argue that the yard's failure to construct the dry dock amounted to repudiatory breach. That might well have been the case, but it did not advance the owner's position against the refund guarantor because the refund guarantee only secured contractual refunds, not damages for repudiatory breach.

The parties are of course free to agree a guarantee which covers more than just contractual refunds, but that would be unusual in my experience, except where it is the yard's parent that is giving the guarantee. We then get into the territory of full performance guarantees where all claims against the builder are generally covered.

For the sake of completeness, I should also make a brief mention of completion guarantees. These are relevant in the context of shipbuilding contracts under which title to the vessel passes to the buyer during construction. In such cases, simply refunding the pre-delivery payments to the buyer makes no sense because it already has title to the vessel. Instead, the buyer may be more interested in having the vessel completed and will want recourse against the guarantor for the costs of doing so if the yard fails. This will require a more bespoke form of guarantee which falls outside the scope of this chapter.

5.4 Nature of guarantor's liability – primary or secondary obligation?

As mentioned in my introduction, the most significant legal issue to address when drafting and negotiating an English law refund guarantee is to determine at the outset whether the refund guarantee is intended to impose a primary or a secondary payment obligation on the guarantor. This distinction can get somewhat blurred from a commercial perspective; but it is crucial from a legal point of view, not least because English law contains a number of provisions which, unless specifically addressed in the document, will serve to release the guarantor from liability where that liability is secondary in nature.

In simple terms, when distinguishing between primary and secondary payment obligations, we are talking about the difference between a guarantee in its strict sense under English law (under which the obligations are secondary) and an instrument in the form of a demand bond[1] or standby letter of credit (under which the obligations are primary).

A guarantee imposes a secondary payment obligation on the guarantor because it is contingent on the existence of an enforceable underlying obligation of the builder to the buyer. The guarantee is dependent on the underlying shipbuilding contract and the guarantor is generally only liable to the buyer if and to the same extent that the builder is itself liable to the buyer.[2] This means that, if there is a dispute, the buyer must first establish that the builder is indeed liable to pay the refund amount, typically through arbitration proceedings under the building contract, and then wait for the builder to fail in making payment before it can call the refund guarantee.

A demand bond or standby letter of credit, on the other hand, imposes a primary payment obligation on the guarantor because its obligation to pay is independent of the position of the parties under the underlying shipbuilding contract. Instead, it is similar in nature to a documentary credit and payment is required to be made simply against presentation of one or more specified documents (typically a written demand for payment accompanied by a statement from the buyer that the refund has fallen due for payment from the builder but has not yet been paid). This allows the buyer to call the refund guarantee as soon as it believes that the builder's obligation to refund the pre-delivery instalments has arisen. Unless the guarantor has notice of fraud, payment has to be made against the specified documents even if the demand is made in circumstances under which the buyer is not, in fact, entitled to the refund under the terms of the building contract. For a straightforward recent example of this, see *Wuhan Guoyu Logistics Group Co Ltd and another v Emporiki Bank of Greece SA*.[3] In this case, payment was validly demanded under a demand bond but it subsequently became clear that the beneficiary was not, in fact, entitled to the guaranteed amount under the underlying contract. The Court of Appeal held that, once a valid demand had been made under the demand bond, the beneficiary acquired

1 Confusingly, demand bonds are often referred to as demand guarantees (as well, of course, as refund guarantees) even though they do not constitute 'guarantees' as such in the strict sense under English law.

2 This is referred to as the accessory nature of a guarantee or the 'co-extensiveness principle'.

3 [2013] EWCA Civ 1679; [2014] 1 Lloyd's Rep 273.

a complete and immediately enforceable cause of action against the guarantor. Tomlinson LJ memorably described the guarantor's argument against this proposition as 'a heretical proposition which, if accepted, would be subversive of the basis upon which international trade is routinely financed'.[4]

As might be imagined, buyers (and their banks) prefer having the right to make an immediate demand under a demand bond or standby letter of credit whereas builders prefer 'true' guarantees[5] because they do not want to risk the buyer being paid out if the refund amount is in dispute (particularly as many buyers are single-purpose companies from whom it would be difficult to recover compensation at a later date if the guarantee is paid in circumstances where the amount claimed exceeds the buyer's actual entitlement under the building contract). That said, even if the refund guarantee takes the form of a demand bond or standby letter of credit, it does not necessarily follow that the guarantor must pay immediately upon receipt of a valid demand from the buyer as language is often included in the refund guarantee which builds in some protection for the builder. In particular, the guarantor's obligation to pay will often be deferred for a specified grace period so that, if the builder disputes the refund and commences arbitration under the building contract within that period, the guarantor's obligation is then suspended pending the outcome of the arbitration.

If it is commercially agreed that the guarantor will only be obliged to pay once any arbitration proceedings have been completed (as is often the case), it is tempting to think that there is in fact little practical difference between having a guarantee on the one hand and a demand bond or standby letter of credit on the other. That may be true in terms of the timing of payment (as opposed to timing of the demand) but, as noted above, there are material differences between the treatment of these instruments under English law. Guarantors whose payment obligations are secondary enjoy various protections under English law which are not afforded to those with primary payment obligations. These differences need to be considered when drafting or negotiating the refund guarantee.

5.5 A comparison between primary and secondary payment obligations

5.5.1 Availability of defences

As mentioned already, one of the key features of a guarantee is that the guarantee is secondary to the building contract and so the guarantor is generally entitled to the same defences as the builder. This contrasts with the position under a demand bond or standby letter of credit where the guarantor has primary liability.

5.5.2 Non-disclosure of 'unusual' facts

Generally, the beneficiary of a guarantee is under no obligation to disclose material facts to the guarantor before the guarantee is issued. However, in some circumstances,

4 *Ibid* at [2].
5 Sometimes referred to as 'see to it' guarantees.

the beneficiary may be under a duty to disclose 'unusual' facts to the prospective guarantor and, if it fails to do so, the guarantee will be void.[6] The circumstances in which a guarantee would be set aside on these grounds are limited and refund guarantees rarely address this issue (although one might insert a non-reliance statement from the guarantor to mitigate this risk). This limited duty of disclosure does not apply to demand bonds or standby letters of credit.

5.5.3 Events discharging the guarantee

Under English law there are several circumstances in which a guarantee will be automatically released if they take place without the guarantor's approval. In the context of shipbuilding transactions, these include:

- material variation of the shipbuilding contract;
- repudiatory breach of the shipbuilding contract by the buyer;
- the giving of time by the buyer; and
- the release or discharge of any other guarantee or security held by the buyer to secure the refund amount.

It is generally possible to contract out of these rules for automatic discharge by including so-called 'anti-technicality' clauses, but only up to a point. Thus, whilst a provision stating that the refund guarantee will survive any future variation of the building contract is effective, that variation must be 'within the general purview of the original guarantee' and the provision will not be construed as capable of creating what is effectively an entirely new obligation it its place.[7]

These rules on discharge do not apply to demand bonds or standby letters of credit given that liability there is primary.

5.5.4 Statute of Frauds 1677

Under section 4 of the Statute of Frauds 1677, a guarantee must be in writing (or evidenced in writing) and signed by the guarantor or a person authorised by the guarantor. These requirements seem easy to satisfy but issues do sometimes arise, particularly regarding signature where guarantees are issued electronically. In this regard, it is accepted that guarantees may be signed electronically.[8] The signature may appear anywhere in the document provided it is inserted with the intention of acting as a signature to give authenticity to the guarantee.[9] The signature may also be as informal as 'a first name, initials, or perhaps a nickname'.[10] However, I have seen a few instances of guarantees being issued in electronic form by SWIFT message where the signature clause has either been omitted completely or incorporated in a

6 See, e.g., *North Shore Ventures Ltd v Anstead Holdings Inc* [2011] EWCA Civ 230; [2012] Ch 31.

7 *Triodos Bank NV v Dobbs* [2005] EWCA Civ 630; [2005] 2 Lloyd's Rep 588 at [14] (Longmore LJ).

8 See, e.g., *WS Tankship II BV v Kwangju Bank Ltd* [2011] EWHC 3103 (Comm); [2012] CILL 3155 at [153] (Blair J).

9 *J Pereira Fernandes SA v Mehta* [2006] EWHC 813 (Ch); [2006] 1 WLR 1543.

10 *Golden Ocean Group Ltd v Salgaocar Mining Industries Pvt Ltd* [2012] EWCA Civ 265; [2012] 1 Lloyd's Rep 542 at [32] (Tomlinson LJ).

form which makes it worthless (including one case where all that was included was a solitary '*Brgds*').

Helpfully, the High Court stated *obiter* in 2011[11] that a header containing the guarantor's name in a SWIFT message will suffice for the purposes of the Statute of Frauds and, if followed, this ought to avoid most cases of SWIFT guarantees otherwise failing for want of a signature. The one caveat to this is that it will only apply where the bank sending the SWIFT message is the same entity as the guarantor; problems could continue to arise where a bank advises a guarantee on behalf of another party.

The Statute of Frauds does not apply to demand bonds or standby letters of credit.

5.6 Key guidelines to interpreting refund guarantees

There have been numerous cases in which the courts have had to determine whether an instrument is a guarantee or a payment bond. Indeed, in *Wuhan Guoyu Logistics Group Co Ltd v Emporiki Bank of Greece SA*,[12] Longmore LJ observed that the judge at first instance had found it necessary to cite no fewer than 20 authorities in coming to his decision that the instrument in question was a guarantee (a decision which the Court of Appeal overturned).[13]

The key points to be borne in mind when interpreting a refund guarantee may be summarised as follows:

- Where the parties have used unambiguous language, the court must apply it, however improbable the result.[14]
- Where an instrument contains a provision that can have more than one possible meaning, it is generally appropriate to prefer the construction which is consistent with business common sense and to reject the other.[15]
- Just because a document purports to label itself as one thing, it may still have the legal characteristics of something else. An instrument of this type must be construed 'by looking at it as a whole without preconceptions as to what it is'.[16]
- Where the instrument is governed by English law but the underlying obligation guaranteed is governed by another law, English law will not apply the rules of interpretation of that other jurisdiction (which may, for example, imply certain good faith obligations with regard to demands). Nor may overseas parties argue that they do not 'understand the subtleties of English law in relation to guarantees/indemnities'.[17]

---

11 *WS Tankship II BV v Kwangju Bank Ltd and another* [2011] EWHC 3103 (Comm); [2012] CILL 3155 at [155] (Blair J).

12 [2012] EWCA Civ 1629; [2014] 1 Lloyd's Rep 266. Note that these were different proceedings from those referred to in n. 3 above.

13 See [2012] EWCA Civ 1629; [2014] 1 Lloyd's Rep 266 at [22].

14 *Cooperative Wholesale Society Ltd v National Westminster Bank plc* [1995] 1 EGLR 97.

15 *Rainy Sky SA v Kookmin Bank* [2011] UKSC 50; [2011] 1 WLR 2900.

16 *Gold Coast Ltd v Caja de Ahorros del Mediterraneo* [2001] EWCA Civ 1806; [2002] 1 All ER (Comm) 142 at [15] (Tuckey LJ).

17 *Vossloh AG v Alpha Trains (UK) Ltd* [2010] EWHC 2443 (Ch); [2011] 2 All ER (Comm) 307 at [55].

- There is a strong presumption that an instrument is a guarantee if the guarantor is not a bank.[18]
- There is a presumption that an instrument is a demand bond where:
  - it relates to an underlying transaction between parties in different jurisdictions;
  - it is issued by a bank;[19]
  - it contains an undertaking to pay '*on demand*' (with or without the words '*first*' and/or '*written*'); and
  - it does not contain any 'anti-technicality' clauses excluding or limiting the defences available to a guarantor (although this fourth limb need not be satisfied).[20]
- The inclusion of 'anti-technicality' clauses excluding or limiting the defences available to a guarantor is a mark of a guarantee[21] although, as just noted, such clauses may still be included in demand bonds without undermining their character as primary obligations.
- Factors that point towards the instrument being a demand bond include reference to the guarantor transacting as '*primary obligor*' and a requirement for payment to be made '*on demand*' (though neither is determinative).
- Incorporation of a conclusive evidence clause (i.e. one under which, in the absence of manifest error, the amount to be paid by the guarantor is to be certified by the buyer or a specified third party) is generally determinative of the instrument being a demand bond.
- Including references in a demand bond to the building contract does not undermine the principle that a demand bond is independent of the underlying contract.[22]

5.7 Drafting refund guarantees – some key provisions

As mentioned in the introduction above, it is important to ensure that the refund guarantee is fully consistent with the terms of the underlying building contract. This means checking all cross-references if the refund amount is described by reference to particular clauses of the building contract. It also means checking details such as amounts, currency, interest rates and relevant dates. These are simple points to get right but I have seen mistakes made.

Other points to be addressed include the following.

18 *Marubeni Hong Kong & South China Ltd v Government of Mongolia* [2005] EWCA Civ 395; [2005] 2 Lloyd's Rep 231.

19 Interestingly, this second point is probably the most important factor here. English courts generally take the view that banks are able to look after themselves (e.g. banks will usually require counter-security before issuing guarantees) and that they need less protection than other classes of guarantor.

20 *Wuhan Guoyu Logistics Group Co Ltd v Emporiki Bank of Greece SA* [2012] EWCA Civ 1629; [2014] 1 Lloyd's Rep 266.

21 *Trafalgar House Construction (Regions) Ltd v General Surety & Guarantee Co Ltd* [1996] AC 199.

22 *Wuhan Guoyu Logistics Group Co Ltd v Emporiki Bank of Greece SA* [2012] EWCA Civ 1629; [2014] 1 Lloyd's Rep 266.

5.7.1 *Primary or secondary obligation; protective language*

As already discussed, the refund guarantee should make clear whether it is intended to take effect as a guarantee or as a demand bond or standby letter of credit. Although the name given to a document will not be conclusive as to its legal character, it does not hurt to have a heading which describes what is intended and to use expressions such as 'this demand bond' consistently throughout the document.

If the guarantor is intended to have primary liability, the refund guarantee should state that it is contracting 'as primary obligor and not merely as a surety' (or words to that effect) and the undertaking should be to pay 'on demand' against specific described documents. The procedure for making demands and the timing when payment has to be made (e.g. after a particular grace period or, if arbitration is commenced, within so many days after publication of the arbitration award) should be clearly stated. An appropriate conclusive evidence clause should be included in respect of the amount demanded. In the case of a standby letter of credit, it would be usual to incorporate the ICC's International Standby Practices (ISP 98). The ICC also publishes Uniform Rules for Demand Guarantees (URDG 758) which may be incorporated in the case of a demand bond, although it is not necessary (or even that common) for them to be incorporated.

If the guarantor is intended to have a merely secondary liability, the refund guarantee should make this clear by avoiding language consistent with a demand bond. The refund guarantee should also contain 'anti-technicality' clauses contracting out of the rules for automatic discharge of the guarantor's liability. In fact, similar provisions are often included in demand bonds and standby letters of credit even though, strictly speaking, they are unnecessary. Although the courts have held that the inclusion of such language is an indicator that the relevant instrument is a guarantee, the courts have also accepted that such language may be included in demand bonds without undermining the guarantor's primary payment obligation. That being the case, it is common to include this language in demand bonds and standby letters of credit just in case the guarantor tries to argue at a later date that he has secondary liability and has been released from that liability by operation of law.

5.7.2 *Irrevocable and unconditional*

The refund guarantee should state that the guarantor's liability is irrevocable and unconditional.

5.7.3 *No set-off or withholding*

Payment should be required to be made without set-off or counterclaim and without any withholding or deduction.

5.7.4 *Notice provisions*

The refund guarantee should state the refund guarantor's address for receiving demands and any other correspondence.

5.7.5 Effective date

The refund guarantee may be expressed to be conditional upon receipt of the instalment to which it relates. This is because building contracts usually require the builder to procure that the refund guarantee is issued before the buyer is obliged to pay that instalment to the builder.

5.7.6 Expiry date

Ideally, from a buyer's perspective, the refund guarantee will not contain any expiry date and the guarantee will simply fall away with the building contract. However, banks typically want to insert a specific long-stop date on which the guarantee will expire if the vessel has not been delivered by that date. This is where my Norwegian owner referred to above got into trouble, because the refund guarantee failed to match properly the provisions of the building contract. In that case, the owner was entitled to rescind for excessive delay if the vessel was not delivered within 270 days of the original contractual delivery date. The refund guarantee was stated to expire on that 270th day. However, the building contract also contained a grace period of five days after the 270th day before the owner was entitled to give his notice of rescission. In effect, rescission was only permitted on the 276th day and the guarantee (which expired on the 270th day) therefore expired before rescission for excessive delay could be made.

It is worth noting that, even without the five-day grace period, there would still have been a timing difficulty because the owner would have had to have waited until the end of the 270th day to be sure that there was a sufficiently long delay to justify his rescission, but still make his demand within that same day. This is typical of the mistakes we sometimes see. It is essential that the buyer has a sufficient number of days to consider his options after he becomes entitled to rescind for excessive delay and that the refund guarantee does not expire as soon as the right to rescind arises.

The other key issue relating to expiry dates is that refund guarantees must provide for the guarantee period to be extended automatically if the guarantor's obligation to pay is suspended pending the outcome of arbitration proceedings under the building contract. Those arbitration proceedings may take several months to conclude, and so there should be a provision extending the guarantee period until an appropriate number of days after the final arbitration award is published.

5.7.7 Assignment

It is common for buyers to raise pre-delivery finance against an assignment by way of security of the building contract and refund guarantee. From a buyer's perspective, the refund guarantee should therefore be capable of assignment to the buyer's financing bank without the need for the refund guarantor's consent. Ideally, all rights under the refund guarantee should be assignable although in some cases, most commonly in China, we see the right of assignment being restricted to the proceeds of the refund guarantee only. This means that the right to make demands must remain with the buyer. In such cases, the financing bank will typically require the buyer to issue it with a power of attorney authorising it to make demands in the buyer's name.

5.7.8 Representations and warranties

Refund guarantees will sometimes contain representations and warranties from the refund guarantor confirming, for example, that it has capacity to issue the guarantee and that the guarantee does not need to be registered with (or approved by) any authority. The appropriateness of these will depend on the guarantor and its jurisdiction and should be considered on a case-by-case basis.

5.7.9 Governing law and forum

The refund guarantee should contain an express governing law clause and specify the forum in which disputes are to be heard. Refund guarantees often provide for disputes to be referred to arbitration in accordance with the shipbuilding contract. That makes sense when the refund guarantee is a 'true' guarantee because liability is secondary and so liability under the guarantee will be contingent on the builder itself being liable under the building contract. However, in the case of demand bonds and standby letters of credit (which are independent of the underlying contract), I would normally recommend that disputes be referred to the English courts. A caveat to this would be if the refund guarantor is domiciled in a jurisdiction which does not recognise English court judgments, in which case arbitration may be advisable instead, particularly if that jurisdiction is a party to the 1958 New York Convention on the Recognition and Enforcement of Foreign Arbitral Awards.

CHAPTER 6

The evolving nature of builders' risks cover

*Professor Barış Soyer**

6.1 Introduction

Builders' risks cover forms a significant part of the contractual matrix in any construction project. It is often a requirement of any shipbuilding contract, standard or otherwise, that such cover is put in place by the time the construction commences.[1] Although it is customarily the shipyard which is the purchaser of this insurance product, the primary role of builders' risks cover is to provide financial security so that the project could continue in the event of a casualty occurring during the process of construction. Under English law, a shipbuilding contract is viewed as a contract for the sale of goods and, unless it is otherwise provided in the contract itself, the risk of loss of or damage to the ship during construction and until delivery to the buyer is borne by the builder.[2] The existence of a builders' risks cover provides peace of mind to the buyer who will operate in the knowledge that in the event of an unexpected casualty, the builder will secure adequate funds to repair the damage and keep the building contract on track.[3]

It is not uncommon to see the buyer being named as an assured in addition to the shipyard in builders' risks policies.[4] This happens usually for two reasons: either to protect the buyer's interest in an instance where the building contract stipulates that

* Professor of Commercial and Maritime Law, Director of the Institute of International Shipping and Trade Law, Swansea University.

1 For example, BIMCO's NEWBUILDCON, cl 38(a), stipulates:

> From the time of first steel cutting or equivalent (or delivery of the Buyer's Supplies, whichever is earlier) until the Vessel is completed, delivered to and accepted by the Buyer, the Builder shall ... effect and maintain at no cost to the Buyer, Builder's Risk Insurance for the Vessel and Buyer's Supplies ...

2 Section 20(1) of the Sale of Goods Act 1979 reads:

> Unless otherwise agreed, the goods remain at the seller's risk until the property in them is transferred to the buyer, but when the property in them is transferred to the buyer the goods are at the buyer's risk whether the delivery has been made or not.

Since passing of ownership is invariably delayed until final handover and acceptance, it follows that risk remains with the builder throughout the build process.

3 In some cases, if the insured vessel suffers an actual or constructive loss, the contract might be terminated if the builder and buyer are unable to agree within a reasonable time on an extension to the delivery date and/or any other amendment to the contract. See, for example, cl. 38(b) of NEWBUILDCON.

4 Article 9 of the Association of West European Shipbuilders and Repairers (AWES) Standard Shipbuilding Contract provides that the insurance policies shall be in the joint names of the builder and the purchaser, that in the event of damage to the ship the insurance money shall be applied by the builder in making good such damage, and that, in the event of loss of the ship, the insurance money shall be paid

ownership of the newly built ship passes progressively to the buyer as the buyer pays instalments of the price, or to provide a security to the buyer for the advances made during the construction period.[5] In both instances, there would be no difficulty for the buyer to demonstrate that he has insurable interest: in the former instance because he owns property which is at risk[6] and in the latter instance because he has paid one or more instalments of the contractual price.[7]

Turning to contemporary practice, it is very common to see standard shipbuilding contracts stipulating that the insurance must be effected on the Institute Clauses for Builders' Risks (ICBR) (1/6/88).[8] The origins of the ICBR 1988 can be traced to a form developed in 1963 and still in frequent use in the London insurance market and beyond.[9] The ICBR 1988 is in many respects similar to the standard forms used to insure hull risks against marine losses, with the exception that insurance provided under the ICBR 1988 is on an 'all-risks' basis.[10] The following perils have been excluded from the scope of the cover provided by the ICBR 1988: war risks (including terrorist or other politically motivated activities),[11] strikes,[12] malicious acts,[13] nuclear risks,[14] earthquake and volcanic eruption,[15] cost of renewing faulty welds[16] and any liability upon the builder by way of workmen's compensation as a result of

to the purchaser to the extent of any instalments of price previously paid to the builder and thereafter any balance of insurance money shall be paid to the builder.

5 The same result can be achieved by assigning the builders' risks policy to the buyer. It should be noted that the mere existence of a loss payee clause in the policy would not be adequate to facilitate an effective assignment. Donaldson J, in *Iraqi Ministry of Defence v Arcepey Shipping Co Ltd SA (The Angel Bell)* [1979] 2 Lloyd's Rep 491 at 497, expressed the view that 'a loss payable clause gives no rights to the loss payee unless it constitutes or evidences an assignment of the assured's rights under the policy or evidences the fact that the designated person is an original assured'. Hence, contemporary builders' risks policies indicate that for assignment to be binding, it is necessary that a dated notice of such assignment signed by the assured, and by the assignor in the case of subsequent assignment, is endorsed on the policy and the policy with such endorsement is produced before payment of any claim or return of premium thereunder.

6 *Anderson v Morice* (1876) 1 App Cas 713.

7 *Ebsworth v Alliance Marine Insurance Co* (1873) LR 8 CP 596.

8 Clause 38(a)(ii) of NEWBUILDCON reads: '[S]uch Builder's Risk Insurance shall: be on terms no less than Institute Clauses for Builder's Risk terms (1/6/88) including Institute War and Institute Strike Clauses.' Similarly, Article XX.1 of the Shipbuilders' Association of Japan (SAJ) Form is usually amended to provide that the insurance coverage should be placed on the terms of the ICBR 1988, rather than the Japanese wording. On the other hand, Norwegian Standard Form Shipbuilding Contract 2000 does not specify the form that should be used for insuring the newbuild apart from stating that customary 'All Risks' terms should be used. Similarly, Article XXVIII of the China Maritime Arbitration Commission Standard Newbuilding Contract does not specify the form that the builders' risks insurance must be placed upon except indicating that the insurance must be with 'a qualified Chinese insurance company'.

9 Cover for builders' risks is also available under the Nordic Marine Insurance Plan of 2013, Chapter 19 and some newbuild ships are insured under the Nordic Plan.

10 The ICBR 1988 provides cover for loss or damage of the subject matter insured, collision liabilities, protection and indemnity liabilities, salvage and general average contributions, launching failure and sue and labour expenses.

11 See cl. 21 of the ICBR 1988.

12 See cl. 22 of the ICBR 1988. The perils of war and strikes may be reinstated into the builder's policy by agreement with the underwriters on the terms of the Builders' Risks War Clauses and Strikes Clauses respectively.

13 See cl. 23 of the ICBR 1988.

14 See cl. 24 of the ICBR 1988.

15 See cl. 6 of the ICBR 1988.

16 See cl. 5 of the ICBR 1988.

accident or illness (which extends to any contractual obligation to indemnify the buyer in respect of liabilities to his own employees).[17]

In 2007, following an extensive review carried out by the Joint Hull Committee, the International Underwriting Association of London published the London Marine Construction All Risks Wording (Mar CAR) (01/09/07) which was intended to revise the ICBR 1988. A number of factors motivated the market forces to undertake this review. The insurers offering builders' risks cover suffered significant losses between October 2002 and January 2004. Some of these losses were caused by hurricanes on the Gulf Coast and were significant in volume.[18] Another motivating factor was the desire to clarify some of the historical ambiguities that were inherent in the ICBR 1988 with the intention of achieving a higher degree of contract certainty. Lastly, the general view in the market was that the ICBR 1988 is not appropriate for the insurance of complex modern constructions, such as high-value cruise vessels and LNG carriers; it was felt that there was a desperate need for a new form that would provide a more flexible and user-friendly approach to construction insurance in a maritime context.[19]

The Mar CAR 2007 is a more comprehensive form than the ICBR 1988. It contains five parts: Part I provides cover on an 'all risks' basis for hull and machinery, collision, liability, war and strikes risks, terrorism, political activity and malicious acts;[20] Part II lists exclusions from cover; Part III deals with issues concerning claims; Part IV provides general provisions and definitions; and Part V includes provisions for a number of 'optional buy-back' extensions of coverage for which an additional premium is payable. The Mar CAR 2007 is drafted in the form of a policy wording rather than a set of clauses; the intention is that it can be used as a template which can then be tailored to an individual client's specific needs if required. This provides the flexibility needed to offer cover for complex modern constructions. The Mar CAR 2007 also includes a special provision[21] which can be used to provide cover for related construction activities such as conversion and repair. More fundamentally, a number of provisions that appeared in the ICBR 1988 have been modified to achieve more contractual certainty.[22]

The Mar CAR 2007 has certainly gained a reasonable degree of industry support but it appears that the ICBR 1988 is still widely used by the insurance market.[23] It is the view of the author that, from a legal perspective, Mar CAR 2007 offers a more precise indemnity package for the assured; but at the same time several key concepts

17 See cl. 19 of the ICBR 1988.

18 For example, the amount of indemnity paid for the loss of *The Diamond Princess* was $310m. The loss of *The Pride of America* cost the sector $228m (figures obtained from *Willis Marine Builders' Risk Review*, Summer/Autumn 2005).

19 For the sake of completeness, it must be stressed that WELCAR is the form that is often used in offshore construction projects. WELCAR was introduced to the market in April 2001 by a Lloyd's Syndicate, Wellington, and was revised in 2011 following a review undertaken by the Joint Rig Committee.

20 The ICBR 1988, on the other hand, excludes war and strikes risks and malicious acts from its scope. Cover for such risks needs to be purchased additionally if the ICBR 1988 is used as the basis of cover when insuring ships under construction.

21 See cl. 43 of the Mar CAR 2007.

22 For a more comprehensive analysis of the changes introduced by the Mar CAR 2007 see S. Curtis, *The Law of Shipbuilding Contracts* (4th edn, Informa Law, 2012), Art. XII, pp. 222–232.

23 See, P. Reed, *Construction All Risks Insurance* (Sweet & Maxwell, 2014) at 23-042.

and definitions remain unclear, and some of the recently introduced provisions could potentially fuel litigation. To this end, in the next part amendments made to the policy wording of the ICBR 1988, which were perceived to have shortcomings, will be analysed. This analysis is significant not only in demonstrating the improvements made by the Mar CAR 2007 form but also raising awareness, amongst those who continue using the ICBR 1988 form as the basis of their builders' risks cover, as to the amendments in the form that they need to negotiate in the contracting process with their underwriters to ensure that their interests are fully protected. In the final part of the chapter, potentially problematic aspects of the Mar CAR 2007 will be discussed in the light of legal authorities and similar provisions that appear in other standard insurance contracts. The author's view is that, despite the improvements achieved by the introduction of the Mar CAR 2007, there remain some significant legal difficulties that would benefit from further clarification.

6.2 Changes introduced by the Mar CAR 2007 – problematic aspects of the ICBR 1988

6.2.1 Position of contractors/subcontractors

Shipbuilding is a rather complex project involving the assembly of materials obtained from a variety of suppliers. Even if a contemporary shipyard possesses all the requisite technical know-how and manpower to put increasingly complex structures in place, it is inevitable that it will need to utilise a number of contractors and subcontractors which would be responsible for supplying and assembling different parts of a new ship under construction.[24] The standard contracts that shipbuilders enter into with their contractors and subcontractors would invariably require the builder to extend the insurance cover obtained for the construction process for their benefit. To achieve this goal, contractors and subcontractors need to be included in the builder's insurance policy as co-assureds, but the ICBR 1988 is surprisingly silent on this matter. To fill this gap, in practice, brokers often add a generic phrase stressing that the policy is extended to 'contractors, subcontractors and affiliates' of the builder. The use of such a standard phrase could potentially give rise to two major legal difficulties.

First, it becomes a matter of construction whether a particular contractor or subcontractor would come under the definition of the assured in each case. If the view adopted by the court is to the effect that the definition cannot be extended to a particular contractor or subcontractor, in case of that contractor or subcontractor acting negligently this will enable the insurer to exercise rights of subrogation to bring an action in the name of the assured builder against that party in respect of loss or damage to property for which the builder is insured.[25] This was the position adopted by the court in *Hopewell Project Management Ltd and Hopewell Energy (Philippines) Corp v Embank Preece Ltd.*[26] There, a firm that was engaged by a project

24 Most of the sophisticated builders use a module or block process. Full blocks of vessels are built separately and often by different contractors and then moved into place for final construction.

25 *BP Exploration Operating Co Ltd v Kvaerner Oilfield Products Ltd and another* [2004] EWHC 999 (Comm); [2005] 1 Lloyd's Rep 307.

26 [1998] 1 Lloyd's Rep 448.

developer involved in the construction of a power station to provide engineering consultancy contended that it was covered by a clause in the CAR policy which defined the assured as, *inter alia*, 'all contractors and subcontractors' and should accordingly be treated as a co-assured. Mr Recorder Jackson QC reached the decision that this phrase was limited to those contractors and subcontractors carrying out 'physical works of construction' and accordingly did not cover a firm providing only professional (i.e. consultancy) services. In similar fashion, the Court of Appeal of British Columbia in *Canadian Pacific Ltd v Base-fort Security Services (BC) Ltd*[27] reached the conclusion that a firm responsible for security on a construction site did not come under the definition of 'General Contractors and/or Contractors and/or Sub-Contractors' in the CAR policy. Hollinrake J viewed only 'those persons whose contributions are an integral and necessary part of the construction process itself' as coming within the definition of 'Insured' in the policy. Others, such as security companies, whose contributions are collateral to the construction process in the view of Hollinrake J would fall short of the definition.[28]

The second difficulty relates to whether contractors and subcontractors, assuming that they are included in the definition of the assured in the policy, would have an insurable interest to enable them to insure the entire works and recover the whole of the loss insured despite having only limited proprietary and possessory rights in the works. If that holds true, then the contractor or subcontractor can recover the whole of the insured sum in case of a total loss (obviously holding the excess over his own interest on trust for the others). Lloyd J was adamant in *Petrofina (UK) Ltd v Magnaload Ltd*[29] that this represented the legal position. There, during the process of extension works to an oil refinery, a gantry came loose and fell to the ground, damaging the works. The insurer, having indemnified the assured (main contractor), brought a subrogated action against the subcontractors who were responsible for the heavy lifting system which had collapsed. The subcontractors argued that they were the co-assureds, as they were insured for loss or damage to the entirety of works, not just their own. The subcontractors' defence was upheld. In reaching this decision, Lloyd J relied on the well-established rule that a bailee could insure for the full value of the goods even though their interest in the goods was limited.[30] He was heavily influenced by a Canadian authority in which it was held on a similar set of facts that the subcontractors had a 'pervasive interest' in the contract works on the basis that they had 'such a relationship with the entire works that their potential liability therefore constituted an insurable interest in the whole'.[31]

---

27 (1991) 767 (BC CA).

28 Conversely, see the judgment of Deputy Judge Mr Anthony Colman QC in *Stone Vickers Ltd v Appledore Ferguson Shipbuilders Ltd* [1991] 2 Lloyd's Rep 288 where it was held that the policy was extended in favour of a subcontractor for work on shipbuilding, even though the subcontractor was not employed to work in the shipyard but to fabricate parts (the propeller and ancillary equipment) on its own premises. The decision of the Court of Appeal [1992] 2 Lloyd's Rep 578 turned on the facts of the case and it was held that there was insufficient evidence of intention that the contractor should insure on behalf of this particular subcontractor. However, the reasoning adopted by the first instance judge on the definition of the assured covering subcontractors that do not work on the premises but engage in the production process at their own premises remains valid.

29 [1984] QB 127.

30 *Tomlinson (A) Hauliers Ltd v Hepburn* [1966] AC 451.

31 *Commonwealth Construction Co Ltd v Imperial Oil Ltd* [1978] 1 SCR 317, at 322–323.

It is debatable, to say the least, whether a parallel can be drawn between the position of a bailee and a subcontractor in the manner Lloyd J has done, given that a bailee has a contractual and legal right over the goods left in his care, whilst working on the works does not, on its own, create similar rights for a contractor over the goods belonging to others. However, the reasoning adopted by the Supreme Court of Canada in *Commonwealth Construction* has an instinctive appeal and seems to have found further support amongst other British judges. Most notably, Colman J in *National Oilwell (UK) Ltd v Davy Offshore Ltd*[32] was prepared to hold that a subcontractor should be able to insure against loss of or damage to property involved in a common project not owned by him and not in his possession, either as a result of potential liability arising from the existence of a contract between the assured and the owner of property or from the assured's proximate physical relationship to the property in question.[33]

The Court of Appeal's judgment in *Deepak Fertilisers & Petrochemicals Corporation v Davy McKie (London) Ltd and ICI Chemicals and Polymers Ltd*,[34] however, has cast some doubt on the correctness of the stance taken by several first instance judges on the matter. The defendant contractor was responsible for designing and supervising the construction of a power plant in India which exploded with tragic consequences after it began operating. An action was brought in the name of the employees against the contractor for negligence. It was argued that the defendant contractor was a co-assured and, accordingly, an action brought in the name of the employers was excluded. The Court of Appeal had no difficulty in finding that the defendant contractor enjoyed an insurable interest in the property, although he had no proprietary interest in it, up to the point of completion and commissioning because he had an expectation of remuneration in respect of the contract works, which was dependent on the continued existence of the property. However, in overruling the decision of the first instance judge on this point, it was held that the defendant contractor would have no insurable interest after the works were completed by reason of his potential liability for them after completion and commissioning, as the effect of this would be to convert the policy from a property policy to a liability policy.[35] This seems at odds particularly with the reasoning adopted by Colman J in *National Oilwell*, who deemed it adequate to establish insurable interest that the contractor has a potential liability stemming from an existing contract with his principal. Although the issue did not arise for decision, in *Feasey v Sun Life Assurance Co of Canada*,[36] the majority of a differently constituted Court of Appeal expressed the view that there is no reason why potential liability for damage to property should not create an insurable interest in the property as long as the assured was in a legal relationship to the property (i.e. a contractor or subcontractor undertaking part of the construction) and the relevant liability fell within the subject of insurance.[37] It is the view of the

32 [1993] 2 Lloyd's Rep 582.

33 *Ibid.*, at 611. See also, *Stone Vickers Ltd* v *Appledore Ferguson Shipbuilders Ltd* [1991] 2 Lloyd's Rep 288 and *Anthony John Sharp and Roarer Investments Ltd v Sphere Drake Insurance plc (The Moonacre)* [1992] 2 Lloyd's Rep 501.

34 [1999] 1 Lloyd's Rep 387.

35 *Ibid.*, at [65]–[67].

36 [2003] EWCA Civ 885; [2003] Lloyd's Rep IR 637.

37 *Ibid.*, at [94]–[97] per Waller LJ. Similarly, Dyson LJ at [122], said: '. . . I can see no useful purpose

author that in the light of the fact that courts have demonstrated a tendency in the last few decades of taking a more liberal stance on the issue of insurable interest, the majority judgment in *Feasley* correctly represents the current state of law, but it is also undisputable that the law on this point is far from settled.[38]

Turning to the Mar CAR 2007, it is pleasing to see that provisions have been added to offer a solution for both of these legal problems. Clause 56.2 provides a definition of people who will be treated as 'additional assured' for the purposes of the policy. Accordingly, a contractor or subcontractor to whom benefit of this policy has been provided for in their written contract with the assured or contractor, respectively, will be treated as an additional assured (in effect a co-assured). Therefore, contractors or subcontractors, regardless of the nature of the function they serve in the construction, will benefit from this policy as long as they have a written contract with the assured or contractor and the benefit of this policy is extended to them in that contract. This would provide a solution to the legal problems that manifested themselves in the case of *Hopewell Project Management Ltd and Hopewell Energy (Philippines) Corp v Embank Preece Ltd*. Similarly, difficulties surrounding insurable interest would not arise under the Mar CAR 2007 as clause 42.1 stipulates that 'additional assureds shall be insured under this insurance to the extent of their respective rights and interest in the Subject Matter Insured, but in no case shall any Additional Assured be insured under this insurance to any greater extent than provided for in their written contract with the Assured'. This provision is supplemented by clause 42.3 which prevents additional assureds from exercising their rights themselves. The assured is authorised to make claims on their behalf under the policy, thus averting the possibility that the additional assured would recover the full insured sum under the policy in case of a total loss.

6.2.2 Matters concerning coverage

6.2.2.1 Damage or loss caused by latent defect

Clause 5 of the ICBR 1988 stipulates that cover is available for loss or damage to the subject-matter insured 'caused or discovered during the period of this insurance including the cost of repairing, replacing or renewing any defective part condemned solely in consequence of the discovery therein during the period of this insurance of a latent defect'. This provision might give the impression that any loss or damage of a defective part is covered even if this is simply a case of discovering a latent defect on that part. However, in the context of hull policies since the decision of the Court

---

in holding that a contractor has an insurable interest in [a] plant (of which he supplies only a small component) up to the time of completion and commissioning, but not thereafter. On the facts of a case like *Deepak*, the subcontractor's commercial interest in the plant as a whole during the construction and commissioning stage lies at least as much in his potential liability for damage caused to the plant by his breach of contract and duty as in his interest in not losing the opportunity to do the work and be remunerated for it if the plant is damaged or destroyed by any of the risks covered by an all risks policy.'

38 Dissenting from the majority, Ward LJ *ibid.*, at [185]–[188], held that the only insurable interest possessed by a subcontractor in the plant is his own pecuniary loss in the event of damage to the plant; his liability for the loss of the plant could not support a property policy as there was no legal or equitable relationship between the subcontractor and the subject matter itself.

of Appeal in *Promet Engineering (Singapore) Pte Ltd* v *Sturge (The Nukila)*[39] it is clear that a distinction should be made between a previous latent defect on a part simply becoming patent and the latent defect operating upon the insured property so as to occasion damage to it.[40] Loss or damage of the former type is not covered under a standard hull policy, whilst if a latent defect in a part progresses to cause damage on that part or any other part, that damage will be covered by the policy.[41]

It would be logical to follow the stance adopted in hull policies in the context of builders' risks policies as well, but due to the manner in which clause 5 is worded, it is not clear under the ICBR 1988 whether cover is provided against simple discovery of a latent defect without that effect causing loss or damage. Clause 3 of the Mar CAR clarifies the position by stating that the cover is extended to the 'cost of repairing physical loss or physical damage caused by latent defect and discovered during the period of insurance'. This means that the basic cover provided under the Mar CAR 2007 will not extend to the cost of rectifying the latent defect or the latently defective part,[42] which is in line with the position adopted by standard hull policies.

6.2.2.2 Cover for losses caused by tsunami

Clause 6 of the ICBR 1988 excludes from the scope of cover loss or damage caused by 'earthquake or volcanic eruption'. This brings to mind a significant question as to whether the assured could recover for losses caused by tsunami if his cover is based on the ICBR 1988. A tsunami is usually caused by earthquakes, volcanic eruptions, other underwater explosions, landslides, glacier calvings, meteorite impacts and other disturbances above or below water. Accordingly, an insurer could plausibly argue that the proximate cause[43] of loss in a case when the insured vessel is lost or damaged as a result of tsunami, is, in fact, 'earthquake' or 'volcanic eruption' either of which is an excepted peril.

On the other hand, there is a considerable body of case law as to which occurrences qualify as 'perils of the seas',[44] and the use of the word 'peril' denotes that something fortuitous or accidental is envisaged.[45] Recently, it has been stressed

---

39 [1997] 2 Lloyd's Rep 146.

40 That said, Hobhouse, LJ, *ibid.*, at 157, acknowledged the potential for evidential difficulty in determining whether a latent defect had progressed so as to have occasioned damage.

41 Clause 2.2.2 of the International Hull Clauses (01/11/03) makes reference to the position established by the Court of Appeal in *The Nukila*: 'This insurance covers loss of or damage to the subject matter caused by any latent defect in the machinery or hull *but does not cover any of the costs of correcting the latent defect*' (emphasis added).

42 It is open to the assured to purchase additional cover under cl. 57 of the Mar CAR 2007 to cover the cost that would have been incurred to replace, repair or rectify a latent defect that is discovered prior to the occurrence of the physical loss or physical damage.

43 In identifying the proximate cause of a loss the court's function is to identify the cause without which the loss would not have occurred. See the House of Lords' judgment in *Leyland Shipping Co Ltd v Norwich Union Fire Insurance Society Ltd* [1918] AC 350.

44 See, e.g., *Canada Rice Mills Ltd v Union Marine & General Insurance Co Ltd* [1941] AC 55; *Baxendale v Fane (The Lapwing)* [1940] P 112; *Samuel (P) & Co Ltd v Dumas* [1924] AC 431 and more recently in *Versloot Dredging v HDI-Gerling Industrie Versicherung AG (The DC Merwestone)* [2013] EWHC 1666 (Comm); [2013] 2 Lloyd's Rep 131.

45 Rule 7 of the Rules for Construction of Policy attached to the MIA 1906 stipulates: 'The term "perils of the seas" refers only to fortuitous accidents or casualties of the seas.'

by the Supreme Court in *Global Process Systems Inc v Syarikat Takaful Malaysia Berhad (The Cendor Mopu)*[46] that there is no threshold in identifying what will amount to a fortuity. The question in each case is whether sea and weather conditions were such as to have caused a fortuitous accident or casualty.[47] This might lend support to the submission that as long as the loss is not attributable to internal failure or deficiencies,[48] any external fortuitous event, including unexpected movement of the sea and waves however caused, could qualify as a peril of the sea. If taken to its natural conclusion, the assured could argue that the proximate cause of loss or damage caused by tsunami is 'perils of the seas' and, therefore, covered by the ICBR 1988. No such difficulty would arise under the Mar CAR 2007 as the 'earthquake and volcanic eruption' exclusion has been removed, meaning that in those instances the assured would obviously be able to make a claim under the policy.

6.2.2.3 Constructive total loss

The relevant provision of the ICBR 1988, which deals with constructive total loss, is modelled on the corresponding provision in standard hull clauses. Therefore, in ascertaining whether the subject matter insured is a constructive total loss, the insured value is taken into consideration and the assured could claim constructive total loss only in instances where the cost of recovery and/or repair would exceed the insured value under the policy.[49] This makes perfect sense in the context of hull insurance where the value of the subject matter of insurance normally remains unchanged throughout the policy period (subject to market fluctuations and reasonable depreciation, of course). However, ship construction is a progressive venture and it is very unlikely that the cost of recovery and/or repair would exceed the insured value unless the loss occurs very late in the project. This, therefore, limits the prospect of claiming constructive total loss under the builders' risk policy although, in reality, the parties may wish to abandon a construction project at an early stage in the event of significant damage or loss.

The Mar CAR 2007 incorporates a new provision[50] allowing the assured to seek abandonment of the works if the cost of repairing the damage to the ship under construction does not exceed the sum insured, but does exceed the 'works value'.[51] This provision, in effect, creates an additional contractual constructive total loss regime allowing the assured to abandon the project part of the way through the construction work. Needless to say, in such instance the assured could not claim the whole of the insured value as indemnity, but only the works value plus the profit percentage thereon at the time of loss. Given that most standard shipbuilding contracts enable the parties to bring the construction project to an end in cases where a significant loss or damage occurs, this new provision brings

46 [2011] UKSC 5; [2011] 1 Lloyd's Rep 560.

47 *Ibid.*, at [104] per Lord Clarke.

48 See, e.g., *Dedgeon v Pembroke* (1874) LR 9 QB 581 and *Lamb Head Shipping Co Ltd v Jennings (The Marel)* [1992] 1 Lloyd's Rep 402.

49 Clause 12 of the ICBR 1988.

50 Clause 31 of the Mar CAR 2007.

51 That is, the cost of the works to date.

builders' risk insurance in line with commercial practice and assists the shipyards by giving them the assurance that they will be able to recover for their losses if they choose to bring the shipbuilding contract to an end for economical or physical reasons.

6.2.2.4 Notice provision

Most indemnity policies would contain a provision requiring the assured to give notice to the underwriters in case of an event or loss or damage occurring that might give rise to a claim. The ICBR 1988 is not an exception in that regard and clause 14 stipulates that in the event of loss, damage liability or expense which may result in a claim the assured shall give 'prompt notice to the underwriters prior to repair'. If the subject matter is under construction abroad, notice should be given 'to the nearest Lloyd's Agent so that a surveyor may be appointed to represent the underwriters should they so desire'.

This provision leaves a lot to be desired in terms of clarity. For example, it will be a question of fact in each case whether prompt notice has been given or not. The word 'prompt' presumably implies 'vigorous action, without any delay'[52] but surrounding circumstances concerning a casualty or event (e.g. the precise location, availability of witnesses, legal or physical restrictions) will be expected to play a key role in each instance in determining whether the notice was given promptly or not. More significantly, the provision fails to identify the consequence of breach. The insurer could attempt to argue that the provision is a 'condition precedent' which discharges himself from liability in relation to that claim in case of non-compliance. The only factor that goes in favour of the insurer minded of contending that the notice obligation has this kind of legal effect is the fact that it has been qualified by the use of the word 'shall'. On the other hand, there is nothing in the clause suggesting that the assured would be deprived of the claim if the notice requirement is not fulfilled. Therefore, one should not be surprised if the judge decides to construe the clause against the interests of the insurer (i.e. *contra proferentem*). A provision of this nature that relates to the claims stage and sets out a procedure that needs to be followed by the assured in case of an event arising that could give rise to a claim, without specifying the consequences in case of its breach, could have been treated as a severable innominate term, following the reasoning of Waller LJ in *Alfred McAlpine plc v BAI (Run-Off) Ltd*,[53] thus allowing the insurer to reject the claim in question if it is seriously breached but leaving the contract in place. However, it is doubtful whether the analysis of Waller LJ survives the judgment of the Court of Appeal in *Friends Provident Life & Pensions Ltd v Sirius International Insurance*.[54] If it does not, this provision is likely to be treated as a mere condition that results in damages only,[55] but it is fair to say that this is far from certain.

52 *R v Berkshire Justices* (1878) 4 LR QBD 469, at 476, per Cockburn CJ.

53 [2000] 1 Lloyd's Rep 437. See also *K/S Merc-Scandia XXXXII v Certain Lloyd's Underwriters (The Mercandian Continent)* [2001] EWCA Civ 1275; [2001] 2 Lloyd's Rep 563.

54 [2005] EWCA Civ 601; [2005] 2 Lloyd's Rep 517.

55 It was held in *Stoneham v Ocean Railway and General Accident* (1887) 19 QBD 237 that a term which required the company to give notice in case of a fatal accident within seven days was a contractual

Fortunately, the new notice provision that appears in the Mar CAR 2007 is much clearer both in terms of the time that the assured is allowed to comply with its requirements and also the legal consequence of its breach. Clause 26 of the Mar CAR requires the assured or his project management to give notice to the underwriters as soon as possible and certainly within 180 days of becoming aware of loss or damage, loss that may result in a claim and in case of physical loss or damage to the subject matter insured prior to the commencement of any repair work. It has been set out expressly in clauses 26.2 and 26.3 that in case of breach of these obligations, no claim shall be recoverable under the policy for such loss, damage or liability or expense, unless the underwriters agree to the contrary in writing.

6.3 Problematic aspects of the Mar CAR 2007

6.3.1 Meaning of physical loss and damage

The coverage provided by the Mar CAR 2007 is against all risks of 'physical loss of or physical damage'.[56] This represents a deviation from the position adopted by the ICBR 1988 which provides indemnity for 'loss of or damage'.[57] The use of the word 'physical' before 'loss' and 'damage' is a positive development which clearly indicates that the policy's coverage could not be extended to include defects and non-physical losses such as economic losses.

Despite the fact that there is consensus amongst the judges that 'physical loss or damage' requires 'some altered state',[58] there still remains difficulty in identifying the degree to which the insured works have to be materially altered to constitute 'physical loss or damage'. It is, for example, debatable whether temporary damage would qualify as 'physical damage' for the purposes of a policy of this nature. In the context of tort law, the view taken by courts is that the adverse change in the physical condition does not need to be permanent to constitute damage. In *Losinjska Plovidba v Transco Overseas Ltd (The Orjula)*[59] drums of hydrochloric acid and sodium hypochlorite shipped in two containers were carried on board a vessel from Felixstowe to Benghazi in Libya. The vessel was expected to make a call to Rotterdam *en route* to load further cargo. On approaching Rotterdam, it was noted that one of the containers containing the acid was leaking. The vessel was ordered to a tank-cleaning berth where the containers were removed and the vessel's contaminated deck and hatch covers were decontaminated by using soda and fresh water. It transpired, following inspection, that the drums of chemicals had not been properly stowed within the containers. The bareboat charterer of the vessel raised claims in contract and/or tort of negligence against the shipper of the cargo, the supplier, freight forwarder and road haulier in the UK. On behalf of the defendants, it was

term, the breach of which imposed an obligation upon the assured's representatives to reimburse the company for extra expenses that they might incur from having to investigate the circumstances of an accident at a long interval after its occurrence.

56 See cl. 2 of the Mar CAR 2007.

57 See cl. 5 of the ICBR 1988.

58 See *Pilkington UK Ltd v CGU Insurance plc* [2004] EWCA Civ 23; [2004] Lloyd's Rep IR 891, at [51], per Potter LJ.

59 [1995] 2 Lloyd's Rep 395.

contended that there could be no duty of care in relation to the contamination of the vessel as the incident did not amount to physical damage but only left a layer of hydrochloric acid over the part of the deck and hatch covers. Mance J (as he then was), gaining some assistance from a number of criminal cases,[60] found in favour of the bareboat charterers and he said:

> Here, specialist contractors were engaged in undertaking the decontamination work using soda to neutralise the acid before washing the deck and hatch covers down with fresh water; further, it is pleaded, not perhaps surprisingly, that the vessel was required to be decontaminated of the hydrochloric acid before she could sail from the special berth to which she had been directed after discovery of the leakage. On these alleged facts, I would have no hesitation in concluding that the vessel should be regarded as having suffered damage by reason of her contamination.

A similar outcome emerged in *British Celanese Ltd v A H Hunt (Capacitors) Ltd*,[61] which involved a claim arising from a power interruption that resulted in some machinery becoming clogged with solidified material that had to be cleaned out before that machinery could be used again. It was the contention of the defendant that the fact that the machinery had to be cleaned did not evidence that the clogging of the machinery constituted an injury to property. Lawton J disagreed and found that the clogging did constitute physical injury.[62]

It is submitted that a parallel can be drawn between these authorities and insurance contracts on shipbuilding risks. It is immaterial whether the condition of the property can be restored to its original status by repair or by replacing the defective part or simply by cleaning it. The key consideration must be whether the functionality of the relevant property is adversely affected or not and the need for work and the expenditure of money has arisen to restore the property to its former usable condition. If both of these conditions are satisfied, there should be 'loss or damage' within the meaning of the policy even if this is temporary in nature.

It is arguable that a similar outcome should follow in a case where the paint or coating applied by a shipyard is excessive. The issue was raised in the *State of Netherlands v Youell*[63] where the shipyard, during the process of constructing submarines for the Dutch Navy, failed to prevent excessive thickness in the primer coating although the Navy had drawn this point to the attention of the shipyard. The yard refused to carry out repairs without prior payment by the Navy. Once the payment was made to the yard, the Navy sought indemnity from the underwriters. Originally the underwriters attempted to argue that the policy in question was a joint policy, meaning that the failings of the yard could prejudice cover in favour of the Navy, but this was rejected. An alternative argument to the effect that the yard was the Navy's agent for the purposes of the statutory duty to sue and labour also did not receive

60 Particularly the unreported Court of Appeal authority of *R v Henderson and Batley* (29 November 1984) cited with approval by the Court of Appeal in *Cox v Riley* (1986) 83 Crim App R 54 and *R v Whiteley* [1991] 93 Crim App R 25.

61 [1969] 1 WLR 959.

62 It is worth noting that a similar stance has been adopted in other common law jurisdictions. For example, in *Ranicar v Frigmobile Pty Ltd* [1983] Tas R 113, at 116, Green CJ said: '.. the phrase "damage to" when used in relation to goods, is a physical alteration or change, not necessarily permanent or irreparable, which impairs the usefulness of the things said to be damaged.'

63 [1997] 2 Lloyd's Rep 440.

the support of the court. The dispute between the parties was settled following the decision of the Court of Appeal,[64] so the need to determine whether the de-bonding of excessive paint could amount to 'damage' did not arise. However, in the light of the authorities discussed above, it is submitted that as long as it can be shown that application of excessive painting has an impact on reducing the functionality of the vessel in construction, that should be adequate to qualify the cost of de-bonding and repainting as 'physical loss or damage'.[65]

At this juncture, one should make reference to a related matter that could give rise to difficulties under the Mar CAR 2007. It is clear that the insurance does not cover the cost of replacing, repairing, or rectifying of a latent defect or defect in material, design or workmanship.[66] Therefore, if painting of a vessel reveals that her frame has been defectively cast so as to cause shrinkage cracks requiring its replacement, there is no 'damage' within the scope of the policy, so the cost of replacing the frame is not recoverable.[67] The interesting question in this context is whether the cost of repainting the frame (and the vessel) after the defective frame is replaced would be recoverable under the policy. Put differently, is this a case of a latent defect or defective workmanship causing physical loss or damage that is covered by the policy? The problem here is that the painting itself is not faulty but it is an ancillary component that needs to be replaced when the faulty component is rectified. Although the Mar CAR 2007 is silent on this point, it is submitted that this should be treated as 'physical loss' caused by a defective part given that replacement of a defective part renders the original painting wasted and necessitates the application of a new coat of paint. Therefore, it is the view of the author that, in this instance, the cost of repainting the new frame should be recoverable under the policy, although the cost of replacing the defective part itself is not covered.

An even more intriguing question is whether there is 'physical loss or damage' if materials suffer from a degree of stress (fatigue) but they can still be used for the purpose for which they are intended. There is authority pointing to the direction that if the stress that the materials suffer causes a reduction in their life expectancy, this amounts to physical damage. *Ranicar v Frigmobile Pty Ltd*[68] concerned a claim in respect of a consignment of shellfish that were found upon delivery to be at a temperature higher than contracted. Expert evidence indicated that this did not cause any apparent damage to the shellfish but had, in fact, shortened their life expectancy. On that basis, it was held that the insured property suffered a 'loss or damage' within the meaning of the Institute Cargo Clauses (All Risks). It is submitted that a similar reasoning can be applied here, and if the life expectancy of materials used in construction are shortened due to their having suffered from stress (or fatigue), this should qualify as a 'physical loss or damage'. Logically, a similar conclusion should

64 [1998] 1 Lloyd's Rep 236.

65 This presupposes that 'paint' could be regarded as a part in the manner described in *The Nukila* [1997] 2 Lloyd's Rep 146. There is no definition of what a part is in the Mar CAR 2007 and, as acknowledged by Mustill J in *JJ Lloyd Instruments Ltd v Northern Star Insurance Co Ltd (The Miss Jay Jay)* [1985] 1 Lloyd's Rep 264 at 273, it may be difficult to find a complete answer to this question and provide a comprehensive definition as to what a part is.

66 See, cl. 3 of the Mar CAR 2007.

67 See, e.g., *Hutchins Brothers v Royal Exchange Insurance Corp* [1911] 2 KB 398.

68 [1983] Tas R 113.

be reached in a case where stress leads to a diminution in the value of the materials used.

A more difficult question is whether we can talk about 'damage or loss' to materials in a case where they undergo a molecular change although they show no clear evidence of any impairment. A molecular change is likely to have an impact on the life expectancy or quality of the materials used so it should be possible to qualify such a change as a 'physical damage or loss' within the meaning of Mar CAR 2007. Support for this can be drawn from a case, *Quorum AS* v *Schramm*,[69] which concerned a fine art insurance on the Degas pastel *La Danse Grecque*. The pastel had been insured in respect of 'direct physical loss or direct damage'. The pastel was stored in a strong room in a specialist warehouse. On 7 October 1991 there was a fire in the warehouse which destroyed all the contents except for the materials stored in the strong room. Although no smoke penetrated the strong room, it was agreed that, on the balance of probability, the insured pastel was exposed to rapid change in heat and humidity within the strong room as a result of fire. Following the fire, the pastel was examined by an expert who noted two tears in the paper which was partially detached from the backing board which itself was bowed and with evidence of recent mould. The expert also believed that the pastel had been damaged by being in that environment but he said that it was difficult to observe the damage. He made recommendation that a number of restoration works be carried out. Underwriters accepted liability for some tears and bowing of the board as direct physical damage caused by fire but the assured's contention was that the pastel suffered further damage not readily observable by the naked eye.[70] Further experts in Paris stated that the picture superficially appeared in good condition, but it had been through severe stresses which had probably shortened its life expectancy and it was impossible to give any guarantee for the duration of the effects of the restoration. Experts appointed by the parties for the purpose of the hearing agreed that there would have been molecular changes to the pastel as a result of the heat and humidity and also sub-molecular changes at a chemical level that were irreversible. Agreeing with the evidence of the experts, Thomas J (as he then was) held:[71]

> . . . there was sub-molecular damage to the pastel caused by fire: that was in my view, damage to the picture. In my view such damage is clearly direct physical damage resulting from fire, even though it might not be visible and its extent could not be determined without testing, which could not be carried out because of its effects on the pastel.

It is appreciated that providing an exhaustive definition for the term 'physical loss or damage' is not an easy task and might not be practical but, in the light of the discussion above, it will be very useful if it can be specified with clarity whether the term includes: (i) loss or damage incurred as a result of temporary causes; (ii) loss or damage when the functionality of the materials or their life expectancy are affected even if there is no apparent physical impairment; and also (iii) loss or damage

69 [2002] 1 Lloyd's Rep 249.

70 There were also disputes between the parties, as to the value of the pastel before and after the fire and whether the policy was a valued policy or not, which the trial judge had to deal with.

71 [2002] 1 Lloyd's Rep 249, at 264.

suffered as a result of molecular change in the structure of materials even though they suffer from no apparent physical impairment.

6.3.2 Due diligence provision

Clause 48 of the Mar CAR 2007, in a similar fashion to other standard contracts used in the energy insurance sector,[72] stipulates:

> 48.1 It is the duty of the Assured to exercise due diligence in the conduct of all operations relating to the Subject Matter Insured and in the utilisation of all safety practices and equipment considered prudent for such operations, such duty including but not limited to:
> 48.1.1 the exercise of due diligence in the selection and employment of Contractors, Subcontractors and others in connection with the Subject Matter Insured; and
> 48.1.2 the exercise of due diligence in relation to such precautions as may be reasonably required to prevent loss, damage, liability or expense in connection with the Subject Matter Insured.
>
> 48.2 In the event of any breach of the duty under Clause 48.1, the Underwriters shall in no case be liable for any loss, damage, liability or expense attributable to such breach.

Although the consequence of breach of this obligation is clearly set out in cl. 48(2), it is submitted that the first part of the clause lacks clarity and has the potential to create legal disputes in many respects. The first difficulty is identifying the appropriate standard of care required of the assured under this provision. A due diligence proviso, which often appears in a standard hull policy, is designed to qualify the cover provided under an Inchmaree clause by stating that cover for loss or damage caused by certain types of perils[73] will be available 'provided such loss or damage has not resulted from want of due diligence by the Assured, the Owners or managers of the property insured'. In two recent English authorities on the subject,[74] negligence has been accepted as the appropriate standard of care required under an Inchmaree clause. This would mean that the underwriter will be liable for loss or damage caused by an Inchmaree peril only if he has not negligently contributed to the occurrence of that peril.

One might be tempted to argue that a similar degree of standard is adequate for the insurer to rely on clause 48 of the Mar CAR 2007 as a defence. It is submitted that the judicial line taken in the context of Inchmaree clauses is not appropriate here for two reasons. First, the due diligence proviso as appears in the context of an Inchmaree clause serves the purpose of defining the scope of the cover provided.

72 See, e.g., cl. 4E of the London Standard Platform Form 2009. Similar provisions exist in control of well and operations policies.

73 Such as bursting of boilers or breakage of shafts; latent defect in the machinery or hull; negligence of master, officers, crew or pilots; negligence of repairers or charterers or barratry of master, officers or crew.

74 *Sealion Shipping Ltd Toisa Horizon Inc v Valiant Insurance Co (The Toisa Pisces)* [2012] EWHC 50 (Comm); [2012] 1 Lloyd's Rep 252 (aff'd on different grounds by the Court of Appeal [2012] EWCA Civ 1625; [2013] 1 Lloyd's Rep 108) and *Versloot Dredging v HDI-Gerling Industrie Versicherung AG (The DC Merwestone)* [2013] EWHC 1666 (Comm); [2013] 2 Lloyd's Rep 131 (aff'd on different grounds by the Court of Appeal [2014] EWCA Civ 1349).

Clause 48, on the other hand, is very different in nature as it creates an obligation on the part of the assured and its breach deprives the assured from indemnity for any loss, damage or liability attributable to such breach. Secondly, if negligence is adequate for the insurer to be able to rely on clause 48, this might thwart the commercial objective of the policy, which is designed to provide cover against all risks and liabilities associated with shipbuilding.

As it stands, clause 48 is more akin to 'reasonable care' clauses commonly found in liability insurance policies. Commenting on a clause of that nature, Diplock LJ in *Fraser v BN Furman (Productions) Ltd* said:[75]

> Obviously, the condition cannot mean that the insured must take measures to avert dangers which he does not himself foresee, although the hypothetical reasonably careful employer would foresee them. That would be repugnant to the commercial purpose of the contract, for failure to foresee dangers is one of the commonest grounds of liability in negligence. What, in my view, is 'reasonable' as between the insured and the insurer, without being repugnant to the commercial object of the contract, is that the insured should not deliberately court a danger, the existence of which he recognises, by refraining from taking any measures to avert it . . . What, in my judgment, is reasonable as between the insured and the insurer, without being repugnant to the commercial purpose of the contract, is that the insured, where he does recognise a danger should not deliberately court it by taking measures which he himself knows are inadequate to avert it. In other words, it is not enough that the employer's omission to take any particular precautions to avoid accidents should be negligent; it must be at least *reckless*, that is to say, made with actual recognition by the insured himself that a danger exists, and not caring whether or not it is averted. The purpose of the condition is to ensure that the insured will not, because he is covered against loss by the policy refrain from taking precautions which he knows ought to be taken.[76]

It is submitted that the assured would be in breach of clause 48 only if he acts 'recklessly' for the reasons discussed above.[77] The next difficulty with this clause is identifying the person who will qualify as the assured in a case where the assured is a corporate entity. Would it be, for example, necessary to assess the actions of the directing mind or will of the insured corporation or could the actions of those lower in the ranks of management be attributable to the assured company? Applying the test laid down by the Court of Appeal in *Meridian Global Funds Management Asia Ltd v Securities Commission*,[78] it will be necessary to determine whose act in the company structure for the purpose of clause 48 was intended to count as the act of the company. Taking into account the language of the clause and its content, it can be suggested that the actions of the person entrusted with the management of the assured company with regard to the day-to-day running of its affairs would possibly count as the act of the assured company. This is obviously a fact-sensitive issue and in each instance the burden will be on the insurers

---

75 [1967] 1 WLR 898, at 905–906 (emphasis added).

76 See also Jackson J in *Tate Gallery (Board of Trustees) v Duffy Construction Ltd & Anor (No 2)* [2007] EWHC 912 (TCC); [2008] Lloyd's Rep IR 159 at [26]: 'In a policy of liability or property insurance a reasonable precautions clause in the conventional form is not breached by negligence. Recklessness is what constitutes a breach of such a clause.'

77 A similar point has been made by G. Leloudas & B. Soyer, 'Standard Contracts Used in the Offshore Insurance Sector' published as Chap. 10 in *Offshore Contracts and Liabilities* (2014, Informa Routledge), at p. 243, with regard to cl. 4E of the London Standard Platform Form 2009 which is drafted in a very similar manner.

78 [1995] UKPC 5.

to prove that the appropriate person failed to act with regard to matters that come under the scope of clause 48.

A related, but equally fascinating, question is what the position will be when the assured company (e.g. a building yard) decides to delegate one of these functions to another organisation. Let us assume that the insured building yard appoints a company which is responsible for the security of the building yard. One evening, an employee of the security company fails to activate the alarm, leading to theft of machinery belonging to the subject matter of insurance. When determining whether clause 48 has been breached or not, is the insurer expected to show that the assured shipyard acted recklessly in engaging this particular security company, or would it be adequate to demonstrate that the security company acted recklessly in appointing the individual whose error led to the loss? Drawing a parallel from other areas of law,[79] it can tentatively be suggested that if management of a particular operation is delegated to a separate corporate entity, for the purpose of clause 48, that entity should be regarded as the assured,[80] but of course it would have been a much better solution had this been explicitly dealt with in clause 48.

6.3.3 Protecting the right to recovery

Clause 39(1) of the Mar CAR 2007, putting the assured under a duty to protect the rights of his insurer, stipulates the assured shall take reasonable steps to:

> 39.1.1 assess as soon as possible whether there are any prospects of a recovery from third parties in respect of matters giving rise to a claim or to a potential claim under this insurance;
> 39.1.2 protect any claims against such third parties if necessary by the commencement of proceedings and the taking of appropriate steps to obtain security from the claim from third parties;
> 39.1.3 keep the Underwriters and the appointed average adjuster (if any) advised of the recovery prospects of any action taken by third parties;
> 39.1.4 co-operate with the Underwriters in the taking of such steps as may be reasonably required to pursue any claims against third parties.

Some hull policies contain an equivalent provision,[81] but the problem is that the consequence of breach of a provision of this nature is not expressly spelt out even though clause 39(2) imposes a correlative obligation on the insurer to pay the reasonable costs incurred by the assured pursuant to clause 39(1). Given the fact that this clause is secondary in nature and deals with the issue of subrogation, one might be tempted to argue along the lines that its breach sounds only in damages.[82]

79 In *The Marion* [1984] 2 Lloyd's Rep 1 (HL), the management company's act or omissions were deemed to be acts or omissions of the shipowning company for the purpose of breaking the limits.

80 There is authority to the effect that failure by an employee to take reasonable precautions would not amount to breach of a reasonable care condition on the part of the assured in liability policies, *Woolfall & Rimmer Ltd v Moyle* [1942] 1 KB 66; but evidently the situation here is rather different given that the management of a particular aspect of operations has been delegated to another entity.

81 Clause 49 of the International Hill Clauses 2009. A similar provision appears in cargo policies; see, e.g., cl. 16(2) of the Institute Cargo Clauses (A, B and C) (1/1/09).

82 *Noble Resources v Greenwood (The Vasso)* [1993] 2 Lloyd's Rep 309. Prior to the decision of the Court of Appeal in *Friends Provident Life & Pensions Ltd v Sirius International Insurance* [2005] EWCA Civ 601; [2005] 2 Lloyd's Rep 517, this provision could have been treated as a severable innominate

However, there is certainly room for arguing that breach of this obligation could deprive the assured from the cover. It can, for example, be argued by drawing a parallel with surety contracts that the breach of clause 39 is repudiatory, as there is a similarity between an assured failing to protect the subrogation rights of his insurer and a creditor releasing a surety by granting indulgence to the principal debtor.[83] In the view of the author, on the basis that breach of this obligation is unlikely to be causative of the loss or damage, it is very plausible that the appropriate remedy will be damages in the shape of a reduction from the claim. Considering the complicated nature of shipbuilding contracts and the various parties that the assureds are expected to deal with in the course of a shipbuilding process (i.e. suppliers and subcontractors not insured under the policy) it is not a distant possibility that this obligation might be breached; it is certainly a cause for concern that the policy fails to identify the applicable remedy with precision.

6.4 Conclusion

The insurance market has proved on several occasions how sophisticated it can be by amending standard forms in use when practice and evolution of case law necessitate such a change. Almost 20 years after its introduction, the ICBR 1988 has been subjected to an extensive review process. The outcome was the Mar CAR 2007, which is a more elegant contractual document.

Despite its superior features, Mar CAR 2007 is not yet in use in practice as much as the ICBR 1988. This chapter, by illustrating the failings of the ICBR 1988, intends to send a strong message to those assureds who continue using it as the basis of their cover. Using the ICBR 1988 without making any changes could create hazardous consequences for the assureds. Although it advances contractual certainty considerably, it has also been highlighted in the chapter that there remain several provisions in the Mar CAR 2007 that need to be reassessed and ideally modified. As far as standard forms used to insure shipbuilding risks are concerned, it is fair to say that the evolution that has naturally taken place in the market represents a significant improvement, but there is still some way to go to achieve the desired contractual certainty.

---

term allowing the insurer to discharge himself from the liability with regard to the claim concerned in case of its breach in a serious fashion, as stipulated by Waller LJ in *Alfred McAlpine plc v BAI (Run-Off) Ltd* [2001] 1 Lloyd's Rep 437. However, it is very doubtful whether the analysis of Waller LJ survives the judgment of the Court of Appeal in *Friends Provident*.

83 *Andrews v Patriotic Assurance Co of Ireland* (1886) 18 LR Ir 355.

PART 2

SHIP SALE CONTRACTS AND PRACTICE

CHAPTER 7

Battle of the sale forms

*Matt Hannaford**

7.1 Contractual setting

Many aspects of shipping commerce are negotiated and documented using standard form contracts which have, over time, earned acceptance from the international maritime community. For some shipping activities, most notably time chartering, numerous standard forms are available to meet the specific requirements of particular vessel types and trading demands. Other core areas such as technical ship management and crewing lend themselves more naturally to a uniform approach and this is reflected by the predominance, in each case, of a single standard form to regulate the contracting of these activities.

A large number of the internationally recognised standard form contracts have been produced by the Baltic and International Maritime Council (more commonly known as BIMCO). BIMCO is the world's largest shipping association with around 2,300 members across 130 countries. One of its key objectives is the development of standard form contracts and clauses for the purpose of promoting free trade and harmonisation within the shipping industry. Of these standard form contracts, few have achieved greater global penetration than Norwegian Saleform, the standard form contract for the sale and purchase of second-hand ships. Norwegian Saleform is in fact the product of the Norwegian Shipbrokers' Association but for many years has been acknowledged as a core component of BIMCO's document suite.[1] In 1967 Singh & Colinvaux made the following observation with regard to the 1956 version of Norwegian Saleform:[2]

> The Norwegian 'Saleform' . . . is a document in common use as a contract for sale and purchase [and] has been evolved to assist both buyers and sellers in an effort to standardise certain phraseology . . .

It is therefore evident that Norwegian Saleform has enjoyed a wide degree of acceptance from the international shipping community for at least half a century. Indeed, the 1956 form was not the first version of Norwegian Saleform – it is stated on the form that it is a 'revision' – so its 'common use' may date back to the first half of the twentieth century. Since 1956, Norwegian Saleform has been periodically updated,

---

* Partner, Curtis Davis Garrard LLP.

1 BIMCO has formally adopted each version of Norwegian Saleform since the 1956 form and works alongside the Norwegian Shipbrokers' Association on each Norwegian Saleform revision.

2 Singh & Colinvaux, *British Shipping Laws: Shipowners (Volume 13)*, (Stevens, 1967).

most recently in 2012.[3] Each revision has been a response to changes in law, regulation and market practice but, despite the passage of almost 60 years, the structure and content of the 1956 Norwegian Saleform would be strikingly familiar to those involved today in the business of buying and selling ships.

Although Norwegian Saleform has enjoyed widespread acceptance from the maritime community, it is not the only standard form available in the market for the sale and purchase of second-hand ships. Since 1965, the Japan Shipping Exchange has produced its own standard form, 'Nipponsale'.[4] This too has been through various revisions, the latest of which bears the code name 'Nipponsale 1999'. However, Nipponsale typically is used only where at least one of the contracting parties is from Japan, its geographical focus being demonstrated by the fact that disputes under Nipponsale are to be submitted to arbitration in Tokyo in accordance with the Rules of the Tokyo Maritime Arbitration Commission (TOMAC).[5] Moreover, in 2011, another standard form for the sale and purchase of second-hand ships was introduced to the market – the Singapore Ship Sale Form, or 'SSF2011'. Commissioned by the Singapore Maritime Foundation, it was launched in January 2011 with the express objectives of seeking '. . . to serve as a viable alternative option next to the other widely used forms . . . and propel Singapore as a leading international maritime centre in the long run'.[6]

Today, therefore, buyers and sellers of second-hand ships have three principal standard form contracts to choose from: Norwegian Saleform, Nipponsale and SSF2011. All three of these standard forms are designed to be used for the sale and purchase of second-hand, commercially trading vessels. They are not suitable for ships under construction as their terms and conditions presume that the ship will have a trading history. Norwegian Saleform is sometimes used for newbuilding sales (on heavily amended terms) but only where the shipyard is contracting with a buyer for the acquisition of a completed ship or where a ship is being on-sold by the seller immediately upon the seller taking delivery of the ship from the shipyard. Furthermore, just as Norwegian Saleform, Nipponsale and SSF2011 presume that the ship will have a trading history, so do they also contemplate that the ship will have a trading future. For this reason, these standard forms are not usable without extensive amendment for the sale of a ship for demolition and recycling at the end of its working life. In recent years, with increased legislative and regulatory measures in place which focus on the environmental impact of shipping, standard form contracts have been developed and revised to meet the particular needs of this industry sector.[7]

A distinction also needs to be made between commercial vessels and pleasure craft. Whilst Norwegian Saleform, Nipponsale and SSF2011 are readily usable (or at least adaptable) for most vessel types within the commercial sphere, other standard

3 Since 1956, Norwegian Saleform has been revised in 1966, 1983, 1987, 1993 and 2012.

4 The Nipponsale Memorandum of Agreement of the Documentary Committee of the Japan Shipping Exchange Inc 1965.

5 Both Norwegian Saleform and SSF2011 offer parties the opportunity to agree on governing law and the seat of arbitration, in the latter case with a default position of London and Singapore respectively.

6 Singapore Maritime Foundation at www.singforms.com (last visited on 31 December 2014).

7 For example, see BIMCO's 'Demolishcon'.

form contracts are available which are specifically tailored for the sale and purchase of second-hand yachts.[8] That said, in the context of very large yachts – where their size, sophistication and complement of passengers and crew demand adherence to many of the regulatory requirements applicable to commercial ships – Norwegian Saleform or SSF2011 may provide a more suitable base contract than a standard form yacht sale contract.[9]

Generally speaking, the form and content of Norwegian Saleform, Nipponsale and SSF2011 are extremely consistent (although Nipponsale perhaps to a lesser extent, because of its age and comparative brevity). Each form demands little by way of vessel description and contemplates a cursory initial inspection by the buyer of the subject vessel and its records. There is a requirement for the buyer to lodge a deposit by way of security for its contractual obligations following which the buyer is entitled to place one or more representatives on board the ship for familiarisation purposes. There is a temporal delivery range and a geographical delivery range. Close to the time of delivery, there is an opportunity for the buyer to arrange for a classification society surveyor to perform an underwater inspection (or, if agreed in the alternative, an inspection whilst the ship is in drydock). There is an agreed date by which the seller must be able to validly tender the ship for delivery. Finally, simultaneously with the physical handover of the ship, there is a documentary closing meeting where the seller's delivery documents are handed over (including, crucially, the bill of sale, transferring title in the ship from the seller to buyer) in exchange for payment of the purchase monies.[10]

7.2 Factors supporting the development of a global and uniform approach to ship sales

The extent to which the global shipping market has embraced a uniform approach to the sale and purchase of second-hand ships is a remarkable achievement and demonstrates how the industry has overcome regional customs and practices in the interests of furthering international commerce. In this regard, the influence of English law and specifically, English law principles relating to the sale of goods, has been significant.[11] However, from a practical viewpoint, the erosion of national (and even local) customs in favour of internationally accepted practices owes much to the development and interplay of three key factors: (i) industry acceptance of the single purpose ship owning company, (ii) influence of the open flag states and (iii) reliance upon the classification societies.

8 Most notably the Mediterranean Yacht Brokers Association, or 'MYBA' form.

9 Yachts of a size and passenger complement sufficient to be termed 'superyachts' will likely need to comply with the International Maritime Organisation (IMO) Conventions – most notably the International Convention for the Safety of Life at Sea, commonly known as SOLAS – unless they are eligible to operate within one of the yacht equivalent regimes such as LY2 or the 'Passenger Yacht Code'.

10 The bill of sale does not record the date and time that title (and risk) in the ship is transferred from the seller to the buyer, it being an instrument which is effective on physical delivery. Accordingly, it is customary for parties to sign a separate document known as a protocol of delivery and acceptance to confirm the time of transfer and all three standard form contracts make express provision for this document.

11 In particular the Sale of Goods Act 1893 as the same was replaced by the Sale of Goods Act 1979 (and as the same has been subsequently amended from time to time).

Critical to the understanding of the make-up of ship sale and purchase, is the long-established practice of ship owners using single ship owning companies. Typically established in no or low tax offshore jurisdictions which have extremely limited reporting requirements, the principal purpose of these companies is to utilise the doctrine of distinctive corporate personality in order to ringfence the liability of the ship owner on an individual ship basis. In second-hand ship sale transactions, the seller will likely be a single ship owning company (to become a company with no assets whatsoever immediately upon sale and delivery of the ship to the buyer). Conversely, the buyer of a second-hand ship will often be a newly incorporated company with no assets (which shall become a single asset company upon taking delivery of the ship from the seller). Of course, there are narrowly defined circumstances where the Courts will be prepared to probe company structures and to impute liability on a parent company by piercing the corporate veil. However, as a general principle, the use of single purpose companies is well accepted by the international shipping community and by the Courts of most sophisticated jurisdictions. In *Atlas Maritime Co SA v Avalon Maritime Ltd (The Coral Rose) (No. 1)* (1991),[12] a case before the English High Court, Staughton LJ observed that:[13]

> The creation or purchase of a subsidiary company with minimal liability, which will operate with the parent's funds and on the parent's directions but not expose the parent to liability, may not seem to some the most honest way of trading. But it is extremely common in the international shipping industry and elsewhere. To hold that it creates an agency relationship between the subsidiary and the parent would be a revolutionary doctrine.

The prevalence of single ship owning companies has had a number of consequential effects on the manner in which second-hand ships typically are bought and sold. First, the absence of any meaningful financial data for the selling company means that very little emphasis is placed on the credibility of the selling entity. Due diligence is focused on the ship, through a combination of log-book entries, class records, port state control history, market knowledge and physical inspections of the ship itself. At the time of a ship's delivery the buyer's key concern regarding the seller is that it holds good title in the ship and has the power to transfer that title to the buyer. From the seller's perspective, the fact that the buyer will not be a company of financial substance has led to the practice of contracts for the sale and purchase of second-hand ships being initiated by the buyer lodging a monetary deposit as security for its performance under the contract. This deposit will be forfeited to the seller in its entirety if the buyer fails to pay for and take delivery of the ship in accordance with the terms of the sale contract. In most cases the forfeited deposit amount will be significantly greater than the seller's actual losses but the market nevertheless has adopted this approach almost universally in recognition of the risks borne by the seller in contracting with a single purpose company.

The development of a system whereby the world's commercial ships are largely owned by single purpose entities registered in offshore jurisdictions has only been possible through the availability of nation states willing to accept ships within their commercial fleet where there is no genuine link between them and the ship's

12 [1991] 1 Lloyd's Rep 563.

13 *Ibid*, at 571.

beneficial owners or managers. Known as 'open flags' or 'flags of convenience', they are typically characterised by tax-friendly arrangements, relaxed crewing rules (as regards nationality) and, in the case of the larger open flag states, administrative efficiency. As a substantial part of the world's commercial tonnage (new and second-hand) is financed through debt lending with the ship as principal security, another critical aspect of a flag state's ability to attract ships onto its register has been the development of lender-friendly legislation and, specifically, a tried and tested system of ship mortgage enforcement. Those handful of flag states which have managed to secure both owner and lender support are today responsible for the registration of over 80 per cent of the world's commercial fleet on a tonnage basis.[14]

The predominance of the major open flags has been a key component in standardising the process of ship sale and purchase. Whilst the open flags typically have limited eligibility conditions as to the integrity of the ship owners, they do impose adherence to the rules and regulations of the major international maritime conventions, particularly those concerned with safety of life at sea.[15] Accordingly, focus is again placed firmly on the integrity of the ships rather than on those who beneficially own them.

The third key element in the development of a uniform approach to the sale and purchase of commercial ships is the critical role played by the classification societies. Classification societies are independent commercial organisations staffed by marine surveyors who work with shipyards, ship owners, insurers and flag state authorities in matters relating to the design, maintenance and repair of ships and other marine units. It is a requirement of the flag state in which a ship is registered that the ship is entered with one of its approved classification societies. When a ship is first classed, the classification society will issue a certificate recording the class rating assigned to the ship. Thereafter, surveyors from that classification society will survey the ship at regular intervals to determine if the ship continues to comply with rules and other requirements of the class rating assigned to it. From the dozens of classification societies operating around the world, thirteen currently are members of the International Association of Classification Societies, or 'IACS'.[16] Together, IACS members are responsible for the surveying and certification of over 90 per cent of the world's commercial shipping fleet and, invariably, ship financiers will insist on their financed ships being classed with one of the IACS members.

Whilst the classification societies may have their origins in a particular jurisdiction they are independent from government and not otherwise 'national'. The IACS members adhere to a uniform approach and actively seek to harmonise their rules and practices. It is difficult to overstate the reliance (or perhaps over-reliance) of the shipping industry on the classification societies.[17] For second-hand ship sales,

14 As at the date of writing, the largest open flag states in terms of tonnage entered on their commercial ships register include Panama, Liberia, Marshall Islands, Bahamas and St Vincent & the Grenadines.

15 For example, the Safety of Life at Sea (SOLAS) Convention 1974.

16 The current members of IACS are: American Bureau of Shipping (ABS), Bureau Veritas, China Classification Society, Croatian Register of Shipping, Det Norske Veritas, Germanischer Lloyd, Indian Register of Shipping, Korean Register of Shipping, Lloyd's Register, Nippon Kaiji Kyokai (NKK), Polish Register of Shipping, Registro Italiano Navale and Russian Register of Shipping.

17 For discussion on this, see I. Goldrein, M. Hannaford and P. Turner, *Ship Sale and Purchase* (6th edn), (Informa, 2012), Chap. 3.

the standard form contracts benchmark the seller's obligations with regard to the ship's condition at delivery by reference to its class status. In its unamended form, Norwegian Saleform stipulates the condition of the ship at delivery as being 'with her Class maintained without condition/recommendation*, free of average damage affecting the Vessel's class, and with her classification certificates . . . valid and unextended without condition/recommendation*'.[18] Norwegian Saleform also provides that the final underwater inspection of the ship is to be carried out not by the buyer's own technical team but by a representative from the classification society. Nipponsale and SSF2011 place similar reliance on the ship's classification society for determining the delivery condition of the ship.

There have, of course, been many other important legal, commercial, political and social factors which have had influence but, taken together, the evolution and interplay of the single purpose company, the open flag states and the major classification societies have been instrumental in developing a universal approach to ship sale and purchase. In part this also explains why the international market has traditionally been slow to accept any change to the practices that it has become accustomed to. For Norwegian Saleform there has been an initial resistance from the market to take on a new version with practitioners preferring the familiarity of the previous form and showing a ready willingness to address its shortcomings through increased amendment and additional rider clauses.[19] This was particularly evident in the case of Saleform 1993 where, for a number of years after its issuance, many ship sale negotiations would commence with a heated debate as to whether the 1987 form or the 1993 form would be used as the base document. It was not until around 2000 that the 1987 form started to disappear from the negotiating table. In contrast, the take-up for the latest version of Norwegian Saleform has been relatively quick. BIMCO confirmed that in the first half of 2014, sales of Norwegian Saleform 2012 had already overtaken those of Norwegian Saleform 1993.[20] This is due, in no small part, to the fact that, after nearly twenty years, most of the substantive amendments made in Norwegian Saleform 2012 had already been accepted by the market (and customarily inserted in the 1993 form by way of amendment to the printed terms). Indeed, when the Norwegian Shipbrokers' Association and BIMCO first approached the market about revising Saleform 1993, the overwhelming response was that it was time for a new version but that the revision should be as light as possible. There was no appetite for a fundamental review of how ship sale and purchase transactions are carried out: simply recognition that specific provisions within Saleform 1993 no longer reflected market practice and that others required amendment following judicial interpretation of their meaning.

18 The asterisk refers to a note on the form which states: '*Notes and memoranda, if any, in the surveyor's report which are accepted by the Classification Society without condition/recommendation are not to be taken into account.'

19 This issue is less pertinent for Nipponsale which has not been revised since 1999 and has not been tested in respect of SSF2011 which is the first version of that standard form.

20 See https://www.bimco.org/ (last visited 31 December 2014).

7.3 Substantial differences in the approach adopted by different sale forms – a case study on the treatment of deposit

One aspect of Norwegian Saleform 1993 which had been subject to judicial examination, regulatory attention and a consequential change in market approach, relates to its treatment of the deposit. For reasons already discussed, it is a long-held custom of the market that the buyer shall lodge a deposit as security for its performance under the contract. Traditionally, the deposit amount has been fixed at 10 per cent of the purchase price of the ship. This percentage amount is expressly stipulated in Norwegian Saleform 1993, Nipponsale 1999 and SSF2011. Accordingly, if the parties contracting on these standard form contracts reach commercial agreement for a different deposit amount, amendment to the printed terms will be necessary. In recent years, however, there has been a noticeable movement away from parties initiating their ship sale and purchase negotiations on the basis that a deposit of 10 per cent is a market-accepted amount which can only be departed from in the most unusual circumstances. Whilst today 10 per cent remains the most commonly adopted percentage amount, deposits of 5, 15 and 20 per cent now regularly appear in ship sale contracts. Saleform 2012 reacts to this market movement by inviting the parties to insert an agreed percentage amount into a designated space (stipulating that the deposit amount shall be 10 per cent only in the unlikely event that the parties fail to insert a figure).[21]

Whilst this amendment to Norwegian Saleform was market driven, a recent English High Court decision, approved by the Court of Appeal, may have a greater impact on the negotiation of the deposit amount than the drafting subtleties of Norwegian Saleform 2012. In *Griffon Shipping Ltd v Firodi Shipping Ltd (The Griffon)*,[22] the buyer under an amended Norwegian Saleform 1993 contract failed to pay the deposit within the specified time and the seller exercised its right to cancel the contract and sue for damages. It was held that the seller was entitled to the full amount of the deposit notwithstanding that its actual losses (deposit aside) were only a fraction of that sum. On a strict interpretation it is submitted that the Court was correct in its findings (and, it is further submitted, that the position would be the same if the contract had been based on Norwegian Saleform 2012 terms). However, this decision was surprising to many market practitioners. It was considered the phrase in the buyer's default clause of Norwegian Saleform referring to the seller being entitled to compensation for its 'actual losses' if the buyer failed to pay the deposit was limited to the normal and foreseeable costs incurred by the seller up to the point at which it cancelled the contract. As a result of the decision in *The Griffon*, buyers may now be reluctant to agree a substantial deposit amount if they are contracting on unamended Norwegian Saleform terms. In relation to this it is interesting that the drafting team of SSF2011 – who followed the Norwegian Saleform terms so closely in many of the clauses – expressly sought to address this point by providing that the buyer's failure to pay the deposit entitles the seller

21 This raises an interesting question as to the extent to which provisions in the standard form contracts influence market practice or even, over time, determine it.

22 [2013] EWCA 1567; [2014] 1 Lloyd's Rep 15; [2013] EWHC 593 (Comm); [2013] 2 Lloyd's Rep 50.

'. . . to claim compensation for their losses and expenses (but with no automatic right to compensation in the amount of the Deposit)'. Nipponsale 1999, however, takes the opposite course, stipulating that the seller's rights in these circumstances are 'to receive an amount equivalent to the Deposit' with a right to claim further compensation if the deposit amount does not cover the seller's actual losses.

For those contracting on Saleform 1993 or Nipponsale 1999 terms, the risk of a buyer not being able to lodge the deposit amount within the prescribed time limit is very real. Under the printed terms of these standard forms the deposit is to be paid within a specified number of banking days following signing of the contract. But these forms were produced at a time when opening a joint escrow account was a simple, quick and relatively informal process. As such, it was rare for a buyer to find itself unable to lodge the deposit monies – with a bank, broker or law firm – within the three to five banking days normally agreed with the seller that it would do so. However, from the late 1990s, this situation changed both rapidly and radically. The requirement for banks (and other entities that traditionally had been willing to act as deposit holder) to carry out full 'know your client' investigations in order to comply with money laundering and proceeds of crime legislation meant that it could take weeks rather than days to open a joint account. For many in the shipping market this sharp focus on corporate transparency caused particular difficulties at first since their single purpose ship owning companies did not lend themselves readily to the justifiably keen scrutiny of the banks who risked severe penalties if they were unable to trace the contracting vehicle through to its ultimate beneficial owners.

SSF2011 addresses these issues in a detailed and specific manner. Whereas Norwegian Saleform uses the term '*Deposit Holder*' and defines the term widely, SSF2011 contemplates that the deposit will be held by a bank nominated by the seller and that the buyer will remit the deposit amount with a value date no later than the date specified in the contract. With regards the opening of the deposit joint account, SSF2011 provides:

> The Sellers are to arrange the opening of the joint escrow account, latest by 2 banking days prior to the Value Date. The Buyers, latest together with their remittance of the Deposit, are to arrange bank-to-bank confirmation from the remitting bank to the [deposit holding bank] (a) that the Buyers, and the remitting party if different, are a known customer of the bank and should it be required by the bank in Box (i)(a), the Buyers will also arrange for the bank-to-bank confirmation to include the confirmation by the remitting bank that they know the source of funds.[23]

Whilst, under the printed terms of SSF2011, the deposit account is required to be in a joint account, it will be seen that it is the seller's sole responsibility to arrange for its opening within the specified time frame. In this regard the deposit-holding bank may be unwilling to open the account without receipt of appropriate 'know your client' documentation. The bank-to-bank confirmations may not be sufficient for this purpose and, in any case, the buyer is not obliged to provide these until the time that it remits the deposit. Conversely, the buyer must take care that its remitting bank will be willing to send the required bank-to-bank confirmations. The first

23 This wording is followed by a general requirement imposed on both seller and buyer to comply with all anti-money laundering laws and regulations of the country in which the deposit holding bank is located.

confirmation should be straightforward as the remitting bank will presumably have an existing relationship with the buyer. However, the second bank-to-bank confirmation relating to the source of funds is only applicable if the deposit-holding bank requires it. As the buyer must arrange for the bank-to-bank confirmations to be sent at the latest when the deposit monies are remitted, it will be necessary to ascertain in advance whether this second confirmation is, in fact, required. If, for any reason, the deposit monies are remitted but the bank-to-bank confirmations are not sent (or are sent but are considered by the deposit-holding bank to be defective) the buyer is placed in a situation where it is in breach of contract but the deposit is no longer in the buyer's sole possession or control. In practice, the parties usually will find ways to deal with these types of eventualities but this particular provision in SSF2011 does highlight the potential difficulties which can arise when a standard form seeks to impose a prescriptive mechanism on the parties.

In contrast, Norwegian Saleform stipulates that the time period for the buyer to lodge the deposit with the deposit holder only starts to run once the contract is signed and the joint account is set up. This is a sensible amendment and one which has featured as an additional rider clause to the previous version of Norwegian Saleform for a number of years. However, whilst the less prescriptive approach adopted in Norwegian Saleform 2012 does give parties greater flexibility in how (and with whom) they establish the deposit account, it does create uncertainty for parties if the account is not set up in a timely manner. Pursuant to clause 2 of Norwegian Saleform 2012, both the seller and the buyer are obliged to 'provide the Deposit Holder all necessary documentation to open and maintain the account without delay'. In circumstances where the deposit holder is asking for documentation unfamiliar in the jurisdiction of the buyer (or its bank) or, alternatively, where a party is dragging its heels, this provision is likely to cause interpretive difficulties.

In the context of the deposit, the decision in *The Griffon* was not the first time that the English courts have held that particular terms within Norwegian Saleform 1993 do not, in fact, have the meaning or effect that the market (or at least a significant part of it) has always assumed and contracted on the basis of. One example of this is the decision in *PT Berlian Laju Tanker TBK and another v Nuse Shipping Ltd (The Aktor)*.[24] In that case, following negotiations for the sale of a product tanker, the agreed terms were set out in a recap which provided for a 10 per cent deposit to be lodged with the seller's nominated bank in Singapore. The parties then signed a sale contract on Norwegian Saleform 1993 terms which provided for closing to take place in Singapore on the same day the buyer remitted the deposit to that bank. The dispute arose when the seller asserted its right to have 100 per cent of the purchase price paid to its bank in Greece, as nominated by the seller under clause 3 of the contract. The buyer agreed to pay 90 per cent of the price to the seller's bank in Greece, but maintained its right to pay the 10 per cent balance by releasing the deposit to the seller in Singapore. The seller took the view that this amounted to an anticipatory breach by the buyer and accepted it as terminating the contract. The dispute was referred to arbitration, with the seller seeking to forfeit the deposit and the buyer claiming for the return of their deposit (with interest) and for loss of profits.

24 [2008] EWHC 1330 (Comm); [2008] 2 Lloyd's Rep 246.

The arbitrators decided in favour of the seller, finding that the seller's entitlement to nominate a bank (in Greece) for the purpose of receiving the full purchase price under clause 3 was unrestricted. Permission to appeal was granted by the High Court as the dispute raised issues of general public importance. The Court dismissed the appeal, holding that payment of the purchase price, and the place where payment is to be made, is governed by clause 3 and that the requirement under clause 3 is for payment to be made in full. In reaching this conclusion, the Court found that – on the wording of Norwegian Saleform 1993 – the purchase price and the deposit are two different things; accordingly, payment of the deposit as security into a separate deposit account does not constitute part payment of the purchase price.

The *Aktor* decision alerted parties to the need for the buyer's payment obligations to be clearly set out in the sale contract. As clause 3 of Norwegian Saleform 1993 was held to operate completely independently of clauses 2 and 8, the failure to provide for the deposit to be released in part payment of the purchase price could leave the buyer with a need to fund an amount greater than 100 per cent of the purchase price with a right to receive the deposit back only at closing. This could cause particular problems for the buyer whose acquisition of a ship is being financed from an external source. This issue is adequately addressed in Nipponsale 1999 which clearly provides for the 'balance' of the purchase price to be paid at delivery simultaneously with the release of the deposit. SSF2011 and Norwegian Saleform 2012 also follow this approach.

The treatment of the deposit in Norwegian Saleform, Nipponsale and SSF2011 is interesting in its own right but also as an illustrative example of how the standard forms react to judicial interpretation of terms within those standard forms which are at odds with market understanding as to their meaning and effect. Following the decision in *The Aktor* in 2008, Norwegian Saleform 2012 included an amendment to reflect market understanding that the deposit forms part of the overall purchase price. It will be interesting to see if the decision in *The Griffon* will result in the next version of Norwegian Saleform being amended to follow SSF2011 in stipulating that failure to lodge the deposit in time does not entitle the seller to claim for the whole deposit amount or whether the market will react by simply reducing the percentage customarily agreed for the deposit.

7.4 Conclusion

Not surprisingly, this issue is not confined to the deposit. In a recent landmark judgment, *Dalmare SpA v Union Maritime Ltd (The Union Power)*,[25] the English High Court held that the wording of Norwegian Saleform 1993 did not operate to exclude the application of the implied terms of the Sale of Goods Act as to the ship being fit for purpose and of satisfactory quality.[26] To the extent that it was pleaded, the Court was not persuaded that there was sufficient unanimity of market understanding as to the meaning of 'as she was at the time of inspection' to meet the requirements of the

25 [2012] EWHC 3537 (Comm).

26 Sale of Goods Act 1979 (as amended), s. 14(2). For a detailed analysis on this case see Chapter 8 written by Simon Rainey QC.

common 'usage' exception under section 55(1) of the Act. At one level this should perhaps not be surprising: shipping remains a uniquely fragmented industry with numerous features particular to different trading sectors and vessel types. However, in the specific context of ship sale and purchase, there has evolved over many decades a remarkable uniformity of understanding as to how second-hand ships are bought and sold and, going forward, it is hoped that this process will continue and will result in greater recognition of this in judicial interpretation of standard form ship sale contracts where genuine market practices can be identified.

CHAPTER 8

'As is' . . . as you were? *The Union Power* and 'as is' provisions in ship sale and purchase contracts

*Simon Rainey QC**

8.1 Introduction: standard form contracts and English law

The business of the sale and purchase of second-hand trading ships is routinely conducted on or by reference to a standard form contract. Various forms are in common currency. Perhaps the most commonly encountered in English legal practice is the Saleform (commonly referred to as the Norwegian Saleform or 'NSF'[1]), first adopted by the Baltic and International Maritime Council (BIMCO) in 1956 and then the subject of periodic revisions[2] culminating in the latest version of the form, Saleform 2012. Others in common use include Nipponsale 93, produced by the Japan Shipping Exchange Documentary Committee. A recent addition to the corpus is the Singapore Ship Sale Form 2011 or 'SSF' launched by the Singapore Maritime Foundation.

Standard form contracts have their advantages: 'if used intelligently, the form will save the parties from going to the trouble and expense of reinventing what may be a relatively complex commercial and legal relationship. For, in addition to setting out the essential commercial elements of a straightforward deal, a good standard form will also contain essential legal components which might be overlooked where the contract is not prepared by an experienced draughtsman.'[3] However, the proviso of intelligent use is important. The use of a standard form needs to take account of the system of law which the parties choose to govern their transaction, entered into on the basis of that form. The BIMCO Saleform 2012, as is typical and as with its predecessors, provides for an optional choice of governing law between English law, alternatively New York law, alternatively such other law as the parties may select and insert.[4] The form is, therefore, drafted as an 'omnibus' form for application under whatever system of law the parties may choose to adopt. While the form shows English common law lineaments, given that it is commonly governed by English law

* Barrister at Quadrant Chambers. London; Honorary Professor of Law at the University of Swansea, Visiting Fellow at the Institute of International Shipping and Trade Law, University of Swansea.

1 The full title of the form explains its origins: 'Norwegian Shipbrokers' Association Memorandum of Agreement for Sale and Purchase of Ships. Adopted by BIMCO in 1956. Codename: SALEFORM'

2 In 1966, 1983, 1987 and 1993.

3 I. Goldrein, M. Hannaford and P. Turner, *Ship Sale and Purchase* 6th edn, (Informa Law, 2012) (hereinafter referred to as Hannaford & Turner). Unfortunately, the new edition was published on 19 December 2012 and, therefore, went to print before judgment in *The Union Power* was delivered on 13 December 2012.

4 See cl. 16.

and has therefore been much litigated in England, leading to particular revisions,[5] it is not itself drafted as an English law form.

Therein lurks a very real danger. It is necessary to test the operation of the standard omnibus provisions against the rules which may be applied to those provisions under the applicable system of law. Where the system of law has mandatory statutory provisions which apply to all sales of goods, then attention must be given to the effect of those provisions and, where it is intended to exclude their operation, the wording must be such as will do the job in accordance with the terms of the statute and any relevant case law as to the requirements for exclusion. If it is not, then a typed rider to or amendment of the standard form will be necessary.

Nowhere is that danger more real than in the context of the exclusion of the statutory conditions applicable to all sales of goods under the Sale of Goods Act 1979.

The decision of Flaux J in *Dalmare SpA v Union Maritime Limited (The Union Power)*[6] caused some apparent surprise to those engaged in the sale and purchase of second-hand ships and to some of the commentators on the standard Saleform 1993.[7] However, the decision represented the application of ordinary and basic general principles applicable to the Sale of Goods Act and the sale of any item of goods, whether new or second-hand, and including the sale of ships.

It is suggested that the argument that, in some sense, the statutory conditions under the 1979 Act should not be allowed to supplement a 'comprehensive' form and that the Court should be reluctant 'to supplement its terms by implying provisions from their national laws' because 'where international forms such as the NSF 1987 and 1993, promulgated by a widely respected trade association and almost universally adopted by a global industry, define a detailed contractual regime', is inappropriate, misplaced and starts from the wrong end of the argument.[8] The starting point, as considered below, is that the parties have chosen English law to apply to their standard form, for good or ill. English law provides for the statutory regulation of sales of goods (and has done so since 1893) as is well known. The standard form is being used as the basis of the contract; the parties can therefore accommodate the statute and cater for English law principles by a suitable rider or amendment. If they do not, then the result may not be what they might have expected from simply looking at the form. Drafting deficiencies in standard forms (and BIMCO standard forms in particular) are well known; compare the fate, over many revisions, of the ineffective 'consequential loss' exclusion in the BIMCO Towcon and Supplytime forms and the failure to address the case law.[9]

While criticism has been levelled at Flaux J's decision on the basis that it is

5 For example the revisions of cl. 11 in the 1983 and 1987 versions of Saleform in consequence of the decision in *The Buena Trader* [1978] 2 Lloyd's Rep 325 as to the non-implication of terms as to notification of Class of matters affecting classification status.

6 [2012] EWHC 3537 (Comm); [2013] 1 Lloyd's Rep 509.

7 See, e.g., M. Strong and P. Herring, *Sale of Ships* 2nd edn (Sweet & Maxwell, 2009) (hereinafter referred to as Strong & Herring).

8 See S. Curtis, 'The Union Power Decision', at p. 6 (Paper produced for the London Shipping Law Centre Seminar on Saleform Issues, 8 May 2013) which contains a robust criticism of the decision (hereinafter referred to as Curtis).

9 Discussed by the present author in S. Rainey, *The Law of Tug and Tow and Offshore Contracts* 3rd edn, (Informa Law, 2013) at, e.g., pp. 180–187 (Towcon) and by C. Kidd and R. Gay in *Offshore Contracts and Liabilities* edited by B. Soyer and A. Tettenborn (Informa Law, 2014) at chaps 6 and 7 respectively.

retrograde because the Court should be 'very reluctant to assume that [parties] have also agreed to overlay that contract with an additional statutory regime which may be entirely unknown to them; this particularly so given that an accurate analysis of their respective rights and obligations will require a detailed knowledge of and the ability to interpret correctly our national case law',[10] which is not the approach adopted in English law. Nor, it is suggested, can it sensibly be, whether in relation to the BIMCO Saleform or any other standard form contract or trade association template. As the Court of Appeal stated in *Bominflot Bunkergesellschaft für Mineraloele mbH & Co v Petroplus Marketing AG (The Mercini Lady)*,[11] in the context of an attempted exclusion of the Sale of Goods Act 1979 under a contract for the sale of gasoil, the parties must (and can only be) taken to be contracting by reference to principles and settled meanings in English law:

> The jurisprudence extends beyond individual decisions and has become expressive of a principle . . . I must consider that the parties to this English law contract, foreign as both of them are and quite possibly ignorant of the consequences of their choice of language, intended to contract by reference to what English law had to say about the language which they have adopted.[12]

As the Court further explained, at the heart of the matter is the need for commercial and legal certainty of result:

> [. . .] there has also been a judicial consensus that such obligations can only be excluded by language which expressly (or perhaps one may add which must necessarily be taken to) refer to conditions: and that such language as has been used in our case falls within that consensus and principle. Lord Diplock was speaking[13] of a judicial consensus which creates implied obligations: but there may also be a judicial consensus which preserves such implied obligations in the face of inadequate exclusions. In such circumstances, there is an importance in the certainty of our commercial law which goes beyond the answer that may be given in a particular case.[14]

Commentaries have treated *The Union Power* as a decision which relegates 'as is' sales effectively to irrelevance or meaninglessness. But it is important to focus on what the true issue in the case was.

The principal issue was whether the contract in fact contained an 'as is' provision at all. It was held that it did not, therefore, the meaning of 'as is' did not arise directly for decision. The Court went on to determine that separate question *obiter* and on the basis of the specific (and limited) arguments as to how the term should be construed and why it should be so construed. Accordingly, the Court did not, in fact, address the case where a particular contract uses 'as is' wording nor did it have to consider any particular form of 'as is' wording; it merely expressed a non-binding and self-avowedly 'provisional', albeit highly persuasive, view on the meaning of 'as is' in general terms.[15]

---

10 See Curtis, at p. 6.
11 [2010] EWCA Civ 1145; [2011] 1 Lloyd's Rep 442.
12 *Ibid* at [61] per Rix LJ.
13 In *Lambert v Lewis* [1982] AC 225 at 273, 276.
14 [2010] EWCA Civ 1145; [2011] 1 Lloyd's Rep 442, at [62] per Rix LJ.
15 [2012] EWHC 3537 (Comm); [2013] 1 Lloyd's Rep 509 at [84].

8.2 The approach to 'as is' clauses under the Sale of Goods Act

As said above, the decision of Flaux J represented the application of ordinary and basic general principles applicable under the Sale of Goods Act 1979 to the sale of a second-hand ship, just as with any other item of goods, new or second-hand, complex or simple. It is necessary to consider those general principles since, irrespective of the decision in *The Union Power*, any argument which seeks to exclude the operation of the Act by reference to an 'as is' provision will necessarily invoke those general principles.

It is convenient to take those principles in logical stages.

8.2.1 Application of the Act to ship sale and purchase

Stage one of the analysis is that, as a matter of English law, the 1979 Act applies to a contract for the second-hand sale and purchase of ships just as it does to any other contract for the sale and purchase of a highly complex engineered chattel (whether bought new or second-hand) or internationally traded high value commodities as well as humbler forms of goods.

The 1979 Act (like its predecessor of 1893) is not a consumer protection statute, nor is section 14 a consumer protection measure: it represents a codification of the law of sale of goods both as it affects commerce and trade and as it involves consumers. The Act codified the general law relating to all sales of goods without distinction and section 14(2) in its previous guise of 'merchantable quality' has a long history. The more recent reformulation of the statutory standard as one of 'satisfactory quality' (with guidelines for evaluating what that may entail)[16] does not alter the general applicability of the Act.

The correct starting point is, therefore, that, as with any contract for the sale of goods, from the most humble to the most complex, the sale of a second-hand ship (like the sale of a second-hand aircraft or any other complex mechanical or other chattel) is subject to the Sale of Goods Act 1979: all who buy and sell goods who contract on the basis of English law are taken in English law to be aware of this and to contract upon this basis. The Court in *The Union Power* was, rightly, dismissive of the argument that the Sale of Goods Act should apply in some diluted way to chattels as complex as second-hand ships.[17]

8.2.2 Application of section 14 to such sales

Stage two is that the 1979 Act imports, as a matter of law, an implied term of satisfactory quality into the parties' contract on the Saleform or any other standard form sale contract by section 14(2) ('where the seller sells goods in the course of a business, there is an implied term that the goods supplied under the contract are of satisfactory quality'). In the case of the sale and purchase of any commercial vessel using the Saleform and in the case of *The Union Power*, the seller will sell the vessel to the buyer in the course of a business, usually that of a trading shipowner or operator.

16 By the amendment of the 1979 Act by the Sale and Supply of Goods Act 1994.
17 [2012] EWHC 3537 (Comm); [2013] 1 Lloyd's Rep 509 at [24]; see also [72].

'Satisfactory quality' is an objective test which applies to goods irrespective of their age and price. It has, for example, been applied to the sale of other second-hand goods, the most frequently cited example being second-hand cars.[18] Not only, therefore, does the Act, on its true construction, apply to a contract for the sale of a second-hand ship, but there is also no reason (theoretical or practical) why the section 14(2) term in particular should not apply to the sale of a second-hand ship (although what constitutes 'satisfactory quality' will, of course, very much vary depending on the precise factual context). There are no 'policy reasons' to restrict the implication of the statutory implied terms into contracts for the sale of second-hand ships, any more than for any other complex second-hand (or new) chattel.

Accordingly, the *prima facie* and indeed the default, position under any contract for the sale of a second-hand ship is that the contract contains an implied term that the vessel sold is of a satisfactory quality.

8.2.3 Section 14 applies unless excluded under section 55

Stage three is that the statutory implied term will only be excluded where the conditions for that exclusion which are contained in section 55 of the 1979 Act are met. Insofar as is material, section 55 provides as follows:

> (1) Where a right, duty or liability would arise under a contract of sale of goods by implication of law, it may (subject to the Unfair Contract Terms Act 1977) be negatived or varied by express agreement, or by the course of dealing between the parties, or by such usage as binds both parties to the contract.
> (2) *An express term does not negative a term implied by this Act unless inconsistent with it.*' [emphasis added].

There are, therefore, three routes by which the implication of the statutory conditions into the contract may be excluded: either (i) by an express agreement; (ii) by the parties' course of dealing (unlikely to arise with any frequency in the context of the sale of second-hand ships); or (iii) by 'usage' or some trade custom or practice.

It is important to note that in *The Union Power*, no case of custom or usage or specially understood binding trade meaning in relation to the contract (an amended version of Saleform 93) was run by the seller before the arbitrators (as was noted by Flaux J).[19] Reliance was placed solely upon the alleged meaning of the express terms on their ordinary language and by reference to ordinary principles of construction. Accordingly, the relevant question in that case, and in any case in which some binding special customary meaning is established, was and is whether there is an express term in the contract that is inconsistent with a term that the vessel should be of satisfactory quality.

The concept of an express term inconsistent with the implied term requires little

18 M. Bridge *et al.*, *Benjamin on Sale of Goods* 8th edn (Sweet & Maxwell, 2008), para. 11-048 (hereinafter referred to as Benjamin).

19 [2012] EWHC 3537 (Comm); [2013] 1 Lloyd's Rep 509 at [16]; see also [71], the judge referring to the absence of 'any cogent evidence of market custom' to the effect that 'second hand vessels sold pursuant to MOAs on Saleform were simply sold to a class standard, not to a standard of satisfactory quality under the SOGA'. Cf. the suggestion by Curtis that 'evidence of custom and other practice was submitted to the arbitration tribunal [but] this was not referred to in the award' which is incorrect.

elaboration. Inconsistency means what it says: the seller must be able to point to some other express term of the contract which cannot be reconciled with a warranty of satisfactory quality, written out side by side with it in the contract in question. The term must necessarily negate or 'negative' the implied term because it cannot stand in the same contract with it.

In this context 'express conditions or warranties will normally be construed as additional to the implied terms'.[20] This is entirely appropriate and orthodox: a clause that warrants that, for instance, stores will 'pass survey of the East India Company's officer' (i.e. that those stores will pass an inspection) does not contradict a *parallel* term that those stores are also to be of satisfactory quality (facts taken from the old case of *Bigge v Parkinson* (1862)[21] where the question was whether the express term excluded the implied term of fitness for purpose).

The emphasis in the case law on the exclusion of section 14 statutorily implied terms is on the need for clear language producing a clear result. This is understandable: the Court is being asked by a seller to conclude that the buyer gave up his statutory rights under s 14. If the language relied upon by the seller is ambiguous or can be read sensibly in some way consistently with the statutory rights, then the seller loses. It is well known that the implied terms apply unless excluded; it is a matter for the seller to exclude them expressly or to structure the bargain such that the terms can be shown necessarily to have been excluded by other terms.

By way of example, in *Air Transworld Ltd v Bombardier Inc*,[22] Cooke J summarised the familiar starting point in this context:

> The court was unlikely to be satisfied that a party to a contract had abandoned valuable rights arising by operation of law, unless the terms of the contract made it sufficiently clear that this was intended. The more valuable the right the clearer the language would need to be. Similarly, the more significant the departure from obligations implied by the law or ordinarily assumed under contracts of the kind in question, the more difficult it would be to persuade the court that the parties intended that result.

The Court, therefore, proceeds upon the basic presumption that basic statutory rights in English law will not lightly be assumed to have been abandoned.

8.2.4 The onus is on the seller to satisfy section 55

Stage 4, which flows from the preceding stage, is that the effect of section 55(2) is to place the onus on the seller to demonstrate the exclusion of the Sale of Goods Act terms.

Given that the statutory implied terms provided for by the Sale of Goods Act 1979 will apply to any contract of sale of goods unless and until excluded, then the onus is upon a seller of a second-hand ship who seeks to contend that the implied term as to satisfactory quality under section 14(2) was excluded to point either (i) to an 'express term' within the contract in question which is 'inconsistent' with the particular implied term: see ss 55(1) and 55(2) or (ii) to the term having been

20 Benjamin, op. cit. at para. 11-068.

21 7 H & N 955. Cited in Benjamin as an example of the general principle.

22 [2012] EWHC 243 (Comm); [2012] 1 Lloyd's Rep 349 at [25] per Cooke J.

negatived 'by such usage as binds both parties to the contract'. In the case of (ii), such usage can either be one which applies generally to the type of contract in question or to the meaning of a particular term or expression used in the contract. Accordingly, it will be for the seller to identify either the express term which as a matter of language is said to be inconsistent with the statutory implied term of satisfactory quality or any relevant customary understanding in the relevant trade which excludes the term.

Where a seller relies upon the express terms of the sale contract, this will be a matter of the construction of the particular terms of the contract relied upon, in the context of the contract as a whole and giving the term or terms their sensible and ordinary objective meaning.

8.2.5 'Custom' and customary or settled trade meaning

Much of the controversy which has attended the decision in *The Union Power* springs from commentators who believe that it ran counter to what everyone considered to be the position and upset what had hitherto been a clear and universally settled meaning.[23] The factual premise for that belief is, or has been for a considerable period of years, highly questionable.

Where the seller seeks to rely on any customary meaning, the position as to custom (or 'usage') of the trade (the two terms are usually treated as synonymous) relied upon by a contracting party as giving to a term which is used by the parties a special or customary meaning has been conveniently summarised as follows. 'Evidence of custom is therefore admissible to explain ambiguous mercantile expressions in a charter or to add incidents, or to annex usual terms and conditions which are not inconsistent with the written contract between the parties, but not for any further purpose.'[24]

Reliance by a seller on a customary meaning, however, sets a high threshold and will require compelling evidence. The requirements of a custom or usage of the trade used for this purpose may be taken for present purposes as follows: 'Customs to be enforced by the Courts must be – (i) reasonable; (ii) certain; (iii) consistent with the contract; (iv) universally acquiesced in; (v) not contrary to law.'[25] The position has been described by Professor Guest as follows: 'To be binding however, the usage must be notorious, certain and reasonable, and not contrary to law and it must be proved to be something more than a mere trade practice.'[26]

In the particular context of a custom or usage said to be sufficient to negative the implication of a term as to satisfactory quality under section 14(2) of the Sale of Goods Act, requirements (i) to (iii) and (v) are unlikely to present any difficulty. There is nothing unreasonable or contrary to law about the exclusion of the Sale of Goods Act term: the Act itself provides expressly for the possibility. In a case where

23 See Curtis at p.3, although he recognised that it would not be a view totally acquiesced in but at least 'the overwhelming majority'.

24 This formulation is set out in S. B. Eder *et al*, *Scrutton on Charterparties and Bills of Lading* 22nd edn, (Sweet & Maxwell, 2011) at art. 10; see also H. Beale *et al*, *Chitty on Contracts* 31st edn, vol I, (Sweet & Maxwell, 2012) at para. 12-058.

25 See *Scrutton on Charterparties*, op. cit.

26 See *Chitty on Contracts* 31st edn at para. 12-018; see also para. 12-129.

the contract does not expressly include or contract upon the basis of section 14(2) and where section 14(2) only arises as a matter of statutory implication, then consistency with the contract will not prevent the operation of the custom: the custom is being relied upon as conferring a special meaning on words used. Further, a custom which is said to exclude the Sale of Goods Act term as to satisfactory quality on the basis, for example, that the seller gives no undertaking as to the condition of the thing being sold is sufficiently precise and 'certain'.

The usually problematic area in proving a binding trade or customary usage lies in establishing that the alleged custom or usage is sufficiently universal and total in its application, viz. requirement (iv).

If there are two views on the point or if, while most in a particular trade adhere to one view or adopt the practice in question, others adhere to another view, or either do not follow or do not always follow the practice, then it is not possible to ascribe to the particular word any particular customary meaning: some may use it in one way, others another and the question will fall to be resolved by what it means as a matter of ordinary language.

As already noted, in *The Union Power*, the seller led no evidence as to the meaning of the relevant words relied upon ('as she was') in the parties' contract and in the version of Saleform 1993 under consideration in that case or as to the existence of a settled custom of the trade that the sellers gave no contractual undertakings as to the condition of the vessel sold such as to oust the implication under section 14(2). However, they nevertheless asserted in argument 'it was well settled that such phrases meant that the buyer takes the goods as he finds them, "warts and all" with no warranty or condition as to quality or fitness for purpose'.[27]

Heavy reliance was placed by the seller in *The Union Power* on passages in a leading textbook (Strong and Herring)[28] that the better or the more accepted view was that the Sale of Goods Act implied terms as to quality were not incorporated and that the sale of second-hand ships was or should be viewed as 'as is' in the sense that the seller under the Saleform 1993 gave no undertakings as to the quality or condition of the ship he was selling, other than the various express obligations as to class assumed under various clauses of Saleform. That argument necessarily failed since the seller in that case had not advanced any case of custom or usage as such or led any evidence to make good the existence of any relevant custom or usage: see, for example, per Flaux J rejecting:[29]

> [. . .] the generalised submissions [the seller] made about how application of an implied term as to satisfactory quality would be contrary to the expectations and intentions of parties to MOAs on Saleform and of 'the market'. In the absence of some market custom or usage to exclude the implied terms, as Rix LJ says [in *The 'Mercini Lady'* [2011] 1 Lloyd's Rep 442], the parties to this English law contract are to be taken to have intended to contract by reference to what English law has to say about the language used. Even if [the seller] were right that the expectation of 'the market' was that no term as to satisfactory quality was to be implied into a Saleform MOA, no market custom or usage was pleaded, let alone established.

---

27 [2012] EWHC 3537 (Comm); [2013] 1 Lloyd's Rep 509 at [11].

28 Malcom Strong and Paul Herring (see footnote 7) are two leading practitioners of Ince & Co, and doyens in the subject, with many years of experience in the field.

29 [2012] EWHC 3537 (Comm); [2013] 1 Lloyd's Rep 509 at [30].

However, the likely, or at least a possible, reason for the seller in that case not leading expert evidence that there was a custom or usage in the second-hand ship sale and purchase trade was that, as the textbooks themselves demonstrated, there was considerable uncertainty about the position in the trade and a number of opposing views prior to the decision in *The Union Power*. In these circumstances, requirement (iv) or Chitty's requirement of notoriety would necessarily be difficult to establish.

The question whether or not the standard Sale of Goods Act section 14(2) implied term applies to Saleform 93 was and is one which has been around for a considerable period, has provoked considerable debate and is widely known in the market as a result. Prior to the decision in *The Union Power*, it has attracted two strongly opposed schools of thought. One school (which Strong and Herring advocated) is that it did not or, perhaps more accurately, should not.[30] The other school (of which a leading exponent was one of the arbitrators in *The Union Power*, Simon Crookenden QC) is that it did.[31] As it was accurately and fairly put by Strong and Herring: 'All of that said, the matter cannot be regarded as settled and there are clearly many arguments which can be advanced on both sides.'[32]

Disputes under Saleform 93 are subject to arbitration (the standard form provides for a choice between London, New York and ad hoc).[33] The question raised in *The Union Power* had divided experienced London commercial and legal arbitrators during the life of the 93 form. The position was summarised by Hannaford and Turner (co-authors of the other leading text in this area) before the decision: 'Whatever the decision, the market will benefit from judicial guidance on an issue which has been the subject of legal debate for many years.'[34]

Each party in *The Union Power* was therefore able, on the application for permission to appeal, to pray in aid reference to the existence of a volume of arbitral awards in favour of the school of thought for which it contended. As a paper by Simon Crookenden QC recorded:[35] 'In London arbitrations some tribunals have held the statutory terms to be implied into Norwegian Saleforms contracts [sic] and some have not. The majority of the awards of which the author is aware, however, have held the statutory terms to be implied.' The leading exponents of the contrary view, Strong and Herring, themselves recognised that the question divides arbitrators: 'It is, however, understood that some arbitrators favour the approach that the implied terms in the Act should apply and that that the standard Saleform wording is not sufficient to prevent them from doing so.'[36]

This was not, therefore, a question where it can be said that either party's

30 See Strong & Herring at p. 39: the authors 'unrepentantly continue to prefer the view that the implied terms in the Act should not apply'.

31 See his paper, *Norwegian Saleform Contracts: Implied Terms of Satisfactory Quality*, unpublished but referred to in the judgment of Flaux J; [2012] 3537 (Comm) [2013] 1 Lloyd's Rep 509 at [66].

32 Strong & Herring at p. 39.

33 See cl. 16.

34 Matt Hannaford of Curtis Davis Garrard and Paul Turner of Clyde & Co are equally doyens in this field with Strong and Herring. As noted above, the new edition was published on 19 December 2012 and therefore went to print before judgment in *The Union Power* was delivered on 13 December 2012 and only after permission had been granted to appeal the underlying award under s. 69 of the Arbitration Act 1996.

35 At para. 21.

36 Strong & Herring at p. 39.

construction leads to an obviously uncommercial result or receives one certain, notorious or universally acquiesced-in answer. Experienced London arbitrators selected by buyers and sellers in Saleform 93 arbitrations felt able to endorse either view as arriving at a sensible and workable commercial bargain between the buyer and the seller of a second-hand ship on Saleform 93 terms. The very real uncertainty over the correct result led BIMCO, in its 2012 revision of the Norwegian Saleform, to include an express exclusion of statutory implied terms as to quality and condition.

Another allied problem with a custom or usage argument is the need to identify the particular 'trade' within which the alleged custom or usage is said to operate or exist.

The wider the trade, the more difficult it may be to establish a settled usage applicable to all transactions in it. In *The Union Power*, while no case on custom or usage was in fact advanced, the submission, as seen above, was that it was the market expectation in *any* sale of *any* second-hand ship or at least of any second-hand ship sold on the Saleform terms that there would be no undertaking by the seller as to the quality and condition of the vessel which was sold on terms as to class and after inspection. That might be said to be an ambitious proposition. The position may be very different where the class of sale is much narrower, for example, where the sale occurs in the specific context of sales of vessels for scrap at scrap prices and destined for scrapping or demolition or where the vessel sold is sold as a wreck for wreck removal or scrap or recycling purposes. In such a case, the particular market and those on that market may very well use a specific term, such as 'As Is, Where Is', to connote a special meaning (whatever it means in ordinary English).

8.3 The decision in *The Union Power*

Against this background of the application of the Sale of Goods Act 1979 to a contract for the sale of a second-hand ship and to 'as is' provisions, the decision in *The Union Power* can be considered.

Under a memorandum of agreement (the MOA) on Saleform 1993, the sellers agreed to sell and the buyers agreed to buy a 15-year-old 1994-built tanker for US$7 million. As she has been described, she was 'a lady in her late middle age, who had almost certainly worked hard all her life and could be expected at this stage to be exhibiting a number of aches and pains'.[37] The MOA provided:

> Clause 11. Condition on delivery. *The Vessel shall be delivered and taken over as she was at the time of inspection*, fair wear and tear excepted. However, the Vessel shall be delivered with her class maintained extended to 30 September 2009 without condition/recommendation, free of average damage affecting the Vessels class. (Emphasis added.)

The buyers inspected the vessel at Piraeus on 18 August 2009 and found the vessel to be 'a quite good vessel' at the time. The vessel was then delivered to the buyers at Tuzla, Turkey on 1 October 2009 for dry-docking there. On 6 November 2009, only some 30 hours after the vessel had departed from Tuzla, the main engine broke down, which, as found by the tribunal, was caused by the ovality of the no. 1

37 See Curtis at p. 3.

crankpin. Further, it was found by the tribunal that the ovality had developed to such a state at the time of delivery that the crankpin bearing was likely to fail within a short period of normal operation of the main engine after delivery of the vessel.

The buyers contended in those circumstances that the sellers were in breach of the MOA either because the ovality was 'average damage affecting class' within clause 11 or because there was a breach of the implied term as to satisfactory quality implied into the MOA by virtue of section 14(2) of the Sale of Goods Act 1979. The sellers denied that any SOGA terms were to be implied into the MOA and argued that the terms of clause 11 were inconsistent with the SOGA implied terms in that the vessel was sold 'as she was'. The tribunal, which was composed of three well-known London maritime arbitrators,[38] rejected the sellers' argument, holding that the implied terms as to satisfactory quality were to be implied into the MOA, the sellers were in breach of that term and the buyers' claim succeeded in full.

It is important to bear in mind the considerable limitations upon the decision of Flaux J in *The Union Power*. As seen, no argument as to market or trade custom or usage was run by the seller in that case (save in the very oblique way referred to above by reference to various statements taken from textbooks). It was a case decided solely upon its particular contract terms and therefore cannot be applied directly to any ship-sale contract which uses materially different terms.

In the case, the only argument advanced was based on clause 11 of the contract as set out above and the particular words 'The Vessel shall be delivered and taken over as she was at the time of inspection, fair wear and tear excepted'. This was relied on as an express term inconsistent with section 14(2) for the purposes of section 55. Reliance was also placed on the words 'the Vessel shall be delivered with her class maintained extended to 30 September 2009 without condition/recommendation, free of average damage affecting the Vessels class'.

The contract in *The Union Power*, therefore, did not use the words 'as is' or 'as is where is' as are used in certain contracts by way of stand-alone provisions.

The seller in that case sought to isolate the words 'as she was' which were used as part of the wider sentence dealing with the vessel's condition on delivery and her condition on inspection and to treat them as having the meaning of an 'as is' clause. The seller then argued that an 'as is' clause ousted section 14(2). The seller's argument accordingly raised two separate questions for the Court: (i) the narrower first question was whether clause 11 in the contract was equivalent to an 'as is, where is' basis of contract, *assuming* that phrase had the effect of excluding the SOGA implied terms; (ii) the wider second question was, if clause 11 was equivalent to an 'as is' provision, does an 'as is, where is' clause exclude section 14(2) in the light of section 55?

The judge rejected the seller's case based on question (i) and held that the wording in clause 11 was not equivalent to an 'as is' clause but was simply part of a sentence dealing with a different matter. As he put it:[39]

> [...] the words 'as she was' in the first sentence of clause 11 are a necessary part of a sentence which is recording the obligation to deliver the vessel in the same condition as

38 Simon Crookenden QC; Michael Baker-Harber and Simon Gault.
39 [2012] 3537 (Comm); [2013] 1 Lloyd's Rep 509 at [64].

> she was when inspected. In other words, they are part of a temporal obligation which arises because, usually, there will be a period of time of weeks or even months between inspection and delivery. However, those words tell one nothing about what the sellers' obligations are, either on inspection or delivery, as regards the quality of the vessel. Hence they do not and cannot exclude the implied term as to satisfactory quality under section 14(2) of SOGA.

In reaching his decision on that first question, the judge referred to the fact that other ship sale and purchase contracts did specifically use separate 'as is' or 'as is where is' provisions; this made it difficult to see the use of the words 'as she was' not on their own but as part of another clause as having the same effect. He stated:[40]

> [. . .] the cases on 'as is' provisions, such as The Morning Watch or The Brave Challenger, do pose a real difficulty for the sellers' case, since they demonstrate that sellers who wish to put themselves into an 'as is' regime, with the intention of excluding the statutory implied terms, use those specific words either on their own (as in The Morning Watch) or by way of addition to the contract terms, for example by adding 'as is, where is' after the reference to the sale being 'outright and definite' in clause 4 of Saleform 93.

He also referred as an example of this approach to another arbitration award which had been put in evidence by the seller in which the arbitrators had said:[41]

> We also mention that the sellers' 'as is' argument has no appeal to us. Many sale and purchase contracts are on this basis but this is made clear by use of the well-known words 'as is' used by those involved in the ship sale and purchase market. No such words were in our contract.

Importantly and, it is suggested, correctly, the judge held that, at its highest for the seller, the argument could only be that the wording 'shall be delivered and taken over as she was at the time of inspection' had two possible meanings and this being the case, was fatal to the exclusion of the Sale of Goods Act since the term could be read in the sense which permitted the Act to apply:[42]

> the sellers cannot establish that that was the only meaning the words were intended to have, since plainly the context indicates the temporal purpose of the words, to make it clear that the vessel is to be delivered in the same condition as when inspected. [. . .] even on the sellers' best case the words must have more than one meaning is fatal to the sellers' case that these words exclude the statutory implied terms. Furthermore, given that an obvious sensible meaning of the words is as part of the temporal obligation to which I have referred, section 55(2) defeats the sellers' argument, since it cannot be said that the first sentence of clause 11 is inconsistent with the implied term in section 14(2).

This is the only aspect of his decision which forms the *ratio decidendi*. The judge then went on, *obiter*, to deal with what would have been the position if clause 11 had been capable of being read as an 'as is' clause. Although *obiter*, the view is plainly entitled to careful consideration, being a decision of a highly experienced and reputed Commercial Court judge.

He decided only as a 'provisional view' that 'that the correct approach is to read down "as is" provisions as excluding the right to reject the vessel, whilst leaving the

40 *Ibid* at [65].

41 *Ibid* at [66]. This award was cited in detail in the paper by Simon Crookenden QC 'Norwegian Saleform Contracts: Implied Terms of Satisfactory Quality', unpublished but referred to in the judgment of Flaux J; [2012] 3537 (Comm); [2013] 1 Lloyd's Rep 509 at [66].

42 *Ibid* at [68].

right to claim damages for breach of the implied terms as to description, satisfactory quality and fitness for purpose unaffected'.[43] In other words, such provisions did not exclude the implication of the section 14(2) implied term into the contract and were effective only to exclude the right of the buyer to reject the vessel for breach of that term, while leaving it entitled to claim damages for that breach.

Being only a judgment at first instance, Flaux J's decision is not binding upon any other judge at first instance. He refused permission to appeal but this confers no additional authority upon his decision.

In arbitration, it could be argued that arbitrators are bound to follow his decision, being a decision of the Court.[44] Leaving the question of principle as to the binding effect of a first instance judgment upon arbitrators to one side, even if this proposition were correct, the judgment would only bind as to its *ratio decidendi*. Accordingly, only if the question in the case were precisely the same question of construction which Flaux J had to decide, namely whether the phrase 'The Vessel shall be delivered and taken over as she was at the time of inspection, fair wear and tear excepted' was an 'as is' provision, would his decision have binding effect. The decision has no binding effect on the meaning of an 'as is' provision as the judge himself made clear.

Flaux J's decision, therefore, leaves open for further argument and decision: (i) the effect of differently worded provisions, either in the context of Saleform or otherwise; (ii) the effect of any specific and express 'as is' provision; and (iii) if desired, any argument by a seller as to allegedly relevant binding market or trade custom as to the meaning of the terms in a particular contract or of a standard form contract, such as Saleform.

Given the dichotomy of views applicable to general second-hand sales at least, in the case of a routine sale of a commercial trading vessel (iii) is perhaps now an even more difficult argument than it was before.

8.4 The meaning of 'as is'

The position can be tested at its simplest by taking any contract for the sale of goods which does not involve a second-hand ship. If the parties contract on terms such as 'sold as is' or 'sold as is, where is' or 'basis of sale: as is', what does their language mean? What did they intend to say? How, if one were writing out their bargain in longhand, would one set out what is comprised within these two (or four) terse syllables?

8.4.1 The ordinary English meaning

As a matter of ordinary language, the term 'as is' in the context of a chattel being sold and delivered or provided means, effectively, to take one dictionary meaning, 'in the state that something is in at the present time.[45] Other dictionaries treat the

43 *Ibid* at [84].

44 However, see the thought-provoking contrary analysis by Andrew Baker QC in 'Arbitrators under English law: Inferior Tribunals or A Law Unto Themselves?', a paper delivered at ICMA XVIII (Vancouver).

45 *Cambridge Dictionary Online* (last accessed 31 December 2014).

words 'as is' as shorthand for 'as it is' which perhaps make the position clearer still when used with the term 'sold'[46] It is a term which in its ordinary sense when used in the context of a sale is used to describe goods which the seller offers in their present, existing condition to a prospective buyer. Read ordinarily, it means that the buyer takes the goods as he finds them because they are sold as they are and in the condition in which they then are.

A brief *tour d'horizon* of dictionaries[47] show firstly that the term is one which has a settled dictionary or ordinary language meaning and that meaning is consistent and settled in both English and American English to the same effect: '*as is*: as it stands; unaltered';[48] '*as is, as it is:* in the existing state, things being what they are';[49] '*as is*: in its present condition; without any repairs, improvements or alterations being made, 'the car was priced at $1000 as is';[50] '*as is*: in the existing state of affairs';[51] '*as is*: in the presently existing condition without modification'.[52]

The decision in *The Union Power* does not address directly what the phrase means, simply as a matter of English. The seller in that case cited no dictionary meaning and did not seek to explain the language of an 'as is' provision simply by reference to plain English, stripped of any shipping or ship sale and purchase case law or textbook commentary.

The judge correctly rejected the argument that the phrase was meaningless.[53] This argument was based on a suggestion (by Simon Crookenden QC)[54] that the term 'as is, where is' is a 'mere truism'. If a contract for the sale of the goods is expressed to be 'Basis: sold as is, where is', it is highly unlikely that the parties intended merely to say nothing. Further, as a matter of ordinary English, 'as is' has a meaning as the extract from the dictionary cited above demonstrates. Therefore, as a matter of construction, absent some special trade meaning, the necessary starting point is thus to ask what the words mean simply as a matter of ordinary English. Once that has been established, that is the meaning of the term in the contract; the next question is then whether that term with that meaning can be read consistently with a section 14(2) term. If it cannot, then the Sale of Goods Act term is negatived under section 55(2).

It might be said that the seller's argument in *The Union Power*, in seeking to confer upon it some special 'second-hand ship sale and purchase received meaning' missed, with respect, a more obvious and far simpler argument as to the ordinary meaning of the words as a matter of ordinary language.

If the term properly construed means that the buyer takes the goods as they are in their present and existing condition, whatever it is, and they are sold to him as they are and in that condition, then it is at least highly arguable that that is flatly

46 *Shorter Oxford Dictionary* 6th edn, (OUP, 2007).

47 I am grateful to Andrew Carruth, barrister of Quadrant Chambers who greatly assisted me in this exercise.

48 *The Chambers Dictionary* 12th edn, (Chambers, 2011).

49 *Shorter Oxford English Dictionary* 6th edn, (OUP, 2007).

50 *Webster's Third New International Dictionary*, (G Bell & Sons, 1961).

51 *Collins English Dictionary* 6th edn, (Collins, 2011).

52 *The Merriam Webster Dictionary*, (online, last accessed 30 July 2014).

53 [2012] 3537 (Comm); [2013] 1 Lloyd's Rep 509, at [82].

54 In the paper referred to above.

inconsistent with the seller, at the same time, undertaking that they will also be of any particular, still less satisfactory, quality etc.

While the decision in *The Union Power* holds, provisionally and without reaching any definitive conclusion, that an 'as is' provision does not exclude any of the Sale of Goods Act statutory implied terms as to description, satisfactory quality and fitness for purpose but restricts the remedy for breach of such implied terms to the right to claim damages for breach but excluding the right to reject the vessel, it is suggested that the reasoning of the judge is open to question.

8.4.2 The cases where the meaning of 'as is' has been considered

The starting point in considering his judgment overall is that no previous case or commentary supports the decision of the trial judge. All of the authorities cited before him and the additional Canadian cases considered by him after oral argument are consistent in their view that an 'as is' provision excludes all and any implied terms as to quality and condition.

The judge carefully analysed the English cases *Mariola Marine Corporation v Lloyd's Register of Shipping (The Morning Watch)*[55] and *The Brave Challenger*[56] and demonstrated that the issue before him and the application of section 55 of the Sale of Goods Act 1979 had not been argued in those cases; these were, therefore, *obiter* expressions of view.

However, while this is undoubtedly so, the relevant point in the present context might be argued differently in that highly experienced judges (respectively Phillips J, as he then was, and David Steel J) have seen the phrase 'as is' in a sale of goods context as having a clear and understood general meaning of negativing the implication of any terms as to quality and condition. Their natural reaction, as a matter of language, was that this meant that the goods were sold and taken as they then were and in their present, existing, condition whatever that was; it followed that there was no room for a warranty as to the quality or condition of the goods. Their quality and condition was defined by how they then were and their present factual condition. This, it is suggested, is the significance of the cases: they explain what the language means when it is 'unpacked'.

This is perhaps clearest from the statement of Phillips J in *The Morning Watch*. He stated, in relation to a contract (for the sale of a yacht) on terms 'sold as is, where is': 'The term "as is" has a clearly recognised meaning in a contract of sale. The purchaser *takes the object sold as he finds it* without any warranty as to quality or condition. It was not open to Mariola having agreed to purchase the Morning Watch "as is where is" to complain that her condition did not justify a classification 100A1 at Lloyd's' (emphasis supplied).[57] The words are interpreted as meaning that the buyer takes the goods as they are and as they are found to be.

A similar instinctive reaction was given to the phrase 'as is' in *Choil Trading SA v*

55 [1990] 1 Lloyd's Rep 547 ('as is where is').

56 [2003] EWHC 3154 (Admlty) ('The vessel with everything belonging to her shall be at the seller's risk and expense until she is delivered to the buyer but subject to the conditions of this contract she shall be delivered and taken over as she is at Hasler Marina Portsmouth').

57 [1990] 1 Lloyd's Rep 547 at 555.

Sahara Energy Resources Ltd[58] by Christopher Clarke J. In that case, the term 'as is' was combined with a quality clause which provided that the seller was to provide 'PHRC naphtha quality'. It was, ambitiously, argued by the seller that there was no term as to quality. This was rejected by the court. But the judge held that the term 'as is' meant the goods were sold as they were, provided they were of that broad quality. He stated: 'The expression "as is" was sufficient to indicate that the naphtha to be received *would have whatever characteristics (within that range) the cargo supplied to Choil ex PHRC happened to have*. But, whatever might have been the position if the words "as is" had stood alone, they cannot in context be taken to signify that the sellers could provide cargo which was not of normal PHRC quality for naphtha because of its very high MTBE content'[59] (emphasis supplied).

Flaux J in *The Union Power* correctly distinguished *Covas v Bingham*,[60] a decision of the Court of Queen's Bench, on which the seller relied as establishing a legal meaning of the term ousting warranties, given the very particular basis of the decision in that case on statements in bills of lading. However, it is again interesting to note the court in that case, considering an analogous term of 'as it stands' expressed views as to the ordinary meaning of the term which are synonymous with those of Phillips J and David Steel J. This case is cited in Stroud as authority for the meaning of the term 'as it stands':

> *AS IT STANDS*. A contract to take, e.g. a cargo 'as it stands' (though it specify a quantity), means that the cargo is 'to be taken by the purchaser for better for worse, for less or for more' (per Campbell L.C., *Covas v Bingham*, 23 L.J.Q.B. 29: see Benj. 565).[61]

It is important to note that, like the English cases, the Canadian cases cited by Flaux J in *The Union Power*[62] similarly reach the result they do by giving the term 'as is' not some special meaning but its ordinary linguistic meaning of the goods being taken in their existing and present condition.

In *MacLeod v Ens*,[63] in an almost direct echo of the wording used by Phillips J in *The Morning Watch* and Lord Chancellor Campbell in *Covas*, the court stated: 'When used with reference to a sale, people generally take the term "as is" to mean that the product *is bought and sold in the condition in which it then exists, for better or for worse*, with altogether no warranties in relation to quality, durability, or fitness, and with the entire risk in those respects to be borne by the buyer'[64] (emphasis added).

No case espousing any contrary or narrower meaning in English of the term 'as is', nor any commentary to like effect, was found in *The Union Power*.

Although not cited by either party before the judge and not referred to in the textbooks, there is additional authority which has been turned up by the author in the preparation of this paper which again supports the view that, simply as language, the term 'as is' has a clear and settled meaning.

58 [2010] EWHC 374 (Comm).
59 *Ibid* at [104].
60 2 El & Bl 836.
61 *Stroud's Judicial Dictionary of Words and Phrases* 8th edn, (Sweet & Maxwell, 2012).
62 Viz. *MacLeod v Ens* (1983) 135 DLR 3d 365 and *Smith v Lasko* [1987] 5 WWR 412,
63 (1983) 135 DLR 3d 365.
64 *Ibid* at [59].

In *Norwest Holdings Ltd (in liquidation) v Newport Mining Ltd*,[65] the Singapore High Court considered a sale of a Chinese mining business on 'as is, where is' terms. The ambitious argument in that case was that this entitled the seller to transfer the mining business in any condition whatsoever, and therefore the adverse effects of an earthquake in Sichuan which caused a considerable loss in value were for the buyer's account and at its risk. The Court considered that where a buyer offers to buy on 'as is where is' terms, that defines the condition of the subject matter as at the time of offer. If that offer is accepted, the condition warranted is not of the goods at the time of acceptance (fair wear and tear excepted, which would exclude an earthquake).

The Court's reasoning built upon what it saw as the ordinary meaning of the phrase. The passage merits citation in full.[66]

> Norwest's liquidator maintained during cross-examination that, since the negotiations and sale proceeded on an 'as is, where is' basis, he was not giving any warranty as to the condition of the Chinese business. This is correct, but merely one side of the coin. In an 'as is, where is' sale, the buyer agrees to take the subject-matters 'as it is' or 'as seen' without any warranty of guarantee as to quality, character, condition, size, weight or kind[67] . . . The buyer cannot ask for the subject matter to be better than it is actually is, but at the same time, and crucially for this case, the buyer had agreed to accept nothing less. Put in another way, it is against all commercial sense to say that an interested buyer who offers to take in existing condition thereby offers to take in any condition henceforth – an agreement to take 'as is' does not operate as an entire assumption of risk. Absent any contrary intention, the condition which the offeror-buyer offers to take is, quite naturally, the condition at the time the offer was made and not the time of acceptance, which is by definition a complete and unconditional agreement to the terms of the offer. Of course, it can well be argued that the buyer would take subject to any change which might occur in the ordinary course of events such as wear and tear, but it can hardly be doubted that the massive Sichuan earthquake and the damage it caused were not events of that nature.

The Court referred to and applied a previous Malaysian decision in *Mensa Mercantile (Far East) Pte Ltd v Eikobina (M) Sdn Bhd*.[68] In that case construction machines were sold on an 'as is, where is' basis. In a short passage the High Court stated:[69]

> The term 'as is where is' appearing in a sale transaction is defined in the Encyclopaedia of Practical usages of Terminology in Business Agreements, at p. 40,[70] as being a sale of goods 'without any warranty of guarantee as to quality, character, condition, size, weight or kind'. In other words, to quote [one of the party's witnesses] 'I take them as I see them'.

In addition, the United States perspective, at least in terms of the language, is the same (which may explain the approach of the Canadian courts).

The Uniform Commercial Code refers to 'as is' as a term which is indicative of the buyer taking the goods as they are and for this reason as excluding warranties or conditions as to the goods. See § 2-316(3)(a):[71] 'unless the circumstances indicate

65 [2010] 3 SLR.

66 *Ibid* at [42] per Belinda Ang Saw Ean J.

67 Citing Krishana, *Words, Phrases and Maxims Legally and Judicially Defined* (LexisNexis, 2008) at A1887.

68 [1989] 2 MLJ 170.

69 *Ibid* at 175–176.

70 This appears to be identical to the text cited by the Singapore High Court in *Norwest* [2010] 3 SLR 956.

71 Headed 'Exclusion or Modification of Warranties'.

otherwise, all implied warranties are excluded by expressions like "as is", "with all faults" or other language which in common understanding calls the buyer's attention to the exclusion of warranties and makes plain that there is no implied warranty'. American law dictionaries unsurprisingly contain similar statements. The relevance for present purposes is that, apart from the asserted secondary meaning that all warranties etc. are excluded, the primary meaning which leads to this is that the goods are sold 'in the existing condition' or (where inspected) 'in the condition in which they are presently found'.[72]

Lastly, and since writing the above, the position has very recently been considered briefly again by the English Courts in a manner which conveniently summarises the state of the authorities on the plain English meaning of the phrase 'as is' in a sale context. The cases can conveniently be concluded by Leggatt J's statement in *Hirtenstein & Il Sole Ltd v Hill Dickinson LLP*[73] where he said:

> The first is that the Yacht was offered for sale to Mr Hirtenstein, and he agreed in principle to buy it, 'as is, where is'. I would regard that phrase as self-explanatory. It clearly signified that the buyer would acquire the Yacht in whatever condition the boat was at the time of purchase with no right to complain subsequently if the boat should turn out to have any defect.[74]

He approved the view of a witness that this meant that he was buying the boat in question 'with its beauty and the warts'.[75] While the issue in that case concerned the parties' personal subjective understanding of that phrase and not the interpretation of it in the contract in which it was used,[76] the Court was dismissive of the particular phrase having any other meaning, quite apart from these parties' understanding of it.

8.4.3 Flaux J's 'reading down' or via media approach

Flaux J correctly rejected the seller's argument that the words are and should be treated as meaningless and just a 'truism' meaning nothing more than 'here are the goods' (which the arbitrators had held).[77] He held that they had to be given

72 See *Black's Law Dictionary* 4th pocket edn, (West, 2011) – '*as is, adv. & adj.* In the existing condition without modification "the customer bought the car as is". Under UCC § 2-316(3)(a), a seller can disclaim all implied warranties by stating that the goods are being sold "as is" or "with all faults". Generally, a sale of property "as is" means that the property is sold in its existing condition, and use of the phrase as is relieves the seller from liability for defects in that condition.' See also *Barron's Law Dictionary* (6th edn, 2010.) – 'AS IS: in the condition in which presently found; commercial term denoting agreement that buyer shall accept delivery of goods in the condition in which they are found on inspection prior to purchase, even if they are damaged or defective. Sale of goods "as is" when based on inspection of a sample requires that the goods delivered be of the same type and quality as the sample. UCC § 2-316. Homes and used cars are often sold "as is".'

73 [2014] EWHC 2711 (Comm); judgment handed down 31 July 2014. This appears to be the first judicial consideration of *The Union Power* in any context.

74 *Ibid* at [51].

75 *Ibid* at [52].

76 See at [56]: 'In the present case, however, the question is not one of interpretation of the sale contract but simply one of the background understanding of the parties against which the contractual documentation was negotiated. I have no doubt that Mr Hirtenstein, Mr Candy, their solicitors, and the broker, Mr Cecil-Wright, all understood that buying the Yacht "as is, where is" meant that Mr Hirtenstein would get the Yacht in its existing condition, good or bad, with no subsequent recourse against the seller for any fault which the Yacht might turn out to have.'

77 Cited by Flaux J [2012] 3537 (Comm); [2013] 1 Lloyd's Rep 509 at [10]: 'The phrase "as is,

substantial meaning. But what meaning is to be given to the words 'as is' other than that the goods are being sold as they are and in their present and existing condition?

The judge went on to hold that substantial meaning could be given by using cases on 'with all faults' clauses to 'read down' the clause limitatively so as to restrict it to the right to reject for breaches of the implied terms as to quality but not as excluding them altogether. That is, undoubtedly, an arguable approach, especially where the phrase 'as is' is used in conjunction with the buyer (e.g. 'taking delivery' (as in the 'with all faults' case of *Christopher Hill Limited v Ashington Piggeries Limited*[78])). However, it is strongly arguable that this is a very strained meaning to give to the ordinary English language sense of 'as is' and imports a level of distinction between remedies which is objectively unlikely to have been intended by comparison with the buyer agreeing to take the goods for all purposes as they are. Certainly, as a matter of ordinary language, that is a meaning which it is very difficult to deduce or derive from the use of the phrase 'as is'. This was the reaction of Leggatt J in the recent decision of *Hirtenstein*,[79] where he stated in relation to Flaux J's *via media* approach:[80]

> In a contract between commercial parties such an interpretation would seem to me to be generous to the buyer. Drawing such a distinction between the right to reject and the right to damages and treating the words "as is" as excluding the former but not the latter seems to me most unlikely to reflect the expectations of ordinary business people or to be an interpretation that would occur to anyone other than an ingenious lawyer.

Further, much will depend on how and where 'as is' is used.

If the way in which it is used differs, then unless the phrase has this settled *via media* meaning, halfway between full exclusion of section 14 and a meaningless truism, its meaning will be sensitive to its linguistic setting. In *The Union Power*, the judge was considering the position on the hypothesis that the phrase 'as she was' in Clause 11 was a species of 'as is' provision. In that case the phrase would have been used as follows: 'The Vessel shall be delivered and taken over [As Is].' It is arguable that where the term 'as is' is used in direct juxtaposition with taking over and making delivery, it can be read as dealing only with the right to reject. If the phrase is differently worded, for example 'Goods sold basis As Is Where Is', it is much more difficult to suggest that this is what the language used means, unless this is what it always means in ordinary English. There is no support for that view.

Flaux J recognised that this *via media* was not reflected in the English cases (although they had not had to address the section 55 question) and was at odds with the Canadian cases which, in his view, excluded the Sale of Goods Act terms 'completely'.

In respect of the Canadian cases, Flaux J accepted that these cases supported the argument that 'as is' excluded Sale of Goods Act terms. He stated:[81]

---

where is", on the other hand is simply a truism (clearly any vessel must be as it is and where it is on delivery) if some other meaning is not to be imputed to the phrase.'

78 [1969] 2 Lloyd's Rep 425.

79 [2014] EWHC 2711 (Comm).

80 [2012] 3537 (Comm); [2013] 1 Lloyd's Rep 509, at [55].

81 *Ibid* at [61].

> I accept that the Canadian cases decide that, as a general rule, a statement that the contract is 'as is' will exclude the statutory implied terms as to merchantability and fitness for purpose, save where consumer protection legislation renders such a statement of no effect or there are other inconsistent terms of the contract. There is no justification for limiting the Canadian cases on 'as is' to exclusion of the implied term as to fitness for purpose, but not that as to merchantable quality, as [the buyer] suggested [. . .].

The judge's reason for not applying the Canadian cases in the same way as the *obiter* expressions of view in the English case law was that, in his view, they failed to grapple with the distinction between warranties and conditions and to recognise that the Sale of Goods Act terms are implied conditions.[82] A difficulty with the judge's approach is that he used the different cases of *The Mercini Lady*[83] and *Air Transworld Ltd v Bombardier Inc*[84] to arrive at a view that if one sought to exclude conditions one needed to refer expressly to conditions or as he put it 'the implied terms are conditions and that (whilst it may not be necessary in a modern context to exclude conditions expressly) clear words are needed to exclude the statutory implied conditions (see for example Air Transworld).'

The line of authority represented by those two cases deals with a specific type of clause which excludes certain specified types of obligation and not others. Accordingly, if a clause in a contract for the sale of goods seeks to exclude 'warranties' (or other terms such as 'guarantees') it will not be read as excluding conditions, therefore the implied conditions under the Sale of Goods Act 1979 including under section 14 will not be excluded thereby. This was the issue in *Wallis, Son & Wells v Pratt & Haynes*[85] and was similarly the issue before the Court of Appeal in *The Mercini Lady*, where it was argued that under modern principles of construction, this approach was archaic and should not be followed. In *Air Transworld*, a very elaborate clause which did not mention the term 'conditions' was, however, held to have excluded the Sale of Goods Act conditions read as a whole.

This line of authority does not address the meaning of 'as is'. Nor does it address the question whether a shorthand term such as 'as is', if it has acquired either a settled legal or popular ordinary meaning as a matter of language of 'no undertakings as to quality and condition' is sufficient to exclude those conditions. This, however, is what the Canadian cases held (and the consistent trend of *obiter* statements in the English cases confirms). The cases of *The Mercini Lady* and *Air Transworld* cited by Flaux J are not inconsistent with the Canadian cases or their reasoning. While the court in the Canadian cases referred to 'warranties', I do not consider that on a fair reading of the cases any distinction was being drawn between 'conditions' and 'warranties'. Flaux J appears to have read them as negativing completely the implication of the statutory terms, however classified.

It is, therefore, suggested that there is a strong argument that the judge's provisional conclusion on the meaning of 'as is' was wrong.

In particular, the judge did not have any real argument addressed to him by the seller in that case as to the 'ordinary English meaning approach'. The seller's focus

82 *Ibid* at [63].
83 [2010] EWCA Civ 1145; [2011] 1 Lloyd's Rep 442.
84 [2012] EWHC 243 (Comm); [2012] 1 Lloyd's Rep 349.
85 [1911] AC 394.

was rather on two matters. First, reliance was placed on what was said to be the special *ship* sale and purchase meaning of the phrase (based on seeking to trace the wording of clause 11 back over previous versions of the Saleform: an exercise which the judge deprecated as 'archaeology of the forms').[86] Second, reliance was placed on older cases on differently worded 'all faults' clauses from which it was unsuccessfully attempted to show that the modern phrase 'as is' is to be derived.[87]

The first argument was summarily rejected. The difficulty with it is that the Saleform has greatly changed in its wording. While the 'Ur form' of Clause 11 in Saleform 1956 and 1966 provided that 'the vessel ... shall be delivered and taken over as she is at the time of delivery after which the Sellers shall have no responsibility for possible faults or deficiencies of any description',[88] the later versions of Saleform have completely departed from this wording. It is obviously impermissible to construe the 93 form by saying that it has the same meaning as a wholly differently worded earlier version, especially where the phrase 'after which the Sellers shall have no responsibility for possible faults or deficiencies of any description' had been dropped and the 'as she is' wording replaced with a new concept of continuity between inspection and delivery only.

The seller's second argument was heavily based on the proposition that 'as is' is modern shorthand for older clauses referring to language such as 'with all faults'. This was not a necessary argument for the sellers to have advanced. The attempt to trace the historical origin of 'as is' is difficult and no guidance is given in the cases or commentaries. However, while interesting in terms of legal research, it is an unnecessary argument since the correct question is: what does the separate term 'as is' *now* mean in current English both generally and when used in a contract of sale where the seller is entitled to deliver and the buyer is bound to take the goods 'as is' or where the sale is described as being on an 'as is' basis? That question is not answered by looking at older cases on different clauses. However, the fact that the older cases shed no light (as the judge held) is neutral and leaves the critical question as to the modern meaning of 'as is' as a matter of ordinary language untouched.

The judge was heavily influenced by the consideration that it would be surprising if the simple words 'as is' could exclude the Sale of Goods Act terms when much more elaborate linguistic formulae had failed to have that effect as cases such as *The Mercini Lady* and *Air Transworld* demonstrated: if that were the case, he asked rhetorically, why were the terms not more commonly used?[89] This reasoning is, with respect, open to question. The sole question is whether the phrase 'as is' has the meaning asserted as a matter of language. That is not answered by asking whether

86 [2012] 3537 (Comm); [2013] 1 Lloyd's Rep 509 at [58]. He noted and applied the distaste of the Court of Appeal for similar exercises in previous BIMCO forms in *Polestar Maritime Limited v YHM Shipping Co Ltd (The Rewa)* [2012] EWCA Civ 153; [2012] 1 Lloyd's Rep 510 (see at [30] per Aikens LJ): 'I would generally discourage such exercises in "the archaeology of the forms". In most cases it makes the task of interpretation of contractual wording unnecessarily over-elaborate and it can add to the expense and time taken in litigating what should be short points of construction.'

87 Founded upon *Lloyd del Pacifico v Board of Trade* (1929) 35 LlL Rep 217. See [2012] 3537 (Comm); [2013] 1 Lloyd's Rep 509, at [37] to [41] per Flaux J. He rejected this at [78] and [79].

88 See Strong & Herring, at p. 154.

89 [2012] 3537 (Comm); [2013] 1 Lloyd's Rep 509 at [81].

and if so why that wording should be more effective than different types of more long-winded clauses which speak only of 'warranties'.

Different terms are commonly used in different trades and in different contexts. The concept of a party buying a chattel outright after an inspection (or blind and without inspection) on terms that he takes it as it is will only apply to certain types of chattel and judging from the decided cases, second-hand ones such as cars and ships. That concept is not common in sales of new chattels or more sophisticated transactions such as bulk sales of commodities (as in *The Mercini Lady*) where a very detailed contractual specification of the goods is given and the seller assumes considerable responsibilities for the condition of the goods sold under agreed express terms: there the concern is to exclude the operation of additional general terms which may be implied so as to operate over and above the express obligations (as the wording of the clauses in both *The Mercini Lady* and *Air Transworld* shows).

8.5 Conclusions

The difficulty encountered in rejecting the ouster of the Sale of Goods Act 1979 implied conditions as to quality is what sensible linguistic meaning to give to the phrase 'as is' or 'as is, where is' (or variations of the same) where it is used by parties in a contract for the sale of goods to define the basis on which the goods are being sold and delivered.

Irrespective of any special trade meaning or legally binding meaning, the term 'as is' is used by buyers and sellers to mean *something* (as the judge accepted). In some simpler cases, it may be the only term which they use. The authorities all suggest that they mean and are commonly understood as meaning that the seller does not give any undertakings as to the quality and condition of the goods because they are being sold in their present condition and as they are at the time of sale and nothing more.

That being the case and if the term 'as is' is opened out from its compressed elliptical form and written out in longhand, the express term which the parties have thereby agreed is necessarily inconsistent with the implication of further terms as to the quality or nature of the goods under section 14(2) of the Sale of Goods Act. Section 55(2) is satisfied because the parties, albeit in a two-word shorthand, have expressly contracted on a clearly defined basis which defines quality solely by reference to the goods as they are at the time of sale and in the condition they are then in, whatever that condition is.

The consequences of *The Union Power* in the immediate context of commercial ship sale and purchase of ordinary trading vessels concluded on Saleform terms as dealt with by the Court are likely to be short lived. A straw poll of a selection of leading London Maritime Arbitrators Association arbitrators[90] suggests that, since the decision in *The Union Power*, the point has not arisen for decision, at least not going to an award. To that extent the decision may finally have brought that measure

90 I am very grateful to Messrs Mike Baker-Harber, Robert Gaisford, Simon Gault, Ian Gaunt, Lindsay Gordon, Bruce Harris, Mark Hamsher and Patrick O'Donovan for taking the time to respond to my enquiries and for their helpful thoughts on this subject.

of certainty which the commentators sought (even if contrary to their adamant viewpoint).

For the future, the matter has been sought to be addressed by BIMCO. As the case was being argued, the new Saleform 2012 was being finalised. It seeks to tackle the point in issue head on by its new 'Entire Agreement Clause' in clause 18, the relevant portion of which provides:[91] 'Any terms implied into this Agreement by any applicable statute or law are hereby excluded to the extent that such exclusion can legally be made.' The BIMCO Explanatory Notes[92] state:

> The purpose of the EAC is to limit the rights of the parties to the written terms of the contract. . . .. Although the Clause is designed to work under any system of law, under English law it should effectively exclude the implied terms of [SOGA]. This should remove the uncertainty concerning the sale of ships under English law and a potential obligation . . . of 'satisfactory quality' . . .

Which side of the line drawn by *The Mercini Lady* this falls may be a matter for argument,[93] although the drafters have avoided the word 'warranty' and gone for a more inclusive term, eschewing 'condition', 'warranty' etc. But the point as to the different results of an omnibus wording under different governing laws is, again, potentially illustrated.

In other shipping sale and purchase contexts and under earlier transactions still using the older Saleforms (an unfailing bad habit),[94] Nipponsale 93 or the new SSF, as well as in general sale of goods contexts quite divorced from shipping, the 'as is' point remains a very live one. The decision in *The Union Power* is therefore likely to give rise to further consideration and discussion of the meaning of the these two, rather pregnant, but in reality and in ordinary English 'self-explanatory'[95] syllables.

---

91 At lines 407–409.

92 Available on the BIMCO website.

93 Cf. for example the view expressed by M. Bek, 'Unsatisfactory Implication – A Case for Revision' [2012] 18 JIML 420 at 422 that the new Saleform 2012 Clause 18 will exclude the Sale of Goods Act terms with a more cautious approach by J. Leonida, '"As is" Is Not What It Used To Be' (2012) *The Superyacht Report* (Issue 148) to the similarly drawn entire agreement/exclusions clause in cl. 34 of the Mediterranean Yacht Brokers Association Memorandum of Agreement (MYBA MOA) upon which most second-hand yacht sale and purchase agreements are concluded.

94 See Curtis at p. 7.

95 In *Hirtenstein & Il Sole Ltd v Hill Dickinson LLP* [2014] EWHC 2711 (Comm) at [51] per Leggatt LJ.

CHAPTER 9

The practicalities of ship sales in the current market

*William Maclachlan**

9.1 Introduction

We are aware of temptation by some in the shipping industry to view the process of ship sale and purchase as a relatively straightforward proposition. This is particularly true where they are using a standard from sale agreement such as the most commonly used form, the Norwegian Shipbrokers' Association Memorandum of Agreement for sale and purchase of ships (1993 or 2012 revision), known in the industry as the NSF and available form BIMCO as part of the suite of standard documents published by them.

However, given the particular nature of the asset in question, the fact that the parties are normally multinational and the large numbers of moving parts involved, the sale of second-hand vessels can become extremely complicated very quickly. Indeed, from a legal point of view, it can be amongst the hardest thing solicitors are asked to assist with. Getting a troublesome deal over the line can sometimes be likened to landing a disabled aircraft at night on an unlit runway in a storm whilst the passengers are fighting and the crew have locked themselves in the cockpit.

The process can be greatly assisted if the memorandum of agreement is carefully drafted in the first place. But the reality is that, even with the finest memorandum of agreement, the parties are not contracting in laboratory conditions and many things can crop up to scupper their best-laid plans.

As the title suggests, the aim of this chapter is to give an insight into the practical aspects of ship sale and purchase which legal practitioners encounter, and a few useful tips on how avoid some of the more common pitfalls. As the author is an English lawyer, the legal analysis must be viewed in the context of English law. This may, in any event, be appropriate, given that English law and London arbitration remain the preferred choice in the vast majority of ship sale and purchase deals, and much of the leading case law on the subject emanates from the English courts.

9.2 The current market

The year 2008 saw the high-water mark for second-hand values across all types and classes of vessel, following which values largely crashed. Whilst they have begun to

---

* Senior Associate, Holman Fenwick Willan LLP.

recover from their 2012/2013 lows and some who bought at low prices are now, at least on paper, making a lot of money, prices remain volatile.

The improved values appear somewhat disconnected from the prevailing freight rates, leading some brokers to believe that it is the big speculators driving prices. At the time of writing, it is not easy to get a clear picture on market sentiment and values are very much asset-specific. The volatility in asset prices and the difficult times being faced by most western ship finance banks make the current ship sale and purchase market a challenging and unique place to do business.

9.3 Pre-contractual negotiations

Once the buyer has identified the right vessel at the right price, his first move is ordinarily to have his surveyors and technical team physically inspect the vessel and its classification records. If he is happy with what he sees, he will need his broker to secure the deal. In our experience, whilst we always encourage our clients to involve us from the outset, it is very unusual for external lawyers to be involved at this point. However, there are pitfalls even at this early stage, to which the parties should be alive.

In most cases, the contract is made when a sale and purchase agreement, known in shipping as the 'memorandum of agreement' or 'MOA', is signed by both parties. However, in shipping it is common for the key terms of the MOA, commercial and otherwise, to be agreed by the parties before they are formally documented.

Negotiations are commonly conducted between the parties' brokers through an exchange of written offers and counter-offers, normally by e-mail. Once most of the elements are agreed, the seller's broker will usually issue a written message, known as the 'recap', to the buyer's broker for approval. The elements of a contract are understood; it goes without saying that an offer should make clear the key terms and conditions on which the party making it is promising to contract, contain a period within which it should be accepted and confirm whether or not, if it is accepted, there is a binding contract at that stage. It is surprising how often this last point is overlooked and brokers should remember that, until they are authorised by their respective principals, correspondence between them should be clearly marked, 'subject to contract' to avoid (i) any suggestion that the principal is immediately bound and (ii) the broker's possible exposure to a claim for breach of warranty of authority.[1]

In the ship sale and purchase context, you may find correspondence marked '*sub details*'. This is broker shorthand for 'subject details' and, as Langley J held in *Thoresen & Co (Bangkok) Ltd v Fathom Marine Company Ltd*,[2] this can be taken as having the same meaning as 'subject to contract'.

The fact that the recap doesn't cover every term of the contract to be made may not be a valid defence against the existence of a binding contract;[3] it should

1 *Manatee Towing v Oceanbulk (The Bay Ridge)* [1999] 2 Lloyd's Rep 227.

2 [2004] EWHC 167 (Comm); [2004] 1 Lloyd's Rep 622.

3 *RTS Flexible Systems Ltd v Molkerei Alois Müller GmbH & Co KG (UK Production)* [2010] UKSC 14; [2010] 1 WLR 753.

be remembered that a recap might be considered as objective evidence of the parties' intention to be bound.[4]

9.4 Know your deal

It is a reality of the second-hand market, particularly in its current volatile state, that things generally move pretty quickly from the point at which terms are agreed to the formal signing of the MOA.

It is common for a vessel to be owned by a special purpose vehicle (SPV) with no assets other than the vessel in question. This can make bringing an action against the owner fruitless, as there may be nothing to enforce an award or judgment against. Parties may rightly, therefore, be expected to look for additional security for their counterparties' obligations. Whilst a seller can take some comfort from the deposit, which is paid as security for due performance of the buyer's obligations under the MOA, it is the buyer who is particularly exposed. Generally, a claim against a seller for a breach of the MOA will not give rise to a right of arrest against the vessel.[5] In addition, should the buyer, following delivery, have cause to bring a claim against the seller for a pre-existing defect or maritime lien, he is likely to be up against the problem of enforcing against an SPV. Given the risks, the buyer must know what it is buying and from whom.

9.4.1 Your counterparty

In ship sale and purchase transactions it is common practice for the parties to exchange copies of their constitutional documents and corporate authorities ratifying the MOA at or just before delivery. By this point it is rather late to discover that your counterparty is not validly incorporated or that the signatory to the MOA or other delivery documents is not duly authorised. We therefore recommend that parties do all that they reasonably can to establish at an early stage the identity and status of their counterparty and whether those signing for the counterparty have due authority to do so. A party undertaking effective due diligence on its counterparty will reduce the risk of any argument that the MOA was not validly executed on its behalf (whether through lack of corporate capacity or otherwise).

Trite as it might sound, the industry grapevine can be a useful starting point for anyone wishing to learn about their counterparty; brokers may well know something of the other side and their business and be in a position to pass that information on to their principal. Equally, if engaged, lawyers generally have access to a number of electronic search services, such as Dun & Bradstreet and World-Check, from which useful information can be obtained.

4 *The Merak* [1976] 2 Lloyd's Rep 250.

5 Arrest of vessels in most states is governed by the 1952 Brussels Convention on the Arrest of Sea-going Ships or the 1999 International Arrest Convention on Arrest of Ships. The former, which does not allow arrest by a buyer, is the dominant convention and is that in force in the UK (where it is enacted via the Senior Courts Act 1981). The 1999 Convention does grant a right of arrest for a dispute arising from a contract for the sale of a vessel, but given the limited numbers of signatories, this is of little practical use in most circumstances at the time of writing.

For each party, it is good practice to do the following in good time. First, it should obtain copies of the counterparty's constitutional documents. Second, if the counterparty is in a jurisdiction where directorships and shareholdings are a matter of public record, it should carry out a search of the companies registry in the counterparty's place of incorporation and any other registry that might be maintained (e.g. in Panama, a party should search both the flag state registry and the public registry). If directorships and shareholdings are not a matter of public record, it may be advantageous to ask the counterparty to provide a certificate of incumbency or equivalent from their companies registry confirming the directors and shareholders of the company, though it is likely that this will be resisted until nearer the time of delivery.

9.4.2 Your vessel

Once a buyer has accepted the vessel and signed an MOA, it generally has very limited scope to later reject the vessel or otherwise make a claim for her condition, particularly if the sale is on an 'as is' basis. There are frequent cases of clients involved in disputes, particularly cash buyers in the demolition market, who have entered into quick and dirty 'as is' deals only to discover, on delivery, that the vessel has a major and very expensive defect preventing its movement and that they have little recourse to the seller.

However, subject to the terms of the MOA, that is not to say that there is no scope for a claim, particularly if using the NSF 1993 with its unamended standard terms. As Simon Rainey QC explains elsewhere, in the case of *Dalmare SpA v Union Maritime Ltd (The Union Power)*,[6] *The Union Power* was sold on the terms of an NSF 1993. The main engine broke down shortly after delivery as a result of a pre-existing defect. The court was asked to consider whether the buyer could rely on the seller's breach of the 'warranty of quality' under the UK's Sale of Goods Act 1979[7] or whether this was excluded by the language of the contract. The court found that, in the absence of express language excluding statutory implied terms, the buyer's right to claim for a breach of the statutory warranty under the Sale of Goods Act remained unimpaired.

If buying, there is no substitute for ensuring that a thorough inspection of any target vessel and its classification records is conducted by qualified surveyors and a senior member of the buyer's technical team. Any defects or issues discovered in such an inspection should be raised with the seller and addressed commercially (often by demanding a reduction in the purchase price) prior to acceptance.

As well as the existence of any physical defects, it is also essential that the buyer is comfortable that there are no maritime liens arising out of any action or inaction of the seller prior to the sale which might still be attached to the vessel following delivery and thereby potentially give a third party a right to arrest her. Under clause 9 of the NSF and again in the bill of sale, the seller warrants that the vessel is free from all mortgages, encumbrances and maritime liens; however, for the reasons set out

6 [2012] EWHC 3537 (Comm); [2013] 1 Lloyd's Rep 509.
7 The Sale of Goods Act 1979 (as amended), s. 14(2)

above, this is of relatively little comfort if the seller is an SPV and the buyer wishes to recover damages from it following the arrest of the vessel and has to post security to get it released.

Whilst it might be obvious, it is also very important that the entity contracting as the seller is, in fact, the registered owner of the vessel and therefore, at least in principle, entitled to pass title in and physical possession of the vessel to the buyer. It must be remembered, however, that registered ownership is not conclusive evidence of title.[8]

The buyer can confirm the identity of the registered owner and whether or not the vessel is subject to any registered encumbrances by obtaining a certificate of ownership and encumbrances or transcript of register (the particular form of certificate varies from registry to registry). Whilst there is a small fee to be paid for such a certificate, a buyer should consider it good practice to obtain this early on rather than wait to receive one from the seller at delivery, by which point it may be too late to deal with any issues.

Unfortunately for buyers, the existence of a maritime lien over a vessel is not a matter of public record, and there is no hard-and-fast means of establishing whether any are likely to exist. If the vessel has been stationary in one place for any length of time it may be prudent to do a local writ search in that jurisdiction, though this will tell you little about any claims beyond that jurisdiction. To obtain some comfort as regards the wider risk, there are, however, some things that can be done. By using information portals such as Lloyd's List Intelligence, talking to brokers and keeping an ear to the ground whilst inspecting the vessel, a buyer should be able to establish, to a degree at least (i) where the vessel has been trading, (ii) whether or not she has been detained for any reason by port state control, (iii) whether she has historically been arrested by creditors, or (iv) whether she has otherwise suffered any casualties or incidents likely to give rise to a claim. With such information, a buyer should be able to take a view on the likely level of risk.

9.5 What can be done if enquiries reveal an issue?

If the buyer has concerns as to the possibility of maritime liens existing over the vessel or the performance of any post-delivery obligations by the seller, he might be well advised to request that the seller procures a guarantee from a trading company within its group. Subject to the parties' respective bargaining positions, such a request may well be resisted by the seller, but there are good arguments in the buyer's favour should he choose to pursue them.

Alternatively, the buyer might like to consider taking out indemnity insurance or pre-existing liens insurance to cover any maritime liens. Whilst such cover is available, it is expensive. Worse still, it may not cover a claim that the buyer became aware of through its enquiries before taking out the insurance (i.e. the very reason the buyer took out the cover in the first place) so its value is therefore questionable.

If a buyer feels exposed for whatever reason, it might like to negotiate the retention of a part of the purchase price in escrow for an agreed period of time. Such a proposal is likely to be unpalatable to the seller; indeed, it may be completely

---

8 *The Bineta* [1967] 1 WLR 121.

unacceptable if the vessel is the subject of finance, as the seller's bank may expect to receive the full purchase price. If it can be agreed in principle, the parties should ensure any escrow agreement is tightly worded to prevent funds being released other than on the escrow agent's receipt of the agreed trigger document (if appropriate) or otherwise on the joint written instructions of the parties or upon receipt of a final and unappealable arbitration award. No escrow agent should ever be left in the position of having to make a decision himself on whether or not to release funds.

9.6 The memorandum of agreement

In an ideal world the negotiation and agreement of the MOA should be a relatively quick and painless process once terms have, in principle, been agreed and the buyer is comfortable with what he is buying. In reality, negotiating the MOA can be surprisingly protracted; it also often happens simultaneously with all of the above. By convention, the MOA is drafted by the seller with comments from the buyer's side. It is not uncommon for parties to seek to save on legal costs by preparing the MOA themselves or through their brokers. Often, it is worth adding, MOAs are negotiated entirely between brokers with no legal input until after the MOA is signed, by which point it is, of course, too late to address any concerns with the drafting. English law is by far the most common choice of law for those using the NSF form and, if the draftsman lacks a clear understanding of the nuances of English law, this approach is not without its inherent risks.

As various recent cases have shown,[9] there are some valid amendments the parties might wish to make to the standard NSF and commercial and technical issues may also need to be addressed. However, as a rule, it is not recommended that the standard terms be amended without a good deal of care and even then, only when necessary. One frequently sees very loose drafting and the amendment of important provisions without apparent good reason or where the draftsman has otherwise failed to follow through with all necessary further amendments to the MOA. Whilst an amendment might appear innocuous to those without any knowledge of the English law approach to the interpretation of contracts, it can have serious consequences if not properly thought through.

As a practical note, for those MOAs prepared on standard NSF 1993 or 2012 terms, the draftsman will have to prepare the document using BIMCO's Idea portal. This is an extremely useful product and source of standard contracts. However, making extensive amendments to documents in Idea can be difficult and parties may find it easier to make all but the most minor amendments in a separate MS Word addendum.

When assisting clients, who often have a very short window in which to secure a deal and are under intense pressure to do so, transactional lawyers must never lose sight of the fact that their purpose is to advise the clients on the terms to which they are agreeing and assist them to secure the deal with as little risk to them as

9 Including, amongst others, *Dalmare SpA v Union Maritime Ltd and another (The Union Power)* [2012] EWHC 3537 (Comm); [2013] 1 Lloyd's Rep 509 and *Firodi Shipping Ltd v Griffon Shipping LLC* [2013] EWCA Civ 1597; [2014] 1 Lloyd's Rep 471.

possible. Faced with the reality of commercial pressure and the general perception by those in the commercial world that their lawyer is the 'deal prevention officer' who really need not worry because 'everything will be all right', it is not always possible to entirely eliminate commercial risk for a client. It is ultimately up to the client whether or not the potential commercial advantage outweighs the risk. Rarely is a contract entirely perfect for all parties; but if it is reasonable and balanced, it is more likely to survive.

Issues often arise after the MOA has been signed which require the parties to re-negotiate certain terms. There is sometimes a temptation to cut corners in this respect, especially if close to delivery; but, as with any contract, all amendments should be made in writing and signed by both parties. If there is a guarantee in place, the recipient should always ensure that the guarantor has been made aware of and confirms its acceptance of any amendment to the MOA; otherwise there is a risk of the guarantee becoming invalid. Further, if a purchase is being financed, the buyer's bank should be informed by the buyer of any change in the purchase price or else risk being in default under its loan agreement.

9.7 The deposit

Once the MOA is signed, the buyer has three banking days in which to pay the deposit to the deposit holder. The deposit is commonly held either on a joint account opened with the seller's bank or in accordance with an escrow agreement by the seller's lawyer.

The process of opening a joint account and satisfying the account-holding bank's requirements is generally more expensive and time consuming than placing funds with a lawyer in escrow. However, either way, the parties cannot escape the modern reality of having to satisfy the deposit holder's compliance requirements, which generally means providing documents confirming the identity of the parties paying and receiving the deposit and any individual owning 25 per cent of more of their respective shares. Whilst the requirements of law firms, particularly those based in the US and UK, are increasingly strict, the banks generally remain harder to satisfy.

When agreeing the terms of an escrow agreement (usually on the firm's standard terms), it is important both for the deposit holder and the parties that the funds can only be released in accordance with the joint written instructions of the parties or a final and unappealable arbitration award. As already noted, no discretion should be left to the deposit holder when it comes to release.

Delays in setting up the mechanism for the holding of the deposit or in the buyer satisfying the deposit holder's compliance requirements are unlikely to provide an excuse for non-payment of the deposit. As the recent case, *Firodi Shipping Ltd v Griffon Shipping LLC*[10] showed, unless the standard NSF terms are clearly amended by express words intending to deprive the seller of the deposit in the event of it not being paid, the seller's right to receive the deposit is unconditional.

---

10 [2013] EWCA Civ 1597; [2014] 1 Lloyd's Rep 471.

9.8 Delivery documents

9.8.1 What is the scope of delivery documents?

Clause 8 of the NSF sets out the documents to be provided by the parties on delivery. However, it is common for the parties to delete the schedule of documents in clause 8 and prepare their own addendum. This is sensible as, on a number of points, clause 8 does not provide sufficient certainty to either side in terms of what they will receive and what they are obliged to provide, particularly with regard to the vessel's technical documents and trading certificates.

However, as with all amendments, care should be taken. The author's firm was recently involved in a case where, without legal input, the parties had deleted clause 8 in its entirety and signed the MOA but then failed to agree an addendum. By the delivery date there was still no agreement on the delivery documents and, whilst the seller was anxious to complete, the buyer was looking to exploit the situation and either get out of the deal altogether or extract a significant price reduction. The potential dispute was resolved through legal involvement; but it highlighted the danger of deferring agreement until a later date. This generally defers, rather than avoids, an argument and may, in fact, increase the chances of a full-blown dispute.

9.8.2 Local formalities

The documentary requirements of flag states differ widely and the buyer must be comfortable that it will be able to get all the documents it needs for registration from the seller, within the timescale and in the form required by the buyer's chosen flag state. Equally, the seller will want to know well in advance of completion what additional documentation, if any, it is required to give and, again, it is recommended that this be expressly set out in the MOA or an addendum.

Whilst the standard language of the NSF in respect of the bill of sale requires the seller to provide a bill of sale in the form 'required by the Buyers' Nominated Flag State' and this language is commonly mirrored in any addendum, it is sensible for the buyer to pre-clear the form of the bill of sale (including any necessary notarial certification) and any other applicable documents with its ship registry in advance of completion.

One of the documents that can cause a problem is the deletion certificate. This is issued by the seller's outgoing registry upon the deletion of the vessel from its flag and is required by the buyer's incoming registry. Not all registries issue their deletion certificate immediately following closing (an example is Panama); and whilst most are flexible and will, at least provisionally, register a vessel without a deletion certificate, some – such as the Canadian flag registry – simply will not. The author's firm recently sold a vessel from Panama to Canada, and this point took weeks to organise and required a complicated escrow arrangement to be put in place dealing with the release of funds and other delivery documentation upon final receipt of the deletion certificate from Panama.

9.8.3 Leave enough time

In some parts of the world it is difficult, if not even impossible, to obtain a useful notarial certificate or have a document legalised or apostilled. There are ways around this but these can sometimes take time. In addition, some registries are quicker than others when it comes to approving documents or issuing certificates; the parties should ensure that the laycan given in the MOA realistically gives them enough time to get everything they need done, including getting the right documents in original form to the place of closing in time for delivery. The seller should never tender notice of readiness until it is absolutely comfortable that it can satisfy its documentary obligations.

The author's firm was involved in a deal recently in which the laycan was impossibly short and its client, the buyer, had no prospect of getting its corporate authorities certified in accordance with the requirements of the MOA, let alone physically in the place of closing on time. Luckily, the seller was desperate and agreed to complete on the basis of virtually no paperwork from the buyer. However, it was under no obligation to be as flexible as this; the buyer could have found itself in a very awkward situation.

9.9 Logistics

Much of the role of lawyers in any transaction is to ensure clear lines of communication between all those involved and coordinate the (often complicated) logistics of closing, particularly when acting for the seller who will have much to achieve, normally in a fairly short space of time.

The parties to a deal and the vessel itself are frequently separated by several time zones which can result in a very short window each day within which to close. It is, therefore, essential that there are no delays in the closing process and that all those involved understand the time zones and their respective responsibilities.

9.9.1 Place of delivery

The MOA should set either a specific place of delivery or a range within which the vessel is to be delivered. If the latter, the parties will need to agree the actual place of delivery as soon as possible so arrangements can be made for taking over the vessel. The seller is contractually obliged under the standard NSF terms to deliver 'at a safe and accessible berth or anchorage', though it is, in any event, common sense that the vessel should be anchored or berthed somewhere with safe reliable access for the exchange of crews. However, the author has, in the past, been faced with situations where relations have broken down between the parties and the seller has refused to cooperate by bringing a vessel into shelter and so put lives at risk.

If acting for the seller, one must be cautious about agreeing to deliver a vessel in the buyer's home port. A little time ago, the author's firm was acting for the seller of a vessel who had agreed to deliver in the buyer's home port. Shortly after arrival the vessel was detained by port state control. It soon became apparent that the buyer, port state control and the local class representative had colluded to gain a commercial advantage for the buyer.

9.9.2 Notice of readiness (NOR)

Once the vessel is 'at the place of delivery and physically ready for delivery' the seller should give the buyer notice of readiness (known as the NOR). If the seller is in any doubt as to the vessel's physical condition or whether or not it can comply with its documentary obligations under the MOA, it should not tender NOR. Following the giving of NOR, the buyer has three banking days in which to take delivery. The seller may find itself in default if it fails to deliver when the buyer comes to take delivery and pay the purchase price. In such a volatile market, this is not a risk generally worth taking.

However, if the seller is in all respects ready and it suspects that the buyer is not or is otherwise playing games, the giving of NOR can be a useful means of flushing the buyer out as, if they fail to take delivery and pay the purchase price within the three banking days they are vulnerable to the seller terminating the MOA for buyer's default and retaining the deposit. In such circumstances, any seller would be well advised to wait until after midnight on the third banking day before making such a claim, or else risk being found to have repudiated the contract itself.

9.9.3 Seller's mortgage

Under the standard NSF terms, the buyer is obliged to pay only upon delivery of the vessel free from all encumbrances. It is clear, therefore, that the seller must discharge any mortgage in advance. In reality, very few sellers will be in a position to obtain the discharge of their mortgage ahead of payment by the buyer and, given the prevailing conditions in most ship finance banks, it is almost unheard-of for a bank to discharge its mortgage before it knows beyond all doubt that it is going to get paid in full. There is good reason for this, given that neither the seller nor its bank has any control over the buyer's actions; if, for whatever reason, the buyer fails to complete as planned, the bank having discharged its mortgage early would find itself as an unsecured creditor. That said, in Italy where it is understood that it is very challenging to obtain a simultaneous discharge of a mortgage at closing because of the processes involved, banks will, on occasion, discharge their mortgage in advance to facilitate a deal and, where possible, take alternative security from their borrower.

Nevertheless, the seller's obligation to deliver free and clear of all encumbrances remains; and the banks' current lack of flexibility can, on occasion, put their selling clients in pretty difficult positions. It is prudent to begin discussing the closing mechanism as soon as possible after the execution of the MOA (it can often take some time to settle and get the bank or banks comfortable with it). Once agreed, each party will need to ensure that its bank is suitably organised and that all appropriate steps have been taken to ensure efficient and timely closing. It is prudent to prepare a full memorandum setting out the closing mechanism but remember this is unlikely to be a binding document and serves only as an *aide-mémoire*.

9.10 Payment of the balance of the purchase price

In most sale and purchase transactions the buyer is required to make only two payments (three if they pay separately for bunkers and lubricating oils remaining on board). Nevertheless, it never ceases to amaze how often mistakes are made by buyers one way or another when making these payments, whether it be because they leave it too late to make payment, send payment to the wrong account (e.g. they credit the stakeholder's account rather than seller's account), fail to send enough, or forget about the deduction of banking charges en route. It must be remembered that unless the seller is prepared to take a lenient view, all amounts owed by the buyer under the MOA must arrive in full and in cleared funds in the designated account for the buyer to have discharged its obligations.

The prevailing currency in shipping remains the US dollar. However, any transfer of US dollars (no matter how routed) must clear through the New York banking system. This can only happen during banking hours on a working day in New York, and can take time. In addition, there is a significantly heightened risk of a US dollar payment being stopped for compliance reasons. It is becoming increasingly common for the innocent use of a name or word in a payment instruction to trigger an automated spot check. When this happens, the banks are unable to tell you whether or why a payment has been stopped and no amount of cajoling, threatening or otherwise will persuade them to do anything that might jeopardise their US licences. Whilst there is no way to avoid this, it is another reason to ensure payments are made well in advance. It has also been found that, if including a vessel's name in the payment instructions, it can help reduce the likelihood of such a stop if you also include its IMO number. In any case, always run a World-Check against the vessel's name before referring to it and, if in doubt limit the amount of information set out in the payment instructions, some prefer to avoid all mention of the vessel or its name.

9.10.1 Payment mechanisms

(a) The old-fashioned approach

The so called 'old-fashioned' approach is for the parties to convene a closing meeting, table and sign all necessary documents and, when ready, have the buyer instruct its bank to make the payment. Thereafter, none of the documents leave the meeting room until the seller receives cleared funds in full in its account. Of course, waiting for funds to clear, particularly given what has been said above, can be time-consuming and sometimes impracticable. If the funds are unexpectedly delayed, the parties need to have a plan in place for the treatment of the documents overnight until the funds are received. Furthermore, if delivery is delayed as a result of missing funds, the buyer is exposed to the risk of default under the MOA as well as further commercial risk with regard to the cost of its crew standing idle and any possible onwards charter it might have arranged.

(b) Pre-position funds?

It is not uncommon for a seller to demand that the buyer pay funds into its bank in advance of closing. This cannot be recommended to a buyer without a clear mechanism for the return of such funds if, for whatever reason, the deal does not complete. However, the effective pre-positioning of funds can greatly speed things up and simplify the closing process. Indeed, sometimes this is the only way to facilitate a difficult closing or one that is to be held outside of US or European banking hours. There are various methods by which funds can be pre-positioned and the buyer can be given comfort that the funds will be returned to it if needs be. Two such methods are as follows.

First, funds can be pre-positioned with the buyer's lawyers, to be held and released to the seller in accordance with the terms of a solicitor's undertaking. A solicitor, it should be noted, should only give an undertaking if it is within his power to perform the task required by it: a breach of an undertaking is a matter of professional misconduct. Indeed, the giving of undertakings is sufficiently serious to see most firms today restrict the right to give them only to partners. However, whilst a solicitor's undertaking represents a very effective means of closing, it is disappointing to note the increasing numbers of law firms advising their clients not to accept them.

Second, funds may be pre-positioned with the seller's bank to be held in a suspense account and released in accordance with the terms of an irrevocable payment instruction sent in the form of an MT199 SWIFT message. Provided the buyer is happy to do this, it can work well; however, in this age of extreme risk-aversion, banks are increasingly reluctant to act on MT199s and a number of well known ship finance banks will not accept them. Furthermore, it can be difficult to get banks who do not have a relationship to talk to one another and agree release terms.

PART 3

LEGAL AND PRACTICAL ISSUES RELATING TO SHIP FINANCE

CHAPTER 10

Ship mortgagees: enforcement and remedies

*Charles Buss**

Unprecedented numbers of shipowners have defaulted on their bank loans since the most recent global economic downturn began in 2008. Six years later, as the world economy makes a very slow and patchy recovery, shipowner loan defaults and insolvencies still continue. However, the arrest and lay-up of ships that was so prevalent in the immediate aftermath of the 2008 downturn, even converting the river Fal in Cornwall, whose sheltered location and deep draught makes it the ideal English location for mothballing ships, into a tourist attraction,[1] has largely given way to alternative ship mortgage enforcement methods. In this chapter, I shall consider the benefits of the Admiralty enforcement process in England and other common law jurisdictions and the extent to which these may have been somewhat eroded by recent developments in the law and by the increasing use of insolvency proceedings by shipowners in financial difficulties.

10.1 Introduction: in what circumstances do ship mortgagees arrest ships?

Ship mortgage enforcement is, for any lender, an option of last resort. If the borrower's default can be resolved by a consensual restructuring of the loan, which may involve an orderly open market sale process through brokers, typically contracted for by memorandum of agreement on Norwegian Saleform or similar terms, then that is infinitely preferable to a protracted and costly court process, involving a distressed sale at a knock-down price. Even if a restructuring is not viable, an MOA sale with the borrower's cooperation is surely preferable. Why then would a lender ever contemplate an arrest and judicial sale process? In practice, lenders will only do this in three scenarios.

The first is where the borrower is refusing to cooperate, hoping either that a market recovery is imminent or that he can pressurise the lender into accepting an unrealistic loan restructuring. This leaves the lender little option but to arrest the borrower's ships and put them through a court sale process. In recent years, many ship-owning businesses have evolved from small family concerns to large corporate conglomerates with many shareholders, raising significant capital and being listed on major stock exchanges. For their part, lenders have flocked to lend to these larger

* Partner, Watson, Farley & Williams LLP.

1 See, e.g., *Marooned on the Fal* (*The Guardian*, 31 January 2009); *Ghostship Creek* (*Daily Mail*, 19 August 2010).

owners. It follows that the scenario in which the individual beneficial owner of a family owned shipping business becomes locked in a battle to the death with his mortgagee lender is much rarer than it used to be as recently as 20 years ago.

Second, there is the situation where the borrower is unable to pay off or secure the claims of other maritime creditors who have already arrested his ship, leaving the mortgagee no option but to finance the ship's release or to accelerate the loan and join in the Admiralty process. Maritime credit is much tighter than it used to be, reflecting the fact that fewer credit lines are available to marine suppliers themselves, such as bunker suppliers. Again, therefore, it is much less frequent to see vessels arrested by a raft of trade creditors for debts running into millions of US dollars.

The third case is where the mortgagee believes that a court sale is required to give the purchaser 'clean' title. Unlike an MOA sale, a court sale will deliver the ship to the buyer free of all pre-existing liens and encumbrances.[2] This will remove the risk of the vessel being arrested in her new ownership for debts of the previous owner which would have survived an MOA sale. In practice, however, this risk is small. The only debts which survive an MOA sale are those attracting maritime liens, and those which do not but have already been protected by the issuance of *in rem* proceedings before the sale.

There should be no concern about maritime liens, at least under English law, where the outgoing crew has been paid their wages and any outstanding collision or salvage claims have been insured or bonded. Regard will have to be had, however, to foreign law in jurisdictions where the ship may trade globally post-sale. For example arrests are sometimes founded post-sale in states signatory to the 1926 Maritime Liens and Mortgages Convention,[3] which gives maritime lien status to 'claims resulting from contracts entered into or acts done by the master, acting within the scope of his authority, away from the vessel's home port, where such contracts or acts are necessary for the preservation of the vessel or continuance of the voyage'[4] where the claim has been brought within six months of its having arisen[5]. Such arrests are nevertheless usually speculative and can often be successfully challenged, even if they may cause nuisance in the process.

Nor should there be concern about *in rem* proceedings as long as searches have been carried out in appropriate court registries to verify whether any protective writs have been issued. For this reason, pre-completion court searches are not uncommonly made in England, Gibraltar, Singapore and Hong Kong. However, even though the risk of arrest post-sale for debts of the seller can be to some extent contained, interested buyers dealing with mortgagees not uncommonly insist on purchasing through a court bill of sale where they perceive the seller to be in serious financial difficulties. In these circumstances, the usual MOA warranty as to freedom from liens and encumbrances given by the seller[6] will be worthless and mortgagees are, unsurprisingly, unwilling to give open-ended guarantees to MOA buyers of their defaulting borrowers' warranty.

---

2 *The Acrux* [1962] 1 Lloyd's Rep 405.

3 See *League of Nations Treaty Series*, vol. 120, p. 187 for the text.

4 Art. 2(5) of the 1926 Convention.

5 Art. 9 of the 1926 Convention.

6 For example that in cl. 9 of Norwegian Saleform 2012.

It can be seen from the above that the circumstances in which a mortgagee will need to arrest its borrower's vessel are becoming less common, even in a time of significant economic difficulty.

10.2 The use of self-help remedies

There may also be circumstances in which the mortgagee prefers to deploy other weapons which it is likely to hold in the armoury of its security package, if its ship mortgage is governed by English law or the law of other common law jurisdictions, including the USA and countries with historic ties to the USA, such as Liberia, Panama and the Marshall Islands. So, for example, the mortgage will typically contain a right to take possession of the vessel and to sell her as mortgagee in possession, together with a power of attorney, by which the owners irrevocably appoint the mortgagee as their attorney to sell the vessel on their behalf. Further, shareholders may have pledged their shares in the registered owner to the mortgagee. In practice, mortgagees will on occasion enforce their security by exercising their powers of sale or enforcing their share pledges. For example, if the owners' bankruptcy is feared, these rights, which may, at least in theory, be quickly enforced, may succeed in removing the mortgaged asset from the bankrupt estate before any such process commences, bringing with it the inevitable costs and delays. Enforcement of a share pledge may also, by transferring ownership of the asset at corporate level, preserve a valuable long-term time charter commitment, which might otherwise be lost if the asset itself were sold. This is because, in the absence of express words in the charter to the contrary, a time charterer may treat a change of ownership of the vessel, but not, usually, a change of ownership of the registered owning company,[7] as a repudiatory breach of charter.[8] However, these contractual rights of the mortgagee exist on paper only and will not work to deliver a ship into the physical possession of the mortgagee or its intended buyer if the shipowner resists their exercise. If the owner does resist, and the mortgagee finds itself having to arrest a vessel to recover possession, then the time advantages offered by these self-help remedies will quickly be lost and the mortgagee may simply find itself embroiled in litigation in an unfavourable foreign jurisdiction. In practice, where the mortgaged vessel is managed by professional third-party ship managers, they can usually be relied on to follow a mortgagee's instructions, but ship managers associated with the owners will probably be less compliant.

10.3 The procedural advantages offered by English and common law jurisdiction for arrest and sale

If arrest and judicial sale is the mortgagee's preferred or only option, the mortgagee may then have to decide in which jurisdiction(s) to strike. In this regard,

7 Sometimes, charters expressly restrict changes in beneficial ownership. For example, the Shelltime 4 (2003) printed form prohibits changes in '*ownership structure*' throughout the charter period (cl. 1(i)), ll. 26–27).

8 See *Omnium d'Entreprises v Sutherland* [1919] 1 KB 618 (CA) (sale); *Fratelli Sorrentino v Buerger* [1915] 3 KB 367 (CA) (restructuring). These are, however, decisions which the authors of *Time Charters* (6th edn, Informa, 2008) at para. 3.13 consider to be outmoded.

the mortgagee may have rather more scope for forum-shopping than may at first appear, whether because he can track the mortgaged ships over a period of weeks or months until they call at ports in favourable jurisdictions for arrest and sale, or because he has the ability to take control and direct the vessels (or otherwise persuade the owners to sail them) to such favourable jurisdictions. To the extent that forum-shopping is possible, English common law jurisdictions, in particular, England, Gibraltar, Hong Kong and Singapore, have, over many years, earned a kind of most-favoured-nation status amongst mortgagees, although, as I shall discuss, a number of recent legal developments have begun to dent their reputation.

The main procedural advantages of these English common law jurisdictions are as follows.

10.3.1 The possibility of sale 'pendente lite'

First, these jurisdictions entitle the mortgagee to obtain an order for sale before judgment, even where the owner contests the claim. In England, at any time after the arrest, the claimant may apply for the vessel to be appraised and sold.[9] The claimant may even apply before the 14-day period for acknowledging service of the claim form has expired.[10] The court will not make an order for appraisement and sale 'except for good reason'.[11] A 'good reason' is usually that the vessel is a wasting asset, in other words that 'unless a sale is ordered, heavy and continuing costs of maintaining the arrest will be incurred over a long period, with consequent substantial diminution in the value of the plaintiffs' security for their claim'.[12] Accordingly, if there is no equity in the vessel and the vessel's outgoings are not being met by the shipowner (especially, if the shipowner has ceased trading or is insolvent), the court, in most cases, will take the view that the vessel is a wasting asset and that a prompt sale is the only sensible solution. In practice, by the time a mortgagee accelerates its loan, there simply will be little or no equity in the vessel.

The court may make an order for sale *pendente lite* even if opposed by the defendant shipowner, although it will examine the merits more critically in such a situation than it would do where the mortgagee has obtained judgment.

By contrast, non-common law jurisdictions, including almost all continental European jurisdictions, invariably have codified rules requiring the mortgagee to obtain judgment or 'executory title' on its claim before the court will order a sale of the ship. This allows owners to cause substantial procedural delay by advancing nuisance defences, which must then be tried, and to apply pressure on the lender to agree to an unfavourable loan restructuring or release of other security, for example a personal guarantee. The advantage to a mortgagee of obtaining an order for sale *pendente lite* therefore cannot be over-estimated in the hostile enforcement scenario. The costs of maintaining a vessel in the period between arrest and auction, including berthing charges, bunkers, crew wages, insurance and supplies, which the

9 Civil Procedure Rules (CPR) 1999, Part 61.10(1).
10 See CPR, 61.10PD, para. 9.3.
11 *The Myrto (No. 1)* [1977] 2 Lloyd's Rep 243, 260 (Brandon J).
12 *Ibid.*

mortgagee will have to pay or reimburse to the court, can be crippling and can significantly erode the mortgagee's recovery.

Second, even if the shipowner does cooperate in the foreclosure, cargo interests and other maritime creditors may not. Any kind of litigation is likely to delay the process of sale and distribution of the fund. Common law jurisdictions, which have specialist Admiralty judges, drawn from the senior ranks of the profession, applying judicial precedents which often give highly detailed guidance, usually produce more certain rulings, and much more quickly, than inexperienced judges in local courts in jurisdictions whose codified laws contain little or no guidance on the sometimes complex issues that tend to arise in this context. For the mortgagee, as the vessel will usually be a wasting asset, every day of delay until the vessel has been sold and converted into cash is likely to be very costly. Berthing charges alone may cost tens of thousands of US dollars per day.

10.3.2 The possibility of court-approved direct sales

Another of the great advantages offered by common law jurisdictions, at least until very recently, was the possibility of the court approving a 'direct' sale, without the need for a formal auction, to a specific buyer (often another customer of the mortgagee, whom the mortgagee then finances to purchase the ship). This fast-track procedure obviates the need for physical appraisal of the vessel to establish a reserve price, advertisement of the auction in the press, marketing through the court's brokers and inspections by bidders, and typically reduces the time from arrest to sale from two to three months to two to three weeks.

Until very recently, the majority of judicial sales in England were direct sales to a specific buyer (typically financed by the mortgagee), for a specific price, without auction. These sales have been given various names, from 'fast-track' sales to '*ex parte*' or 'private treaty' sales, the private treaty in such cases being with the court, rather than the owners or the mortgagee. Not only did such sales save lenders hundreds of thousands of dollars in ship operating expenses, but they also avoided liability for the court-appointed brokers' fees (usually 1 per cent of the purchase price). Court sales without auction accounted for 23 of 31 mortgagee sales in England in 1999–2000.

In England, the Admiralty Marshal has for many years had the flexibility to sell either by public auction or by private treaty.[13] Whichever way the court sells the ship, the Admiralty Marshal's bill of sale delivers the ship to that purchaser free of pre-existing liens and encumbrances.

To avoid an auction, the mortgagee usually had to satisfy the court that the intended purchaser was unrelated to the defendant owner and that the sale was for a purchase price at least as high as a 'willing seller, willing buyer' price for the ship, if sold on the open market. Mortgagees were often amenable to financing above-market purchases as it permitted them to roll forward all or a substantial part of the old borrower's loan debt (which very often exceeded the ship's open market value) into a new loan to the purchaser. To support its application, the mortgagee

13 See, e.g., *The Everilda* (1927) 29 Ll LR 238.

would adduce in evidence two or more desk valuations from reputable shipbrokers to support the price offered. The court was amenable to such applications on the basis that judicial auctions tend to achieve depressed 'forced sale' prices, such that a sale for a full market price without the need for an auction benefited all maritime claimants by swelling the fund and reducing priority arrest and sale costs.

Not all English common law jurisdictions have adopted this approach. In Hong Kong, for example, as long ago as 1998, the court held in *The Margo L*[14] that there must exist '*powerful special features*' to justify such an order. In that case, the judge, Waung J, was sceptical of the evidential value of the three short brokers' desk valuations relied on by the bank and, further, was not persuaded that the vessel might not be sold at auction for 'considerably more' than the US$3.4 million offered, despite that offer exceeding the highest of the three valuations.

Direct sales continued thereafter to be approved in England and Singapore until 2013, when the courts of both these countries delivered judgments whose effect has been to bring the practice almost to a grinding halt. In England, in *The Union Gold*,[15] Teare J ruled that a judicial sale would only be ordered to a direct buyer without appraisement and auction if there were 'special circumstances' to justify such an order. An example of 'special circumstances' given by the judge would be where it was shown that a delay in the sale would cause the loss of a particularly valuable employment fixture.

Like Waung J before him, Teare J was clearly uncomfortable with the brokers' valuation evidence supporting the mortgagee bank's application. Indeed, in *The Union Gold*, which concerned four small cargo ships, the Admiralty Marshal's own brokers acknowledged that they had little experience of valuing ships of this kind. The various brokers' valuations obtained by the bank varied quite considerably. Further, Teare J expressed doubts that a court auction would realise a lower price than a private sale through brokers, given that the court could give the purchaser title free of liens and encumbrances, unlike a shipowner, especially one in financial difficulty. The judge also noted that, where a bank sold a ship under its 'power of sale' contained in its mortgage, it would usually be unwilling to give the purchaser any warranty that the ship was free of liens and encumbrances.

Ultimately, the judge was unconvinced that the cost of the short delay involved in appraisal and advertising an auction outweighed the risk of a 'direct' sale not achieving the best possible price. The judge relied on dicta in a number of older cases which emphasised that (i) the Admiralty Marshal should decide the timing of an auction[16] even if this meant further costs being incurred to maintain the ship and (ii) the court had previously disapproved of parties marketing ships for private sale, even if done under the Admiralty Marshal's supervision, once an order for appraisal and sale had been made.[17]

The judge was less perturbed that the bank intended to finance a 'newco' owned by the former managing director of the defaulting borrower to purchase one of the

14 [1998] 1 HKC 217

15 *Bank of Scotland Plc v The Owners of the Union Gold* [2013] EWHC 1696 (Admlty); [2014] 1 Lloyd's Rep 53.

16 See *The Halcyon Isle* [1975] 1 Lloyd's Rep 525.

17 See *The APJ Shalin* [1991] 2 Lloyd's Rep 62.

four ships. That was the only direct sale which the judge ordered, given that there was evidence, in that case, that the ship would lose a valuable employment contract if she were not promptly sold.

In Singapore, in *The Turtle Bay*,[18] Belinda Ang Saw Ean J followed and approved *The Margo L* and *Union Gold* decisions. Later, in *The Sea Urchin*,[19] the same judge declined to order the direct sale of a vessel laden with soya beans, despite the fact that the intended purchaser had agreed to deliver them to their intended Chinese destination, saving US$700,000 in transhipment costs. She did not accept that this saving would benefit the general body of creditors, given that the transhipment costs would be for cargo interests' account, and not a sheriff's expense. She was also unimpressed by the short-form desk valuation of a Greek broker relied on by the bank and by the absence of a formal, written offer by the intended purchaser, which was neither a party to nor intervener in the proceedings, on the sheriff's terms and conditions of sale. She also referred to the Canadian case of *The Nel*,[20] in which the court had ordered the direct sale of a vessel laden with a bulk sulphur cargo, and noted that, unlike in the case before her, in *The Nel* the costs of discharge had been relatively prohibitive – US$1.3m as against a ship value of US$4m; worse still, there had been a risk of the cargo corroding the ship's holds unless shipped and discharged quickly.

Whilst the judges in each of the above cases may have raised some legitimate concerns, their judgments make no reference to the fact that, in such cases, it will make no difference whatever to maritime creditors if an auction delivers a higher price. In *The Margo L*, the judge had already given judgment to the mortgagee in the sum of US$12 million, over three times the sum offered by the direct purchaser. Likewise, in *The Sea Urchin*, although not apparent from the judgment, the mortgagee's claim exceeded the price offered several times over. So even if an auction produces a higher price, the mortgagee, not the other creditors, will invariably scoop the benefit. However, if the auction produces a lower price, the additional cost and delay involved will invariably fall on the bank, not the other creditors.

To conclude, it seems that a number of judges have been irritated by the cursory brokers' valuation evidence presented to them and have, at least for the time being, closed off this useful enforcement procedure route. Unfortunately, banks and their legal advisers may, over the years, have become rather too complacent about direct sale applications. Judges have always been concerned to obtain the best possible price for a ship and whilst in some cases it is relatively straightforward to identify a ship's open market value by reference to recent, comparable sales, in others, it is not so simple. Moreover, whilst brokers' desk valuations may routinely be provided to and relied on by banks, they hardly qualify as expert evidence to be relied on by a court. Had these judges been presented with fully reasoned expert valuation reports, supported by evidence of precedent sales, they might have made different decisions. A further problem is that, whilst it may appear to judges that the auctions they order often achieve respectable prices, practitioners know that this is rarely the case unless

18 [2013] SGHC 165; [2013] 4 SLR 615.
19 [2014] SGHC 24; [2014] 2 SLR 646.
20 (1997) 140 FTR 271.

the mortgagee has financed an offer by a direct sale buyer or an auction bid with a view to boosting the price and recouping more of its debt.

In the meantime, other countries are catching up with the fast-track procedure that England has now all but abandoned. In Malta, for example, court sales without auction are now available to mortgagees under the Code of Organisation and Civil Procedure. The effect of such a sale is to give clean title to the purchaser.[21]

All in all, the near loss of the direct sale procedure will probably have less practical impact than feared and will not suffice of itself to remove the competitive edge of England and Singapore for judicial sales of ships. In practice, where a mortgagee arrests a ship, the process of identifying, and agreeing a financing package with, a potential purchaser, of replacing or re-engaging the crew and of conducting repairs to make the ship ready for future trading usually takes several weeks, such that an auction timetable will usually cause no, or no significant, overall loss of time. Further, in the current market, many mortgagees are exiting the shipping market for good and are content to rely on an auction process to realise their security, rather than to finance a new owner to purchase the ship. Finally, the ability of mortgagee banks to finance purchases at 'above market' levels has been considerably curtailed by an increasingly strict regulatory environment in which such transactions must be accorded a high risk weighting and may even find their way onto the bank's balance sheet.

10.4 The position of the mortgagee towards cargo interests

Whenever a mortgagee arrests a ship, the mortgagee may discover that she is laden and not in her discharge port, either because she is loading, taking on board bunkers or conducting repairs, effecting a crew change or discharging at one of several intermediate discharge ports. Container ships, for example, usually remain laden at all times, even if the number of empty boxes on board may vary considerably. To some degree, a mortgagee may be able to surmise whether the ship is laden, by reference to the ship's trading pattern, or to what the shipowner may have informed the mortgagee, or by monitoring freights received into an account which the mortgagee operates. Sometimes it may be possible to ascertain through investigative agents whether the ship is laden, although this may take time and risks information leaking back to the defaulting shipowner.

Even if the ship is laden or believed to be laden, the mortgagee may want to arrest the vessel in a jurisdiction perceived to favour its interests more than the jurisdiction in which the ship is due to discharge. However, before taking a decision whether to arrest the laden vessel, the mortgagee will need to have regard to (a) its potential exposure in tort to cargo owners; and (ii) its potential liability to bear the costs of cargo discharge. I shall consider these issues in turn.

10.4.1 Tortious liability of a ship mortgagee

If the mortgagee arrests a laden vessel, can cargo owners claim damages from the mortgagee on the grounds that, by arresting the vessel, the mortgagee tortiously

21 COCP, s. 3064. See *Lloyd's Maritime and Commercial Law Quarterly* [2012] at p.356 where this new procedure is discussed in detail.

interfered with their contractual rights under the contract of carriage, to which contract the mortgagee is a stranger?

In *The Myrto (No. 1)*,[22] the court held that the mortgagee was entitled to exercise its rights under the mortgage without regard to any contract made by the owner with a third party for the employment of the ship, where either (a) the contract of carriage was of such a kind, and/or made or performable in such circumstances, that the security of the mortgagee was impaired; or (b) the owner was unwilling and/or unable to perform the contract of carriage. *The Myrto* was later confirmed by Christopher Clarke J. in *Anton Dürbeck GmbH v Den Norske Bank ASA*.[23]

Impairment of security is a question of fact[24] which can be shown in a number of ways:

1. It may be clear that if the voyage continues, the ship will be at risk of arrest by other creditors in a jurisdiction with a less favourable law of priorities than English law: '. . . if the ship should be arrested by any creditor, the others would be driven to protect their own interests by entering caveats against release or otherwise. If this happened during the voyage at a foreign port, where the law relating to priorities might be different from English law, the bank's security would clearly be endangered'.[25] So in the Hong Kong case of *The CC Orient*,[26] Waung J found nothing inequitable in the mortgagee's conduct where, having taken possession of a vessel in Japan laden with cargo bound for the Philippines, it directed cargo interests to discharge it there, before sailing the vessel to Hong Kong for judicial sale. The judge refused to confer priority on cargo interests' claim in the circumstances.
2. It may be that 'the evidence shows that the owner is impecunious, that he can only perform the voyage to which the contract relates, if at all, on credit, and that the ship is subject to pressing liabilities and charges'.[27]
3. It is possible that if the voyage continued, priority maritime liens would accrue, because, for example, the owner was not paying his crew, even if the crew was not then pressing for immediate payment.[28]
4. It may also be that if the voyage continued, the ship would be uninsured for P&I risks.[29]

Inability of the owners to perform the contract of carriage is also a question of fact.[30] It may be shown, for example, by evidence that the owners are unable to bond or to pay off other arrests;[31] or that they or cannot fund the costs of the voyage from earnings (which may extend to the situation where the earnings are not available to

22 [1977] 2 Lloyd's Rep 243. See esp at 254 (Brandon J).
23 [2005] EWHC 2497 (Comm); [2006] 1 Lloyd's Rep 93.
24 See *The Myrto (No. 1)* [1977] 2 Lloyd's Rep 243, 254 (Brandon J).
25 *Ibid.* at 258 (Brandon J).
26 [1985] HKC 115.
27 *The Myrto (No. 1)* [1977] 2 Lloyd's Rep 243, 254 (Brandon J).
28 *Ibid.* at 256 (Brandon J).
29 See *Anton Dürbeck GmbH v Den Norske Bank ASA* [2005] EWHC 2497 (Comm); [2006] 1 Lloyd's Rep 93 at [79] (Christopher Clarke J).
30 *The Myrto (No. 1)* [1977] 2 Lloyd's Rep 243, 254 (Brandon J).
31 *Ibid.* at 258.

the owner simply because the mortgagee has applied them to the loan debt under its assignment).[32]

In *The Myrto (No. 2)*,[33] the vessel was initially arrested by bunker suppliers, whose claim the owners were unable to pay or to bond. Sheen J concluded that the loading of multiple cargo parcels without any stowage plan indicated that the owners had no intention of completing the voyage but rather of pocketing freights, which they hid from the mortgagee, even presenting a forged time charter to the mortgagee, before engineering the loss of the vessel on a winter voyage through the Bay of Biscay in order to make a substantial insurance claim.[34] The owner's misconduct in *The Myrto* was egregious. Usually it will only be the mortgagee's arrest which will prevent the vessel from continuing on her voyage. Even if other creditors have filed claims or issued cautions against release, it is often only as a result of their learning of the mortgagee's arrest. *The Myrto* indicates that the mortgagee cannot safely rely on its own arrest but only on other arrests which have occurred and would have occurred despite the mortgagee's claim and which the ship owner is unwilling or unable to bond or to secure. Accordingly, this leg of the *Myrto* criteria would appear to be difficult for a mortgagee to satisfy in most foreclosure situations.

Since the decision in *Anton Dürbeck*, the law of economic torts was comprehensively reviewed and revised by the House of Lords in *OBG Ltd v Allan*.[35] In that case, the House of Lords made clear that there exist two economic torts.

The first tort is committed by a third party who induces a party to a contract to breach its contract with its contracting counterparty. This might in theory apply if a mortgagee (for example) persuaded a borrower to sail its vessel to a favourable arrest port, breaching a charter in the process. For liability to accrue, the mortgagee would have to be shown to have intended to induce its borrower to breach its contract with the charterer. It would not be necessary to show that the bank had used unlawful means. However, crucially, it was also made clear in the *OBG* case that a party was allowed to interfere with a third party's contract in order to protect an equal or superior right of his own.[36] The House of Lords cited with approval an earlier (land) mortgage case, namely *Edwin Hill & Partners v First National Finance Corporation Plc*.[37] Here, a bank had a mortgage over some freehold property as security for repayment of a loan. The owner could not repay the loan or fund a development for which it had engaged a firm of architects. The bank agreed to advance further funds on condition that the architects were removed from the intended development. The Court of Appeal held that the bank, which could have brought the development (and therefore the architects' contract) to an end by exercising its power of sale or appointing a receiver, was justified in procuring a breach of the architect's contract in defence and protection of its equal or superior right. Put another way, the alleged tort assumed a positive obligation on the bank to fund the development.

32 *Ibid.* at 259.
33 [1984] 2 Lloyd's Rep 341.
34 See *The Myrto (No. 2)* [1984] 2 Lloyd's Rep 341, 347.
35 [2007] UKHL 21; [2008] 1 AC 1.
36 [2007] UKHL 21; [2008] 1 AC 1 at [193] (Lord Nicholls).
37 [1989] 1 WLR 225.

The second tort is where a party's unlawful acts prevent another party from performing its contract. Unlawful means is a vital ingredient of this tort. It is not unlawful for a mortgagee to arrest a vessel under an accrued legal and contractual right to do so. In *OBG v Allan*, Lord Nicholls emphasised that for these purposes 'unlawful means' encompassed only acts not permitted under the civil and criminal law.[38]

However, in *Anton Dürbeck*, Christopher Clarke J held that *The Myrto* criteria were still good law as regards ship mortgages, and distinguished *The Myrto* from *Edwin Hill* on the basis that a ship mortgagor, unlike a real property mortgagor who could not bind his mortgagee by a lease, could bind his mortgagee by a charterparty.[39]

It remains for a mortgagee to challenge the decision in *Anton Dürbeck* in reliance on the House of Lords' decision in *OBG v Allan*. Until then, however, *Anton Dürbeck* remains good law, and the exposure of a ship mortgagee to tortious liability is, somewhat anomalously, greater than that of all other *in rem* claimants, whose liability would be determined under the principles in *OBG v Allan*, despite their lower ranking (with the exception of maritime lien holders) to the mortgagee. Indeed, *in rem* trade creditors often prefer to arrest vessels when laden to increase the pressure on the shipowner to settle their claims.

It is submitted that, whether or not mortgages of real property and ships can be distinguished on the grounds cited by the judge in *Anton Dürbeck*, the *Edwin Hill* test should be applied to ship mortgagees. Any charterer or cargo owner is aware of, and surely accepts, the risk of the contract of carriage not being performed where the owner is unable to pay his debts, which must surely include those owing to the mortgagee. Likewise, any charterer or cargo owner could, if he has any concern, seek and obtain comfort or quiet enjoyment from the mortgagee. It is surely anomalous for a ship mortgagee to be in a worse position than not only lower ranking *in rem* claimants but also all other tortfeasors. It will be of little comfort to a mortgagee held liable in tort to a cargo owner that the mortgagee is not also liable to the mortgagor shipowner or in breach of the loan agreement.[40]

In assessing mortgagee risk, reference must be made to both the law of the arrest court and the law of the country of domicile of the mortgagee and perhaps other jurisdictions in which the mortgagee has assets or a presence and may be sued because different countries decide the applicable law of a tort under different rules. It may be by reference to where the harmful event occurred (*lex loci delicti commissi*) or where the damage occurred (*lex loci damni*) or the place with which the tort is most closely connected.

In England, until 2009, the applicable law of a tort was the law of the place in which 'the significant elements of [the events constituting the tort] occurred'.[41] In *Anton Dürbeck*,[42] a Norwegian mortgagee, acting through its London branch, arrested a ship, the *Tropical Reefer*, in Panama when laden with bananas, which subsequently rotted and became a constructive total loss. The mortgagee was then sued

38 [2007] UKHL 21; [2008] 1 AC 1 at [162].
39 See [2005] EWHC 2497 (Comm); [2006] 1 Lloyd's Rep 93 at [64].
40 See the comment of Longmore LJ in *The Tropical Reefer* [2004] 1 Lloyd's Rep 1, 7.
41 Private International Law (Miscellaneous Provisions) Act 1995, s. 11(2)(c).
42 [2005] EWHC 2497 (Comm); [2006] 1 Lloyd's Rep 93.

in England by the cargo owners. The court held that Panamanian law governed any tortious liability because Panama was where the significant elements of the tort had occurred (rather than England, where the mortgagee had given the instruction for the arrest to take place).[43]

For proceedings commenced after 11 January 2009, the law of a tort is presumptively 'the law of the country in which the damage occurs'.[44] If the claimant and mortgagee are both domiciled in the same EU member state, the law of that state – *lex domicilii* – will apply.[45]

Civil law jurisdictions may be similar to Panama as regards economic torts. In *Anton Dürbeck*, it was held that for the mortgagee to be liable under Panamanian law, they would have had to have arrested the vessel intentionally to harm the cargo interests/charterers or otherwise to have arrested in bad faith.[46] Neither criterion was made out on the facts where the mortgagee had a clear right of arrest (*qui iure suo utitur nemini facit injuriam*).

The cargo owner can bring a claim *in rem* against the arrested vessel, for the owner's breach of the contract of carriage, but such a claim will almost certainly rank below the mortgagee in priority.

The cargo owner may, therefore, prefer to bring an action *in personam* against the mortgagee in tort, either in the mortgagee's home jurisdiction or possibly elsewhere, for example, where the claimant can show that he has suffered loss.

In *Anton Dürbeck*, a German fruit importer and bill of lading holder, who had paid some US$1.2m for the cargo, sued the mortgagee in England, relying on the Brussels I Regulation[47] which provides that a defendant to a tort action may be sued not only in his country of domicile[48] but also in the courts were the 'harmful event' occurred[49] and, where the dispute arises out of the operations of a branch, in the place where the branch is situated.[50] In proceedings prior to those mentioned above, the judge found that the 'harmful event' had occurred in Panama.[51] It is certainly possible to imagine courts in other countries holding that the harmful event occurred in the claimant's home jurisdiction, where he will typically suffer financial loss. On appeal, the Court of Appeal held that the dispute did arise out of the operation of the branch, and so the mortgagee could be sued in England, although the court stayed the proceedings in favour of those in Panama.[52]

10.4.2 Liability of a ship mortgagee to cargo discharge costs

Where a ship is under arrest, and persons interested in the ship or cargo wish to discharge the cargo, they may request the Admiralty Marshal to authorise steps to

43 See [2005] EWHC 2497 (Comm); [2006] 1 Lloyd's Rep 93 at [32] (Christopher Clarke J).
44 See EC Regulation 864/2007 (hereafter 'Rome II Regulation'), art. 4.
45 *Ibid.*, art. 4(2).
46 [2005] EWHC 2497 (Comm); [2006] 1 Lloyd's Rep 93 at [33]–[49].
47 Alias EC Regulation 44/2001.
48 *Ibid.*, art. 2.
49 *Ibid.*, art. 5(3).
50 *Ibid.*, art. 5(5).
51 See [2002] EWHC 1173 (Comm).
52 See *Anton Dürbeck GmbH v Den Norske Bank ASA* [2003] 2 All ER 411 (CA).

discharge it.[53] For this purpose, a mortgagee is a person interested in the ship.[54] The Marshal will then apply, if he considers the request reasonable and the party making the request has given the Admiralty Marshal an acceptable undertaking to pay his fees and expenses on demand, for an order to permit discharge.[55] The Admiralty Marshal will then arrange for discharge. If there are no crane facilities to discharge the cargo or no storage facilities, the vessel may have to be moved under arrest to a different port.[56]

Who bears the costs of discharge? Under English law, the cargo owners will be responsible provided that *The Myrto* criteria have been met: '. . . when a shipowner is unable to perform a contract of carriage the owner of the cargo laden in his ship is entitled to take his cargo out of that ship at his own expense or abandon the cargo'.[57] If the mortgagee has put the Admiralty Marshal in funds to discharge the cargo, the mortgagee is entitled to reimbursement by cargo interests, together with interest.[58] If cargo interests wish to take their cargo away upon discharge, they should either pay the discharge costs there and then or provide appropriate security.[59]

The rationale is that, if the costs of discharge were treated as an Admiralty Marshal's expense, the cargo owners' claim for damages against owners for failing to discharge the cargo would acquire a priority over other claims to which they are not entitled.[60] Further, 'when freight is paid in advance the cargo-owners accept the commercial risk that the voyage will not be completed for whatever reason',[61] thus leaving them either to bear the loss if the contract of carriage is frustrated[62] or to claim it from the carrier if the contract of carriage is repudiated.

Alternatively, if the cargo owners abandon the cargo, the Admiralty Marshal can sell the cargo and recover his expenses from the sale proceeds.[63]

If the mortgagee arrests the vessel, and this alone prevents the owner from performing the contract of carriage, should the cost of cargo discharge still fall on cargo interests? In *The Myrto*, as already discussed *supra*, the owners were found to have been unable to perform the contract of carriage. It is submitted that the outcome would be no different where the court was satisfied that continuance of the voyage would impair the mortgagee's security. The test, it is submitted, is the corollary of that for tortious interference (see *supra*). Again, it is submitted that the *Edwin Hill* criteria, rather than *The Myrto* criteria, should be applied in this situation.

The position is the same in Singapore,[64] Hong Kong[65] and Canada.[66] By contrast,

53 CPR, 61.8(8).

54 *The Gulf Venture* [1985] 1 Lloyd's Rep 131.

55 CPR, 61.8(9)

56 See *The Myrto (No. 2)* [1984] 2 Lloyd's Rep 341, 348 (Sheen J).

57 *Ibid.* at 352.

58 *Ibid.* at 354. Interest was awarded at 12 per cent p.a.

59 See the CA decision in *The Myrto (No. 1)* [1978] 1 Lloyd's Rep 11, 15 (Roskill LJ).

60 *The Myrto (No. 2)* [1984] 2 Lloyd's Rep 341, 352 (Sheen J).

61 *The Jogoo* [1981] 1 WLR 1376, 1380 (Sheen J).

62 The Law Reform (Frustrated Contracts) Act 1943 does not apply to contracts for the carriage of goods by sea – see s. 2(5)(a).

63 See N. Meeson & J. Kimbell, *Admiralty Jurisdiction and Practice* (4th edn, Informa, 2011), para. 4.68; see also *The Myrto (No. 2)* [1984] 2 Lloyd's Rep 341, 352 (Sheen J).

64 *The Nagaski Spirit* [1994] 2 SLR 621 at [30].

65 *The Mingren Development* [1979] HKCU 19

66 *The MCP Altona* (2013) 225 ACWS (3d) 292

US courts cargo rank discharge costs above a mortgage claim as priority *custodia legis* expenses on the basis that they benefit the fund by preserving and maintaining the vessel whilst under arrest.[67] In most jurisdictions, the courts are simply unprepared to make orders compelling cargo interests to discharge their cargo in a timescale which is of use to the mortgagee.

10.5 The cost of a judicial sale

England has certainly lost some ground to the Netherlands as the jurisdiction of choice in northern Europe. Although England still retains the advantage of the *pendente lite* sale, which is not available in the Netherlands when a ship owner defends the claim, the Netherlands arguably outscores England in two respects.

First, in England, the purchaser of a ship must pay the purchase price into court.[68] It is not possible for the mortgagee to 'bid its debt', that is to set off the amount of its claim against the purchase price or provide a guarantee to cover priority claims. Accordingly, a mortgagee which finances the purchaser must effectively 'double fund' its debt, having to outlay new cash before it may recoup and apply the funds in court, usually several weeks later, to repay its old loan.

By contrast, in the Netherlands, a buyer may secure the purchase price by way of providing an on-demand guarantee rather than paying cash down.[69] This is particularly attractive to a mortgagee which has financed the purchaser.

Whether the English court might ever be persuaded to accept a bank guarantee from a mortgagee which has funded the successful bid remains to be tested. In Gibraltar, the Admiralty Marshal currently accepts bank letters of undertaking for the 90 per cent balance of the purchase price where the mortgagee has funded the successful bid and where he is satisfied (on a case-by-case basis) with the mortgagee bank's credit. In England, the Court of Appeal has recently accepted in *The Atlantik Confidence*[70] that a limitation fund could be validly constituted by a P&I Club guarantee, despite contrary indications in the rules of court and long-established practice, although the court there relied heavily on the fact that guarantees were expressly contemplated under Article 11.2 of the 1976 Convention. It is submitted that there is no good or fundamental reason why the purchase price for a vessel should not be guaranteed rather than paid into court. When a ship is arrested, the court is empowered to, and invariably will, order security in the form of a bank guarantee or P&I Club LOU to secure the claim of an *in rem* claimant.[71] Where a ship is judicially sold, although the sale proceeds will constitute a fund available to all *in rem* claimants, and not just one, the reality is that the mortgagee will invariably be the party entitled to those proceeds. If the ship is sold for more than the mortgagee's debt, then there is no reason why the mortgagee could not simply pay the surplus in cash and guarantee the rest.

In other jurisdictions, credit bidding is also permitted. For example, in the USA, a

67 *The Emilia*, 1963 AMC 1447, quoted by Sheen J in *The Myrto (No. 2)* at 350.
68 CPR, 61.10(5); and CPR, 61PD, para. 9.5.
69 See art. 577 of the Dutch Code of Civil Procedure.
70 [2014] EWCA Civ 217; [2014] 1 Lloyd's Rep 586.
71 CPR, 61.5(10).

mortgagee bank may 'credit bid' where it is buying the ship in through a nominee.[72] The bidding bank will simply be required to put up a bond for the amount of other claims against the ship. As another example, in South Africa, the court's Conditions of Sale permit the 90 per cent balance of the purchase price to be paid in cash or secured by bond, letter of credit or other acceptable security.

Secondly, whilst the court and brokerage fees payable in England are rather more competitive than those of many other jurisdictions, they exceed those payable in the Netherlands.

In England, the mortgagee will have to pay *ad valorem* court fees of approximately 0.5 per cent of the sale price[73] together with a commission of approximately 1 per cent of the sale price to the Marshal's brokers. In Gibraltar, the court fees and brokerage are approximately 1 per cent. In Singapore, a 2.5 per cent sheriff's commission is payable on sale of the vessel.[74]

UK Government policy has, for many years, been for the civil courts to be self-financing. That leaves little prospect of the court fees payable on sale of ships being reduced unless, perhaps, a surge in ship sales were predicted to ensue, producing an overall increase in fee revenue. Such a prediction would not be wholly without precedent. On 23 January 2003, the Gibraltar House of Assembly resolved to reduce its 1 per cent court fee on sale down to 0.75 per cent on the excess portion of the sale price above US$15 million, and also to discount the fees payable on fleet sales. At the time, there had been a number of foreclosures in the cruise sector, which had been in turmoil since the 9/11 terrorist outrages. Undoubtedly the Gibraltar Government hoped to encourage mortgagees of cruise ships to use, or at least not to deter them from using, Gibraltar as their preferred enforcement jurisdiction.

By contrast, in the Bahamas, the court fees payable on sale were increased relatively recently from 0.8 to 3 per cent,[75] to which must also be added the 1 per cent brokerage payable to that court's brokers. This will undoubtedly act as a significant disincentive to mortgagees to foreclose in that jurisdiction, something which may well not have been appreciated by the legislators when approving this increase.

It is submitted that the *ad valorem* fees payable on judicial sales of ships are wholly disproportionate to all other categories of court fees and are difficult, on any view, to justify. They hark back to a bygone era when ships were generally much more modest in size and value and the Admiralty Marshal himself took a share of the fees. The same applies to the court brokers' fees. A 1 per cent sale and purchase broker's commission can only be justified on the basis that for each successful transaction, many will be aborted.[76] When a court instructs a broker to market an arrested ship for sale, a sale will invariably result, quite unlike when a shipowner instructs a broker to market a ship for sale on the open market. That aside, many vessels are today valued at in excess of US$10 million, and often considerably more. Even ship sale and purchase brokers operating in the open market rarely command

72 *Jefferson Bank & Trust Co v Van Niman*, 722 F2d 251, 252 (5th Cir 1984).
73 Supreme Court Fees Order 1999.
74 Item 57 of the Rules of the Supreme Court of Singapore, Appendix B (Court Fees).
75 Supreme Court (Amendment) Rules, 2010.
76 As recognised by May LJ in *The Manifest Lipkowy* [1989] 2 Lloyd's Rep 138, 140.

such handsome commissions when instructed by owners to circulate for sale ships of high value.

By contrast, in the Netherlands, court and bailiff costs, including the cost of advertisements, usually run to around €7,500. In this respect, the competitive edge of that jurisdiction is undoubted.

10.6 The increasing use of international insolvency procedures

Whilst evolving court procedures and ever higher court fees may offer ever fewer incentives to mortgagees to avail themselves of Admiralty procedures for ship mortgage enforcement, the greatest encroachment upon such procedures in recent years has undoubtedly been the trend towards the use of international insolvency regimes.

The reasons for this are twofold. First, the business of ship owning and ship operating has been brought increasingly 'onshore' for taxation purposes. Most shipowners, even if their ships still fly flags of convenience, now operate their vessels under the framework of an onshore taxation regime, and are correspondingly less reluctant to use the insolvency regimes of the onshore jurisdictions where they have their 'centre of main interest' (COMI) to obtain protection from their creditors. Second, the international framework for recognition of foreign bankruptcies, in the form of the UNCITRAL Model Law on Cross-Border Insolvency and, within Europe, the EU Insolvency Regulation[77] 1346/2000, has enabled such bankruptcy regimes to give effective protection to businesses such as shipping whose assets and liabilities are globally dispersed.

Previously, cross-border insolvency was an uncertain affair, which in England, relied on common law comity principles for the recognition of foreign bankruptcy orders,[78] save in those cases where to do so offended against English bankruptcy procedures or, for example, favoured foreign over local creditor litigants.[79]

In the ship mortgage context, there was, however, one important exception to that prior uncertainty, which deserves mention at this point. Where a defaulting shipowner sought protection from the US courts under Chapter 11 of the US Federal Bankruptcy Code, an automatic stay would apply, freezing all pre-petition debts. As far as most mortgagees were concerned, that stay would have immediate worldwide application through what is commonly referred to as the US courts' 'long arm' jurisdiction. This referred not only to the automatic stay's worldwide application but also, crucially, to the fact that violations were punishable by contempt proceedings against all US-domiciled persons, including officers of the US branches and/or US affiliates of non-US mortgagees, whom US bankruptcy judges regarded as subject to their jurisdiction. In practice, almost all international ship finance lenders have some such US presence through which they may be so held to account.

---

77 Council Regulation 1346/2000.

78 For example, a US bankruptcy stay, in *Banque Indosuez Ferromet Resources* [1993] BCLC 112.

79 As was the case in *Felixstowe Dock & Railway Co v United States Lines Inc* [1989] QB 360, in which Hirst J refused to give effect to a US bankruptcy stay by discharging Mareva injunctions over English assets of the debtor company. The judge was strongly influenced by the fact that the debtor's Chapter 11 reorganisation plan envisaged them closing down their European operations, leaving the English creditors to object fruitlessly in the US bankruptcy to the injuncted funds being consumed in the reorganisation.

Defaulting shipowners, therefore, could always resort to Chapter 11 to obtain an effective breathing space against their foreign lenders. Further, the fact that the shipowner's connection with the USA was tenuous would not usually disbar him from founding jurisdiction to commence such proceedings. US bankruptcy judges are predisposed to assume jurisdiction over both foreign creditors and foreign debtors. So whilst, to be eligible, a debtor must have property or a place of business in the USA, it has been held that the mere presence of a credit balance in a US bank account will suffice. So, for example, in *Re Marco Polo Seatrade BV*,[80] a New York bankruptcy court held in 2011 that (i) the presence of funds in the client account of the debtor company's US lawyers; and (ii) the presence of earnings held in trust by the pool operator, OSG, for those of the debtor's vessels that were entered in the OSG Aframax pool, each sufficed to establish jurisdiction.

These *de minimis* jurisdictional entry requirements, together with the perceived 'debtor friendly' nature of the Chapter 11 regime, have seen a spate of Chapter 11 motions since the 2008 downturn. In some cases, shipowners seeking a 'pre-pack' rehabilitation process to assist them to disclaim onerous charters have gone into Chapter 11.[81] In other cases, ship owners seeking to prolong their life in the hope or expectation of a market upturn and/or to give leverage in debt restructuring negotiations with their mortgagees have filed for Chapter 11.[82] In the latter category, it is well known that US bankruptcy judges tend to go to considerable lengths to explore whether a debtor company or group may be saved as a going concern. That aside, the Chapter 11 procedure is rightly perceived to be 'debtor friendly' in three key respects.

First, the debtor's existing management stays in possession of the debtor's assets and operations as a 'debtor-in-possession'. This stands in stark distinction to the 'practitioner-in-possession' bankruptcy regimes, such as England, where management is devolved to a court-appointed insolvency practitioner. Secondly, the debtor has a 120-day exclusive period to prepare a reorganisation plan, a period which may be, and often is, extended. This gives a generous breathing space from creditors. Thirdly, the automatic nature of the stay removes the element of judicial scrutiny at the outset to weed out hopeless cases. In such cases, a creditor's motion to lift the automatic stay on grounds that there is no possibility of the debtor's reorganisation plan succeeding is unlikely, where opposed by the shipowner, to be heard for at least 60 days.

Although in the past a shipowner could successfully resort to the Chapter 11 regime for protection against his mortgagee, he could not usually rely on it to prevent his remaining non-US maritime creditors from arresting his ships in non-US ports to which they sailed. Now, the shipowner may have the US automatic stay reciprocally enforced in those nations which have adopted the UNCITRAL Model Law on Cross-Border Insolvency, including the UK, where it has been enacted in the Cross-Border Insolvency Regulations 2006,[83] so as to prevent mortgagees and lienholders from taking steps to enforce any security.[84] A number of recent Far Eastern charterer

80 SDNY Bankruptcy Court, 21October 2011, ref: 11-13634 (JMP).

81 Examples include General Maritime Corp (Genmar), OSG and TBS International Plc.

82 For instance, Omega Navigation Enterprises Inc, Marco Polo Seatrade BV, and TMT.

83 SI 2006/1030.

84 Defined in art. 2(n)(i) of the Cross-Border Insolvency Regulations 2006 as 'any mortgage, charge, lien or other security'.

insolvency stays have been reciprocally enforced in this way. For example, in 2012, the English Companies Court enforced a Korean court stay granted to a Korean charterer, Korea Line Corporation, under the Korean Debtor Rehabilitation and Bankruptcy Law.[85] Indeed, the English court has reciprocally enforced a number of stays granted to Korean and Japanese charterers, including not only the Korea Line but also Sanko Steamship[86] and STX Panocean,[87] on the application of foreign bankruptcy administrators seeking to stymie arrests, the issuance of protective writs and the prosecution of London arbitrations.

In the event that a shipowner's indebtedness to his mortgagee is the sole or principal motivation for his Chapter 11 filing, the eventual outcome will be either that the shipowner's business is saved, after his mortgagee accepts to restructure his debts, or it is not saved, in which case the US bankruptcy court will typically approve the sale or transfer of the mortgaged ships free of liens and encumbrances to the mortgagee in settlement of their secured debt in the context of a Chapter 7 liquidation. In this way, enforcement through an Admiralty process will be avoided.

Chapter 11 remains a popular avenue of resort for shipowners locked in a fight to the death with their mortgagees. That said, it is perceived, with some justification, to be considerably more expensive than equivalent corporate rescue regimes available in other jurisdictions which shipowners perceive to be debtor-friendly. These include remedies in a number of EU jurisdictions, such as *concordato preventivo* in Italy,[88] *sauvegarde* in France,[89] and corporate reorganisation under Greek law.[90] In the case of stays granted by the courts of EU member states, under the Insolvency Regulation[91] all other EU courts must recognise them[92] without further investigation.[93]

A number of shipowners having their COMI in EU member states, particularly Italian shipowners, have availed themselves of these onshore bankruptcy regimes since the 2008 downturn. EU-wide recognition can be obtained very quickly by straightforward applications in those countries where the stay needs to be enforced. In England, the application is made by the foreign administrator/liquidator to the Companies Court[94] and will typically be published to notify creditors in Lloyd's List.

As with Chapter 11, the obtaining of enforceable court-ordered stays is perceived by ship owners to give them leverage in restructuring talks with their mortgagees. Like all perceptions, this one has a kernel of truth to it, but, in reality, such proceedings usually lead to full-blown liquidations unless there is a realistic restructuring to be done, or the shipping market makes an unexpected recovery during the period of the bankruptcy.

85 Unreported, Companies Court, February 25, 2012 (Norris J). These proceedings were referred to by Andrew Smith J in *The Bulk Chile* [2012] EWHC 2107 (Comm); [2012] 2 Lloyd's Rep 594 at [31].

86 Unreported, Newey J, 30 July 2012 ('*kaisha–kosei*' under Corporate Reorganisation Law of Japan).

87 Unreported, Warren J, 25 June 2013.

88 Under s. 160 of RD 267/1942.

89 See *Code de Commerce*, art. L620.

90 See Bankruptcy Law No. 3588/2007.

91 Council Regulation (EC) No. 1346/2000 of 29 May 2000 on insolvency proceedings.

92 Insolvency Regulation, art. 16(1).

93 *Ibid.*, art. 17(1).

94 Cross-Border Insolvency Regulations SI 2006/1030, Sched 1, art. 15.

Either way, the opening of insolvency proceedings in an EU member state will operate to curtail the ability of European mortgagees to enforce their security using an Admiralty process.

The UK administration procedure has had limited uptake in this context, for two main reasons. First, not many shipowning groups have a UK COMI. Second, ship owners tend to be unwilling to surrender their businesses to their lenders or to favour their mortgagees over their unsecured, trade creditors. Thus, resort to a pre-pack administration procedure by which the debtor company cooperates in the sale of the business and novation of the secured debt to a 'newco' (often owned by the existing management or lenders), leaving all or part of the unsecured debt behind in the old shell company, has not proved attractive.

10.7 EU Insolvency Regulation, Article 5(1)

The enforceability of EU stays under the Insolvency Regulation is subject to a possible small exception in Article 5(1), which provides as follows:

> The opening of insolvency proceedings shall not affect the rights in rem of creditors or third parties in respect of tangible assets ... belonging to the debtor which are situated within the territory of another Member State at the time of the opening of proceedings.

These words suggest that, if a mortgagee arrests a vessel in one EU member state before the defaulting shipowner files for protection in another EU member state, the mortgagee may proceed with its enforcement. What then, is the position if a ship is located outside the EU altogether at the time the EU insolvency proceedings are commenced? Is an EU-domiciled mortgagee then precluded by the Regulation from enforcing against such an asset? Probably it is, although the Regulation is unclear.

What then if the insolvency administrator in the main proceedings opens secondary proceedings in the member state where the vessel is arrested? Probably the Article 5(1) protection will fail, provided that local law recognises the concept of an insolvency moratorium or stay.[95] In the 2011 decision of *Foreign Cooperation Company of China v MV Beluga Sydney*,[96] a Maltese court declined to order the release of that vessel on the application of her owner's German insolvency administrator, placing reliance on Article 5(1). No secondary proceedings had at the time been opened in Malta.

Thus far, the impression may have arisen that insolvency regimes are only resorted to by shipowners in order to delay or to frustrate feared ship mortgage enforcement, by seeking sanctuary in corporate rescue procedures. If so, this impression was unintended. In many cases, insolvent shipowners will take action to wind up their businesses in an orderly fashion, usually in cooperation with their lenders.

For example, in Germany, where large numbers of vessels are owned by German private equity investors through a limited liability partnership,[97] a managing director[98]

95 This is the view of the editors of R. Olivares-Caminal & Ors, *Debt Restructuring* (OUP, 2011), para. 2.46.

96 Unreported, Civil Court of Malta, 30 December 2011.

97 I.e. *Kommanditgesellschaft* (KG).

98 Known as a *Geschäftsführer*.

of a KG's general partner will typically resort to German insolvency procedures as the most convenient asset realisation process in a mortgage default situation for the sale of the ship and any other assets, and the ranking and distribution of the sale proceeds by an insolvency administrator according to German insolvency law. Such managing directors can rely on the insolvency process to protect themselves from personal liability for wrongful trading[99] by requesting the opening of insolvency proceedings no later than three weeks after the KG becomes over-indebted.[100] From the mortgagee's perspective, this process confers the advantage of placing control of the asset swiftly into the hands of an insolvency administrator.

In these orderly winding-up scenarios, the Admiralty enforcement process may yet retain a small, residual role. In the case of a corporate rehabilitation or winding-up procedure, the administrator/liquidator and the mortgagee may prefer the mortgaged ships to be sold by way of an Admiralty Court bill of sale than through the insolvency proceedings. An Admiralty Court bill of sale is recognised worldwide as delivering a ship to its purchaser free of liens and encumbrances. It is certainly perceived to confer the best protection against arrests post-sale. As discussed earlier, it will defeat protective *in rem* writs issued against the ship. By contrast, an administrator's or liquidator's sale may not have the same legal effect. As such, an Admiralty sale will usually deliver a better price.[101] Further, the administrator or liquidator may need to apply to foreign bankruptcy courts in the jurisdictions where the ships are located to assist him to complete the sale. Such assistance may be available, for example under Article 21 of the UNCITRAL Model Law, or by commencement of secondary proceedings under the Insolvency Regulations. However, the terms on which such sales may be approved by the local bankruptcy court may be less clear. Nor will such approval be given if it offends against local law.[102]

An administrator may, for example, wish to release only the mortgagee from the stay to assist the mortgagee to arrest and to apply for the judicial sale of the mortgaged ship. If so, should the remaining maritime creditors be restrained from bringing their claims *in rem* against the sale proceeds, so that, after satisfaction of the mortgagee's claim, the residue will be repaid to the administrator for use and/or distribution in the bankruptcy? This was what happened in the recent case of *The Sanko Mineral*[103] in which the Japanese bankruptcy trustee of Sanko Steamship Co Ltd authorised a Japanese bank to enforce its ship mortgage over a bulk carrier by way of arrest and judicial sale in England, whilst requiring other *in rem* creditors to prove in the Japanese corporate reorganisation proceedings. Following the judicial sale of the vessel by the English court for a sum which substantially exceeded the mortgage debt, the trustee applied for payment out of the residue of the sale proceeds, after

99 Under art. 64 of the German Limited Liability Company Law (*Gesetz betreffend die Gesellschaft mit beschränkter Haftung*).

100 See art. 15A of the German Insolvency Law (*Insolvenzordnung*).

101 A point recognised by Arden J in *Re Lineas Navieras Bolivianas SAM* [1995] BCC 666.

102 For example, in *Fibria Cellulose SA v Pan Ocean Co Ltd* [2014] EWHC 2124 (Ch); [2014] Bus LR 1041, Morgan J refused to give relief to Pan Ocean's administrator under art. 21 to restrain the claimant from terminating a contract in reliance on an *ipso facto* clause, conferring the right to terminate on the counterparty's insolvency, which was invalid under the law of Korea, where the insolvency was proceeding, but lawful under English law, the law governing the contract.

103 *Bank of Tokyo-Mitsubishi UFJ Ltd v MV Sanko Mineral (Owners)* [2014] EWHC 3927 (Admty).

satisfaction of the mortgagee's judgment. The trustee's application was opposed by Glencore Ltd, an *in rem* creditor with a substantial damages claim arising from delay to a cargo in the USA. On the facts, the court dismissed Glencore's claim as time-barred under the Hague-Visby Rules. Furthermore, having regard to Article 20(4) of the Cross-Border Insolvency Regulations 2006, which permits proceedings to be taken to the extent necessary to preserve a claim, the judge held that the stay ordered under the 2006 Regulations had not excused Glencore's failure to commence proceedings on time. Nevertheless, the judge, having regard to the fact that the Japanese reorganisation proceedings had been recognised by the Companies Court as the main foreign proceeding under the 2006 Regulations, ordered the trustee to hold the proceeds to abide the outcome of a petition that Glencore had filed in the Tokyo District Court to have its claim admitted as a secured claim in Sanko's reorganisation, on the basis of its attracting a maritime lien under US and/or Japanese law. Accordingly, it is clear that, even if the court had not dismissed Glencore's claim, the outcome would have been the same: the court would have deferred to the foreign insolvency court to determine Glencore's claim. Indeed, the Tokyo District Court's decision would ordinarily be enforceable against the trustee in England as a judgment *in personam*, given the parties' submission to the jurisdiction of the Tokyo District Court.[104] So whilst the judgment leaves open the question whether an *in rem* creditor who has not participated in a foreign insolvency could claim against sale proceeds in Admiralty despite the existence of a stay under the 2006 Regulations, it is highly likely that the attitude of the English Court would be no different. It is submitted that there should be no fundamental objection to an Admiralty Court conducting an Admiralty sale on such terms, provided at least that it is satisfied that any maritime lien claims will be met in the bankruptcy. Were the Admiralty Court not so satisfied, it might well regard the approach of the foreign insolvency court as offending against English law.

104 See *Rubin v Eurofinance SA* [2012] UKSC 46; [2013] 1 AC 236 at [156]–[167] (Lord Collins).

CHAPTER 11

Mortgagees' interest insurance

*Peter MacDonald Eggers QC**

11.1 Introduction

When ships are built or purchased, the purchaser – the shipowner – very often uses funds borrowed from a lender who will secure the loan over the ship which is the subject of the shipbuilding contract or the contract of purchase. The mortgage advance in respect of a ship will be based on mortgagee documentation and agreements designed to protect the lender's interest against the various risks to that interest. As regards ship's mortgages, the risks arise by reason of the mobility of the vessel, the exposure of the vessel to the political risks associated with the various jurisdictions and legal systems which the vessel will encounter, the effect of maritime law, the full range of maritime perils to which the vessel will be exposed,[1] and the financial risks associated with the credit of the shipowner, the vessel's earning potential and charter engagements, the vessel's value and condition and the amount of the loan.

These risks are managed by the agreements between the shipowner and the mortgagee, including (a) the loan agreement and deed of covenants, regulating the parties' rights and obligations (relating, amongst other things, to the vessel's registration, classification, maintenance and repair and the shipowner's obligations to discharge other debts and liens, to trade the vessel lawfully, to report changes in the risk to the lender, and not to sell or encumber the vessel), (b) the mortgage over the vessel creating the mortgagee's security interest in the vessel, (c) an assignment of the vessel's receivables (in particular, charter hire or freight), (d) personal and corporate guarantees, (e) an assignment of the shipowner's insurances (including hull and machinery and war risks policies and protection and indemnity club entries), and (f) insurance policies designed to protect the mortgagee's interest.

An essential element of the loan agreement, deed of covenants and mortgage agreement is the procurement of insurance by the shipowner.[2] Similarly, the shipyard and vendor will have insured the vessel up until the time of delivery

* Barrister, 7 King's Bench Walk; Visiting Fellow, Institute of International Shipping and Trade Law.

1 Defined by s. 3 of the Marine Insurance Act (MIA) 1906 to mean: 'the perils consequent on, or incidental to, the navigation of the sea, that is to say, perils of the seas, fire, war perils, pirates, rovers, thieves, captures, seisures, restraints, and detainments of princes and peoples, jettisons, barratry, and any other perils, either of the like kind or which may be designated by the policy.'

2 *The Good Luck* [1988] 1 Lloyd's Rep 514, at 521. For a comprehensive analysis of standard forms used in the London market for insuring risks associated with ship building see Chapter 6 written by Professor B. Soyer.

and, where the vessel is the subject of a mortgage or other form of finance, the shipyard and vendor will have taken out insurance. Independently of the insurance procured by the shipowner, the lender will often take out a further policy to insure either its entire interest or to insure its interest insofar as it is uninsured under the hull insurance by reason of some breach of duty or lapse on the part of the shipowner.

These insurance arrangements and their impact on the relationship between the shipowner and the mortgagee are irregularly considered by the courts and have given rise to a number of issues over the years. With this in mind, this chapter will address some of the issues which arise with respect to the insurance of the mortgagee's interest:

- The mortgagee's insurable interest.
- The mortgagee's participation in the shipowner's insurance policy.
- Mortgagees' interest insurance policies.
- Rights of subrogation.

11.2 The mortgagee's insurable interest and its valuation

Whether the insurance contract insures a vessel which navigates the seas or is under construction, the MIA 1906 will apply to the policy.[3] In order to consider the nature of the mortgagee's interest in respect of a mortgaged vessel and the insurability of that interest, one needs to consider the nature of the marine insurance policy which is agreed to protect that interest. In this context, one is concerned only with indemnity insurance policies, as a marine insurance contract is defined to mean indemnities against 'marine losses'.[4]

Section 3 of the MIA 1906 defines (albeit not exhaustively) a marine adventure by reference to (a) the exposure of insurable property, including a ship, to maritime perils, (b) 'the earning or acquisition of any freight, passage money, commission, profit, or other pecuniary benefit, or the security for any advances, loan, or disbursements, is endangered by the exposure of insurable property to maritime perils', and (c) the incurring of any liability to a third party may be incurred by the owner of, or other person interested in or responsible for, insurable property, by reason of maritime perils. These marine adventures roughly translate to property insurance, financial loss insurance and third party liability insurance. A mortgagee may wish to protect or insure its own interest in the vessel, the financial losses associated with the mortgage over the vessel, and/or the liability which the mortgagee may incur in respect of its interest in the vessel.

The mortgagee's interest in its own loan advance and its potential liability is self-evident. In such cases, there is no meaningful reason to analyse a separate requirement of an insurable interest. The mortgagee's interest in a mortgaged ship as property is the subject of a separate requirement of an insurable interest.[5] The mortgagee has an insurable interest in the vessel: it arises essentially by reason of its

3 Sections 2 and 3.
4 Section 1.
5 Defined inclusively in s. 5 of the MIA 1906.

security interest in the vessel. An *in rem* or secured creditor in respect of a ship has an insurable interest in that ship.[6] By contrast, an unsecured creditor is unlikely to have an insurable interest,[7] although this is a questionable proposition.

The mortgagee's interest in the vessel is one which has been recognised by the common law and by statute. Section 14 of the MIA 1906 provides that 'the mortgagee has an insurable interest in respect of any sum due or to become due under the mortgage'.[8] In addition, the section allows the shipowner and mortgagee to place insurance for the benefit of others with an insurable interest as well as for themselves: 'A mortgagee, consignee, or other person having an interest in the subject-matter insured may insure on behalf and for the benefit of other persons interested as well as for his own benefit'.

Accordingly, a mortgagee may insure the vessel for the full extent of its financial interest, even if this exceeds the market value of the vessel.[9] By contrast, if the mortgagee's financial interest is less than the market value of the vessel, the mortgagee may insure the vessel in its own right up to the full value of the vessel,[10] but in that event, if the mortgagee recovers an indemnity reflecting the full value of the vessel, after receiving an indemnity for its own limited interest, it then will be subject to a duty (albeit not as a trustee) to account to the shipowner for the balance, even if the shipowner itself is not a party to the insurance contract.[11]

11.3 The mortgagee's participation in the shipowner's policy

The mortgagee will very often wish to protect its interest in a vessel being purchased or under construction by means of insurance, amongst other things. Such protection may be procured by the mortgagee's participation in the shipowner's own policy. Such participation may be manifested in one of three ways: (a) as a loss payee, (b) as an assignee, and (c) as a co-assured. Indeed, the mortgagee's interest may be protected by some or all of these methods. That said, there are significant differences between these approaches.

11.3.1 Loss payee provisions

The owner or prospective owner of a ship may insure his interest.[12] As security for any advance or loan made by a mortgagee over the vessel, the policy of marine

6 *Moran, Galloway & Co v Uzielli* [1905] 2 KB 555; *O'Kane v Jones* [2003] EWHC 2158 (Comm); [2004] 1 Lloyd's Rep 389.

7 *Macaura v Northern Assurance* [1925] AC 619; *contra Constitution Insurance Co of Canada v Kosmopoulos* (1987) 34 DLR (4th) 208 (Sup Ct Canada).

8 See also s. 10: 'The lender of money on bottomry or respondentia has an insurable interest in respect of the loan'.

9 *The Maira (No. 2)* [1986] 2 Lloyd's Rep 12, at 17; *Arnould's Law of Marine Insurance and Average* 18th edn, (Sweet & Maxwell, 2013), paras 12-24, 12-36.

10 Section 14(2) of the MIA 1906.

11 *Lonsdale & Thompson Ltd v Black Arrow Group plc* [1993] 2 WLR 815; *DG Finance Ltd v Scott* [1999] Lloyd's Rep IR 387 at 392; *Talbot Underwriting Ltd v Nausch Hogan & Murray* [2006] EWCA Civ 889; [2006] 2 Lloyd's Rep 195, at [68–72] (CA).

12 The prospective owner, i.e. a purchaser, of a vessel has an insurable interest under s. 7 of the MIA 1906, at least if the vessel is at its risk or if it has paid for the vessel in advance: *Arnould: Law of Marine*

insurance may provide that, in the event of a loss, the claims proceeds will be paid by the insurer to the mortgagee. In such a case, the mortgagee is often described as a loss payee. This is, in effect, an agreement between the insurer and the shipowner that in the event that the shipowner has a recoverable claim for an indemnity, the insurer will pay the proceeds of claim not to the shipowner, but to the mortgagee. If, however, the mortgagee is not a party to the insurance contract contained in the policy, at common law the mortgagee is not entitled to enforce the insurer's obligation to pay the mortgagee.[13] The mortgagee will be reliant on the shipowner enforcing the insurance contract and the mortgagee may be able to compel the shipowner to take steps to enforce that obligation, if the contract between the shipowner and the mortgagee so provides.

The limitations of this position at common law were swept away by the Contracts (Rights of Third Parties) Act 1999, by which the mortgagee – as a third party to the insurance contract – is entitled to enforce the insurer's promise to pay the loss payee directly, provided that (i) the insurance contract confers a benefit on the mortgagee, (ii) the mortgagee is identified by name or description or as belonging to a particular class, and (iii) on the proper interpretation of the insurance contract, there is no provision by which the parties did not intend to confer such a right on the mortgagee.[14] In many marine policies, the policy will contain a Contracts (Rights of Third Parties) Act 1999 exclusion clause, by which the parties to the insurance contract – the insurer and the shipowner – intend that no third party such as a mortgagee will enforce any right to a benefit conferred on that third party.[15] Accordingly, such provisions may deprive the mortgagee of rights of direct action by failing to fulfil the third requirement set out above.

Assuming there is no such exclusion clause and no such intention, the mortgagee's statutory right to enforce the insurer's obligation to pay the mortgagee will exist if the other requirements of the 1999 Act are satisfied. Any such right derives from the terms of the contract between the insurer and the shipowner. Accordingly, the mortgagee's right of direct action is subject to the terms of that contract and therefore will only be as valid and effective as that contract allows.[16] Accordingly, if there is a relevant exclusion applicable to the shipowner's right to an indemnity under the policy, it will be equally effective against the loss payee. Further, if the insurer is entitled to avoid the insurance contract by reason of an unfair presentation of the risk and does avoid the contract, the mortgagee will have no right to enforce a valid contractual benefit.

Insurance and Average 18th edn, (Sweet & Maxwell, 2013), paras 11-29, 11-58; F. Rose, *Marine Insurance Law and Practice* 2nd edn (Informa Law, 2012), para. 3.27. See also *Ebsworth v The Alliance Marine Insurance Company* (1873) LR 8 CP 596, at 611–612, 637.

13 *The Angel Bell* [1979] 2 Lloyd's Rep 491, at 497.

14 Section 1. See also *Arnould's Law of Marine Insurance and Average* 18th edn, (Sweet & Maxwell, 2013), para. 8-22.

15 See, e.g., cl. 36 of the International Hull Clauses (01/11/2003): 'No benefit of this insurance is intended to be conferred on or enforceable by any party other than the Assured, save as may be expressly provided herein to the contrary.' The question will inevitably arise whether the mere existence of a loss payee clause is a provision to the contrary.

16 Section 1(4) of the 1999 Act.

11.3.2 Assignments

The shipowner's rights under a marine insurance policy may be assigned to the mortgagee bank.[17] Upon such an assignment, the right to enforce a claim under the policy belongs to the mortgagee and may be enforced directly by the assignee mortgagee. If the assignment is a legal assignment, the assignor ceases to have any directly enforceable right against the insurer. If the assignment is an equitable assignment, the assignor retains the right to sue the insurer for the benefit of the assignee, but the assignee also has the power or title to sue the insurer.

There are three principal ways in which rights under a marine insurance policy may be assigned: (a) under section 50 of the MIA 1906, (b) under section 136 of the Law of Property Act 1925, and (c) in equity.

(a) Section 50 of the MIA 1906

Under section 50 of the MIA 1906, a policy of marine insurance may be assigned before or after a loss, although different requirements and effects arise depending on whether the loss has yet occurred.[18] Indeed, any marine policy is assignable unless there is a provision prohibiting such an assignment in the policy. In order that the policy is effectively assigned pursuant to section 50, the following requirements must be satisfied:

1. The assured must have an insurable interest in the subject-matter insured at the time of the assignment and, if the assignment takes place after the loss, at the time of the loss.[19] If the policy is assigned to a third party before loss, that third party will require an insurable interest, at the time of the loss, in order to be entitled to be indemnified for a loss.
2. The assignment must be of the 'whole beneficial interest' in the policy.[20] If the assured purports to retain some interest in the policy (and in the vessel if the assignment takes place before loss),[21] there will have been no effective assignment for the purposes of section 50. The normal position as a matter of law is that a right, not an obligation, may be assigned without the consent of the obligor (i.e. the person against whom the right may be enforced), unless of course the contract provides otherwise. Therefore, the obligation of A owed to B under a contract cannot be assigned to C without B's

17 See, e.g., *Swan v Maritime Insurance Company* [1907] 1 KB 116.

18 *Raiffeisen Zentralbank Österreich AG v Five Star Trading LLC* [2001] EWCA Civ 68; [2001] QB 825, [63–66].

19 Sections 6 and 51 of the MIA 1906; *Williams v Atlantic Assurance Co Ltd* [1933] 1 KB 81, at 105; *Raiffeisen Zentralbank Österreich AG v Five Star Trading LLC* [2001] EWCA Civ 68; [2001] QB 825, [64–65]. Merely transferring the assured's interest in the subject-matter of the insurance does not automatically amount to an assignment of the policy: s. 15 of the MIA 1906.

20 *Williams v Atlantic Assurance Co Ltd* [1933] 1 KB 81; *The Evelpidis Era* [1981] 1 Lloyd's Rep 54; *Raiffeisen Zentralbank Österreich AG v Five Star Trading LLC* [2001] EWCA Civ 68; [2001] QB 825, [63], [67–73]. Even if the policy purports to assign the whole beneficial interest, there may be other provisions within it which still inure to the benefit of the assignor (see e.g. *The Surf City* [1995] 2 Lloyd's Rep 242). Ordinarily, notice is not a requirement of an assignment under s. 50 of the MIA 1906 (*Baker v Adam* (1910) 15 Com Cas 227, at 234).

21 *Raiffeisen Zentralbank Österreich AG v Five Star Trading LLC* [2001] EWCA Civ 68; [2001] QB 825, [63–66].

consent.[22] However, an assignment of the policy under section 50 probably carries with it all of the rights and obligations pertaining to it,[23] allowing the assignee to present a claim under the policy in respect of his own loss.[24] If all of the rights and obligations under the contract are also assigned or transferred, in that case the assignee would also bear obligations of good faith towards the insurer, insofar as such duties are applicable (many such duties are discharged at the time of the conclusion of the contract so that any post-contractual assignment is not likely to impose fresh duties of pre-contractual disclosure).[25]

3. The assignment must be effected by indorsement on the policy or in some other customary manner.[26]
4. The policy must not provide that the policy is not assignable.[27]

Taking these requirements into account, it is unlikely that most assignments of a policy to a mortgagee, before loss, will be effective under section 50 of the MIA 1906, because the shipowner will invariably retain an insurable interest in the insured vessel and, more importantly, because it is unlikely that the whole beneficial interest in the policy will have been transferred to the mortgagee. This conclusion is even more likely where the hull policy insures not only the vessel as property but also provides liability indemnity insurance in favour of the shipowner (such as collision liability or protection and indemnity cover).[28]

If the assignment is effective under section 50 of the 1906 Act, the assignee is entitled to sue on the policy in its own name. However, as the assignee's title is derivative, the assignee can acquire no greater right from the assignor than the assignor itself had. Accordingly, the insurer is entitled to raise any defence as against the assignee as could have been raised had the claim been made by the assignor (the original assured).[29] Accordingly, if the insurer is entitled to avoid the insurance contract by reason of the assignor's unfair presentation of the risk, the insurer can still avoid the contract even if the claim is presented by the assignee.[30] Similarly, if the original assured scuttled the insured vessel or procured the loss by some other

22 H. Beale *et al*, *Chitty on Contracts* 31st edn, (Sweet & Maxwell, 2012), paras 19-077–19-078.

23 *Arnould's Law of Marine Insurance and Average* 18th edn (Sweet & Maxwell, 2013), para. 8-42.

24 *Landauer v Asser* [1905] 2 KB 184, at 189–190.

25 *The Good Luck* [1988] 1 Lloyd's Rep 514, at 546–547; [1989] 2 Lloyd's Rep 238, at 264–265.

26 *Quaere* whether mere delivery of the policy without more would suffice: *Baker v Adam* (1910) 15 Com Cas 227; *Safadi v Western Assurance Co* (1933) 46 Ll L Rep 140, at 144.

27 It may be that the policy will impose conditions before an effective assignment can take place (e.g. notice to the insurer: see Institute Time Clauses 1/11/95, clause 21; International Hull Clauses, clause 23). There is no such notice requirement under s. 50: *Raiffeisen Zentralbank Österreich AG v Five Star Trading LLC* [2001] EWCA Civ 68; [2001] QB 825, at [62].

28 *Raiffeisen Zentralbank Österreich AG v Five Star Trading LLC* [2001] EWCA Civ 68; [2001] QB 825, at [70].

29 Section 50(2) of the MIA 1906. Of course, once the original assured ceases to have any interest in the policy, the conduct of that original assured which takes place after the assignment is unlikely to affect the assignee's rights.

30 *William Pickersgill & Sons Ltd v London & Provincial Marine & General Insurance Co Ltd* [1912] 3 KB 614; *The Litsion Pride* [1985] 1 Lloyd's Rep 437, at 517–519. However, the assignee does not bear any independent duty of the utmost good faith to the insurer, unless any such obligation is also transferred to the assignee (*The Good Luck* [1988] 1 Lloyd's Rep 514, at 546–547; [1990] 1 QB 818, at 889–890).

means of wilful misconduct, the assignee cannot recover under the policy.[31] If the policy is illegal because of the assured's nationality, for example because he or she is an enemy national or because of the existence of sanctions, the policy does not become valid in the hands of an assignee whose nationality does not cause the same difficulties.[32]

(b) Section 136 of the Law of Property Act 1925

An assignment under section 50 of the MIA 1906 operates as a legal assignment and therefore allows for some priority. A legal assignment of a right or chose in action under the insurance contract may also be effected under section 136(1) of the Law of Property Act 1925. That section provides that:

> Any absolute assignment in writing under the hand of the assignor (not purporting to be by way of charge only) of any debt or other legal thing in action, of which express notice in writing has been given to the debtor . . . is effectual in law (subject to equities having priority over the right of the assignee) to pass and transfer from the date of such notice – (a) the legal right to such debt or thing in action; (b) all legal or other remedies for the same; and (c) the power to give a good discharge for the same without the concurrence of the assignor.

Whereas section 50 allows an assignment only of a policy of insurance, or at least all of the rights under the policy, an assignment under section 136(1) is valid in respect of an individual contractual right or a bundle of such rights. However, in order that the assignment under section 136(1) is effective, it must comply with the following statutory requirements: (a) the assignment must be absolute;[33] (b) the assignment must be in writing; and (c) notice of the assignment must be given to the obligor.[34] That said, an assignment may nevertheless be valid under section 136(1) if it is given by way of security and is subject to an equity of redemption.[35] As with an assignment under section 50, the title transferred by an assignment under section 136(1) is derivative and is subject to any defences which are available to the insurer.

(c) Equity

An assignment which is valid in equity, but not recognised at common law, may result in the transfer of rights to a mortgagee which are directly enforceable by the mortgagee against the insurer.[36] There are no formal requirements which must be satisfied for an assignment in equity. All that appears to be required is that the assignor informs the assignee of the assignment or by informing the insurer that the right under the policy is to be discharged by payment to the assignee, provided that there is an intention to assign. Whilst notice to the insurer is not a requirement of an equitable

31 *Graham Joint Stock Shipping Co Ltd v Merchants' Marine Insurance Co Ltd* (1922) 13 Ll L Rep 509; [1924] AC 294.

32 *Bank of New South Wales v South British Insurance Co Ltd* (1920) 4 Ll L Rep 266, at 384.

33 *Raiffeisen Zentralbank Österreich AG v Five Star Trading LLC* [2001] EWCA Civ 68; [2001] QB 825, [74–75]; *Bexhill UK Ltd v Razzaq* [2012] EWCA Civ 1376.

34 *Williams v Atlantic Assurance Co Ltd* [1933] 1 KB 81, at 105–106.

35 *Raiffeisen Zentralbank Österreich AG v Five Star Trading LLC* [2001] EWCA Civ 68; [2001] QB 825, [74].

36 *Colonial Mutual General Insurance Co Ltd v ANZ Banking Group (New Zealand) Ltd* [1995] 1 WLR 1140; *Raiffeisen Zentralbank Österreich AG v Five Star Trading LLC* [2001] EWCA Civ 68; [2001] QB 825, [76–83].

assignment, the debtor who has had no such notice is entitled to discharge the debt by payment to the assignor.[37] An equitable assignee can sue the insurer in his or her own name, but there is a rule of practice that the assignor (who retains the legal title) should be joined if there is a risk that the assignor might bring his or her own claim.[38]

As with an assignment under the MIA 1906 and the Law of Property Act 1925, any defence available to the insurer in respect of any claim which could have been brought by the assignor is available to the insurer in respect of any claim brought by the equitable assignee. That is, the assignee's title is derivative in the same way as any right it may have by reason of a loss payee provision.

11.3.3 The mortgagee as co-assured

The policy may provide insurance cover for the benefit of the mortgagee and the mortgagee may itself be a party to the insurance contract as an original assured or co-assured. That is, if the policy insures the shipowner, it may also additionally and separately insure the interest of the mortgagee. Such additional insurance may be agreed by identifying the mortgagee by name or by class. Alternatively, the shipowner may enter into the insurance contract on its own behalf but also on behalf of the mortgagee as an undisclosed principal. Whether or not a mortgagee is an assured by name or class is a matter of construction and whether or not a mortgagee is a co-assured by class or as an undisclosed principal depends on (a) the named assured having the authority to enter into the insurance contract on behalf of the mortgagee at the time of the contract,[39] and (b) the named assured subjectively intending, at the time of the contract, that the cover will extend to the mortgagee as co-assured.[40]

Where a policy insures the interests of more than one person, and where the interests are distinct or severable, even if equally extensive,[41] the policy is described as composite.[42] Just as there is a separate insurance contract with each of the insurers subscribing to the relevant risk,[43] there is a separate contract between each insurer and each of the assureds whose interests are separate.[44] As a co-assured, the

37 *Colonial Mutual General Insurance Co Ltd v ANZ Banking Group (New Zealand) Ltd* [1995] 1 WLR 1140.

38 *Three Rivers District Council v Bank of England* [1996] QB 292, at 304, 313, 315; *Raiffeisen Zentralbank Österreich AG v Five Star Trading LLC* [2001] EWCA Civ 68; [2001] QB 825, [60].

39 If the named assured lacks authority at the time of the contract, the contract may be ratified by the principal even after loss (s. 86 of the MIA 1906). However, no such ratification is permitted if the principal is undisclosed: *Keighley, Maxsted & Co v Durant* [1901] AC 240.

40 *National Oilwell (UK) Ltd v Davy Offshore Ltd* [1993] 2 Lloyd's Rep 582; *O'Kane v Jones* [2003] EWHC 2158 (Comm); [2004] 1 Lloyd's Rep 389; *BP Exploration Operating Co Ltd v Kvaerner Oilfield Products Ltd* [2004] EWHC 999 (Comm); [2005] 1 Lloyd's Rep 307; *Talbot Underwriting Ltd v Nausch Hogan & Murray* [2005] EWHC 2359 (Comm), [28–34] (Cooke, J); *affd* [2006] EWCA Civ 889; [2006] 2 Lloyd's Rep 195.

41 In such cases, the interests are described as 'pervasive': *Petrofina (UK) Ltd v Magnaload Ltd* [1983] 2 Lloyd's Rep 91, at 95–96; *The State of the Netherlands v Youell* [1997] 2 Lloyd's Rep 440, at 448–450.

42 By contrast, policies may be 'joint' where the interests are 'inseparably connected so that a loss or gain necessarily affects' all of the co-assureds in the same way and to the same extent (*General Accident Fire & Life Assurance Corp Ltd v Midland Bank Ltd* [1940] 2 KB 388, at 404–406; *Direct Line Insurance plc v Khan* [2001] EWCA Civ 1794; [2002] Lloyd's Rep IR 364).

43 Section 24(2) of the MIA 1906.

44 See, e.g., *Samuel & Company Limited v Dumas* [1924] AC 431; *Tomlinson (Hauliers) Ltd v Hepburn*

mortgagee's rights under the policy are not derived from the shipowner's rights, but must be assessed independently as belonging to a distinct party to the insurance contract.

11.3.4 Distinction between loss payees, assignees and co-assureds

As should be apparent from the above discussion, the interest and right of a loss payee and of an assignee represent a derivative title which is no better than the interest and right of the original assured. Any defence to an insurance claim by the original assured under the marine policy which is available to the insurer is also available in respect of a claim presented by a loss payee or by the assignee. Accordingly, the loss payee or assignee will suffer if the original assured has failed to observe his contractual duties or his duties in connection with the presentation of the risk or the presentation of a claim.[45] Moreover, any exclusion or breach of warranty which arises by reason of the original assured's conduct will likewise affect the assignee mortgagee, unless the policy provides otherwise.

If, however, the mortgagee is a co-assured, the mortgagee has a distinct title in that the mortgagee can maintain a claim under the policy independently of any breach of duty or conduct on the part of the original assured, unless such breach or conduct can legitimately be attributed to the mortgagee as well, for example if the original assured unfairly presents the risk to the insurer not only on its own behalf but also on behalf of the mortgagee. Thus, in the case of a composite policy, a breach of the duty of utmost good faith by the shipowner will not necessarily affect the insurance contract as regards any other co-assured, including the mortgagee.[46] That is, the insurer will be entitled to avoid the insurance contract only as regards the assured who has acted in breach of duty, not as regards any other assured,[47] probably because each co-assured has a separate insurance contract with the insurer.[48] Further, in the event that a loss is procured by the wilful misconduct of one assured under a composite policy, the insurer is not liable to indemnify that assured, but the insurer will remain liable to indemnify any other co-assured in respect of the

[1966] AC 451; *Petrofina (UK) Ltd v Magnaload Ltd* [1983] 2 Lloyd's Rep 91; *Arab Bank plc v Zurich Insurance Co* [1999] 1 Lloyd's Rep 262, 276–277; *Eide UK Ltd v Lowndes Lambert Group Ltd* [1999] QB 199; *New World Harbourview Hotel Co Ltd v Ace Insurance Ltd* [2011] Lloyd's Rep IR 230, [81–89]; *Parker v National Farmers Union Mutual Insurance Society Ltd* [2012] EWHC 2156 (Comm); [2013] Lloyd's Rep IR 253, [165–167].

45 *William Pickersgill & Sons Ltd v London and Provincial Marine and General Insurance Co Ltd* [1912] 3 KB 614, at 619–620; *Bank of New South Wales v South British Insurance Co Ltd* (1920) 4 Ll L Rep 266, at 384; *Graham Joint Stock Shipping Co Ltd v Merchants' Marine Insurance Co Ltd* [1924] AC 294; *Samuel & Company Limited* v *Dumas* [1924] AC 431, at 444–445, 450–451, 460.

46 There may be complications in that knowledge belonging to the shipowner may be attributed to the mortgagee as information which the mortgagee ought to know in the ordinary course of business (s. 18(1) of the MIA 1906). The mortgagee's deemed knowledge will become a great deal more expansive when the Insurance Act 2015 comes into force in August 2016: section 4(6) of the Act provides that 'an insured ought to know what should reasonably have been revealed by a reasonable search of information available to the insured (whether within its own organisation or held by others, for example by its agent, and whether the search is conducted by making enquiries or by any other means)'.

47 *New Hampshire Insurance Company v MGN Ltd* [1997] LRLR 24; *Arab Bank plc v Zurich Insurance Co* [1999] 1 Lloyd's Rep 262, at 276–277. See also *FNCB Ltd v Barnet Devanney (Harrow) Ltd* [1999] Lloyd's Rep IR 459.

48 *Arab Bank plc v Zurich Insurance Co* [1999] 1 Lloyd's Rep 262, at 276–277.

loss (assuming of course it is otherwise covered by the policy).[49] Of course, if the mortgagee is independently in breach of any relevant duty, such breach may operate to prejudice the mortgagee's own claim under the policy.

11.4 Mortgagees' interest policies

11.4.1 Mortgagees' interest insurance

Although many mortgagees benefit from participating in a shipowner's insurance policy on a ship, whether as loss payee, assignee or co-assured, ship financiers will very often take out additional insurance to protect their specific interest as a mortgagee, namely mortgagees' interest insurance policies. In theory, such policies may insure the mortgagee's interest without qualification against the same insured perils as are insured under any shipowner's hull, war risks or protection and indemnity policy.

However, in practice, mortgagees' interest policies very often are worded so as to provide supplementary cover to the shipowner's policy (the benefit of which may inure to the mortgagee by one means or another) in that (a) the mortgagees' interest insurers will pay only insofar as the shipowner's insurers do not pay,[50] and (b) there is a warranty by the mortgagee that the shipowner has taken out policies of hull or war risks insurance (and protection and indemnity insurance).[51] Such policies are commonly placed on the basis of the Institute Mortgagees' Interest Clauses 1986[52] or 1997,[53] although there are other market forms and policies may well be tailored to suit the particular requirements of the mortgagee or the adventure.

There are few English authorities considering the nature and effect of mortgagees' interest policies. Such authorities as exist and the terms of the Institute Clauses give some clue to their nature. First, they are frequently worded as a policy of insurance against property loss or damage or liability. In *The Captain Panagos DP*, as regards a mortgagees' interest policy, Mustill, J said:[54]

> I have come to the conclusion that its general tenor and shape speak more of an insurance against physical damage to (and liability of) the ship, than against a financial damage to the mortgagee's interests.

Mortgagees' interest policies are not commonly expressed to provide insurance against financial loss. That said, the measure of indemnity is often calculated by reference to the mortgagee's 'net loss',[55] which is defined to mean the mortgagee's loss under the relevant loan agreement to the extent secured by a mortgage on

---

49 *Samuel & Company Limited v Dumas* [1924] AC 431, 444–445, at 450–451, 460; *The State of the Netherlands v Youell* [1997] 2 Lloyd's Rep 440, at 446–450; *Parker v National Farmers Union Mutual Insurance Society Ltd* [2012] EWHC 2156 (Comm); [2013] Lloyd's Rep IR 253, [171–175].

50 See, e.g., *The Alexion Hope* [1988] 1 Lloyd's Rep 311; *The Good Luck* [1988] 1 Lloyd's Rep 514, at 521.

51 See Institute Mortgagees' Interest Clauses – Hulls (1/3/97), cll. 1 and 4.

52 (30/5/1986).

53 (1/3/1997).

54 [1985] 1 Lloyd's Rep 625, at 630. See the marine adventures defined in s. 3 of the MIA 1906. See also *The Frio Chile* [1995] 1 Lloyd's Rep 182, at 190; *Shell UK Ltd v CLM Engineering Ltd* [2000] 1 Lloyd's Rep 612, [26]; cf. *The Joint Frost* [1998] 1 Lloyd's Rep 310.

55 Institute Mortgagees' Interest Clauses – Hulls (1/3/97), cl. 2.3.

the mortgaged vessel (net of all proceeds collected or collectible under all security arrangements) and the sum insured under the shipowner's policy. In other words, in the event of an insured peril, the mortgagee will be entitled to an indemnity quantified by reference to the outstanding indebtedness, provided that it does not exceed the value of the vessel (which may be an agreed value).[56]

Second, the mortgagees' interest policy indemnifies the mortgagee against loss, damage or liability which in the absence of a defined 'insured peril' would *prima facie* be covered by the shipowner's hull, disbursements, war risks or protection and indemnity insurance, and where the underwriters of such policies do not pay as a result of any insured peril.[57] The insured perils in the Institute Mortgagees' Interest Clauses 1997 are defined to include defences based on or avoidance of the owners' policy by reason of:[58] (a) misrepresentation or non-disclosure, (b) a breach of the duty of good faith, (c) a breach of a promissory warranty (including warranties as to legality, seaworthiness, trading, classification or condition surveys, ownership, management or flag), (d) the failure to exercise due diligence insofar as it is required under the owners' policy, (e) the expiry of a time limitation period and, even, (f) in the case of a total loss where there is a judgment or award holding that the shipowner's claim is not recoverable under the owners' hull or war risks policies on the grounds that the loss has not been proved to have been proximately caused by a peril insured against, but is not otherwise excluded by any exclusion or provision.[59]

Third, there are express exclusions in the Institute Mortgagees' Interest Clauses, including where the loss or expense arises from (a) the termination, cancellation or suspension of the shipowner's policy or non-payment of claims under the shipowner's policy by reason of the non-payment of premiums or calls, (b) the insolvency or financial default of any of the underwriters subscribing to the shipowner's policies, (c) the inability of any party to transmit funds, (d) any fluctuation in exchange rates, and (e) the operation of any franchise or deductible.[60]

Fourth, there are warranties in the Institute Clauses 1997 requiring the mortgagee to promise that the shipowner's policies have been taken out and shall be maintained throughout the policy period for an amount not less than the insured amount under the mortgagees' interest policy or the amount of the loan outstanding, that the shipowner's policies are 'endorsed to the extent of the [mortgagee's] interest',[61]

56 *Ibid.* cl. 1.2.

57 *Ibid.* cl. 1.1.

58 *Ibid.* cl. 2.1. See also *The Good Luck* [1988] 1 Lloyd's Rep 514, at 521. *Cf. FNCB Ltd v Barnet Devanney (Harrow) Ltd* [1999] Lloyd's Rep IR 459.

59 In *The Aliza Glacial* [2002] EWCA Civ 577; [2002] 2 Lloyd's Rep 421, a claim was made under a policy based on the Institute Mortgagees' Interest Clauses 1986 for an indemnity in respect of a vessel detained in Australia on the grounds of illegal fishing. There was, in the shipowner's war risks policy, an exclusion in respect of 'infringement of any customs or trading regulations' and there was a separate warranty 'no illegal fishing'. The Court of Appeal held that illegal fishing constituted a breach of warranty and not an operative exclusion.

60 Institute Mortgagees' Interest Clauses – Hulls (1/3/97), cl. 3.1. There are also the usual nuclear, etc exclusions: cl. 3.2.

61 The precise meaning of such an endorsement is unclear. It may mean nothing more than an acknowledgement by the underwriters that the vessel is subject to a mortgage, but reference to the mortgagee's 'interest' might suggest something more, such as a loss payee or co-assured provision or an acknowledgement of an assignment to the mortgagee.

and the mortgagee has procured and registered a valid first mortgage.[62] Moreover, the Institute Clauses 1997 provide that in the event of a change of ownership, flag, management or control of the mortgaged vessel or a change of the vessel's classification, the mortgagees' interest policy shall automatically terminate, unless the mortgagee gives prompt written notice of the change to the mortgagees' interest insurers and agrees to pay an additional premium.[63]

Fifth, the Institute Clauses require the mortgagee to prove a claim in one of two ways: (a) by demonstrating to the satisfaction of the mortgagees' interest insurers that by reason of the insured perils under the mortgagees' interest policy there is no reasonable prospect of the shipowner and/or the mortgagee succeeding in the claim against the insurers of the shipowner's policies, or (b) in the event of disagreement between the mortgagees' interest insurers and the mortgagee by referring the issue to arbitration or on final court judgment or arbitration award delivered in favour of the insurers of the shipowner's policies.[64]

11.4.2 Other policies protecting mortgagees

In addition to mortgagees' interest insurance policies, there are a number of other insurance products which are designed to protect mortgagees against the risks which may materialise to the prejudice of their interests in respect of the mortgaged vessel, including Mortgagees Additional Perils insurance, Mortgage Rights Insurance, Pre-existing Liens insurance, Ship Mortgage Indemnity policies, mortgagee financial loss insurance, and refund guarantee insurance.

Mortgagees Additional Perils policies were introduced into the market after the *Exxon Valdez* oil spill. In the case of pollution risks, the financial limits of a shipowner's own protection and indemnity or other liability policies might be exceeded especially in the case of a catastrophic loss. For example, a court may order damages to be paid by the shipowner well in excess of its liability policy limits. This may result in the insured vessel being detained or arrested by third party claimants to the detriment of the mortgagee's interest. Where the vessel is mortgaged, the mortgagee may obtain Mortgagees' Additional Perils insurance to protect its security interest in the vessel. Such policies are available not only against pollution risks but also other protection and indemnity risks.

Mortgage Rights Insurance provides protection to a mortgagee where a foreign government acts against a mortgaged vessel so as to prejudice the mortgagee's security over the vessel. For example, because of sovereign debt, a foreign country's government may confiscate property belonging to nationals of the defaulting state or vessels flagged in the defaulting state. Pre-existing Liens Insurance is available

62 Institute Mortgagees' Interest Clauses – Hulls (1/3/97), cl. 4.

63 Institute Mortgagees' Interest Clauses – Hulls (1/3/97), cl. 5. As to the operation of such 'held covered' clauses, see *Liberian Insurance Agency Inc v Mosse* [1977] 2 Lloyd's Rep 560.

64 *Ibid.*, cl. 7.1. As to the meaning of '*final*' court judgment, see the following US decisions concerning other types of policy: *Unencumbered Assets Trust v Great American Insurance* 817 F Supp 2d 1014 (2011), 1033; *Farkas v National Union Fire Insurance Company of Pittsburgh* 861 F Supp 2d 716 (2012), 721 (US District Court); unreported, 11 April 2013 (US Court of Appeals for Fourth Circuit). Some policies may require proof of non-recovery against the hull underwriters in other ways: see, e.g., *The Alexion Hope* [1988] 1 Lloyd's Rep 311, at 318–319, 321–323.

especially where a second-hand ship is sold and financed by a mortgagee. That vessel may be subject to pre-existing maritime or statutory liens which accrued by reason of the conduct of the vendor or a former owner. Where such liens prejudice or take priority over the mortgagee's security over the vessel, the Pre-existing Liens Insurance provides cover.

Ship Mortgage Indemnity is a type of insurance cover which indemnifies a primary mortgagee where there is a failure to repay on the part of the owner of the mortgaged vessel the indebtedness due in respect of a mezzanine loan agreement, which underwrites a subordinate or mezzanine level of debt. This cover provides assistance to the primary mortgagee because the servicing of subordinate debt can prove costly to a shipowner and affect the viability of the vessel's trading which plainly can have negative consequences for the primary mortgagee.

Mortgagee financial loss insurance may be available to indemnify the mortgagee where the shipowner fails to repay the mortgage debt. This is a form of credit insurance. Refund guarantee insurance may be taken out in respect of a default on the part of guarantors who guarantee the refund of advance payments made under shipbuilding contracts. There has been much litigation in recent years in respect of the terms and effect of refund guarantees.[65] In the case of default under such guarantees, the mortgagee who has financed the building of the mortgaged vessel may be indemnified by refund guarantee insurance policy.

11.5 Subrogation

As a mortgagee of a ship has a severable and distinct insurable interest from that of a shipowner, issues of subrogation may arise in a number of different ways. Before considering the ways in which thorny problems of subrogation may manifest themselves, it is worth remembering the cardinal rationale of the doctrine of subrogation. This doctrine allows the insurer who has indemnified the assured to 'take advantage of any means available to the assured to extinguish or diminish the loss for which the insurer has indemnified the assured'.[66] The insurer's rights of subrogation are set out in section 79 of the MIA 1906.[67]

The purpose of the doctrine of subrogation is therefore to permit the insurer to diminish or extinguish the loss it has had to pay to the assured. If, prior to receiving an indemnity from the insurer, the assured receives a benefit from a third party which operates to diminish the indemnifiable loss, the assured is permitted only to recover from the insurer his or her net loss (i.e. after taking into account the benefit received); this is the principle of indemnity.[68] After the insurer has paid the assured an indemnity, the insurer may exercise the rights of the assured

65 See, e.g., *Sea Cargo Skips A/S v State Bank of India* [2013] EWHC 177 (Comm); [2013] 2 Lloyd's Rep 477. See also Chapter 5 written by J. Forrester.

66 *MacGillivray on Insurance Law* 12th edn, (Sweet & Maxwell, 2012), para. 23-001.

67 *Caledonia North Sea Ltd v British Telecommunications plc* [2002] UKHL 4; [2002] 1 Lloyd's Rep 553, [11].

68 *Burnand v Rodocanachi Sons & Co* (1882) 7 App Cas 333, at 339; *Castellain v Preston* (1883) 11 QBD 380, at 386–387; *Goole and Hull Steam Towing Co Ltd v Ocean Marine Insurance Co Ltd* [1928] 1 KB 589.

to diminish its loss and any recoveries will inure to the benefit of the insurer with a view to diminishing that insurance payout. There is another purpose behind the doctrine of subrogation, namely to ensure that an assured who has been paid an insurance indemnity will not be over-indemnified by any recoveries from third parties.

Accordingly, in the standard case, the insurer pays the assured under a marine policy and the insurer can exercise the assured's rights against any third party with a view to passing the burden of the insured loss on to that third party. Where mortgagees are involved, difficulties can arise in three specific cases: (a) where the indemnified assured has rights of action against a co-assured; (b) where the underlying contract between the co-assureds contains provisions for joint insurance; and (c) where under a mortgagees' interest policy, the insurer acquires rights of subrogation.

11.5.1 The exercise of rights of subrogation against co-assureds

In some jurisdictions, there is an absolute bar on an insurer exercising rights of subrogation against a co-assured. Not so in England and Wales. By a number of decisions culminating in the decision of Colman J in *National Oilwell (UK) Ltd v Davy Offshore Ltd*,[69] the law (based on an implied term in the insurance contract) was that an insurer who has indemnified an assured can, via the doctrine of subrogation, exercise that assured's right of action against a co-assured insofar as the loss sought to be imposed on the defendant co-assured is itself not indemnifiable under the insurance policy in question. The position has been complicated by the Court of Appeal's decision in *Tyco Fire & Integrated Solutions (UK) Ltd v Rolls-Royce Motor Cars Ltd*,[70] where Rix LJ expressed the opinion that there is no such implied term in the insurance contract; rather, the insurer can exercise rights of subrogation against a co-assured, unless there is an express waiver in the insurance policy or unless the underlying contract between the co-assured exempts the defendant co-assured from liability to the indemnified assured. Recently, in *Rathbone Brothers plc v Novae Corporate Underwriting*,[71] the Commercial Court applied the analysis adopted by Colman J in *National Oilwell (UK) Ltd v Davy Offshore Ltd*. However, on appeal in *Rathbone Brothers plc v Novae Corporate Underwriting*,[72] the Court of Appeal held that, on the facts before the Court, there was an implied term both in the policy and in the underlying contract between co-assureds, restricting the underwriters' right to pursue rights of subrogation in the name of one co-assured against another co-assured, irrespective of the extent of cover available under the policy to the latter. The Court does not appear to have rejected the existence of the *NOW v DOL* implied term.

Accordingly, assuming there is such an implied term, this means that if the defendant co-assured is not entitled to an indemnity by reason of an operative

69 [1993] 2 Lloyd's Rep 582, at 603–604, 614–615.
70 [2008] EWCA Civ 286; [2008] Lloyd's Rep IR 617, [77].
71 [2013] EWHC 3457 (Comm); [2014] Lloyd's Rep IR 203, [60–72]. This is currently subject to appeal.
72 [2014] EWCA Civ 1464.

exclusion in the policy,[73] a breach of a condition precedent,[74] wilful misconduct,[75] or by a difference in the scope of cover between the two co-assureds,[76] the paying insurer can exercise the indemnified assured's rights of action against the defendant co-assured. Such a claim is permitted because the defendant co-assured cannot then turn around and seek an indemnity from the insurer.[77] However, this analysis may be subject to the imposition of an overriding implied term (or of course an express term) restricting such rights of subrogation, as the Court of Appeal held in *Rathbone Brothers plc v Novae Corporate Underwriting*.

Accordingly, if a mortgagee is a co-assured under a composite marine policy taken out by the shipowner and the shipowner is not entitled to an indemnity in respect of loss of or damage to the insured vessel, for example because of the operation of an exclusion or because of wilful misconduct, the insurer, having indemnified the mortgagee, may be entitled to exercise any rights of action belonging to the mortgagee against the shipowner.[78] Whether that right of action is limited to rights of action relating to causes of the loss of the vessel or extends also to the repayment of the mortgage debt is unclear, although the latter seems less likely.[79]

11.5.2 The provision for joint names insurance in the underlying contract

The loan agreement between the shipowner and the mortgagee will invariably contain provisions requiring one or both of the parties to take out insurance jointly in the name of both (i.e. as composite co-assureds).[80] In that event, the question arises whether any liability owing by the shipowner to the mortgagee (or *vice versa*) is excluded or exempt insofar as the relevant loss is covered by the policy. There are a host of decisions which consider these issues and come to different conclusions depending on the terms of the contract.

It seems relatively clear that the mere obligation of one party to take out insurance for the benefit of the other party (in particular the defendant co-assured) is not on its own sufficient to lead to the conclusion that the parties intended that there will be no liability between them insofar as any loss falls within the scope of the policy.[81] Indeed, there must be a clear intention on the part of the parties to the underlying contract, objectively expressed, that there will be no liability insofar as the loss falls within the scope of the policy which the claiming co-assured was obliged to take

73 *Rathbone Brothers plc v Novae Corporate Underwriting* [2013] EWHC 3457 (Comm); [2014] Lloyd's Rep IR 203, [60-72]; cf. [2014] EWCA Civ 1464.

74 *Lambert v Keymood Ltd* [1999] Lloyd's Rep IR 80, at 88–89, 91–92; *Board of Trustees of the Tate Gallery v Duffy Construction Ltd* [2007] EWHC 361 (TCC); [2007] Lloyd's Rep IR 758, [62–64].

75 *Samuel v Dumas* [1924] AC 431, at 445.

76 *National Oilwell (UK) Ltd v Davy Offshore Ltd* [1993] 2 Lloyd's Rep 582, at 603–604, 612–615.

77 This was previously referred to as an instance of circuity and the shortcomings of the concept of circuity have led the courts to avoid the concept. Strictly, this is not circuity.

78 *Samuel v Dumas* [1924] AC 431, at 445.

79 *The WD Fairway (No. 3)* [2009] EWHC 1782; [2009] 2 Lloyd's Rep 420, [16]; *cf. Castellain v Preston* (1883) 11 QBD 380.

80 Indeed, it may be that the mortgage agreement requires the mortgagee to take out mortgagees' interest insurance: *The Good Luck* [1988] 1 Lloyd's Rep 514, at 521.

81 *The Ocean Victory* [2013] EWHC 2199 (Comm); [2014] 1 Lloyd's Rep 59, [199–204].

out for the benefit of the defendant co-assured.[82] For example, in *Mark Rowlands Ltd v Berni Inns Ltd*,[83] the Court of Appeal held that the insurer could not exercise the landlord's rights against a tenant in circumstances where under the contract between the landlord and the tenant, it was agreed that (a) the tenant would pay an insurance rent, (b) the landlord would keep the building insured, (c) the landlord would use the insurance proceeds to repair the building, and (d) the tenant would be relieved of any liability in the event of fire. Nevertheless, the facts of the case may be such that, even if the contract between co-assureds makes no reference to insurance or an exemption from liability, there will be an implied term restricting the rights of suit by one co-assured against another co-assured to the extent that insurance is available.[84]

11.5.3 Mortgagees' interest insurance

Under a mortgagees' interest insurance policy, there is no complication arising by reason of any co-insurance, because the mortgagee – not the shipowner – will be a party to the mortgagees' interest insurance policy. Under the Institute Mortgagees' Interest Clauses 1997, it is provided that: 'Upon payment to the Assured of a claim hereunder, the Underwriters shall be subrogated to all the rights and remedies of the Assured in respect of such payment.'[85] The loss to the mortgagee will be the outstanding indebtedness secured under the mortgage granted to the mortgagee over the vessel. In the event of a claim, the mortgagee is entitled to pursue the shipowner, using the mortgagee's rights of action e.g. under the loan agreement, to recover the mortgagee's indemnified losses. Indeed, the Institute Clauses 1997 also provide – although this seems unnecessary[86] – that any payments by the insurer will not be applied by the indemnified mortgagee towards the discharge or satisfaction of the outstanding indebtedness.[87]

However, it is possible that the loan agreement between the parties may require the mortgagee to take out such mortgagees' interest insurance. That provision – perhaps with more – might serve as the foundation for an argument that the parties intended that the shipowner would not be liable to the mortgagee in the event that the mortgagee is covered by the mortgagees' interest policy (see above).

Some mortgagees' interest insurance policies may respond to a claim by the mortgagee in the event that the shipowner is unable to recover under its own policy, even though the mortgagee may be able to assert an independent valid claim under

82 *Scottish and Newcastle plc v GD Construction (St Albans) Ltd* [2003] EWCA Civ 16; [2003] Lloyd's Rep IR 809, [57–58]; *Rathbone Brothers plc v Novae Corporate Underwriting* [2013] EWHC 3457 (Comm); [2014] Lloyd's Rep IR 203, [67–72]; *cf.* [2014] EWCA Civ 1464. The mere incidence of the obligation to pay a premium is not on its own relevant to establish any such exemption: *Arab Bank plc v John D Wood Commercial Ltd* [2000] Lloyd's Rep IR 471, [93–95], [101]. However, see *Gard Marine & Energy Ltd v China National Chartering Co Ltd (The Ocean Victory)* [2015] EWCA Civ 16, [2015] Lloyd's Rep 381, [78–83].

83 [1986] 1 QB 211, at 233–234.

84 See, e.g., *Rathbone Brothers plc v Novae Corporate Underwriting* [2014] EWCA Civ 1464.

85 Clause 8.1. ATE policies often include provisions allowing the ATE underwriter to exercise rights of subrogation using the assured's rights under another policy: see, e.g., *Greene Wood McLean LLP v Templeton Insurance Ltd* [2010] EWHC 2679 (Comm); [2011] 1 Lloyd's Rep 557.

86 Because any such step would probably amount to a breach of the implied duty on the part of the assured not to prejudice the insurer's right of subrogation: *Horwood v Land of Leather Ltd* [2010] EWHC 546 (Comm); [2010] Lloyd's Rep IR 453, [56–70].

87 Clause 8.2.

the same policy. Further, the mortgagee may be able to recover under another policy which has been placed in its favour. In such cases, the mortgagees' interest insurers, having paid a claim to the mortgagee, may have rights available either in subrogation or contribution arising by reason of the mortgagee's rights under the other policies. In such cases, the question will arise whether any of the competing policies are intended to be the primary source of indemnity or whether they are co-ordinate. If one policy is primary, the other insurers may be subrogated to the mortgagee's rights under the primary policy.[88] If the policies are co-ordinate, the indemnifying insurer will be entitled to a contribution from the other co-ordinate insurers.[89]

Finally, there is the question of the right to salvage, as provided for in section 79(1) of the MIA 1906. That section provides that: 'Where the insurer pays for a total loss, either of the whole, or in the case of goods of any apportionable part, of the subject-matter insured, he thereupon becomes entitled to take over the interest of the assured in whatever may remain of the subject-matter so paid for.'[90] In the ordinary case of a hull and machinery policy, if a shipowner claims for a total loss under the policy, the insurer is entitled to take over the interest of the shipowner upon payment of the claim. This is referred to as 'abandonment', whereby the shipowner's title and right to the vessel, and liabilities associated with the vessel, are transferred or ceded to the indemnifying insurer, as from the date of the casualty.[91]

Where the insurer under a mortgagees' interest policy pays a claim for a total loss of the vessel, in circumstances where the hull and machinery underwriters do not pay, is there a similar right of abandonment available to the mortgagees' interest insurer? As the policy has been analysed, in the ordinary case, as a property policy,[92] it should follow that the mortgagees' interest insurer should be entitled to take over the mortgagee's interest in the vessel (together with assuming responsibility for all liabilities as from the date of the casualty). Accordingly, the security interest of the mortgagee becomes the insurer's upon the latter's payment of a claim. More difficult questions may arise where both the hull and machinery insurer and the mortgagees' interest insurer (assuming that the former does not also insure the separate interest of the mortgagee) both settle a claim for a total loss, but in those circumstances one may imagine that their respective rights would follow the respective status of the policies and the respective rights of the shipowner and the mortgagee.[93]

88 *Caledonia North Sea Ltd v British Telecommunications plc* [2002] UKHL 4; [2002] 1 Lloyd's Rep 553; *Rathbone Brothers plc v Novae Corporate Underwriting* [2013] EWHC 3457 (Comm); [2014] Lloyd's Rep IR 203, [71–72]; *cf.* [2014] EWCA Civ 1464.

89 Section 80 of the MIA 1906. The policies are less likely to be co-ordinate if the assureds of each policy are different, in which case the court will have to determine which policy is primarily liable: see *North British and Mercantile Insurance Co v London, Liverpool and Globe Insurance Co* (1877) 5 ChD 569, at 583–584; *O'Kane v Jones* [2003] EWHC 2158 (Comm); [2004] 1 Lloyd's Rep 389, [166–176]; *cf. Nichols & Co v Scottish Union and National Insurance Company* (1885) 14 R 1904.

90 See also s. 63 of MIA 1906.

91 *Kastor Navigation Co Ltd v AGF MAT* [2004] EWCA Civ 277; [2004] 2 Lloyd's Rep 119, [76]; *The WD Fairway* [2009] EWHC 889 (Admlty); [2009] 2 Lloyd's Rep 191.

92 *The Captain Panagos DP* [1985] 1 Lloyd's Rep 625, at 630.

93 Cf. *The WD Fairway* [2009] EWHC 889 (Admlty); [2009] 2 Lloyd's Rep 191, [61].

11.6 Conclusion

The mortgagee's interest in a mortgaged vessel is one which has traditionally and rightly been recognised as insurable. That insurance may be taken out in a variety of forms and in a variety of ways, allowing a great deal of commercial flexibility to the mortgagee in the protection of that interest. It is the mortgagee's participation in the shipowner's policies which most commonly has arisen for consideration and by reason of such participation the light has seldom shone on the nature and effect of mortgagees' interest policies. Nevertheless, they are an important form of insurance requiring greater analysis, if for no other reason than to understand their commercial appeal.

CHAPTER 12

Lease finance and demise charters – lessors' risks and liabilities

*Professor Simon Baughen**

12.1 Introduction

An alternative to financing the purchase of a ship by a loan secured by a mortgage is through lease financing. This will generally be used for tax reasons whereby the lender can obtain the benefit of capital allowances on the vessel and is thereby able to reduce the cost of the loan.[1] Under UK taxation rules this was possible for finance leases entered into prior to 1 April 2006 unless there was an option for the borrower to purchase the vessel before the end of the loan.[2] The new rules which apply to leases entered into on or after 1 April 2006 allocate capital allowances under long funding leases to the lessee, not the lessor, so that long funding lessors are taxed in a similar way to which the way in which they would have been taxed had they made a loan, and long funding lessees in a way that is similar to the way in which they would be taxed had they purchased the asset.[3] Under a finance lease, legal title to the vessel being bought will be transferred to the lender who will then execute a bareboat charter to the borrower for the term of the loan.[4] This will transfer possession of the vessel for the term of the charter to the borrower, who will crew the vessel. The lender's primary security will be its ownership of the vessel, although this will be supplemented by guarantees from various parties associated with the borrower.

* Professor of Shipping Law, Institute of International Shipping and Trade Law, Swansea University.

1 In the US there was a boom in finance leases in the 1970s to take advantage of federal income tax incentives and many of the first generation of Alaska crude carriers were financed in this way. See 'Lease Financing for Vessels Engaged in the Coastwise Trades'. Available at: www.marinemoneyoffshore.com/node/5684 (last accessed 31 December 2014).

2 A finance lease with an option for the lessee to acquire the asset falls within s. 67 of the Capital Allowances Act 2001 if the option to acquire the asset is below the expected market value of the asset at the date the option is exercised. In this case capital allowances will be claimable by the lessee not the lessor because the payments under the lease contain an element that is regarded as capital. If there is no option to purchase, the charter will still be a finance lease unless ownership passes automatically to the lessee at the end of the lease. It could be a finance lease if the total rentals over the initial lease period represent at least 90 per cent of the value of the equipment. Chapter 11 of the Capital Allowances Act 2001 'Overseas Leasing' is directed at limiting the extent to which the benefit of 25 per cent UK writing-down allowances on the acquisition of capital assets can flow through to non-UK residents.

3 Finance leases of more than five years will constitute long funding leases, but a lease of between five and seven years will constitute a short lease if: (a) it is treated as a finance lease under generally acceptable accounting principles; (b) the residual value of the equipment is not more than 5 per cent of its market value at the commencement of the lease; and (c) the rentals in each year must generally not vary more than 10 per cent from the value of the rentals due in the previous year.

4 Under a 'synthetic lease', additional finance may be obtained by the lender who will then execute a mortgage on the vessel.

The bareboat charter will reflect the fact that this is a conditional sale. Loan repayments will be made through payments of hire throughout the life of the charter at the end of which the borrower will obtain ownership of the vessel. Alternatively, a ship may, upon expiry of the financing charter, be scrapped or disposed of at her then current scrap or market value or at a price driven by tax considerations. The charter may include an option to acquire ownership before the end of the charter term. It may also provide the lender with a right to mortgage the vessel to obtain additional finance, together with an assignment of the ship's earnings, insurances and requisition compensation as security for any loan that the owners take, subject to a direct undertaking from the mortgagee bank that they will allow the charterers quiet enjoyment of the vessel. The owners will usually be given the express right to sell or otherwise transfer their rights in the ship together with the charter to a third party provided that, in so doing, they do not increase the charterers' obligations. The charter may also provide for a set fee to be paid on termination reflecting the remaining hire instalments due under the charter, discounted for early payment, and providing that on sale of the vessel any excess over this sum is to be paid to the charterers as a refund or rebate of rental.

This chapter will examine three issues arising out of the use of bareboat charters as a means of financing ship purchase. First, what are the potential liabilities to which the lender will be exposing itself by taking ownership of the vessel during the course of the charter? How will these be addressed by charter provisions as to indemnity and insurance? Second, what are the lender's rights in the event of default by the charterer? Third, what is the lender's position in the event of the borrower's insolvency? I shall look at these issues from the perspective of UK law and US law and will refer to the standard bareboat form, BARECON 2001.

12.2 Owners' liabilities

12.2.1 In personam *claims*

Under a bareboat charter possession of the vessel will be transferred to the charterer who will crew the vessel. Accordingly, there will generally be no contractual liability under bills of lading issued during the currency of the demise charter as the charterers will have no actual or ostensible authority to bind the owners. In *Baumwoll Manufactur Von Carl Scheibler v Furness*,[5] the bill of lading was signed for the master, but the master being the servant of the charterers, the bill of lading constituted a contract with the demise charterers not the owners. Nor were the owners bailees of the cargo. It is also extremely unlikely that the owners will incur any liability in negligence for the operation of the vessel during the charter, except possibly as regards damage caused by unseaworthiness existing at the time the vessel was delivered and of which the owners were, or should have been aware.[6] Where damage is caused

5 [1893] AC 8.

6 Part III of Barecon deals with purchase of a new build and contains a provision in cl. 1(d) which precludes charterers from raising claims in respect of the vessel's performance or specification or defects arising except as regards defects which manifest themselves within the first 12 months from the date of delivery of the vessel. In this event, the owner's obligation is to endeavour to compel the builders to

by concurrent causes of initial unseaworthiness and negligence/unseaworthiness arising after delivery, the owners would still be liable if the initial unseaworthiness was a material cause of the damage.[7]

However, it is possible for the owners to incur personal liabilities in the interim period between termination of the charter and redelivery of the vessel. In *The Chem Orchid*[8] the charter was terminated in April 2011 and bills of lading were issued in June and July 2011 before the physical redelivery of the vessel. The High Court of Singapore held that there was an arguable case that the vessel had been constructively redelivered in mid-April 2011 which meant that from that moment on the possession and control of the vessel passed to the owners, even though the crew continued to be employed by the charterers. The owners had placed the demise charterers in the position as their representatives in respect of bills of lading issued and indirectly represented to shippers that the demise charterers were authorised to bind the owners to the bills of lading; and the shippers had relied on this conduct.[9] After the termination of the charter, the owners could have made it clear that the vessel no longer had their authority to take on cargo and issue bills of lading.[10] Although the owners had repeatedly demanded the return of the vessel after the termination of the charter, they took no steps to recover possession of the vessel as expressly provided for in article 26(5) of the Lease Agreement whereby in the event that the charterer delayed return of the vessel, the owner was to be allowed unilaterally to retrieve the vessel at the charterer's expense.

The lender's exposure to owner's liabilities during the currency of the charter may be limited by vesting ownership in a subsidiary of the lender. Under UK law the courts will not pierce the corporate veil to make the parent company liable,[11]

---

repair, replace or remedy any defects or to recover from the builders any expenditure incurred in carrying out such repairs, replacements or remedies. Lines 35 to 44 expressly limit owners' liability to charterers to the extent that the owners are able to make a recovery against the builders under the guarantee clause of the newbuilding contract. Lines 39 to 44 provide that charterers are to accept such sum as the owners are reasonably able to recover from the shipyard, and not to make any further claim on the owners for the difference between the amount recovered and the true cost of the repairs. These provisions will not affect owners' liability to third parties who suffer loss or damage as a result of the vessel's unseaworthiness at the time of delivery.

7 In *IBA v EMI and BICC* (1980) 14 BLR 1 a television aerial mast collapsed due to oscillation during high winds. The collapse was caused by two forces of stress on the mast but the party who designed the mast was liable only in respect of one of them and the other stress, for which they were not liable, was 'by far the more important cause' of the collapse. However, as the first stress 'materially contributed to the collapse' the House of Lords held that the designer was liable in full. Lord Fraser, at 37 and 38, applied the statement of Lord Reid in *McGhee v National Coal Board* [1973] 1 WLR 1, 4: 'It has always been the law that a pursuer succeeds if he can show that fault of the defender caused or materially contributed to his injury. There may have been two separate causes but it is enough if one of those causes arose from the fault of the defender. The pursuer does not have to prove that this cause would of itself have been enough to cause him injury.' Although the House of Lords in *Fairchild v Glenhaven Funeral Services Ltd* [2002] UKHL 22; [2003] 1 AC 32 cast doubt on certain dicta in *McGhee*'s case, it is suggested that it remains authority for the proposition just stated.

8 [2014] SGHCR 1; [2014] 1 Lloyd's Rep 520.

9 A similar finding had been made in Australia in *CMC (Australia) Pty Ltd v The Ship Socofl Stream* [1999] FCA 1419; (1999) 95 FCR 403; (CA) [2001] FCA 961.

10 This argument was only raised by two of the three bill of lading claimants, but could have applied equally to the third claimants and also to the other claimants who had supplied bunkers to the vessel in June 2011 after the termination of the demise charter.

11 *Adams v Cape Industries plc* [1990] Ch 433 (CA).

although there may be a greater chance of this happening in the US. However, most personal liabilities that can be incurred by a shipowner will be subject to mandatory insurance provisions, which are discussed later in this paper.[12]

12.2.2 Liability under international conventions

The owners of a vessel will also come under two strict liability regimes in relation to oil pollution and bunker spills by reason of two international conventions which the UK has ratified. The first is the 1992 International Convention on Civil Liability for Oil Pollution Damage (1992 CLC) which imposes strict liability on all shipowners,[13] irrespective of their nationality or flag, in respect of discharges or escapes of 'persistent oil' from laden bulk oil tankers and to escapes of oil from oil tankers on ballast voyages following the carriage of oil, unless it is proved that no residues from the carriage of any such oil remain in the ship. Liability is also imposed in respect of escapes from combination bulk carriers, for example, oil, bulk, ore carriers (OBOs). The Convention applies:

(a) to pollution damage[14] caused:
  (i) in the territory, including the territorial sea, of a Contracting State, and
  (ii) in the exclusive economic zone of a Contracting State, established in accordance with international law, or, if a Contracting State has not established such a zone, in an area beyond and adjacent to the territorial sea of that State determined by that State in accordance with international law and extending not more than 200 nautical miles from the baselines from which the breadth of its territorial sea is measured;

(b) to preventive measures,[15] wherever taken, to prevent or minimise such damage.

Liability is strict, but the shipowner may escape liability if it can prove that the discharge or escape or the threat of contamination:

(a) resulted from an act of war, hostilities, civil war, insurrection or an exceptional, inevitable and irresistible natural phenomenon; or
(b) was due wholly to anything done or left undone by another person, not being a servant or agent of the owner, with intent to do damage; or
(c) was due wholly to the negligence or wrongful act of a government or other

---

12 *The Amoco Cadiz* [1984] 2 Lloyd's Rep 304.

13 Defined in art. 1.3 as 'the person or persons registered as the owner of the ship or, in the absence of registration, the person or persons owning the ship'.

14 Defined in art. 1.6 as '(a) loss or damage caused outside the ship by contamination resulting from the escape or discharge of oil from the ship, wherever such escape or discharge may occur, provided that compensation for impairment of the environment other than loss of profit from such impairment shall be limited to costs of reasonable measures of reinstatement; (b) the costs of preventive measures and further loss or damage caused by preventive measures'.

15 Defined in art. 1.7 as 'any reasonable measures taken by any person after an incident has occurred to prevent or minimize pollution damage'. 'Incident' is defined in art. 1.8 as 'any occurrence, or series of occurrences having the same origin, which causes pollution damage or creates a grave and imminent threat of causing such damage'.

authority in exercising its function of maintaining lights or other navigational aids for the maintenance of which it was responsible.[16]

Article III.4 provides: 'No claim for compensation for pollution damage may be made against the owner otherwise than in accordance with this Convention.'

The shipowner is entitled to limit its liability as follows:[17]

- for vessels not exceeding 5,000 gross tons, a flat rate of 4.51 million SDRs;
- for vessels between 5,000 and 140,000 gross tons, 4.51 million SDRs plus an additional 631 SDRs for each additional gross ton above 5,000 gross tons; and
- for vessels of 140,000 gross tons and above, the limit is 89.77 million SDRs.

Under Article 7, all ships carrying in bulk a cargo of more than 2,000 tons of persistent oil are required to supply a certificate confirming liability insurance covering their CLC liabilities. Each Contracting State must ensure that insurance or other security is in force in respect of any such ship wherever registered, entering or leaving a port in its territory, or arriving at or leaving an offshore terminal in its territorial sea, if the ship actually carries more than 2,000 tons of oil in bulk as cargo.

The second international convention is the 2001 International Convention on Civil Liability for Bunker Oil Pollution Damage. This came into force on 21 November 2008. This applies a similar liability regime to that imposed by the CLC in respect of spills of bunker oil outside the vessel, wherever they occur, which cause damage in the territory of State Parties. The Convention, however, defines 'shipowner' in terms wider than those used by the CLC, as 'the owner, including the registered owner, bareboat charterer, manager and operator of the ship'. There is no separate provision for limitation of liability. Instead, claims under the Convention will fall to be limited under either the 1957 or 1976 Limitation Conventions. Article 7 provides that for vessels over 1,000 gross tons, the registered owner, but not the other persons falling within the definition of 'owner', must maintain insurance equal to the amounts of liability under the national or international limitation regime applicable in the flag state, but not exceeding the limits in the 1976 Convention, as may be amended. In the UK section 168 of the Merchant Shipping Act 1995 provides that any liability under the Convention shall be deemed to be a liability to damages in respect of such damage to property as is mentioned in Article 2(1)(a) of the 1976 Convention on Limitation of Liability for Maritime Claims (LLMC).

Strict liability is also imposed on shipowners under two other international conventions, neither of which are yet in force: the 2010 International Convention on Liability and Compensation for Damage in Connection with the Carriage of Hazardous and Noxious Substances by Sea in respect of damage caused by carriage of hazardous and noxious substance;[18] and the 2007 Nairobi International Convention

16 Article 3.2.

17 Subject to art. 4.2 which provides: 'The owner shall not be entitled to limit his liability under this Convention if it is proved that the pollution damage resulted from his personal act or omission, committed with the intent to cause such damage, or recklessly and with knowledge that such damage would probably result.'

18 A public law liability for water pollution by ships carrying hazardous substances is imposed on the 'operator' of the ship by the Environmental Liability Directive 2004/35/EC, as an activity falling within

on the removal of wrecks which makes shipowners liable for the costs of locating, marking and removing wrecks located beyond the territorial sea but within the exclusive economic zone of a State Party. The Convention, which has been ratified by the UK, will come into force on 14 April 2015.[19]

12.2.3 Liability under statute

The demisors may also become liable under a UK statute imposing liability on the 'owners' of a vessel. In *BP Exploration Operating Co Ltd v Chevron Shipping Co*,[20] the House of Lords considered the meaning of section 74 of the Harbours, Docks, and Piers Clauses Act 1847 under which strict liability was imposed upon the owner of a vessel which caused damage to any harbour, dock, pier or quay. It was held that for the purposes of section 74 of the 1847 Act, the 'owner' of a vessel meant the registered owner and did not include a charterer, even a bareboat charterer.

Liability for breach of statutory duty may also arise under section 100 of the Merchant Shipping Act 1995 which places a duty on the owner of a ship to take all reasonable steps to secure that the ship is operated in a safe manner and creates a criminal offence if there is a failure to comply. In *Littlejohn v Wood & Davidson Ltd*,[21] the Scottish Outer House considered that section 31(1) of the Merchant Shipping Act 1988 (since replaced by section 100 of the Merchant Shipping Act 1995) created a civil liability; hence a crew member who was injured while carrying out repairs while the ship was docked was entitled to sue the owner of the ship. The section provided, 'It shall be the duty of the owner of a ship to which this section applies to take all reasonable steps to secure that the ship is operated in a safe manner. . . .' and subsection (4) indicated that in addition to the owner of a ship, a charterer or manager could also be involved in criminal conduct if involved in the operation in question. At the time the vessel was being operated under a contract that was probably a demise charter. It was held that for the purposes of section 31(1), a ship was being operated when at sea, afloat in port, under repair, or loading and unloading cargo.[22] Although failure to comply with the section was

Annex III. Directive 2013/30 on safety of offshore oil and gas operations amends the Environmental Liability Directive so that the ambit of damage to surface water extends from coastal waters (all water up to one nautical mile seaward from the baseline) to all EU waters including the exclusive economic zone (about 370 km from the coast) and the continental shelf where the coastal Member States exercise jurisdiction. The Environmental Liability Directive defines 'operator' as 'the person who operates or controls such an activity, including the holder of a permit or authorisation relating to that activity, or the person registering or notifying such an activity'. Where there is a bareboat charter the 'operator' would be the demise charterer who is the person who operates or controls the relevant activity, the carriage of dangerous cargo, and would not cover the registered owner.

19 The convention has also been ratified by: Bulgaria, Denmark, Germany, India, Iran, Malaysia, Morocco, Nigeria, and Palau.

20 [2001] UKHL 50; [2003] 1 AC 197.

21 1997 SLT 1353.

22 Lord Johnston noted that 'where an owner had simply appointed a manager or had let the ship on demise charter and was not to any extent involved in its management, it would be a sufficient defence for him to aver that he took all reasonable steps as required by subs (1) by either effecting such a management arrangement or by effecting the charter. On the other hand, one can envisage situations where both an owner and a manager could be involved in the event or matters to which the subsection related and therefore joint responsibility was possible'. However, the defendants did not advance this as a defence.

a criminal offence, civil claims for damages could also be allowed as the legislation was aimed at protecting a particular class of persons, that is, all those on board ship, whether at sea or not.[23] Another potential source of liability in the Act is to be found in section 49 which provides that the owner and master shall be guilty of an offence, and the ship subject to detention if in the UK, if a ship goes to sea or attempts to go to sea without carrying such officers and other seamen as it is required to carry under section 47.

In contrast, section 98 of the Merchant Shipping Act 1995 will not apply to the owner where the vessel is demise chartered. The section provides that the owner and master of a ship which is in a port in the UK or is a UK ship and is in any other port will be guilty of an offence if the ship is dangerously unsafe. Subsection 2, however, provides:

> (2) Where, at the time when a ship is dangerously unsafe, any responsibilities of the owner with respect to the matters relevant to its safety have been assumed (whether wholly or in part) by any person or persons other than the owner, and have been so assumed by that person or (as the case may be) by each of those persons either—
> (a) directly, under the terms of a charter-party or management agreement made with the owner, or
> (b) indirectly, under the terms of a series of charter-parties or management agreements, the reference to the owner in subsection (1) above shall be construed as a reference to that other person or (as the case may be) to each of those other persons.

Section 42 implies an obligation into every contract between the owner of a UK ship and the master or any seamen employed on the ship that the owner of the ship shall use all reasonable means to ensure the seaworthiness of the ship for the voyage at the time when it commences and to keep the ship in a seaworthy condition for the voyage during the voyage. This must apply to the demise charterer as 'owner' as there will be no contract of employment between the registered owner and the master and any seamen.

12.2.4 In rem *Admiralty jurisdiction*

An *in rem* action may be brought in the Admiralty Court against the vessel in respect of maritime liens. If a claim is a maritime lien it will attach to the *res* from the date of the claim and will be unaffected by subsequent changes in its ownership. The maritime lien also attaches notwithstanding that at the time of the incident the vessel was on demise charter.[24] Therefore, *in rem* proceedings may be brought against a vessel in respect of a collision notwithstanding that it has been sold to purchasers, without notice of the claim, before the issue of the writ. However, the owners of the demise chartered vessel will not be liable *in personam* and their

---

23 Lord Johnston stated: 'With regard to the secondary question of civil liability, I am clearly of the view that this legislation and particularly s. 31 falls firmly within the exception clearly stated in the law for over 100 years where particular legislation is conceived for the protection of a particular class of people. Here the class is enormous, namely all those at sea, including those on board a ship in port, but is nevertheless a specific and definable class totally akin to those who work in factories wherever they may be. I am therefore firmly of the view that this section admits a claim for civil liability in principle and the defenders' attack in this respect also fails.'

24 *The Father Thames* [1979] 2 Lloyd's Rep 364 (QB).

liability will be limited to the value of the *res*. Further, a maritime lien will be lost if the *res* is sold by an order of the court. A maritime lien may be exercised only against the vessel against which the claim arose and not against any other vessel in the same ownership. At common law the following claims have been established as giving rise to maritime liens:

(a) damage caused by a ship – this will cover claims arising out of a collision but will not cover cargo claims unless they are brought against a vessel that has collided with the vessel on which the cargo was being carried;
(b) salvage
(c) seamen's wages
(d) master's wages and disbursements
(e) bottomry and respondentia.

An *in rem* claim may also be brought against the vessel in respect of the claims listed in section 20(2)(e)–(r) of the Senior Courts Act 1981:[25]

(e) any claim for damage done by a ship . . .
(f) any claim for loss of life or personal injury sustained in consequence of any defect in a ship or in her apparel or equipment, or in consequence of the wrongful act, neglect or default of:
  (i) the owners, charterers or persons in possession or control of a ship; or
  (ii) the master or crew of a ship, or any other person for whose wrongful acts, neglects or defaults the owners, charterers or persons in possession or control of a ship are responsible, being an act, neglect or default in the navigation or management of the ship, in the loading, carriage or discharge of goods on, in or from the ship, or in the embarkation, carriage or disembarkation of persons on, in or from the ship;
(g) any claim for loss of or damage to goods carried in a ship;
(h) any claim arising out of any agreement relating to the carriage of goods in a ship or to the use or hire of a ship;
(j) any claim in the nature of salvage:
  (i) under the Salvage Convention 1989;
  (ii) under any contract for or in relation to salvage services;
  (iii) in the nature of salvage not falling within (i) or (ii) above, or any corresponding claim in connection with an aircraft.
(k) any claim in the nature of towage in respect of a ship or an aircraft;
(l) any claim in the nature of pilotage in respect of a ship or an aircraft;
(m) any claim in respect of goods or materials supplied to a ship for her operation or maintenance;
(n) any claim in respect of the construction, repair or equipment of a ship or in respect of dock charges or dues;
(o) any claim by a master or member of the crew of a ship for wages (including any sum allotted out of wages or adjudged by a superintendent to be due by way of wages) . . .
(p) any claim by a master, shipper, charterer or agent in respect of disbursements made on account of a ship;
(q) any claim arising out of an act which is or is claimed to be general average;
(r) any claim arising out of bottomry.

25 In the case of any such claim as is mentioned in s. 20(2)(a), (c) or (s) or any such question as is mentioned in s. 20(2)(b), an action *in rem* may be brought in the High Court against the ship or property in connection with which the claim or question arises.

The right to proceed *in rem* against the vessel in respect of such claims is subject to section 21(4) of the Senior Courts Act 1981 which provides:

> In the case of any such claim as is mentioned in section 20(2)(e) to (r), where—
> (a) the claim arises in connection with a ship; and
> (b) the person who would be liable on the claim in an action in personam ('the relevant person') was, when the cause of action arose, the owner or charterer of, or in possession or in control of, the ship, an action in rem may (whether or not the claim gives rise to a maritime lien on that ship) be brought in the High Court against—
> (i) that ship, if at the time when the action is brought the relevant person is either the beneficial owner of that ship as respects all the shares in it or the charterer of it under a charter by demise; or
> (ii) any other ship of which, at the time when the action is brought, the relevant person is the beneficial owner as respects all the shares in it.

Therefore, if the demise charterer as 'the relevant person' is liable on the cause of action when the cause of action arose, the *in rem* action can be brought against the ship. If security is not provided for the claim, the ship can then be sold. An *in rem* action can also be brought against a sister ship of the demise charterers, but not against a sister ship of the owners. For there to be an *in rem* action against the vessel for a statutory lien the demise charterer who is liable *in personam* must have been in possession or control of the ship when the cause of action arose.

This provision, or its equivalent, has given rise to litigation in some common law jurisdictions when the demise charterer incurs a liability and subsequently the vessel is withdrawn from the demise charter with proceedings being commenced after the date of withdrawal. The issue is whether when the action is brought the relevant person is still the demise charterer of the vessel. This depends on whether redelivery of the vessel is required for the relevant person to cease to be the demise charterer of the vessel. The authorities on this are mixed. Some cases have required redelivery for termination:[26] others have not.[27]

In *The Chem Orchid*,[28] the High Court of Singapore has recently held that redelivery is required, but a symbolic or constructive redelivery will suffice. The essence of a demise charterparty was that the owners conferred possession and control of the ship on the charterer for the duration of the charter, thus placing charterers in the same position as the owners. An effective termination of a demise charter required the withdrawal of both possession and control of the ship. The guiding principle was that termination and redelivery was required to bring the charter to an end but this was subject to the provisions of the charter itself. In *The Rangiora*,[29] redelivery was

26 Australia: *Patrick Stevedores No. 2 Pty Ltd v MV Turakina* May 1998, Federal Court unreported (*The Turakina*). New Zealand: *Mobil Oil New Zealand Ltd v The Ship Rangiora* [2000] 1 NZLR 49; [2000] 1 Lloyd's Rep 36 (*The Rangiora*).

27 Hong Kong: *Gulf Marine & Industrial Supplies Inc v The Demise Charterer of the Ship or Vessel MV Trident Dawn* [1992] HKCFI 273 (*The Sea Empire*). Australia: *CMC (Australia) Pty Ltd v The Ship Socofl Stream* [1999] FCA 1419; (1999) 95 FCR 403 (*The Socofl Stream* First Instance) and on appeal [2001] FCA 961 (*The Socofl Stream* Appeal); *ASP Holdings Ltd v Pan Australia Shipping Pty Ltd* [2006] FCA 1379; (2006) 235 ALR 554; *Ships Hako Endeavour, Hako Excel, Hako Esteem and Hako Fortress v Programmed Total Marine Services Pty Ltd* [2013] FCAFC 21; (2013) 296 ALR 265 (*The Hako Fortress*).

28 [2014] SGHCR 1; [2014] 1 Lloyd's Rep 520.

29 [2000] 1 NZLR 49; [2000] 1 Lloyd's Rep 36.

held to be required as it was specifically stated in clause 10(a) of the charterparty that hire was to be paid until the day and hour of redelivery of the possession of the vessel which meant that possession of the ship continued until it was redelivered to the owner. However, the charterparty in *The Socofl Stream*[30] did not contain any clause requiring hire to be paid until redelivery. The charterparty contained a clear contractual right to terminate the charterparty. The charterparties in *ASP* and *The Hako Fortress*[31] were on Barecon 2001 form where the position is even clearer with cl. 29 providing that once a notice of termination is issued, the charterer shall hold the vessel as gratuitous bailee pending physical repossession by the owner. Accordingly, once the notice of termination was issued, both control and possession of the ship were terminated and the charterer held the ship on bailment for the owner.

The Sea Empire[32] was the odd man out in these cases, as there, the charterparty provided for hire to be paid until the day and hour of delivery of possession of the vessel, as had been the case in *The Turakina* and *The Rangiora*, but the Hong Kong court held that redelivery was not required and the charterparty was terminated once the notice of termination was issued. The Singapore High Court declined to follow the decision noting that *The Sea Empire* had been decided before the full analysis of the law relating to demise charters by Tamberlin J in *The Turakina*. However, there would be constructive or symbolic redelivery once the charterers acknowledged their intention to surrender the ship, and this had happened before the issue of proceedings by the claimant.[33]

12.2.5 Compulsory insurance

The Merchant Shipping (Compulsory Insurance of Shipowners for Maritime Claims) Regulations 2012 SI 2012/2267[34] came into force on 5 October 2012.[35] The Regulations provide that ships may not leave or enter UK ports unless the shipowner has insurance which must cover at least maritime claims subject to limitation under the 1996 Protocol to the 1976 Convention.

Where the Hague-Visby Rules are incorporated into the contract for the carriage of goods, and that contract imposes liability for loss resulting from delay in the carriage by sea of cargo, the insurance must cover maritime claims in respect of loss resulting from delay in the carriage by sea of cargo. The insurance must cover maritime claims in respect of loss resulting from delay in the carriage by sea of

30 [1999] FCA 1419; (1999) 95 FCR 403; (on appeal) [2001] FCA 961.

31 [2013] FCAFC 21; (2013) 296 ALR 265.

32 [1992] HKCFI 273.

33 In *The Rangiora*, a letter from the liquidators acknowledging receipt of the notice of termination and accepting the termination was sufficient to constitute constructive redelivery of the ship. In *The Turakina* (at [125]) Tamberlin J indicated that a 'statement of intent by the charterer to the effect that possession was surrendered or redelivered' would be an act of symbolic delivery of possession and also (at para. 126) that 'some step or acknowledgement by the charterer to give effect to the redelivery' was required to show constructive redelivery.

34 Implementing Directive 2009/20/EC of the European Parliament and of the Council of 23 April 2009 on the insurance of shipowners for maritime claims.

35 Except for reg. 3(3)(b) which comes into force on the date of commencement of Part 9A and Schedule 11ZA(1) of the Merchant Shipping Act 1995 which were inserted by the Wreck Removal Convention Act 2011 (cl. 8).

passengers or their luggage, but only where the delay is consequent upon – (a) an incident involving a collision, stranding, explosion, fire or other cause affecting the physical condition of the ship so as to render it incapable of safe navigation to the intended destination of the passengers and their luggage; or (b) any other incident involving a threat to the life, health or safety of passengers.

The Regulations apply to seagoing ships of 300 gross tonnes or more but do not apply to warships, auxiliary warships or other State-owned or operated ships used for a non-commercial public service.

12.2.6 US Law

(a) Liability at common law

The vessel owner owes an absolute duty to seamen[36] and those doing seamen's work[37] to provide a seaworthy ship. The warranty continues when the vessel is under a demise charter, if the injury was caused by an unseaworthy condition present when the charter was made.[38] There are dicta in the Fifth Circuit in *Baker v Raymond*[39] that the owner of a demise chartered vessel will be liable in respect of injuries caused by an unseaworthy condition arising after delivery under the charter, but the general consensus of circuits is against this.[40] In *The Marine Sulphur Queen*, the proximate cause of loss of vessel and its crew could be inferred from several aspects of unseaworthiness, some occurring prior to delivery of the vessel to the demise charterer and some occurring after the delivery.[41] This raised a difficult question of causation which the Second Circuit did not have to resolve due to its finding that the owner was the wholly owned subsidiary of the charterer and was therefore held to be liable along with the demise charterer.[42]

The owners will not be liable in tort in respect of loss or damage arising out of

36 *The Osceola*, 189 US 158 (1903) in which the Supreme Court held that an owner of a ship is liable to indemnify seamen in his employ for injuries caused by the unseaworthiness of the vessel or its appurtenant appliances and equipment.

37 In *Seas Shipping Co v Sieracki*, 328 US 85, 89 (1946) the Supreme Court held that 'seamen' are individuals who perform work on board vessels traditionally performed by seamen and who are not covered by the Longshore and Harbor Workers' Compensation Act (33 USC §§ 901–950).

38 *Rodriguez v Coastal Ship Corp*, 210 F Supp 38 (SDNY 1962).

39 *Baker v Raymond International, Inc.*, 656 F 2d 173, 184 (5th Cir 1981). The decision in the case was that the owner could not make a valid bareboat charter relinquishing control.

40 However, a claim *in rem* may be made against the vessel itself in respect of injuries caused by unseaworthiness, whenever arising. *In the Barnstable*, 181 US 464 (1901); *Reed v Steamship Yaka*, 373 US 410 (1963).

41 460 F 2d 89 (2nd Cir 1972).

42 Per Judge Anderson at 100–101: 'Where, as here, the court inferred proximate cause from several aspects of unseaworthiness, some occurring prior to delivery of the vessel to MTL and some occurring after the delivery, we might appear to be faced with an insoluble problem whether MSTC, as owner-demisor, can properly be held liable to the Death Claimants. However, on the peculiar facts of this case, we need not decide that interesting question. Where, as here, the owner-demisor is a wholly owned subsidiary of the charterer, the stock of which would clearly be available to satisfy the claims of the Death Claimants if MTL's other assets were insufficient, we see nothing which, as a practical matter, bars us from affirming liability as to MSTC. We emphasize that we do not decide whether in a case of unexplained disappearance where the owner-demisor is not a wholly owned subsidiary of the parent-demise charterer, the owner could properly be held liable when it was responsible for only some of a group of defects which, in their totality, were inferred to be the proximate cause of the loss.'

the operation of the vessel during the period of the demise charter[43] unless the loss or damage is due to unseaworthiness that was present when the charter was made. *Gabarick v Lauren Maritime*[44] examined a collision involving a barge that had been demise chartered by its owner ACL to DRD, who had then time chartered it back to ACL. A basis for ACL's liability would be in negligence for knowingly placing an unsafe vessel into the hands of unsafe vessel operators and that this placement caused the collision. Although the vessel was in a poor state at the date of the demise charter, this was not causative of the collision. This was caused by the fault of DRD's employees. DRD had knowledge of its employees' misconduct which created an environment conducive to repetitive safety violations by its employees, especially the captain and steersman. The owners, ACL, however, had acted reasonably in the vetting process, although its vetting was imperfect and needed improvement.

If the owners are liable *in personam* they can limit to the amount of the vessel's value absent privity or negligence on their part. With a corporate owner this depends on privity or knowledge on the part of managerial employees.[45] An owner can be vicariously liable for negligence of non-managerial employees and still be entitled to limit.

However, the vessel may still be liable *in rem* up to the amount of her value if service can be effected on her and if the claim is a maritime lien. Maritime lien claims in the US are wider than in the UK and include: ship repairs; ship supplies; towage; use of dry dock or maritime railway or other necessities to any ship; crew wages; tort claims arising from a collision; personal injury claims (excluding Jones Act claims against employer); wharfage; stevedoring; cargo damage/loss; certain maritime contracts (e.g. breach of charter party); preferred ship mortgages; salvage; and claims for maritime pollution.

12.2.7 Statutory liabilities

Section 107 of the Comprehensive Environmental Response, Compensation and Liability Act of 1980 (CERCLA)[46] imposes liability on the owner and operator of a vessel or facility from which there is a release, or a threatened release, which causes the incurrence of response costs, of a hazardous substance. There are three defences available to the person who would otherwise be liable if they can establish by a preponderance of the evidence that the release of the hazardous substance was caused *solely* by:

(1) an act of God, or
(2) an act of war, or
(3) an act or omission of a third party other than an employee or agent of the defendant, or than one whose act or omission occurs in connection with a contractual relationship, existing directly or indirectly, with the defendant. The defendant must establish by a preponderance of the evidence that (a) he

43 *The Lotus Maru*, 615 F Supp 78 (SDNY 1985) where it was held that the owners were under no personal liability in tort in respect of an alleged shortage of a cargo of gasoline.
44 900 F Supp 2d 669 (ED La 2012).
45 *In Re Norfolk Dredging Co*, 2006 WL 3182761 (ED Va).
46 42 USC § 9607.

> exercised due care with respect to the hazardous substance concerned, taking into consideration the characteristics of such hazardous substance, in light of all relevant facts and circumstances, and (b) he took precautions against foreseeable acts or omissions of any such third party and the consequences that could foreseeably result from such acts or omissions.

The limitation figure for vessels carrying hazardous substances is the greater of $5 million or $300 per gross ton. For vessels not carrying hazardous substances as cargo or residue, the limitation is the greater of $500,000 or $300 per gross ton. For incineration vessels, defined as 'any vessel which carries hazardous substances for the purpose of incineration of such substances, as long as such substances or residues of such substances are on board', the limitation is the total of all costs of response plus $50 million. The right to limit will be lost if:

> (A)(i) the release or threat of release of a hazardous substance was the result of willful misconduct or willful negligence within the privity or knowledge of such person, or
>
> (ii) the primary cause of the release was a violation (within the privity or knowledge of such person) of applicable safety, construction, or operating standards or regulations; or
>
> (B) such person fails or refuses to provide all reasonable cooperation and assistance requested by a responsible public official in connection with response activities under the national contingency plan with respect to regulated carriers subject to the provisions of title 49 or vessels subject to the provisions of title 33 or 46, subparagraph (A)(ii) of this paragraph shall be deemed to refer to Federal standards or regulations.[47]

Section 108 then provides: 'The owner or operator of each vessel (except a non self-propelled barge that does not carry hazardous substances as cargo) over three hundred gross tons that uses any port or place in the United States or the navigable waters or any offshore facility, shall establish and maintain, in accordance with regulations promulgated by the President, evidence of financial responsibility of $300 per gross ton (or for a vessel carrying hazardous substances as cargo, or $5,000,000, whichever is greater) to cover the liability prescribed under paragraph (1) of section 9607(a) of this title.'

A similar regime was introduced by the Oil Pollution Act 1990 which imposes liability on the 'responsible party', for a vessel or a facility from which oil is discharged, or which poses the substantial threat of a discharge of oil, into or upon the navigable waters or adjoining shorelines or the exclusive economic zone.[48] The 'responsible party', which in the case of a vessel means the owners, operators, and demise charterers, will be liable for the removal costs and damages specified in OPA 90 that result from such incident, up to prescribed limits of liability, subject to the

47 In addition: 'If any person who is liable for a release or threat of release of a hazardous substance fails without sufficient cause to properly provide removal or remedial action upon order of the President pursuant to sections 9604 or 9606 of this title, such person may be liable to the United States for punitive damages in an amount at least equal to, and not more than three times, the amount of any costs incurred by the Fund as a result of such failure to take proper action.'

48 'Oil' is widely defined and will include bunker oil.

three defences set out in CERCLA 1980. OPA 90 33 USC 2716(a) requires owners and operators, including demise charterers, of certain vessels to establish and maintain evidence of financial responsibility sufficient to meet the maximum amount of liability to which they could be subjected under OPA. These requirements apply to responsible parties for any vessel over 300 gross tons (except a non-self propelled vessel that does not carry oil as cargo or fuel) using any place subject to the jurisdiction of the United States; and any vessel using the waters of the exclusive economic zone to transship or lighter oil destined for a place subject to the jurisdiction of the United States. The current limits of liability for vessels are:

(1) For an oil cargo tank vessel greater than 3,000 gross tons with a single hull, including a single-hull tank vessel fitted with double sides only or a double bottom only: the greater of $3,200 per gross ton or $23,496,000.
(2) For a tank vessel greater than 3,000 gross tons, other than a vessel referred to in (1), the greater of $2,000 per gross ton or $17,088,000.
(3) For an oil cargo tank vessel less than or equal to 3,000 gross tons with a single hull, including a single-hull tank vessel fitted with double sides only or a double bottom only: the greater of $3,200 per gross ton or $6,408,000.
(4) For a tank vessel less than or equal to 3,000 gross tons, other than a vessel referred to in (3): The greater of $2,000 per gross ton or $4,272,000.
(5) For any other vessel: the greater of $1,000 per gross ton or $854,400.

12.2.8 Insurance and indemnity

Barecon 2001 addresses the shipowner's exposure to liabilities arising out of the charterer's operation of the vessel in three ways; by insurance; by indemnity; and by a non-lien clause.

12.2.9 Insurance

Clause 10(iii) requires charterers to 'maintain financial security or responsibility in respect of third party liabilities as required by any government, including federal, state or municipal or other division or authority thereof, to enable the Vessel, without penalty or charge, lawfully to enter, remain at, or leave any port, place, territorial or contiguous waters of any country, state or municipality in performance of this Charter without any delay'. This will cover mandatory insurance, such as that required under CLC or under OPA. The necessary arrangements are to be made at charterers' sole expense and charterers 'shall indemnify the Owners against all consequences whatsoever (including loss of time) for any failure or inability to do so'.

Clause 13 is the general insurance and repairs provision which requires charterers to keep the vessel insured against hull and machinery, war, and P&I risks, as well as any financial security required under clause 10(iii). The insurances are to be arranged to protect the interests of both owners and charterers and the insurance policies 'shall cover the Owners and the Charterers according to their respective interests'. Charterers 'shall effect all insured repairs and shall undertake settlement and reimbursement from the insurers of all costs in connection with such repairs as well as insured charges, expenses and liabilities to the extent of coverage under the insurances herein provided for'. Charterers must also make the necessary documents available to the owners in order for the latter to comply with the insurance provisions of the financial instrument. If the charterers fail to arrange and keep any of the insurances the owners are given a right to withdraw under clause 28(a)(ii) which allows the owners the option of giving the charterers a respite of a specified number of days to rectify the position. Barecon 2001 contains an alternative provision in clause 14 whereby the registered owner will pay for hull insurance, with the charterer still being responsible for P&I cover, with a specific provision that the registered owner and insurers will have no rights of recovery or subrogation against the demise charterer, a provision that does not appear in clause 13.

The effect of clause 13 on the charterer's obligations to the owners and the right of owners' insurers to proceed against charterers by way of subrogation for breach of charter was recently considered in *The Ocean Victory* which concerned a demise charter on Barecon 1989 and the effect of the equivalent clauses, 12 and 13.[49] The vessel had become a total loss and a claim was being pursued by the owners' hull insurers against the time charterers for breach of the safe port warranty. The time charterers argued that the demise charterers had no claim against them as the demise charterers had incurred no liability under the safe port warranty under the head charter because of clause 12. It was submitted that clause 12 of the demise charterparty contained a complete code for the treatment of insured losses as between the parties in the event of a total loss. Accordingly, it could not have been intended by the parties that the demise charterer would have been liable to the registered owner for breach of the safe port warranty in respect of losses covered by the hull insurance taken out by the demise charterer at its expense in the joint names of both the registered owner and demise charterer.[50] The charterers pointed to the express waiver of subrogation in deleted clause 13 which did not appear in clause 12.

Teare J held that clause 12 was not comparable to the war risks clause in *The Evia (No. 2)* and it did not codify the rights and liabilities of the parties with regard to insured risks:

> It provides for the provision of insurance and who is to pay for it, for the demise charterers to be responsible for insured repairs and to reimburse themselves from the proceeds of the insurance policy, for the demise charterers to be responsible for other repairs and for the claims on a total loss to be paid to the mortgagee for distribution to the registered owner and demise charterers in accordance with their respective interests. I do not consider that this 'codifies' in the required sense the rights and liabilities of the parties with regard to breach of the safe port warranty where the casualty caused by the breach has given rise to a claim on the insurance.[51]

Teare J then went on to consider the effect of the provision that the owners and demise charterers were to be co-assureds. Did this bring into play the principle that, generally, an insurer cannot exercise rights of subrogation to pursue a claim in

49 [2013] EWHC 2199 (Comm); [2014] 1 Lloyd's Rep 59. These clauses are now cll. 13 and 14 of Barecon 2001.

50 The value of the vessel was US$88.5 million and the amount paid by the insurers was US$70 million which would mean that the owner's claim would have to be reduced by that amount if the time charterer's argument was correct.

51 At [198].

the name of one co-assured against another? He referred to Rix LJ's explanation of the basis of this principle in *Tyco Fire & Integrated Solutions (UK) Ltd v Rolls-Royce Motor Cars Ltd*[52] that the true basis of the contract between the parties needed to be construed. The contract's terms may provide 'a highly detailed and developed code which made it completely clear that they were designed to supplant any possible liability for negligence on the part of the contractor'[53] or 'the underlying contract envisages that one co-assured may be liable to another for negligence even within the sphere of the cover provided by the policy'. In the latter case Rix LJ was 'inclined to think that there is nothing in the doctrine of subrogation to prevent the insurer suing in the name of the employer to recover the insurance proceeds which the insurer has paid in the absence of any express ouster of the right of subrogation . . .'[54]

Looking at the contract in the instant case, there was an express safe port warranty by the demise charterers, there was no code of rights and obligations in clause 12 with regard to insured losses caused by a breach of the safe port warranty and there was no express ouster of the right of subrogation in clause 12. These features all pointed to an intention that the demise charterer would be liable to the owner for breach of the safe port warranty, notwithstanding that they were joint assured and could take the benefit of the insurance in the manner set out in clause 12. There would be no double recovery by the owner because their recovery from the charterers would go to the insurers by way of subrogation. The decision has since been overruled by the Court of Appeal, [2015] EWCA Civ 16. 22 January 2015, who held that the port was not unsafe and then went on to express the view that clause 12 did constitute a complete code for the treatment of insured losses as between the parties in the event of a total loss. The parties intended there to be an insurance funded result in the event of loss or damage to the vessel by marine risks (and war risks) so that even if the charterers had been in breach of their safe port obligation the owners, and their subrogated insurers, would not have been able to recover from them.

12.2.10 Indemnity

Clause 17 provides a scheme of mutual indemnity whereby charterers are to indemnify the owners against 'any loss, damage or expense incurred by the Owners arising out of or in relation to the operation of the Vessel by the charterers, and against any lien of whatsoever nature arising out of any event occurring during the Charter Period'. In the event of arrest or detention by reason of claims or liens arising out of the vessel's operations, the charterers at their own expense 'take all reasonable steps to secure that within a reasonable time the Vessel is released, including the provision of bail'. Without prejudice to the general indemnity charterers are to indemnify against all consequences or liabilities arising from the master, officers or agents signing bills of lading or other documents.[55] The indemnity does not cover liabilities arising out of

52 [2008] Lloyd's Rep IR 617; [2008] 2 All ER (Comm) 584.
53 At [36].
54 At [77].
55 Claims under the indemnity would be made where the owners became liable under bills of lading signed by the master in jurisdictions in which such bills were regarded as evidencing a contract with the registered owner, notwithstanding that the master was the servant of the demise charterer.

negligence in the operation of the vessel or by reason of its unseaworthiness, and the clause should be amended to make specific reason to loss or damage arising thereby. As a co-assured, the owners will be liable for unpaid calls under the P&I cover and it would be prudent to make specific provision for this in the indemnity provisions. The indemnity should also extend to any fines incurred by the owners due to the charterers' operation of the vessel. The principle of *ex turpi causa non actio oritur* would generally render unenforceable a provision to be indemnified in respect of a fine. However, an indemnity against a strict liability fine where there is no element of fault on the part of the party subject to the fine, has been held to be enforceable.[56]

If the vessel is arrested or detained by claims against owners, the owners, at their own expense, shall 'take all reasonable steps to secure that within a reasonable time the Vessel is released, including the provision of bail' and to indemnify the charterers against 'any loss, damage or expense incurred by the Charterers (including hire paid under this Charter) as a direct consequence of such arrest or detention'.

Clause 20 provides for a further indemnity in relation to wreck removal, as follows: 'In the event of the Vessel becoming a wreck or obstruction to navigation the Charterers shall indemnify the Owners against any sums whatsoever which the Owners shall become liable to pay and shall pay in consequence of the Vessel becoming a wreck or obstruction to navigation.' The indemnity will not arise until the owners have become liable to pay a sum and have paid a sum in consequence of the vessel becoming a wreck or obstruction to navigation. This will be after the charter has come to an end. In consequence, the owners will not be able to secure their claim by arresting a sister ship of the charterers under section 21(4)(ii) of the Senior Courts Act 1981.[57]

12.2.11 Non-lien

Clause 16 is an attempt to prevent maritime liens from arising by giving notice to parties dealing with charterers, such as suppliers, that charterers have no authority to create a maritime lien on the vessel in respect of claims by such parties against the charterers. It provides:

> The Charterers will not suffer, nor permit to be continued, any lien or encumbrance incurred by them or their agents, which might have priority over the title and interest of the Owners in the Vessel. The Charterers further agree to fasten to the Vessel in a conspicuous place and to keep so fastened during the Charter Period a notice reading as follows: 'This Vessel is the property of (name of Owners). It is under charter to (name of Charterers) and by the terms of the Charter Party neither the Charterers nor the Master have any right, power or authority to create, incur or permit to be imposed on the Vessel any lien whatsoever.'

12.3 Owners' remedies for charterers' default

In the event of default by charterers, the owners will want to get their vessel back. This will require termination of the charter and then repossession of the vessel.

56 *Osman v J Ralph Moss Ltd* [1970] 1 Lloyd's Rep 313 (CA); more recently, *Griffin v UHY Hacker Young & Partners (A Firm)* [2010] EWHC 146 (Ch); [2010] PNLR 20.

57 *The Faial* [2000] 1 Lloyd's Rep 473.

Clause 11 provides that hire is to be paid punctually and time shall be of the essence. Hire is to be paid continuously throughout the charter period and there is no provision for off-hire.

12.3.1 Clause 28 – Termination

This provides for withdrawal of the vessel in the event of:

(a) charterers' failure to pay hire in accordance with clause 11, subject to an anti-technicality provision whereby owners are to give charterers a set number of clear banking days within which to rectify the failure, where the failure is due to oversight, negligence, errors or omissions on the part of the charterers or their bankers

(b) the charterers' failure to comply with the requirements of:

  (i) Clause 6 (Trading Restrictions)

  (ii) Clause 13(a) (Insurance and Repairs)

  Under this sub-clause the Owners 'shall have the option, by written notice to the Charterers, to give the Charterers a specified number of days grace within which to rectify the failure without prejudice to the Owners' right to withdraw and terminate under this Clause if the Charterers fail to comply with such notice;'

  (iii) the Charterers fail to rectify any failure to comply with the requirements of sub-clause 10(a)(i) (Maintenance and Repairs) as soon as practically possible after the Owners have requested them in writing so to do and in any event so that the vessel's insurance cover is not prejudiced.[58]

The clause also provides for the charter to be deemed to be terminated if the vessel becomes a total loss or is declared as a constructive or compromised or arranged total loss.

Both charterers and owners are entitled to terminate this charter with immediate effect by written notice to the other party 'in the event of an order being made or resolution passed for the winding up, dissolution, liquidation or bankruptcy of the other party (otherwise than for the purpose of reconstruction or amalgamation) or if a receiver is appointed, or if it suspends payment, ceases to carry on business or makes any special arrangement of composition with its creditors'.

The clause concludes by providing that 'the termination of this Charter shall be without prejudice to all rights accrued due between the parties prior to the date of termination and to any claim that either party might have'. The obligation to pay hire will be a condition as time is expressed to be of the essence. Should the charter be terminated on this ground the owners will be entitled by way of damages to the outstanding instalments discounted for early payment.[59] However, the other clauses

58 The clause also provides for termination due to owners' default, as follows:

> If the Owners shall by any act or omission be in breach of their obligations under this Charter to the extent that the Charterers are deprived of the use of the Vessel and such breach continues for a period of fourteen (14) running days after written notice thereof has been given by the Charterers to the Owners, the Charterers shall be entitled to terminate this Charter with immediate effect by written notice to the Owners.

59 This is assumed to be the measure of damages for a termination of a bareboat charter for breach

which will trigger a right to terminate are not expressed to be conditions but, rather are innominate terms, and a termination on these grounds will not entitle the owners to damages for future sums. For this reason, it is common to include a term providing for a termination fee to be paid in the event of any termination. As a liquidated damages provision, the sum stipulated will become payable without any enquiry into the nature of the losses sustained by the owners or without reference to the duty to mitigate. However, the clause must provide for a genuine pre-estimate of loss in the event of default, otherwise it will be classed as a penalty and will be unenforceable.

The classification of a termination clause in a demise charter used as a finance lease was recently considered by the tribunal in *London Arbitration 14/13*.[60] The charterers and their guarantors argue that the termination sums were penalties because they were 'extortionate and extravagant in amount' and were not a genuine pre-estimate of any loss suffered by owners, relying partly on the fact that the termination sum was always 5 per cent greater than the option price by which the charterers could purchase the vessel before the end of the charter. The owners submitted that the law of penalties only applied where the relevant sums were payable upon a breach of contract and the termination clause provided for payment after any termination event which included a series of events unrelated to any breach by charterers. The tribunal rejected this argument because the owner had terminated because of the charterer's failure to pay hire, which was a breach of contract. The charterers bore the burden of showing that the termination sums were unlawful penalties[61] and the modern cases suggested that the court or tribunal was to ask whether there was a 'commercial justification' for the clause in question.[62] Recent cases such as *Philips Hong Kong Ltd v Attorney General of Hong Kong*[63] indicated that the courts should be slow to strike down clauses as penal, especially in commercial contracts between parties of equal bargaining power.

The tribunal considered what the owners would have been entitled to by way of damages for breach of the obligation to pay hire on time under which time was expressed to be of the essence. Damages would be equal to the unpaid and remaining future hire payments under the charters plus the option price which the charterers would have been required to pay. To account for accelerated receipt, those future payments would need to be discounted to present value. This gave a figure of US$31,875,356. In comparison, the total of the termination sums payable as at the termination date was US$24,681,963.10. In the circumstances, the tribunal would accept the owners' submission that the termination sums were 'commercially justified' and were not intended to be a deterrent to the charterers from breaching the charters. The termination provisions in the charters were negotiated by two commercially sophisticated parties of equal bargaining power. Although, on any

of condition, London Arbitration 14/13. However, the measure of damages for termination of a time charter for breach of condition is the difference between the charter hire rate and that obtained under a substitute charter for the unexpired period of the charter. *The Astra* [2013] EWHC 865 (Comm); [2013] 2 Lloyd's Rep 69.

60 (2013) 880 LMLN 3.

61 *Murray v Leisureplay* [2005] EWCA Civ 963; [2005] IRLR 946.

62 See *Lordsvale Finance plc v Bank of Zambia* [1996] QB 752; *Murray v Leisureplay* above; and *Alfred McAlpine Capital Projects Ltd v Tilebox Ltd* [2005] EWHC 281 (TCC); [2005] BLR 271.

63 (1993) 61 BLR 41.

given date the termination sum would generally be 5 per cent higher than the option price, such a margin, where the termination of the charter was the result of a default in payment by the charterers (or other circumstances evidencing the weakening of the credit underlying the transaction) was 'commercially justified' and consistent with market practice for such transactions. The termination sum was not intended as a 'deterrent', but was a reasonable, if necessarily unscientific, pre-estimate of the internal and other costs which might be incurred by the owners in such a default situation and which were not otherwise recoverable under the charters.

In the event of default, charterers may be able to obtain relief from forfeiture by applying to the court and tendering the outstanding sums owing to owners. In *The Jotunheim*,[64] Cooke J held that relief from forfeiture was available when an owner exercised a right of withdrawal under a demise charter. In *Shiloh Spinners Limited v Harding*,[65] Lord Wilberforce identified two well-recognised heads of the jurisdiction. The first was 'where it is possible to state that the object of the transaction and of the insertion of the right to forfeit is essentially to secure the payment of money'. The second was the case of fraud, accident, mistake or surprise. In *On Demand Plc v Gerson Plc* [66] Walker LJ identified a third head of jurisdiction arising out of Lord Wilberforce's judgment, where the forfeiture provision is security for the production of a stated result which can effectively be attained on the hearing of an application for relief. Applying these principles, Cooke J concluded that the withdrawal clause was either within the first or the third of these categories and that the court did have jurisdiction to give relief from forfeiture.

In exercising its discretion the court would look at the conduct of the party seeking relief from forfeiture, in particular whether his default was wilful, the gravity of the breaches, and of the disparity between the value of the property in respect of which forfeiture is claimed as compared with the damage caused by the breach. In the present case the demise charterers' case was not meritorious in that they had deliberately flouted their obligation to pay hire on time over three months despite the owners' constant protests. The court also had to consider the gravity of the breach as compared with the disparity between the value of the property of which forfeiture is claimed and the damage caused by the breach. The demise charterers had paid three months' hire and a $25,000 deposit in order to earn the freight under contracts under 33 bills of lading, which they had apparently received, as well as incurring some £12,000 expenditure on the ship at the loading port. This was not a case where they had paid the majority of the hire or price for the vessel since the transaction was still at a very early stage. What they would lose, however, was the right to pay further hire over 45 months and to pay further deposits of $200,000 in order to purchase the ship at the end of that period; they would also lose the use of the ship as a profit-earning chattel for the duration of that period. In addition, sums already expended would be wasted and potential liabilities would have been

64 [2004] EWHC 671 (Comm); [2005] 1 Lloyd's Rep 181. In *The Scaptrade* [1988] 2 Lloyd's Rep 253, the House of Lords had held that relief from forfeiture was not available where an owner exercised a right of withdrawal under a time charter, but had left open the position so far as demise charters were concerned.

65 [1973] AC 691.

66 [2000] 4 All ER 734.

incurred, either under the bills of lading or under the sub-charterparty or both. Taking all these factors into account, and in particular the attitude of the charterers towards their payment obligations, Cooke J declined to give relief from forfeiture.

Clause 29 deals with repossession following termination. Termination under clause 28 gives owners the right to repossess the vessel from the charterers at her current or next port of call, or at a port or place convenient to them without hindrance or interference by the charterers, the courts or local authorities.[67] Pending physical repossession, the charterers hold the vessel as gratuitous bailee of the owners. This provision means that charterers' legal possession of the vessel ceases on termination of the charter and any claims against charterers that arise out of the operation of the vessel between termination and physical redelivery will fall outside the provisions of section 20(4) of the Senior Courts Act 1981. However, owners need to take all possible steps to recover physical possession of the vessel after termination if they are not to risk being liable on contracts made by the charterers in this interim period by way of ostensible authority.

Following repossession the lender will then seek to recover the sums outstanding under the sale contract by sale of the vessel, proceedings under guarantees, and attachment of the borrower's assets. If the borrower owns other ships the lender may then consider an arrest. The fact that the demise charter is, in fact, a contract of sale by instalments will probably prevent an arrest of a sister ship of the demise charterer. The issue has recently arisen in the US in *The Icon Amazing*[68] where the owner was unable to attach a sister ship of the demise charterer under supplemental rule B of the Federal Rules of Civil Procedure's (FRCP) Supplemental Rules for Certain Admiralty and Maritime Claims (the Supplemental Rules) that govern the procedure for arresting or attaching a vessel in the United States. Rule B allows a party to obtain quasi *in rem* jurisdiction over a defendant's property for any debt arising out of a maritime claim, when the defendant 'cannot be found within the district'. The Texas court held that the owner's claim was not a maritime claim.[69] The charter required the charterer to purchase the vessel at the end of the term, the charter hire payments were not market-based but rather instalments of the full purchase price for the vessel, and the claim was not only for unpaid charter hire, but also for additional security in the event of a depreciation in the value of the vessel under a loan to value clause. Taking these factors into account, the court held that this was not a 'conventional maritime charter party' but, rather, an 'inseparable component of a larger non-maritime vessel sale/financing transaction'. In particular, the claim for nearly $3 million under the loan to value clause distinguished the case from a previous decision in *Jack Neilson, Inc v Tug Peggy*,[70] in which the charterparty

67 The clause continues:

> The Owners shall arrange for an authorised representative to board the Vessel as soon as reasonably practicable following the termination of the Charter. The Vessel shall be deemed to be repossessed by the Owners from the Charterers upon the boarding of the Vessel by the Owners' representative. All arrangements and expenses relating to the settling of wages, disembarkation and repatriation of the Charterers; Master, officers and crew shall be the sole responsibility of the Charterers.

68 951 F Supp 2d 909 (2013 SD Texas).

69 The general rule is that contracts for the sale of vessels are not maritime claims: *The Ada*, 250 F 194 (2nd Cir 1918).

70 428 F 2d 54, 58 (5th Cir 1970).

was found to be severable from the sale agreement as the plaintiff had amended its complaint to eliminate claims related to the purchase price and only requested damages specifically related to the 'lease'. In this respect it is worth noting that, from March 2013, it has become possible for finance charters over vessels registered in the Marshall Islands to be registered as a preferred mortgage.[71] This would give the owner an independent ground of access to the US District Courts for claims against the demise charterer.

12.3.2 Insolvency

(a) UK law

The compulsory or voluntary winding-up of a bareboat charterer will not normally prohibit or otherwise prevent the finance lessor/owner from exercising its right to repossess its own property (i.e. the ship) then in the hands of the company;[72] and this is also the case when the charterers are in administrative receivership.[73] It is unlikely that the conditional sale contained in the bareboat charter will be recharacterised as a security interest.[74] The provisions of the Insolvency Act 1986 (as amended by the Enterprise Act 2002) relating to administration covers all hire purchase agreements and chattel-leasing agreements such as bareboat charters. Paragraph 111 of Schedule B1 defines 'hire-purchase agreement' as including a conditional sale agreement, a chattel-leasing agreement[75] and a retention of title agreement. Under section 43(3) of Schedule B the owners will need the consent of the administrator or the court to take possession of the vessel. If the hire remaining to be paid under the bareboat charter is considerably less than the current market value of the ship and the bareboat charterers are ready, willing and able to carry on paying the charter hire it is possible that the court will prevent the finance lessor/owner from exercising its right to recover possession of the ship[76] and will permit the bareboat charterers to carry on performing their obligations under the charter, so as to preserve their existing rights under the charter.

Under paragraph 72(1) of Schedule B, an administrator has the power, exercisable with the court's agreement, to dispose of goods (including a ship) in the possession of the company in administration under a chattel-leasing agreement (such as a bareboat charter) 'as if all the rights of the owner under the [chattel leasing] agreement were vested in the company'. The court is charged with the responsibility of ensuring that this provision is only invoked in promoting one or more of the purposes for which the administration order was made. The disposal proceeds must

71 The Maritime (Amendment) Act (No. 1), 2013 which amended certain portions of the Marshall Islands Maritime Act 1990.

72 *General Share & Trust Co v Wetley Brick and Pottery Co* (1882) ChD 260 and *Re Blue Jeans Sales Ltd* [1979] 1 All ER 641.

73 Since the introduction of the Enterprise Act 2002, the holder of a floating charge cannot appoint an administrative receiver unless that floating charge was created before 15 September 2003 or if a statutory exemption applies, none of which is likely to apply in the case of a bareboat charter. See ss 72A to 72G of the Insolvency Act 1986 (as amended by the Enterprise Act 2002) and Order 2003, SI 2003/2095, art. 2.

74 See, e.g., *Welsh Development Agency v Export Finance Co Ltd* [1992] BCC 270 (CA).

75 Defined in s. 251 of the Insolvency Act 1986.

76 Following the decision in *Transag Haulage Ltd v Leyland DAF Finance Ltd* [1994] BCC 356.

also be used to pay down or discharge all amounts outstanding under the charter. This could mean that the administrator is not liable to account for the full value of the ship if (as is likely) it exceeds the amounts due under the charter (unless an order to account for the full value is obtained from the court).

Under paragraph 13 of Schedule 1 of the 1986 Act an administrator is given 'power to make any payment which is necessary or incidental to the performance of his functions'. This would enable it to pay charter hire and to grant leave to the finance lessor/owner to take possession of the ship. The discretion must be exercised as quickly as possible[77] and, where a ship is retained without the full hire (or any hire) being paid, the administrator is obliged to give precise reasons for the exercise of his discretion in such a manner.[78]

There is little English law authority as to whether a clause in a financing bareboat charter which purports automatically to terminate the charter on presentation of a petition for a winding-up or an administration order is effective.[79] Davis[80] suggests that the automatic termination should be expressed to be effective before presentation of the petition, since on presentation of the petition the automatic stay on rights comes into effect (including the right to take possession).[81]

(b) US law

Lease financing for ship finance is not in favour in the US due to the risk of a 'recharacterisation' of the charter so that the ship is treated as the property of the demise charterer in the event of the charterer's insolvency being determined before the US bankruptcy courts. Under US law, an agreement in the form of a lease which vests most of the economic benefits in the lessee may be recharacterised by the courts, particularly in bankruptcy, as in reality a security agreement. If the courts determine that the named lessor does not truly retain the economic benefits of the property ownership, but rather appears to be a creditor secured by its possession of title to the assets, the nominal lessee may be deemed the owner of the asset.[82] It is possible that a security interest in the bareboat charterer's rights to vessel earnings can be perfected by a UCC Financing Statement and perhaps an account control agreement (UCC §9-102(a)(2)(vi)), but the security interest is of low value and will yield to maritime liens, tax liens and other tort liens in any vessel foreclosure.

---

77 In order to avoid attracting other claims against him such as wrongful detention or interference with goods. See *Barclays Mercantile Business Finance Ltd v Sibec Developments Ltd* [1992] 1 WLR 1253 (ChD). This judgment considered the wording of s. 11 of the Insolvency Act 1986, which has now been replaced by Schedule B1 which contains different wording.

78 See *Re Atlantic Computer Systems plc* [1992] Ch 505. Nicholls LJ sets out guidelines for the exercise of an administrator's discretion at p. 529.

79 But see *Transag Haulage Ltd v DAF Finance plc* (cited above) where the court exercised its equitable jurisdiction to grant relief against forfeiture of a hire purchase agreement which had been terminated in accordance with a provision making it terminable upon the appointment of a receiver.

80 M. Davis, *Bareboat Charters* 2nd edn, (Informa Law, 2005) 35–38.

81 See *Re David Meek Plant Ltd* [1994] 1 BCLC 680.

82 See *In re Lykes Bros Steamship Co Inc* (US Bankruptcy Court for the Middle District of Florida, 1996 AMC 1488 decided 18 April 1996), in which the court held that the economic structure of the arrangements was a financing transaction, not a lease and that consequently ownership remained in the bankrupt charterer and the lessors were held to be unsecured creditors.

Until recently, no flag state legal system afforded a means for a ship lessor to perfect a security interest in vessels to which it held title.[83] However, on 6 March 2013, the Marshall Islands became the first flag state to adopt such a system for the lease finance of vessels. Nitijela Bill No 25 amends and adds to the Republic of the Marshall Islands Maritime Act (the Act) to create new definitions for 'documented owner',[84] 'financing charter'[85] and 'finance charterer'.[86] A new section 302A allows either party to file a financing charter with the Marshall Islands registry. A properly filed financing charter will enjoy the status of a preferred mortgage on the vessel in favour of the documented owner. The new law provides that, even if the owner and charterer agree in the charter that the charter constitutes a 'financing charter' this declaration alone is not binding on third parties or courts hearing the question. Under the new law, the documented owner may itself grant conventional preferred mortgages on the vessel, subject to whatever restrictions may exist in prior agreements entered into by the documented owner. It should be possible to continue to argue that the charter is a true lease but, in the alternative, if it is determined in any court or tribunal or by any governmental authority not to be a true lease, then the charter is agreed to be a financing charter entitled to the status of a preferred mortgage under Marshall Islands law. Enforcement of this new preferred mortgage outside the Marshall Islands will depend on the jurisdiction of arrest. Under US law, foreign mortgages are enforceable provided they meet certain minimum requirements of central filing and availability for inspection.

83 Unlike the position with commercial aircraft finance, in which lessors have long been able to record leases or memoranda of leases with governmental authorities for purposes of establishing the priority of rights in the aircraft.

84 A 'documented owner' is the person who appears in the application for documentation, even if he may hold only legal title and not the full benefits and indicia of ownership.

85 A 'financing charter' is 'a demise or bareboat charter, regardless of duration, between the documented owner and the finance charterer, which is agreed by the parties to be or is determined in judicial or arbitral proceedings to create in favor of the documented owner a security interest in the vessel' in favour of the finance charterer.

86 A 'finance charterer' is the person identified as the charterer in the financing charter.

CHAPTER 13

Financing newbuilding vessels and Barecon 2001: a fair deal?

*Dr Theodora Nikaki**

13.1 Introduction: setting the scene

Building and purchasing a vessel is a multi-million-dollar investment on the part of shipowners. As such, it requires careful planning in terms of negotiation and drafting of both the shipbuilding contract and the related financing arrangements, to be effected through either ship mortgages[1] or capital leases, such as finance-based bareboat charterparties.[2] In particular, financing bareboat charterparties of newbuilds, on which this chapter focuses, constitute rather complex commercial transactions, involving the negotiation of at least three types of contract between different players in the shipping and financing industry, namely the shipbuilding contract,[3] the bareboat charterparty and the conditional sale agreement.[4] There may also be other contractual agreements involved, such as a contract between the shipbuilders and their bank for the shipbuilders' financing of the project or the shipbuilders' risk insurance contract[5] but their discussion falls outside the scope of this chapter, which centres on the buyers' financing deal of the vessel.

The cornerstone of financing bareboat charterparties of newbuilds is the shipbuilding contract between the shipyard and the buyers. It governs the respective rights and duties not only of the shipbuilders and the buyers under the shipbuilding contract but also, to a great extent, of the parties to the bareboat charterparty.[6] Once the financial arrangements are settled, the shipbuilding contract will then be novated to the buyers' financiers.[7] This novation is a tripartite agreement between the shipbuilders, the buyers and the new buyers (the financing institution), under which, in consideration of the discharge of the original shipbuilding contract, the

* Associate Professor, Institute of International Shipping and Trade Law, College of Law, Swansea University.

1 Ship mortgages are discussed in Chapter 10 of this book.

2 Other terms used include 'full pay-out' lease or charterparty and 'bareboat lease'. See *Bridge Oil Ltd v Owners and/or Demise Charterers of the Ship Guiseppe di Vittorio (No. 1) (The Guiseppe di Vittorio)* [1998] 1 Lloyd's Rep 136 (CA), 156 (Evans LJ).

3 There are a number of different standard forms in regular use, such as the BIMCO-sponsored NEWBUILDCON and the much older Shipbuilders' Association of Japan (SAJ) Shipbuilding Contract form. See Chapter 1 of this book.

4 See generally the BIMCO BARECON 2001 form, Parts I–IV.

5 See further discussion on such contracts in Chapter 6 of this book.

6 See BARECON 2001, Part III.

7 For an example of a novation clause included in a shipbuilding contract, see *CMA CGM SA v Hyundai Mipo Dockyard Co Ltd* [2008] EWHC 2791; [2009] 1 Lloyd's Rep 213 at [5].

shipbuilders and the original buyers agree to replace it with a new building contract between the shipyard and the financing institution.[8] The novation of the shipbuilding contract will thus divest the original buyers from any responsibility for the performance of their obligations, while the financing institution, as new buyers, will undertake all rights and the duties arising out of the shipbuilding contract,[9] including their right to obtain title to the vessel upon its delivery and acceptance.[10] The financing transaction will not, however, be complete until a financing bareboat charterparty is concluded between the financing institution as owners of the vessel and the original buyers as charterers, which includes either a conditional sale agreement under which property to the vessel will pass to the charterers upon expiration of the charterparty provided certain conditions are met,[11] or an option to purchase in favour of the charterers.[12] Under such a financing charterparty, the charterers (that is, the buyers) bear the obligation to pay hire throughout the stipulated charterparty period, which covers either all of the financiers' capital paid to the shipyard under the terms of the shipbuilding contract plus their profit, or part of it if also combined with a lump sum deposit.[13] Moreover, unlike time charterparties, the risk of the vessel being out of service falls entirely on the charterers who have an operational interest in the vessel, meaning that their obligation to pay hire throughout the agreed duration of the charter will not cease in case of any contingency which may deprive them of the use of the vessel (in other words, 'hell or high water' payment terms).[14]

It comes as no surprise that financing bareboat charters are used in the shipping industry as means of financing the purchase of newbuilds or secondhand vessels, as they are designed to promote the interests of all parties involved in the transaction. From the financing lessors' perspective, the main advantage of such agreements is that they retain the ownership of the vessel as a matter of security, which means that they will be able to retake possession if the bareboat charterers become insolvent.[15] In addition, if all goes well, the lessors will make a profit out of the transaction without getting involved with the operation or the maintenance of the vessel, since

8 *Scarf v Jardine* [1882] All ER Rep 651 (HL), 655 (Selborne LC); *Bradford Old Bank Ltd v Sutcliffe* [1918] 2 KB 833 (CA), 841 (Pickford LJ), 849 (Scrutton LJ). The original parties to the shipbuilding contract may simplify the novation process by inserting clauses such as 'and/or for a company or companies to be nominated by them in due course', which have been construed by the courts as evincing an intention to novate the contract upon the nomination of the new purchasers (i.e. the financing institution, provided that the nominated party also agrees). See *Damon Compania Naviera SA v Hapag-Lloyd International SA (The Blankenstein, Bartenstein and Birkenstein)* [1985] 1 WLR 435 (CA), 446–447 (Fox LJ), 454–455 (Goff LJ).

9 *Bradford Old Bank Ltd v Sutcliffe*, ibid.

10 See, e.g., NEWBUILDCON, art. 31 and SAJ, art. VII.5.

11 See, to that effect, BARECON 2001, Part IV, lines 1–7 and the pertinent discussion under 13.4.

12 E.g. a hire-purchase agreement. See, for instance, the financing agreement in *Petroleo Brasileiro SA v Petromec Inc* [2013] EWCA 150 at [3].

13 On the calculation of hire in bareboat charters, see further M. Davis, *Bareboat Charters* 2nd edn, (LLP, 2005), §§35.5, 35.9 and 35.10. The parties may also agree that the financiers' capital and profit will be paid through lump sum deposit(s) at agreed times and monthly hire. See, e.g., the financing agreement in *More OG Romsdal Fylkesbatar AS v The Demised Charterers of the Ship 'Jotunheim' (The Jotunheim)* [2004] EWHC 671; [2005] 1 Lloyd's Rep 181 at [8].

14 BARECON 2001, Part II, cl. 13 (a) or 14(f), Part III, cll.3. With the exception, for instance, of the vessel being lost or missing. See BARECON 2001, Part II, cll. 11(e), and 28(c). See also Davis, ibid., at §§35.5 and 35.9.

15 On the procedures applicable to the different types of insolvency, see Davis, ibid., at §§35.28–35.36.

such obligations fall into the sphere of the charterers' responsibility (subject to the repairs falling into the scope of guarantee works).[16] Moreover, the lessors' financial interests are further protected because they are entitled to terminate the charterparty if the charterers fail to fulfil any of their core obligations under it, such as their duty to pay hire punctually, to comply with the agreed trading restrictions, and to maintain and repair the vessel.[17]

In a similar vein, the charterers (buyers) also benefit from such an arrangement as the bareboat charterparty constitutes a *contract of hire of the ship*, not for the provision of services.[18] Thus, they are entitled to possession of the ship and also have full control over the master and the crew, the master being the servant of the charterer, not of the owner.[19] In other words, the charterers/buyers can use the vessel as they deem appropriate to maximise their earnings,[20] which will, in turn, be used towards the payment of her purchase price via the payment of monthly instalments in the form of hire.[21]

Nevertheless, attention needs to be paid to any restrictions applying to lease financing agreements under the law of the flag of the vessel. For example, until 1996, it was unlawful for foreign banks to use this means to finance the purchase of a vessel employed in the US coastwise trade as they did not satisfy the rigid citizenship precondition imposed on this type of vessel, namely the qualification that at least 75 per cent of the owners of such vessels had to be a US citizen.[22] To boost the coastwise market, in 1996, the US Government relaxed the financing requirements for such vessels in order to allow foreign lessors to finance their acquisition, introducing the 'foreign lessor exemption', which eliminated the strict citizenship prerequisite.[23] To ensure, however, that foreign investors did not actually operate vessels in the coastwise trade, the foreign lessor exemption was made subject to other strict conditions, such as the requirements that the owner must only have a financial interest in the vessel (a 'passive investment') and also that the vessel be demise chartered for a period of at least three years to a US citizen who was qualified to engage in the coasting trade.[24]

This chapter focuses on a number of legal issues arising out of the financing arrangements related to the acquisition of newbuilding vessels under the BIMCO

16 See BARECON 2001, Part II, cl. 10(a) and Part III, cl. 1(d), lines 29–51.

17 Ibid., at Part II, cl. 28(a).

18 *Sea & Land Securities v Dickinson* [1942] 2 KB 65 (CA), 69–70 (Mackinnon LJ); *Bridge Oil Ltd v Owners and/or Demise Charterers of the Ship Guiseppe di Vittorio (No. 1) (The Guiseppe di Vittorio)* [1998] 1 Lloyd's Rep 136 (CA), 156 (Evans LJ).

19 See BARECON 2001, Part II, cl. 10(a)(i) and (b). See also *Baumwoll Manufactur von Carl Schleibler v Furness* [1892] 1 QB 253 (CA), 259 (Esher LJ), affirmed at [1893] AC 8 (HL), 16 (Herschell LC); *The I Congreso del Partido* [1978] QB 500 (QB), 537–538 (Goff J).

20 The owners' consent is only required if the charterers want to enter into a sub-charter on bareboat basis. See BARECON 2001, Part II, cl. 22(a).

21 BARECON 2001, Part IV, lines 1–7. Also, for a brief discussion on the tax benefits conferred on bareboat charterers under capital lease, see Chapter 12.

22 46 USCA §50501 (2006).

23 46 USCA §12106 (1996) as amended by the Coast Guard and Marine Transportation Act of 2004, PubLNo 108–293, §608, 118 Stat 1028, 1054 (2004).

24 Ibid. See further discussion in A. Hugunine, 'The Emergence of Lease Financing for Vessels Engaged in Coastwise Trade' (2006) 30 *Tul Mar LJ* 411; C. Papavizas, 'US-Flag Vessel Financing and Citizenship Requirements' (2007) 32 *Tul Mar LJ* 35; and D. Michaeli, 'Foreign Investment Restrictions in Coastwise Shipping: A Maritime Mess' (2014) 89 *NY ULRev* 1047.

BARECON 2001 Standard Bareboat Charter, a form which is widely used in the shipping industry. In particular, it analyses the owners' and charterers' respective rights and obligations as regards delivery and repairs. It also investigates the position under a conditional sale agreement (called 'Hire/Purchase' in BARECON 2001, Part IV). Considering the role each party plays in such a financing transaction – owners as financiers and charterers as ultimate buyers of the vessel – it critically evaluates their rights and duties with a view to ascertaining whether the distribution of the parties' rights and obligations under the BARECON 2001 charterparty is fairly balanced. Where appropriate, solutions will be also proposed to restore the balance between the parties' rights. As a final note, this chapter will evaluate the parties' position in the context of English law, being one of the applicable systems of law (and indeed the default) in BARECON 2001.[25]

13.2 Delivery

13.2.1 Overview

The delivery of a vessel to the charterers and her acceptance mark the commencement of the charterparty period and the time from which the duty to pay hire arises.[26] Generally, the vessel tendered for delivery must correspond with the description recorded in the charter form, as negotiated between the owners and the charterers.[27] The position is somewhat different when it comes to financing bareboat charters of newbuilding vessels, as the owners bear the obligation to deliver a ship already fully described in the shipbuilding contract. To that end, it is expressly provided in BARECON 2001 that the owners will discharge their delivery obligation as long as the vessel tendered for delivery is constructed and completed in accordance with the description set out in the shipbuilding contract, and the specifications and plans annexed to it[28] (including subsequent amendments), as approved by the charterers.[29] Thus, as the financing bareboat charterparty is drafted back-to-back with the shipbuilding contract, good delivery of the vessel under the shipbuilding contract amounts to good delivery at one and the same time under the bareboat charterparty,[30] relieving, in turn, the owners from any liability with respect to the vessel's unseaworthiness, specifications, performance and defects, except from those covered by guarantee works.[31]

Furthermore, taking into account the fact that that shipbuilding contracts normally allow for a grace period,[32] it is expressly provided in BARECON 2001 that the owners' duty to deliver the vessel arises only at the date of her actual delivery by the shipbuilders, which may not always coincide with the estimated delivery time.[33] In

25 See BARECON 2001, Part II, cl. 30(a) and (e).
26 See, e.g., BIMCO NYPE 93, cl. 1 and 10, BARECON 2001, Part II, cl. 11(a)–(b).
27 See, e.g., BIMCO NYPE 93, lines 10–20.
28 See, e.g., NEWBUILDCON, art. 2, SAJ, art. I.1–2,
29 BARECON 2001, Part III, cl. 1(a)–(b), 1(d), lines 19–25 and 2(a), lines 53–61.
30 *BW Gas AS v JAS Shipping Ltd* [2010] EWCA 68; [2010] 2 Lloyd's Rep 626 [39] (Rix LJ).
31 BARECON 2001, Part III, cl. 1(d), lines 25–51 and cl. 2(a).
32 See, e.g., NEWBUILDCON, arts 13 and 39(a)(iii) and 39 and SAJ, art. VIII.
33 BARECON 2001, Part III, cl. 1(d), lines 21–25 and cl. 2(a), lines 61–67.

addition, the bareboat charterparty imposes a corresponding duty on the charterers to accept a vessel built in accordance with the terms of the shipbuilding contract at the time she is in fact tendered for delivery.[34] Bearing this leeway in mind, charterers who plan to sub-charter the newbuilding vessel should include flexible cancelling clauses in their sub-charterparties, giving them the right to provide an alternative delivery date – also to be accepted by the sub-charterers – if they cannot meet the cancelling date because the vessel will be delivered later than the originally estimated date.[35] When entering into a sub-charter, the charterers should also provide a realistic 'estimated time of arrival' at the delivery port or 'expected ready to load date'. Otherwise, they will be held liable for damages as, given their involvement in the building process, it would be difficult to prove that they honestly expected that the vessel would be ready to load on the estimated date and that their expectation was based on reasonable grounds.[36]

Such delivery arrangements reflect the nature of the bareboat charterparty of newbuilding vessels as merely a financing tool. On the one hand, the owners are only there to facilitate the purchase of the newbuilding vessel and have practically no involvement in either the construction or the operation of the vessel. It would therefore be unfair to impose on them any liability with respect to the description of the vessel, her seaworthiness or any delay in her delivery. On the other hand, it is the charterers who have the operational interest in the vessel, being her operators and ultimate buyers. In that capacity, they are afforded the opportunity to negotiate and approve the specifications of the new vessel at the time of the conclusion of the shipbuilding contract, which is then novated to the owners, and also at any time during the construction process, as their consent is a prerequisite to any amendments to the terms of the shipbuilding contract and the vessel specifications/plans.[37] In a similar vein, the charterers are entitled to check the vessel's conformity with the agreed specifications throughout the construction process as the bareboat charterparty confers on them the right to send their representative in the shipyard.[38] It should be noted, though, that the bareboat charterparty terms do not automatically grant them such rights, as after the novation of the shipbuilding contract, the charterers have ceased to be a party to it. It is therefore necessary that a special provision to that effect is included in the shipbuilding contract to ensure that they will be allowed to exercise any of these rights. Moreover, it is the charterers who have the final say in deciding whether or not to reject the vessel tendered under the shipbuilding agreement if there is a valid reason to do so. BARECON 2001 imposes an obligation on the owners/lessors to consult with them before deciding on whether or not to accept delivery of such a vessel.[39] More importantly, it confers on the charterers the right to sue their contractual counterparty for damages if the shipbuilders terminate the shipbuilding contract for breach of contract on the part of the owners. All in all, it is

34 Ibid.

35 See, e.g., the cancelling clause in BIMCO NYPE 93, cl. 16, lines 209–218 (Extension of Cancelling).

36 *Maredelanto Compañia Naviera SA v Bergbau-Handel GmbH (The Mihalis Angelos)* [1971] 1 QB 164 (CA), 194 (Denning MR).

37 BARECON 2001, Part III, cl. 1(b).

38 Ibid., at cl. 1(c).

39 Ibid., at cl. 2(c)(i)–(ii).

safe to conclude that, subject to the exceptions discussed below, the bareboat charterparty confers on the charterers essentially the same delivery rights they would have had had they been parties to the shipbuilding contract.

Based on the above discussion, one might say that overall, the delivery rights and obligations of the parties under the financing bareboat charterparty are balanced, reflecting the role each one of them plays. The next subchapter will discuss issues related to the scope of the 'description' of the vessel as agreed in the shipbuilding contract to demonstrate how the different wording employed in standard shipbuilding contracts may impact the protection of the charterers' interests. It will then focus on the assessment of damages for non-delivery arising out of the owners' default with a view to analysing the complexities in calculating damages for breach of the delivery obligation under the financing bareboat charterparty.

13.2.2 The scope of the vessel description

We have so far established that delivery of a newbuilding vessel which conforms to the description provided in the shipbuilding contract simultaneously discharges the owners' delivery obligation under the bareboat charterparty.[40] It is, therefore, important to explore the scope of the vessel description as the charterers' core right to reject the vessel is dependent upon her compliance with it.[41]

It is trite law that full compliance with the terms of the vessel's description set out in the shipbuilding contract does not amount to a strict condition precedent of the charterers' obligation to accept the vessel.[42] The applicable test to the compliance with the contractual description is one of 'mercantile character', assessing whether, according to 'tests men in the market would apply', the vessel tendered for delivery is of a different kind from that which the owners have agreed to buy under the terms of the novated shipbuilding contract.[43] To that end, the description of the vessel encompasses only items amounting to a substantial ingredient of her *'identity'* and does not extend to articles which simply aim to *label* or *identify* the vessel so that the parties can locate her and, should they wish, sub-dispose of her.[44] For example, the description of the vessel entails articles such as her speed, fuel consumption and cubic and deadweight capacity but not the vessel number indicated in the shipbuilding contract (i.e. 'to be built by Osaka Shipbuilding and known as Hull number no. 354'), which is deemed to only form part of her identification under the shipbuilding contract.[45]

Nevertheless, it is at least arguable that the owners will not fulfil their delivery

40 Ibid., at cl. 1(d), lines 19–24 and at cl. 2(a), lines 53–61. See also *BW Gas AS v JAS Shipping Ltd* [2010] EWCA 68; [2010] 2 Lloyd's Rep 626 [39] (Rix LJ).

41 BARECON 2001, Part III, cl. 1(d), lines 19–25 and cl. 2(a).

42 *Reardon Smith Line Ltd v Yngvar Hansen Tangen (The Diana Prosperity)* [1976] 2 Lloyd's Rep 60 (CA), 72 (Denning MR), approved by Lord Wilberforce in the House of Lords at [1976] 1 WLR 989, 998–999.

43 *Christopher Hill Ltd v Ashington Piggeries Ltd* [1972] AC 441 (HL), 489 (Wilberforce LJ), 503 (Diplock LJ).

44 *Couchman v Hill* [1947] KB 554 (CA), 559 (Scott LJ); *The Diana Prosperity* [1976] 1 WLR 989 (HL), 999 (Wilberforce LJ); *Sanko Steamship Co Ltd v Kano Trading Ltd* [1978] 1 Lloyd's Rep 156 (CA), 158–159 (Roskill LJ), 160–161 (Bridge LJ), 161 (Stephenson LJ).

45 *The Diana Prosperity*, ibid. *Sanko Steamship Co Ltd v Kano Trading Ltd*, ibid.

obligation under the charterparty if the tendered vessel is built in a different shipyard from the one(s) expressly stipulated in the shipbuilding contract. This issue may well arise if the vessel is built under the NEWBUILDCON terms. NEWBUILDCON, clause 2, titled 'Description', explicitly provides in paragraph (a) that the vessel shall be built 'at the shipyard and shall have the Builder's Hull Number stated in Box 6'. The shipyard is further defined in NEWBUILDCON as 'the place or places stated in Box 5 where the vessel will be assembled and/or constructed'.[46] In fact, there was a comparable provision in *Sanko*,[47] where clause 41 of the charterparty made reference to a 'vessel to be built *at* Osaka Shipbuilding Co. Ltd. Hull Number 352', whilst the preamble referred to a vessel 'known *as* Osaka Shipbuilding Co. Ltd., Hull No. 352 until named' (emphasis added).[48] The Court of Appeal read the two provisions together to conclude that, in the light of the principles enunciated in *The Diana Prosperity*,[49] the words in the charterparty were not apt to create a contractual obligation requiring the owners to tender a vessel built at the Osaka Shipbuilding Co Ltd's yard at Osaka, therefore the owners were entitled to tender the ship built at Oshima and bearing the yard number 352.[50]

One might argue, however, that *Sanko* is distinguishable for a number of reasons and therefore is not applicable to the wording employed in NEWBUILDCON. NEWBUILDCON, article 2(a), clearly provides for the assembly and construction of the vessel at the agreed place(s), as opposed to the clause at issue in *Sanko*, which merely made reference to a vessel 'to be built *at Osaka Shipbuilding Co., Ltd.*' (emphasis added), wording which Roskill LJ found to be incomplete and unclear.[51] Moreover, as Roskill LJ found himself bound to construe that clause together with the preamble, he left open the issue of the interpretation of the clause had it stood alone, when it might well have meant that the vessel was to be built at the Osaka Shipbuilding Co's own yard.[52] More importantly, it transpires from the text of NEWBUILDCON that it is the intent of the parties to the shipbuilding contract, and in turn, to the bareboat charterparty, to qualify the shipyard where the vessel is built as a substantial ingredient of her identity, as the undertaking to construct the vessel at the agreed shipyard in NEWBUILDCON falls within the scope of the vessel description clause (clause 2). This is also reinforced by the argument that, on the true construction of the shipbuilding contract, and in turn of the bareboat charterparty which makes express reference to the vessel description under it, the express enumeration of such places in the shipbuilding contract demonstrates the parties' reliance on the stipulated contracting and subcontracting arrangements made in the course of the shipbuilding. It is thus the parties' understanding that the vessel they have bargained for may only be assembled and constructed at the place(s) mentioned at the shipbuilding contract.[53] Thus, it is submitted that, unlike

---

46 NEWBUILDCON, lines 48–49.

47 *Sanko Steamship Co Ltd v Kano Trading Ltd* [1978] 1 Lloyd's Rep 156 (CA).

48 Ibid., at 158.

49 [1976] 1 WLR 989 (HL), 999 (Wilberforce LJ).

50 *Sanko Steamship Co Ltd v Kano Trading Ltd, at* 160 (Roskill LJ), 160–161 (Bridge LJ) and 161 (Stephenson LJ).

51 Ibid., at 159 (Roskill LJ).

52 Ibid., at 159 (Roskill LJ) and 161 (Bridge LJ)

53 Such an argument was rejected in both *The Diana Prosperity* and *Sanko* on the basis that the true

the loose language of the clauses at issue in both *The Diana Prosperity* and *Sanko*,[54] the wording employed in NEWBUILDCON, article 2(a), is to be treated as creating a contractual obligation to deliver a vessel constructed in the agreed shipyard(s) and not as introducing merely 'labelling words'.

Although shipbuilding contracts include detailed provisions on items falling into the description of the vessel, such as the guarantees related to the vessel's speed, fuel consumption and cubic and deadweight capacity,[55] there will be cases where, given the complexity of the shipbuilding process, the final product may not fully conform to the agreed specifications. As this may happen quite often, shipbuilding contracts include clauses providing for tolerated breaches in terms of speed, fuel consumption and cubic and deadweight capacity, which entitle the owners to only claim liquidated damages but not to reject the vessel.[56] In turn, the owners will tender such a vessel to the charterers, who are under the obligation to accept her, even though she falls short of the agreed characteristics, as the charterers may reject the vessel only if the owners would have had the same right under the shipbuilding contract.[57] Nevertheless, depending on the charterparty terms, the charterers may be entitled to claim from the owners the stipulated proportion of the liquidated damages the latter will recover from the shipyard.[58]

What may complicate things and also disadvantage the charterers is the provision in shipbuilding contracts for the valid delivery to the owners of a newbuilding vessel with 'minor defects'. Such a clause is found in NEWBUILDCON, which obliges the owners to take delivery of a vessel with 'defects of minor importance, not affecting her class or her operation in its intended trade', which may not be rectified within reasonable time and in any event before the accrual of the owners' rights to terminate the shipbuilding contract.[59] For example, under NEWBUILDCON, the owners/lessors as buyers would be required to take delivery of a container ship even though the shipyard had not completed the decoration of her interior but would not be obliged to accept such a vessel if she had faulty operating cranes. Assuming that the defect is minor, the owners are thus obliged to accept delivery of the vessel on the condition that the shipbuilders agree not only to rectify the defects at their own cost and as soon as possible but also to indemnify them for any consequential loss they incurred, including any loss of time.[60] Then, the charterers in turn have no other option but to accept the tendered vessel, as the owners were not entitled to reject her under the terms of the shipbuilding contract either,[61] and must also

construction of the pertinent clauses did not support such a conclusion. *The Diana Prosperity* [1976] 1 WLR 989 (HL), 1000 (Wilberforce LJ); *Sanko Steamship Co Ltd v Kano Trading Ltd* [1978] 1 Lloyd's Rep 156 (CA), 160 (Roskill LJ).

54 *The Diana Prosperity*, ibid., at 994; *Sanko Steamship Co Ltd v Kano Trading Ltd*, ibid., at 158.

55 NEWBUILDCON, art. 2, SAJ, art. I.2.

56 NEWBUILDCON, cl. 8–13, SAJ, art. III.

57 BARECON 2001, Part III, cl. 2(c)(iii). Similarly, the delivery of a vessel built 'without buyer's supplies' amounts to good delivery if the shipbuilding contract provides that the builder is entitled to construct the vessel without them. See *BW Gas AS v JAS Shipping Ltd* [2010] EWCA 68; [2010] 2 Lloyd's Rep 626.

58 BARECON 2001, Part III, cl. 1(d), lines 45–51.

59 NEWBUILDCON, art. 27(d)(iv).

60 Ibid., at art. 27(d)(iv) (1)–(3).

61 BARECON 2001, Part III, cl. 2(c)(iii).

wait for the shipyard to remedy the defect. It is nevertheless clear that the charterers' interests are not adequately protected as the privity of contract doctrine prevents them from suing the shipbuilders directly if they fail to remedy the minor defect within a reasonable period of time. To make things worse, the charterers have no legal means of forcing the owners to take a legal action against the shipyard that does not comply with its obligation to rectify such faults. The owners' obligation to endeavour to compel the builders to repair the vessel extends only to deficiencies falling into the scope of guarantee works appearing *after the delivery* of the vessel and not to those defects which are in existence *at the time of the delivery*.[62] Even if the necessary repairs are effected by the shipbuilders, it will be the charterers who may incur losses, such as damages for loss of time, if the vessel is taken out of service to effect the repairs, since the hire is agreed on 'hell or high water payment' terms. This is, in fact, an example of how the financing agreement may jeopardise the charterers' interests, as had they been party to the shipbuilding contract, they would have been entitled to sue the shipyard directly and also claim damages for any time lost. Therefore, if the vessel is built under the NEWBUILDCON terms, the parties to the financing bareboat charterparty should consider including additional typed clauses in either the shipbuilding contract or their charterparty agreement to circumvent these hurdles. To that effect, the owners should for example, negotiate with the shipyard the incorporation of a clause in the shipbuilding contract granting the charterers the right to sue the shipyard directly for minor defects or conferring them the benefit of the 'minor defects' clause.[63] Or, if the shipyard is not agreeable, the parties to the bareboat charterparty may agree on the inclusion of two additional clauses in their contract, one imposing on the owners a duty to use best endeavours to compel the builders to repair minor defects, as well as an off-hire clause for the time lost during repairs of the minor defects. Such clauses will certainly advance the charterers' interests without at the same time disadvantaging the owners as any damages for loss of time will be passed on to the shipbuilders who have anyway agreed to pay them for such damages.[64]

13.2.3 Liability for non-delivery

Unlike other types of charterparty, financing bareboat charterparties are often fixed for newbuilding vessels which are then only at the construction stage. Therefore, one should not underestimate the possibility that the vessel may never be delivered at all for a number of reasons, including the termination of the charterparty by either the builders or the buyers for any of the events enumerated in the shipbuilding contract, or because the vessel is lost prior to delivery.[65] Again, bearing in mind the role the owners play in the transaction as financiers, it would be unjust to impose on them any liability if the vessel is not delivered for reasons beyond their control. This is, in fact, reflected in the BARECON 2001, which expressly exonerates them for any such liability if the builders are discharged from their obligation to deliver the vessel

62 Ibid., at cl. 1(d), lines 29–34.
63 Contracts (Rights of Third Parties) Act 1999, s. 1(1).
64 NEWBUILDCON, art. 27(d)(iv)(2)–(3).
65 See, e.g., NEWBUILDCON, art. 39, SAJ, arts X–XI.

for any reason other than the owners' own default,[66] such as the insolvency of the guarantor providing the performance guarantee on behalf of the owners as buyers.[67] They will also be relieved from any liability for non-delivery if the charterers instruct them to reject the vessel for non-compliance with the terms of the shipbuilding contract.[68]

Nevertheless, the owners would not escape liability for non-delivery of the vessel under the financing bareboat charterparty if it was their own default which prevented the delivery. For instance, standard shipbuilding contracts grant the shipyard the right to terminate the building contract if the buyers fail to meet their obligation to pay any of the instalments of the contract price or take delivery of the vessel.[69] If this is the case, the charterers will be entitled to treat the owners' default as anticipatory breach of the combined hire/purchase agreement. In particular, the owners will be guilty of anticipatory breach by renunciation of both the bareboat charterparty and the conditional sale contract in so far as their acts or defaults taking place before the time for delivery would lead a reasonable person to the conclusion that they would not perform their side of either of the contracts.[70] In addition, the charterers may establish anticipatory breach by self-induced impossibility as it is the owners' default which has disabled them from performing their obligation to deliver the vessel to the charterers, depriving them in turn of substantially the whole benefit of the contract (i.e. the use of the newbuilding vessel).[71]

If the charterers opt to act upon the anticipatory renunciation and adopt it as a rescission of the combined hire/purchase agreement,[72] then they will be discharged of all of their obligations, including (most significantly) their duty to pay hire. Moreover, they will be entitled to sue the owners for damages forthwith,[73] without even having to prove that they would have been ready and willing at the time of the renunciation to perform their part of the contract.[74] In assessing their damages, the principle of mitigation will also come into play, meaning that the charterers must take all reasonable steps to reduce their loss naturally flowing from the owners' failure to deliver,[75] for example by going to the market to obtain a substitute vessel,

---

66 BARECON 2001, cl. 2(b) and (iv).

67 See, e.g., NEWBUILDCON, art. 39(b)(i).

68 BARECON 2001, Part III, cl. 2(c)(i) and (iv).

69 BARECON 2001, ibid., at cl. 2(b), NEWBUILDCON, art. 39(b)(ii)–(iii), SAJ, art. XI. 1.

70 *Freeth v Burr* (1874) LR 9 CP 208 (CCP), 213, (Coleridge CJ), 214–215 (Keating J); *Decro-Wall International SA v Practitioners in Marketing Ltd* [1971] 1 WLR 361 (CA), 368 (Salmon LJ), 374 (Sachs LJ), 380 (Buckley LJ); *Universal Cargo Carriers Corporation v Citati* [1957] 2 QB 401, 436–438 (Devlin J); *Geden Operations Ltd v Dry Bulk Handy Holdings Inc (The Bulk Uruguay)* [2014] EWHC 885; [2014] 2 Lloyd's Rep 66 [15]–[21] (Popplewell J).

71 The breach of the financing charterparty is indeed inevitable as the defaulting owners/lessors cannot fulfil their obligation to deliver the vessel to the charterers/buyers. *Universal Cargo Carriers Corporation v Citati*, ibid.; *Geden Operations Ltd v Dry Bulk Handy Holdings Inc (The Bulk Uruguay)*, ibid., at [15]–[18]. For the advantages of establishing anticipatory breach by renunciation, rather than impossibility see *Universal Cargo Carriers Corporation v Citati*, ibid., at 436–437 and 443 (Devlin J).

72 *Johnstone v Milling* (1886) 16 QBD 460 (CA), 467 (Esher MR), 470 (Cotton LJ).

73 Ibid., at 467 (Esher MR).

74 Nevertheless, if they cannot prove that they would have been in a position to complete in any case, then they may be limited to nominal damages: see, e.g., *Cooper Ewing & Co Ltd v Hamel & Horley Ltd* (1922) 13 LIL Rep 590 (KB), 593 (Sankey J) and *Flame SA v Glory Wealth Shipping PTE Ltd (The Glory Wealth)* [2013] EWHC 3153; [2014] QB 1080 [81]–[86] (Teare J).

75 *British Westinghouse Electric & Manufacturing Co Ltd v Underground Electric Railways Co of London*

if a reasonable opportunity arises,[76] and will be debarred from claiming damages for any such loss which they unreasonably failed to avoid.[77] Hence, the measure of the charterers' damages will depend on whether there is an available market for a 'like-for-like' replacement, as well as on whether they have met their duty to mitigate their loss.

This raises the question of what amounts to an 'available market' for a substitute newbuilding vessel. A number of cases have considered the meaning of 'available market' in the context of sales contracts and concluded that such a market exists if there is either at least one actual buyer that day at a fair price or, if there is no actual offer for sale, if on that day there are in the market a sufficient number of traders potentially in touch with each other to establish a market in which the actual or notional seller could sell the goods should he wished to do so.[78] Although this is a definition of the term elaborated and adopted for the purpose of defining 'available market' in the context of sale of goods, courts have found it to be applicable also to charterparty cases,[79] whether the charterers can replace the broken charter with exactly what they bargained for by entering into a replacement charter for the unexpired term on materially equivalent terms (other than freight/hire but including trading limits, hire period and type of cargo).[80] Other cases have stressed that there may be no available market for goods which are too specialised, including goods which had to be manufactured to the specifications requested by the buyers.[81] This is exactly the difficulty the charterers will encounter in finding a readily available substitute vessel, since the newbuilding vessel is constructed in accordance with the specifications agreed in the shipbuilding contract. Such a difficulty raises, in turn, a number of issues with respect to the actions expected to be taken by the charterers to fulfil their obligation to mitigate their loss. The starting point should be that the mitigation rule requires the charterers to only act reasonably, taking an action which a reasonable and prudent person might in the ordinary conduct of business have taken.[82] To that end, reasonable steps to be taken by the charterers to reduce the loss arising from the non-delivery of the newbuilding vessel would not entail a

Ltd [1912] AC 673 (HL), 689 (Viscount Haldane LC); *Fulton Shipping Inc of Panama v Globalia Business Travel SAU of Spain* [2014] EWHC 1547; [2014] 2 Lloyd's Rep 230 [18] (Popplewell J) referring to the summary of the mitigation rules in *McGregor on Damages* 18th edn (Sweet & Maxwell, 2011) [7.003–7.006]. It should also be noted that such a duty does not arise if the anticipatory repudiation is not accepted by the injured party. *Shindler v Northern Raincoat Co* [1960] 1 WLR 1038 (Ass Manch), 1048 (Diplock J).

76 Sale of Goods Act 1979, s. 51(3), embodying a mitigation principle; *Melachrino v Nickoll & Knight* [1920] 1 KB 693 (KB), 697 (Bailhache J).

77 *British Westinghouse Electric & Manufacturing Co Ltd v Underground Electric Railways Co of London Ltd*, ibid.; *Fulton Shipping Inc of Panama v Globalia Business Travel SAU*, ibid.

78 *Shearson Lehman Hutton Inc v Maclaine Watson & Co Ltd (No. 2)* [1990] 3 All ER 723 (KB), 730 (Webster J).

79 *Zodiac Maritime Agencies Ltd v Fortescue Metals Group Ltd (The Kildare)* [2010] EWHC 903; [2011] 2 Lloyd's Rep 360 [56]–[58] (Steel J).

80 Ibid., at [57]–[58], applying *Golden Strait Corporation v Nippon Yusen Kubishka Kaisha (The Golden Victory* [2007] UKHL 12; [2007] 2 AC 353. See also *Spar Shipping AS v Grand China Logistics Holding (Group) Co Ltd* [2015] EWHC 718 [220] (Popplewell J).

81 *Hinde v Liddell* (1874–75) LR 10 QB 265, 269 (Blackburn J). See also *Hirtenstein v Hill Dickinson LLP* [2014] EWHC 2711 [118]. For a detailed discussion on the definition of 'available market', see M.G. Bridge, *The Sale of Goods* 3rd edn (OUP, 2013), §§12.70–12.73.

82 *Dunkirk Colliery Co v Lever* (1878) 9 ChD 20 (CA), 25 (James LJ); *British Westinghouse Electric*

duty to commission the construction of a new vessel[83] or buy a vessel which does not meet the contracted specifications,[84] such specifications to be interpreted as including any tolerated breaches agreed in the shipbuilding contract.

A more challenging question will arise in cases where the construction of the newbuilding vessel at issue was completed in accordance with the shipbuilding specifications but the vessel was not delivered because the owners failed to pay the last instalment or accept delivery. One might argue here for the availability of a market for a substitute vessel in the form of the original vessel herself, as the shipbuilders would usually exercise their right under the shipbuilding contract to sell her elsewhere.[85] Moreover, the charterers will normally be aware of the opportunity created as, having doubtless exercised their rights to monitor the building of the vessel throughout the construction stage, they will clearly know that the vessel is complete and ready. It is, therefore, submitted that the charterers' duty to mitigate their loss requires them to at least approach the shipyard and negotiate the purchase of that particular vessel and, if they are successful, to also enter into a new financing agreement with another financial institution. It is also submitted that the charterers will be deemed to have acted reasonably for the purposes of the avoidable loss rule even if they settled the financing arrangements of the vessel at a higher interest rate, resulting in a higher purchase price.[86] Therefore, they should be entitled to claim any such additional amounts from the owners, as such an extra cost fell on them entirely through their owners' breach.[87] The charterers would not, however, be expected to pay any extortionate price for the repurchase of the vessel, if the shipyard attempted to take advantage of the situation, as this is not an action a reasonable and prudent person might in the ordinary conduct of business have been expected to take.[88] In any case, even if unsuccessful in reaching an agreement with the shipyard to repurchase, the charterers will still be entitled to claim the expenses incurred in taking reasonable steps to repurchase and refinance the acquisition of the newbuild, including legal and brokers' fees.[89]

Assuming that there is an available market, the charterers will be entitled to damages, which will be subject to the usual rules on quantification, including the

---

& Manufacturing Co Ltd v Underground Electric Railways Co of London Ltd [1912] AC 673 (HL), 690 (Viscount Haldane LC).

83 *Sealace Shipping Co Ltd v Oceanvoice Ltd (The Alecos M)* [1991] 1 Lloyd's Rep 120 (CA), 125 (*per curiam*).

84 As a buyer generally is not obliged under the mitigation rules to accept from the seller goods that do not conform with the contract specifications as per *Kersterton v Albiac*, it seems to follow that the buyer is not obliged under the mitigation rules to buy goods that do not meet the contract description from third parties in order to fulfil their duty to mitigate their damages. *Heaven & Kesterton v Établissements François Albiac et Cie* [1956] 2 Lloyd's Rep 316 (QB), 321 (Devlin J).

85 NEWBUILDCON, art. 39(f), SAJ, art. XI.4.

86 See the reasoning in *Bacon v Cooper (Metals) Ltd* [1982] 1 All ER 397(QB), 401–402 (Cantley J).

87 Ibid.

88 *Dunkirk Colliery Co v Lever* (1878) 9 ChD 20 (CA), 25 (James LJ); *Melachrino v Nickoll & Knight* [1920] 1 KB 693 (KB), 697 (Bailhache J), referring to the duty to mitigate 'if a reasonable opportunity arises'.

89 *Fulton Shipping Inc of Panama v Globalia Business Travel SAU of Spain* [2014] EWHC 1547; [2014] 2 Lloyd's Rep 230 [18] (Popplewell J), referring to the summary of the mitigation rules in *McGregor on Damages* 18th edn (Sweet & Maxwell, 2011) [7.003–7.006].

compensatory principle[90] and the mitigation rule.[91] When calculating such damages, one should take into account the interrelationship between the bareboat charterparty and the conditional sale agreement. Neither of these two contracts stands alone as they operate together towards financing the purchase of the vessel. In that respect, the hire instalments paid to the owners throughout the charterparty period and any lump sums agreed (if any) cover the consideration for both the use of the vessel during the hire period and the vessel price. Bearing this in mind, the damages which will place the charterers in the same financial position as if the vessel had been delivered to them and her ownership transferred, will be assessed based on the difference between the agreed contract rate of the financing bareboat charter period (including any lump sums agreed) and the market rate for the chartering in and financing of a substitute vessel for the same period.[92]

Moreover, the charterers' damages will be assessed based on the assumption that, after accepting the anticipatory renunciation of the combined hire/purchase agreement, the charterers acted reasonably to minimise their loss by entering as soon as reasonably possible into a new financing bareboat charterparty for a substitute vessel on the best terms available.[93] Thus, the charterers should be afforded a reasonable period of time after their acceptance of the repudiation to decide how to mitigate,[94] taking into consideration the time required to negotiate the purchase of another vessel and her financing. If, nevertheless, they fail to conclude a new combined hire/purchase agreement within such a reasonable time and in the meantime the market price has gone up, they will be debarred from claiming the difference between the lower price on the date they ought to have entered into a new financing agreement and the higher price they actually paid to hire/purchase a substitute vessel.[95] Certainly, their additional loss is to be attributed to their neglect to mitigate the damages within reasonable time, which breaks the chain of causation between their loss and the owners' breach.[96]

---

90 *Robinson v Harman* (1848) 1 Ex 850 (Exch Ct), 855 (Parke B); *Watts, Watts & Co Ltd v Mitsui & Co Ltd* [1917] AC 227 (HL), 241 (Dunedin LJ); *Golden Strait Corpn v Nippon Kubisha Kaisha (The Golden Victory)* [2007] UKHL 12; [2007] 2 AC 353 [9] (Bingham LJ), [29] (Scott LJ) and [57] (Brown LJ).

91 *British Westinghouse Electric & Manufacturing Co Ltd v Underground Electric Railways Co of London Ltd* [1912] AC 673 (HL), 689 (Viscount Haldane LC); *Fulton Shipping Inc of Panama v Globalia Business Travel SAU of Spain* [2014] EWHC 1547; [2014] 2 Lloyd's Rep 230 [18] (Popplewell J).

92 *Koch Marine Inc v d'Amica SaRL (The Elena d'Amico)* [1980] 1 Lloyd's Rep 75 (QB), 87 (Goff J); *Kaines (UK) v Oesterreichische Warenhandelsgesellschaft Austrowaren GmbH* [1993] 2 Lloyd's Rep 1 (CA), 10 (Bingham LJ).

93 *British Westinghouse Electric & Manufacturing Co Ltd v Underground Electric Railways Co of London Ltd* [1912] AC 673 (HL), 689 (Viscount Haldane LC); *Fulton Shipping Inc of Panama v Globalia Business Travel SAU of Spain* [2014] EWHC 1547; [2014] 2 Lloyd's Rep 230 [18] (Popplewell J).

94 *C Sharpe & Co Ltd v Nosawa & Co* [1917] 2 KB 814 (KB), 821 (Atkin J); *Golden Strait Corpn v Nippon Kubisha Kaisha (The Golden Victory)* [2007] UKHL 12; [2007] 2 AC 353 [10] (Bingham LJ), [57] (Carswell LJ), [80] (Brown LJ).

95 *Melachrino v Nickoll & Knight* [1920] 1 KB 693 (KB), 699 (Bailhache J); *Jamal v Moolla Dawood Sons & Co* [1916] 1 AC 175 (PC), 179 (Wrenbury LJ); *Koch Marine Inc v d'Amica SaRL (The Elena d'Amico)* [1980] 1 Lloyd's Rep 75 (QB), 88–90 (Goff J). See also *Gebruder Metelmann GmbH & Co KG v NBR (London) Ltd* [1984] 1 Lloyd's Rep 614 (CA), 632 (Donaldson MR), holding that if the reasonably anticipated consequence was that the losses would increase day by day, subject to minor unpredictable and short-lived upward movements in the price, the innocent party is under an obligation to take all reasonable measures to prevent this increase in loss.

96 *Kaines (UK) v Oesterreichische Warenhandelsgesellschaft Austrowaren GmbH* [1993] 2 Lloyd's Rep 1 (CA), 10–11 (Bingham LJ).

It is also worthy to note that, in addition to the damages for the breach of the financing bareboat charterparty, the charterers will be entitled to recover all expenses incurred in their reasonable attempts to avoid loss[97] (i.e. any legal or broker's fees incurred to repurchase the vessel). What they will not be allowed to recover, though, is any loss of profits they would have earned if they had gone to the market to conclude a new financing bareboat charterparty for a substitute vessel.[98] Their decision not to enter the market is independent of the owners' default and therefore breaks the causative link between the owners' default and charterers' loss of the opportunity to earn the profits through the use of the vessel.[99]

On the other hand, it is more likely than not that there will no available market for a 'like-for-like' replacement of the newbuilding vessel at the date of the anticipatory renunciation by the owners, which constitutes the breach date of both contracts.[100] If this is the case, the charterers would be again entitled to claim damages under the compensatory principle.[101] But as there will be no available market to measure their damages against, the only way of putting them, as far as money can do so, in the same position as if the financing bareboat charter agreement had been performed, is to allow them to claim their actual loss – to be calculated under the general contract law rules on assessment of damages.[102] The proper measure of such damages is normally damages that can fairly and reasonably be regarded as having been in the contemplation of the parties at the time when the contract was entered into, provided that such damages are likely to happen in the ordinary course of things.[103] When assessing the damages for the breach of contract, one should also consider the nature and object of the business transaction, including the commercial context in which it had been made.[104] Applying this rule to the breach of the financing bareboat charterparty, one would conclude that it was within the reasonable contemplation of parties at the conclusion of the hire/purchase contract that the charterers would exploit the vessel to generate profits. Therefore, reasonable parties to the financing bareboat charterparty would have understood

97 *British Westinghouse Electric & Manufacturing Co Ltd v Underground Electric Railways Co of London Ltd* [1912] AC 673 (HL), 689 (Viscount Haldane LC); *Fulton Shipping Inc of Panama v Globalia Business Travel SAU of Spain* [2014] EWHC 1547; [2014] 2 Lloyd's Rep 230 [18] (Popplewell J).

98 *Jamal v Moolla Dawood Sons & Co* [1916] 1 AC 175 (PC), 179 (Wrenbury LJ); *Koch Marine Inc v d'Amica SaRL (The Elena d'Amico)* [1980] 1 Lloyd's Rep 75 (QB), 88–90 (Goff J).

99 *Koch Marine Inc v d'Amica SaRL (The Elena d'Amico)*, ibid., approved by Bingham LJ in *Kaines (UK) v Oesterreichische Warenhandelsgesellschaft Austrowaren GmbH* [1993] 2 Lloyd's Rep 1 (CA), 10. The issue of causation in cases of mitigation is thoroughly analysed in *Fulton Shipping Inc of Panama v Globalia Business Travel SAU of Spain* [2014] EWHC 1547; [2014] 2 Lloyd's Rep 230.

100 *Frost v Knight* (1871–72) LR7 Ex. 111 (Exch Ct), 114 (Cockburn CJ).

101 *Johnson v Agnew* [1980] AC 367 (HL), 400 (Wilberforce LJ). See also Sale of Goods Act 1979, s. 51(2).

102 *Glory Wealth Shipping Pte Ltd v Korea Line Corp (The Wren)* [2011] EWHC 1819; [2011] 2 Lloyd's Rep 370 [31] (Blair J); *Spar Shipping AS v Grand China Logistics Holding (Group) Co Ltd* [2015] EWHC 718 [226] (Popplewell J) and Sale of Goods Act 1979, s. 51(2).

103 *Hadley v Baxendale* (1854) 9 Ex 341 (Exch Ct), 355 (Alderson B); *Koufos v C Czarnikow Ltd (The Heron II)*, [1969] 1 AC 350 (HL), 382–383, 385 (Reid LJ), 394–395 (Morris LJ), 410–411 (Hodson LJ), 416–417, 421 (Pearce LJ); *Tharros Shipping Co Ltd v Bias Shipping Ltd (The Griparion) (No. 2)* [1994] 1 Lloyd's Rep 533 (QB), 537 (Rix J); *Transfield Shipping Inc v Mercator Shipping Inc (The Achilleas)* [2008] UKHL 48, [2009] 1 AC 61 [15], [26] (Hoffmann LJ), [30], [36] (Hope LJ), [52] (Rodger LJ), [69], [78], [79], (Walker LJ), [91], [93] (Baroness Hale).

104 *Transfield Shipping Inc v Mercator Shipping Inc (The Achilleas)*, ibid., at [78]–[79] (Walker LJ).

that the owners' default in delivering the vessel could result in the charterers' losing the profits they would have expected to earn not only during the hire period but also throughout the expected life of the vessel.[105] It is thus submitted, that the proper measure of damages for the breach of the financing bareboat charterparty is the charterers' likely profit loss calculated from the date of the breach as accrued throughout the expected life of the vessel.[106] The assessment of such damages is a particularly difficult exercise for the courts since they not only have to calculate the damages without a market, but they also must project such losses for a considerably longer period of time (the expected life of the newbuilding vessel). Against such a challenge, the court is expected to make the best assessment that it can, making use of whatever information is available to it, to construct a market value or otherwise calculate the charterers' loss.[107]

As a final note, should the charterers decide to commission the building of a vessel with the same specifications to replace the one which was not delivered, then their damages for the breach of the financing bareboat charter party will again be assessed by reference to their actual loss, although they will not be as extensive.[108] In particular, the measure for assessing their damages is any additional amount they paid to purchase the substitute vessel, as well as any loss of profits incurred between the time the newbuilding vessel should have been delivered and the date the replacement vessel was in fact delivered to them under the new financing bareboat charterparty.[109]

13.3 Repairs

13.3.1 Overview

Bareboat charterparties constitute a distinct category of charters. They qualify as contracts of hire, effectively meaning that full possession and control of the vessel is passed on to the bareboat charterers.[110] Such a full possession and control of the vessel comes also with additional responsibilities imposed on the bareboat charterers,[111] including the duty to maintain the vessel, which normally falls on the owners of a voyage or time-chartered vessel.[112] Under the BARECON 2001 terms, the charterers thus undertake to maintain the vessel and her machinery in a good state of repair, in efficient operating condition and in accordance with good commercial maintenance practice, as well as to keep her class fully up to date with the

105 See the discussion on the calculation of damages of profit earning chattels in *Cullinane v British Rema Manufacturing Co Ltd* [1954] 1 QB 292, 301–302 (Evershed MR), 308 (Jenkins LJ). See also Bridge, n. 81 above, at §§12.179, 12.123–12.132.

106 Ibid.

107 *Deutsche Bank AG v Total Global Steel Ltd* [2012] EWHC 1201; [2012] EnvLRD7 [167] (Smith J) relying on A. Tettenborn, *The Law of Damages* 2nd edn, § 22.99.

108 *Glory Wealth Shipping Pte Ltd v Korea Line Corp (The Wren)* [2011] EWHC 1819; [2011] 2 Lloyd's Rep 370 [31] (Blair J).

109 Ibid., holding that the revival of the (charterparty) market at a later date does not in itself provide the correct measure of damages but may be a factor to take into account in calculating future loss.

110 *Bridge Oil Ltd v Owners and/or Demise Charterers of the Ship Guiseppe di Vittorio (No. 1) (The Guiseppe di Vittorio)* [1998] 1 Lloyd's Rep 136 (CA), 156 (Evans LJ).

111 BARECON 2001, Part II, cll. 10(a)(i), 13(a) or 14 (d)–(f).

112 See, e.g., NYPE 1993, cl. 1, lines 37–38.

classification society and maintain all other certificates in force at all times.[113] The charterers thus have to carry out the required preventive and remedial maintenance of the vessel throughout the hire period, by performing not only regular inspections and surveys but also, subject to the exception discussed below, by repairing the vessel any time she becomes operationally inefficient during the charter period.[114] In terms of the level of their duty, the maintenance obligation is not absolute as it is qualified by 'good commercial maintenance practice'. The bareboat charterers are thus expected to take reasonable steps within a reasonable time, using reasonable skill and care, to remedy any deficiency they have discovered or which they ought to have discovered.[115] In judging whether the bareboat charterers have exercised such reasonable care and skill, regard should be made to factors such as the nature and the complexity of the required remedial work, taking also into account any flexibility available to the charterers about when, where and how the repairs will be effected.[116] Nonetheless, unless the bareboat charterparty is deemed frustrated, the charterers cannot avoid the performance of repairs which may be too expensive, as the proportionality in terms of financial expenditure does not come into play.[117]

The charterers' obligation to perform remedial maintenance does not however encompass repairs, replacements or defects appearing during the first 12 months from her delivery by the builders, if such matters fall within the scope of the guarantee clause in the shipbuilding contract. Specifically, it is expressly agreed between the parties to the financing bareboat charterparty that in case of such repairs, the owners will 'endeavour to compel the Builders to repair, replace or remedy any defects or to recover from the Builders any expenditure incurred in carrying out such repairs, replacements or remedies'.[118] Such an exception aims at protecting the owners from claims they cannot pass on to the shipyard. It also purports to place the charterers in the same position as if they were parties to the shipbuilding contract since, in such a case, they would not have been required to bear the cost of repairs covered by the guarantee clause. Nonetheless, as it will be shown below, the owners' obligation only to 'endeavour to compel the Builders' may not always adequately protect the charterers' interests.

Although guarantee clauses have been discussed elsewhere in this book,[119] it is necessary to make a few points which are of particular relevance to financing bareboat charterparties. The first observation relates to the interrelationship between the guarantee clause and the charterers' recovery for repairs covered by the clause. It is clearly stated in BARECON 2001 that the owners' liability to the charterers for defects appearing within the first 12 months from the vessel's delivery by

113 BARECON 2001, Part II, cl. 10(a)(i). See also Sale of Goods Act 1979, s. 20(3) imposing on the charterers/buyers the duty to take reasonable care of the vessel as bailees.

114 On the maintenance obligation see generally T. Coghlin *et al.*, *Time Charters* 7th edn, (2014), §§11.5–11.14.

115 *Golden Fleece Maritime Inc v ST Shipping & Transport Inc (The 'Elli' and The 'Frixos')* [2007] EWHC 1890; [2008] 2 Lloyd's Rep 119 [59] (Cooke J) (time charterparty case) relying on *'Snia' Societa di Navigazione Industria e Commercio v Suzuki & Co* (1923) 17 LlL Rep 78 (KB), 88 (Greer J).

116 *Golden Fleece Maritime Inc v ST Shipping & Transport Inc (The 'Elli' and The 'Frixos')*, ibid.

117 Ibid.

118 BARECON 2001, Part III, cl. 1(d), lines 29–34.

119 See the discussion in Chapter 1.

the shipyard is contingent upon the validity of the owners' own claim against the shipbuilders under the guarantee clause.[120] This, in turn, means that the charterers' prospects of recovery are dictated by the scope of the guarantee clause, which qualifies as 'a comprehensive provision forming part of what is a complete code for dealing with defects discovered after the delivery of the vessel'.[121] For example, the owners are not obliged to endeavour to compel the shipyard to repair parts which have been replaced or repaired by third parties after her delivery or defects caused by ordinary wear and tear, as it is common in shipbuilding contracts to exclude such repairs from the scope of the guarantee clause.[122] It is unclear though whether the language used in BARECON 2001 is wide enough to cover any claims under the additional guarantee period for repairs or replacements carried out under the guarantee clause, as it is debatable whether they qualify as 'repairs, replacements or defects which appear within the first twelve months from delivery by the builders'.[123] One may argue that they do, as, though they are not literally repairs or replacements appearing during the first 12 months from delivery, additional guarantee works are not independent of the guarantee repairs or replacements. They only become relevant if further action needs to be taken with respect to any of the guarantee repairs and replacements.

And even if the replacements/repairs/remedy of defects are indeed covered by the guarantee clause, it is also expressly agreed that the owners will not bear any greater liability than the shipyard under the shipbuilding contract, as a different approach would be inconsistent with their role as financiers of the vessel purchase. BARECON 2001 thus provides that the owners' liability is limited 'to the extent' they have a valid claim against the shipbuilders under the guarantee clause.[124] In the same vein, it expressly debars the charterers from claiming the difference between their actual expenditures and the amount recoverable under the guarantee clause.[125] Such a scenario may arise if, for example, the guarantee repairs are carried out in an alternative shipyard for any of the reasons stated in the guarantee clause. In such cases, the guarantee clause in standard shipbuilding contracts, such as NEWBUILDCON and SAJ, provides for the reimbursement of only the reasonable costs of making the same repairs or replacement at the original shipyard[126] and not of the actual amount spent on the repairs. One should not, however, jump to the conclusion that the charterers will be worse off for paying the difference between the actual cost of repairs and the amount recovered under the guarantee clause as they would have had to bear such an expense even if they were parties to the shipbuilding contract. Furthermore, BARECON 2001 expressly excludes claims submitted by the charterers for any loss of time. Such a provision is superfluous though as this head of damages is excluded anyway under the guarantee clause and thus is not recoverable

120 BARECON 2001, Part III, cl. 1(d), lines 35–39.

121 *China Shipbuilding Corp v Nippon Yusen Kabukishi Kaisha & Anor (The Seta Maru, The Saikyo and The Suma)* [2000] 1 Lloyd's Rep 367 (QB), 372 (Thomas J)

122 See, e.g., NEWBUILDCON, art. 37(d), SAJ, art. IX.4(a).

123 BARECON 2001, Part III, cl. 1(d), lines 29–34; NEWBUILDCON, art. 35(e).

124 BARECON 2001, Part III, cl. 1(d), lines 35–39.

125 Ibid., at cl. 1(d), lines 39–44.

126 NEWBUILDCON, art. 35 (c)–(d), SAJ art. IX. 3(b). For the difficulties in construing the term 'reasonable costs', see S. Curtis, *The Law of Shipbuilding Contracts* 4th edn (Informa Law, 2012), 179.

under the terms of the bareboat charterparty.[127] Moreover, the charterers are barred from claiming against the owners other types of damages which are excluded by the guarantee clause, even though they are not prescribed in BARECON 2001, such as any consequential losses, damages or expenses, including but not limited to, loss of profit or earnings or demurrage, directly or indirectly caused by the defects.[128]

As a final note, it should be also mentioned that NEWBUILDCON, article 35(g), provides for the possibility of the assignment to the buyers or the entity they nominate, of the benefit of the guarantees or warranties provided by the sub-contractors or suppliers of the materials or equipment used in the newbuilding vessel. Such guarantees or warranties may be assigned to the owners as buyers without the sub-contractors' or the suppliers' consent[129] but only after the expiration of the guarantee period or if the shipbuilders breach their obligation to remedy defects covered by the guarantee clause.[130] Given that the owners have no operational interest in the vessel, it is suggested that the charterers insist on the assignment of such guarantees to them, through the exercise of the owners' express right to nominate a third party. The assignment will then confer on them the right to enforce the terms of the sub-contractors' or suppliers' guarantee or warranty against them in their own right. It should be also noted that, absent such an assignment, the charterers would have no means of pursuing the performance of any works covered by the sub-contractors' or the suppliers' warranty, unless the benefit of such a clause is assigned to the owners, who must also agree to bring an action against such sub-contractors or suppliers. Nonetheless, in such a case, the charterers have to rely on the owners' cooperation on the matter as they cannot force them to start proceedings against such sub-contractors or suppliers since their guarantees fall outside the scope of the guarantee clause.

The next two subchapters will focus on the analysis of the owners' obligation to 'endeavour to compel' the builders to perform the repairs and/or rectify defects under the guarantee clause in the shipbuilding contract. They will examine whether such a duty adequately protects the charterers' interests, or whether their interests would be better protected if the shipbuilding contract confers the benefit of the guarantee clause to them under the Contracts (Rights of Third Parties) Act 1999, or if the owners assigned the benefit of the guarantee clause to them.

13.3.2 The owners' obligation to 'endeavour to compel' the shipbuilders to effect repairs

Under the doctrine of privity of contract,[131] it is only the owners who have the right to enforce in their own right the guarantee clause against the shipbuilders, as novation has extinguished the original shipbuilding contract, preventing thus the

127 BARECON 2001, Part III, cl. 1(d), lines 41–44. See, e.g., NEWBUILDCON, art. 37(b)(ii), SAJ art. IX.4(a).

128 See, e.g., NEWBUILDCON, art. 37(b)(ii), SAJ art. IX.4(a). On the construction of these clauses, see Curtis, n. 126 above, at pp. 185–186.

129 Contrast with BIMCO NEWBUILDCON, art. 45 (b)(ii), which permits assignment of the shipbuilding contract only if the shipbuilders give their consent, 'which shall not be unreasonably withheld'.

130 NEWBUILDCON, art. 35(g).

131 Described in its pre-1999 form in *Dunlop Pneumatic Tyre Co Ltd v Selfridge & Co Ltd* [1915] AC 847 (HL), 853 (Viscount Haldane LC).

charterers from suing the shipyard for any guarantee defects.[132] To ensure that the owners, who have only a financial interest in the vessel, will pursue such a claim against the shipyard, the financing bareboat charterparty imposes on them the obligation to 'endeavour to compel' the shipbuilders to effect repairs, replace or rectify any defects, or to recover from the shipyard any expenses incurred in performing such repairs, replacements or remedies. It should be, however, noted that the owners will not necessarily be in breach of the bareboat charterparty if the shipyard does not perform its duties under the guarantee clause as their obligation is not stated in terms of achievement of the object, namely the performance of the works covered by the guarantee clause or the recovery of their costs, but only of use of (reasonable) endeavours to that effect. There is no doubt that that the duty to 'endeavour to compel' the builders amounts to an enforceable obligation, as the object of the owners' endeavours can be defined with sufficient certainty.[133] Put simply, the object intended by their endeavours is capable of giving rise to a binding legal obligation[134] in the form of the owners' duty to enforce the guarantee clause against the shipyard. In addition, the parties to the bareboat charterparty have provided sufficient objective criteria on the basis of which it will be possible to assess whether the endeavours have been or can be used,[135] as the shipbuilding contract, to which the charterparty refers, defines with great clarity both the scope of the repairs/replacements/defects covered by the guarantee clause and the procedures for claiming under it.[136] This remains the case despite possible difficulty in defining in advance the exact steps the owners are expected to take to meet their obligation (to be discussed further later in this chapter), as a binding obligation is enforceable even if it is difficult to delineate its precise limits beforehand, as long as it may be given practical content.[137]

One would also notice that the owners' obligation to 'endeavour to compel' the builders is not qualified by terms such as 'best' or 'all reasonable' endeavours. Therefore, their obligation will be equated to a duty to use reasonable endeavours to force the shipbuilders to carry out the required repairs/replacements/remedies, an obligation which is less stringent than the one to use best endeavours.[138] Unlike 'best endeavours', which would have required the owners to take all reasonable courses they can[139] or 'to leave no stone unturned, within the bounds of reason',[140]

---

132 *Scarf v Jardine* [1882] All ER Rep 651 (HL), 655 (Selborne LC); *Bradford Old Bank Ltd v Sutcliffe* [1918] 2 KB 833 (CA), 841 (Pickford LJ), 849 (Scrutton LJ).

133 *Little v Courage Ltd* [1995] CLC 164 (CA), 169–170 (Millett LJ); *Jet2.com Ltd v Blackpool Airport Ltd* [2012] EWCA 417; [2012] 1 CLC 605 [29] (Moore-Bick LJ), [69] (Longmore LJ); *Dany Lions Ltd v Bristol Cars Ltd* [2014] EWHC 817; [2014] 2 All ER 403 [36] (Andrews J).

134 *Little v Courage Ltd*, ibid.; *Jet2.com Ltd v Blackpool Airport Ltd*, ibid.; *Dany Lions Ltd v Bristol Cars Ltd*, ibid.

135 See *Bower v Bantam Investments Ltd* [1972] 1 WLR 1120 (Ch D), 1126 (Goff J); *Phillips Petroleum Co UK v Enron Europe Ltd* [1997] CLC 329 (CA), 343 (Potter LJ); *Jet2.com Ltd v Blackpool Airport Ltd*, ibid., at [69] (Longmore LJ); *Dany Lions Ltd v Bristol Cars Ltd*, ibid.

136 See, e.g., NEWBUILDCON, art. 35(a)(ii), SAJ art. IX.4(a).

137 *Jet2.com Ltd v Blackpool Airport Ltd* [2012] EWCA 417; [2012] 1 CLC 605 [29] (Moore-Bick LJ).

138 *Rhodia International Holdings Ltd v Huntsman International LLC* [2007] EWHC 292; [2007] 2 Lloyd's Rep 325 [34]–[35] (Flaux J).

139 *IBM UK Ltd v Rockware Glass Ltd* [1980] FSR 335 (CA), 339 (Buckley LJ); *Rhodia International Holdings Ltd v Huntsman International LLC*, ibid., at [33].

140 *Sheffield District Railway Co v Great Central Railway Co* (1911) 27 TLR 451.

'reasonable endeavours' oblige them to take only one reasonable course, not all of them, to achieve the aim, in this case to enforce the guarantee clause against the shipyard, without having to sacrifice their own commercial interests.[141] To accomplish the desired result, owners must continue using endeavours until the point is reached when it will start repeating itself to go through the same matters again.[142]

The 'reasonable endeavours' test leads next to the question of what steps the owners need to take to meet this obligation. The first thing they are required to do is to give a proper notice to the shipbuilders under the terms of the shipbuilding contract (i.e. one given within the time limits prescribed, and describing the defect).[143] Failure to tender such a notice within the agreed period is likely to be fatal to the charterers' recovery under the guarantee clause as the notice is a condition precedent to the exercise of any rights under the guarantee clause.[144] Next, if for any reason the builders refuse to repair or remedy the guarantee defects, 'reasonable endeavours' entails the initiation by the owners of proceedings for the dispute resolution under the terms of the shipbuilding contracts, usually involving expert determination, mediation or arbitration.[145] Moreover, if an arbitral award is issued and the shipyard fails to comply with it, then the owners should take the necessary steps to enforce the award against the shipbuilders.[146]

It is nevertheless dubious whether the duty to endeavour to compel imposes on the owners an obligation to take additional actions, such as the duty to apply to the court to remove an arbitrator, or after the issuance of the arbitral award, to challenge the arbitral award or appeal against it.[147] There is no clear-cut answer to this issue given that the owners are not obliged to subordinate their financial interests under the contract to enforce the guarantee clause. Whether such additional actions fall into the scope of reasonable endeavours will thus depend on the facts of the case at issue and in particular, on what the prospects of such an appeal/application to the court may be and/or on the division of costs of pursuing the claims against the builders under BARECON, Part III, cl. 1(d). If, however, the charterers are successful in establishing that such additional actions should have been taken by the owners to fulfil their obligation, then they will be entitled to damages (the cost of the repairs/replacements/remedies) provided that they can prove that, on the balance of probabilities, the use of reasonable endeavours would (not could) have resulted in achieving the object of the clause, namely the enforcement of the guarantee clause against the shipyard.[148]

---

141 *Phillips Petroleum Co UK v Enron Europe Ltd* [1997] CLC 329 (CA), 342 (Kennedy LJ); *Yewbelle v London Green Developments* [2006] EWHC 3166 [122]–[123] (Lewison J); *Rhodia International Holdings Ltd v Huntsman International LLC* [2007] EWHC 292, [2007] 2 Lloyd's Rep 325 [35] (Flaux J).

142 *Yewbelle v London Green Developments* [2006] EWHC 3166; [2007] 1 EGLR 137 [123] (Lewison J) (reversed on other grounds at [2007] EWCA 475); [2008] 1 P&CR 17.

143 NEWBUILDCON, art. 35 (a)(ii), SAJ art. IX.4(a).

144 Ibid.

145 See, e.g., NEWBUILDCON, art. 42, SAJ art. XII (arbitration only).

146 Arbitration Act 1996, ss 99–104.

147 Ibid., at ss 24 and 67–69.

148 *Dany Lions Ltd v Bristol Cars Ltd* [2014] EWHC 817; [2014] 2 All ER 403 [62] (Andrews J).

13.3.3 Adequate protection?

In their capacity as operators and ultimate buyers of the vessel, the charterers have an interest in a vessel in a good state of repair and in efficient operating condition. Accordingly, as already discussed, they are required to take reasonable steps to carry out preventive and remedial maintenance of the vessel under the terms of the charterparty throughout the hire period. They nevertheless have to rely on the owners, who are parties to the shipbuilding contract, for any remedy under the guarantee. As shown above, it is questionable whether their interests are adequately protected by this, as, on the one hand, the owners' obligation is only one of reasonable endeavours to enforce the guarantee clause, and furthermore there may be some difficulty in ascertaining in advance the steps the owners should take to meet their obligation. Hence, the charterers would thus have been in a better position if they could overcome the obstacle of the privity of contract to sue the shipyard in their own right and not through the instrumentality of the owners.

If the shipbuilding contract is subject to English law, the parties may rely on the Contracts (Rights of Third Parties) Act 1999 to circumvent the rule of privity of contract. This may be achieved under the 1999 Act, if the shipbuilding contract expressly provides that the charterers have the right to bring an action directly against the shipyard under the guarantee clause,[149] or, alternatively, if it confers the benefit of the guarantee clause on them.[150] The charterers will only be allowed to enforce the guarantee clause in their own right against the shipbuilders if they are expressly identified in the shipbuilding contract by name, as a member of a class or as answering a particular description, such as 'bareboat charterers'.[151] It is suggested that such a clause should be incorporated into the novated shipbuilding contract, as, at the time of novation, the parties to the shipbuilding contract would know that any operational interest in the vessel would lie with the charterers rather than the owners (as buyers of the vessel). Careful drafting is also required to ensure that the relevant clause also confers on the charterers the right to sue the shipyard directly for any claims under the terms of the additional guarantee period clause (if any)[152] by expressly stating that their entitlement to sue the builders in their own right extends to any claims under its terms. It is also worthy of note that such a solution will not jeopardise the shipyard's position as the shipbuilders will still be entitled to rely on the same defences, such as the builders' exclusions under the guarantee clause, that would have been available to them if the proceedings had been brought against them by the buyers (owners).[153]

The second alternative, which is not as straightforward as the first solution, is the assignment of the benefit of the guarantee clause – including the additional guarantee period terms – of the shipbuilding contract to the charterers. If valid, such an assignment will effect an immediate transfer of the owners' rights under the guarantee clause to the charterers, including their right to sue the shipbuilders directly for

149 See also NEWBUILDCON, art. 48 on 'Third Party Rights'.
150 Contracts (Rights of Third Parties) Act 1999, s. 1(1).
151 Ibid., at s. 1(3).
152 E.g., NEWBUILDCON, art. 35(e).
153 Contracts (Rights of Third Parties) Act 1999, s. 3.

breach of the guarantee clause.[154] Similarly, as with the first proposed solution, the assignment will not disadvantage the shipyard, as the charterers will be exposed to all defences set out in the guarantee clause which the shipbuilders would have been entitled to bring against the owners as purchasers of the newbuild.[155]

What, however, complicates things is the qualified prohibition of assignment usually found in shipbuilding contracts. For example, NEWBUILDCON permits assignment only if the shipbuilders give their consent, 'which shall not be unreasonably withheld', whilst the SAJ form makes any assignment under the shipbuilding contract subject to prior consent of the other party and the approval of the Japanese Government.[156] Accordingly, for a valid assignment, the owners are required to request the shipyard's consent first, irrespective of whether its consent could not have been reasonably refused.[157] Furthermore, if the owners ask the shipbuilders to give their consent to the assignment of the guarantee clause to the charterers and the shipyard refuses to consent to it, it will be the owners who will bear the burden of proving that the shipbuilders have unreasonably withheld their consent (question of fact), by establishing that a reasonable shipyard would not have refused to give its consent to such an assignment.[158] While judging whether the shipyard has been unreasonable in withholding its consent, the courts will not rule against a shipyard simply because it considered only its interests when refusing its consent, unless it would be disproportionate to do so having regard to the effects on the shipyard and the owners.[159] In any case, the shipyard should not withhold its consent for reasons which are 'wholly extraneous and completely dissociated from the subject matter of the (shipbuilding) contract',[160] namely reasons which are unrelated to the charterers' ability to perform their part of the deal.[161] Against such criteria, it would be difficult for the shipyard to withhold its consent to the assignment of the guarantee clause, given that the shipbuilders will have in any case to collaborate with the charterers to arrange the performance of the guarantee works.[162] Also, given the operational interest the charterers have in the vessel, one would expect that they would be very keen to cooperate with the shipyard to arrange any guarantee work.

If nonetheless the owners assign the benefit of the guarantee clause to the

154 *Re Armstrong's Will Trusts* (1857) 3 Kay & J 486 (Ch), 495–496 (St Leonards LJ).

155 *Cockell v Taylor* (1852) 15 Beav 103 (Ch), 118–119 (Romilly MR); *Edward Nelson & Co Ltd v Faber & Co* [1903] 2 KB 367 (KB), 375 (Joyce J).

156 NEWBUILDCON, art. 45 (b)(ii). SAJ, art. XIV. There is no issue of validity of such a qualified prohibition of assignment. *Linden Gardens Trust Ltd v Lenesta Sludge Disposal Ltd* [1994] 1 AC 85 (HL), 106–107 (Browne-Wilkinson LJ).

157 *Hendry v Chartsearch Ltd* [1998] CLC 1382 (CA), 1393 (Henry and Millett LJJ).

158 *International Drilling Fluids Ltd v Louisville Investments (Uxbridge) Ltd* [1986] Ch 513 (CA), 520–521 (Balcombe LJ); *Ashworth Frazer Ltd v Gloucester City Council* [2001] UKHL 59, [2001] 1 WLR 2180 [5] (Bingham LJ); *Iqbal v Thakrar* [2004] EWCA 592, [2004] 3 EGLR 21 [26] (Gibson LJ); *Falkonera Shipping Co v Arcadia Energy Pte Ltd (The Falkonera)* [2014] EWCA 713; [2014] 2 Lloyd's Rep 406 [38] (Clarke LJ).

159 *International Drilling Fluids Ltd v Louisville Investments (Uxbridge) Ltd*, ibid., at 521; *Iqbal v Thakrar*, ibid.

160 *Houlder Bros & Co Ltd v Gibbs* [1925] Ch 575 (CA), 583–584 (Pollock MR), 587 (Sargant LJ); *Ashworth Frazer Ltd v Gloucester City Council* [2001] UKHL 59; [2001] 1 WLR 2180 [3] (Bingham LJ).

161 See Curtis, n. 126 above, at 249–250, discussing *British Gas Trading Ltd v Eastern Electricity Plc & Ors*, 18 December 1996 (CA) (unreported).

162 BARECON 2001, Part III, cl. 3.

charterers without requesting the shipyard's prior consent or if for any reason the shipyard has reasonable grounds to refuse its consent, the assignment of the guarantee clause will be of no effect between the charterers and the shipbuilders but will still be valid between them and the owners.[163] This means that the charterers will not be entitled to sue the builders directly, and if they have done so, their claim will be rejected.[164]

13.4 Hire/purchase agreement

13.4.1 Overview

As a financing tool, the bareboat charterparty is combined with a conditional sale agreement of the newbuild, which provides that the ownership of the vessel will be transferred to the charterers if certain conditions are satisfied.[165] For example, under BARECON 2001, property to the vessel will pass from the owners/lessors to the charterers/buyers upon expiration of the charter period provided that the charterers/buyers meet their obligations under the charterparty, including the duty to pay the final instalment of hire.[166]

Once again the division of the duties allocated to the parties to the conditional sale agreement seems fairly balanced, since each party is expected to perform a number of duties which are consistent with the role each of them plays in the financing. On the one hand, the owners, as sellers, are required to transfer the vessel to the charterers on the expiry of the charterparty, also guaranteeing that, at the time of her delivery, she is free from all encumbrances, maritime liens and debts other than those arising from anything done or not done by the buyers.[167] They also bear the obligation to provide the charterers/buyers with the necessary documents (i.e. the bill of sale, classification certificates, plans, etc.), as well as to take the necessary steps, including paying the pertinent taxes and charges, to delete the vessel from the ship's register and to deliver the relevant certificate of deletion to the charterers.[168] On the other hand, the charterers are expected to pay any taxes or charges related to the vessel purchase and her registration under their own flag.[169] More importantly, as they had full operational control of the vessel throughout the hire period, placing them in a better position to assess the vessel's condition, they are obliged to accept delivery of the vessel as 'she is' at the time of delivery.[170] It is also agreed that such delivery also relieves the owners as sellers from any responsibility for 'possible faults or deficiencies of any description'.[171]

---

163 *Linden Gardens Trust Ltd v Lenesta Sludge Disposal Ltd* [1994] 1 AC 85 (HL), 108 (Browne-Wilkinson LJ).

164 Ibid., at 95 (Keith and Bridge LJJ), 96–98 (Griffiths LJ), at 98 (Ackner LJ), and 109 (Browne-Wilkinson LJ).

165 BARECON 2001, Part IV.

166 Ibid., at lines 1–11.

167 Ibid., at lines 10–17 and 45–46.

168 Ibid., at lines 25–38.

169 Ibid., at lines 22–25.

170 Ibid., at lines 42–46.

171 Ibid., at lines 47–48.

The duties imposed on the parties to the hire/purchase agreement are relatively straightforward, with the exception of the charterers/buyers' obligation to take delivery of the vessel 'as she is'. This duty may prove to be problematical and apt to spark litigation as it is contentious whether the language used in BARECON 2001 can exclude the conditions of the vessel's correspondence with the contractual description and satisfactory quality[172] implied into the conditional sales contract by the Sale of Goods Act 1979.[173] The next subchapter will therefore focus on the analysis of the terminology used in BARECON 2001 with a view to examining whether the relevant clause is clearly worded to negate such implied terms, or if it is not, whether the owners/lessors may rely on other defences to avoid such liabilities.

13.4.2 Condition of the vessel on delivery: 'As she is'?

The Sale of Goods Act 1979 imports by implication a number of statutory terms into the conditional sale of the vessel, including that of the correspondence with the description and of satisfactory quality,[174] but it also allows the parties to contract out of any of these terms by including an express agreement to that effect into their contract.[175] Such an agreement will be effective though only if the terms of the contract 'made it sufficiently clear' that the intention of the charterers/buyers was to abandon their rights arising out of the implied terms.[176] This then raises the question of whether the term 'as she is' meets this test.[177]

When BARECON was drafted back in 2001, there was no debate over the meaning of the term 'as she is'. Although there was no case law decided exactly on the point, there was at least the *obiter* expression of view of Phillips J in *The Morning Watch*[178] that the phrase 'as is' means that the buyer takes the goods 'as he finds them without any warranty as to quality or condition'.[179] The same approach was also taken in 2003, again as *obiter*, in *The Brave Challenger*[180] by Steel J who stated that the express term in the contract of sale providing for the acceptance of the vessel 'as is where is' results in negating the implication of the terms set out in section 14 of the 1979 Act.[181] The case which has muddied the waters is *The Union Power*,[182]

172 Sale of Goods Act 1979, ss 13–14.

173 The 1979 Act applies as a conditional sale qualifies as an 'agreement to sell', under s. 2(3) and (5). In addition, there is no doubt that a vessel qualifies as 'goods' for the purposes of the Sale of Goods Act 1979, being a 'chattel personal'. See *Behnke v Bede Shipping Co Ltd* [1927] 1 KB 649 (KB), 659–660 (Wright J); *Dalmare SpA v Union Maritime Ltd* [2012] EWHC 3537; [2013] 1 Lloyd's Rep 509 [24] (Flaux J).

174 Sale of Goods Act 1979, ss 13–14.

175 Ibid., at s. 55(1).

176 *Andrews Brothers (Bournemouth) Ltd v Singer & Co Ltd* [1934] 1 KB 17 (CA), 23 (Scrutton LJ); *Gilbert Ash (Northern) Ltd v Modern Engineering (Bristol) Ltd* [1974] AC 689 (HL), 717 (Diplock LJ); *Air Transworld Ltd v Bombardier Inc* [2012] EWHC 243; [2012] 1 Lloyd's Rep 349 [25] (Cooke J).

177 See further discussion on the interpretation of 'as is' in Chapter 8 of the present work.

178 *Mariola Marine Corp v Lloyd's Register of Shipping (The Morning Watch)* [1990] 1 Lloyd's Rep 547 (QB).

179 Ibid., at 555–556.

180 *Indigo International Holdings Ltd v Owners &/or Demise Charterers of the Vessel Brave Challenger (The Brave Challenger)* [2003] EWHC 3154.

181 Ibid., at [177].

182 *Dalmare SpA v Union Maritime Ltd (The Union Power)* [2012] EWHC 3537; [2013] 1 Lloyd's Rep 509.

decided in 2012, as Flaux J's interpretation of the term 'as is', though once again expressed as *obiter*,[183] was diametrically opposite to the *obiter* expressions of view of the judges in both *The Morning Watch*[184] and *The Brave Challenger*.[185] In particular, his provisional view on the matter was that the phrase 'as is' only excludes the right to reject the goods for breach of the implied conditions as to description, satisfactory quality and fitness for purpose under the 1979 Act.[186] He further stated that, subject to evidence of a contrary custom or market meaning, such a term is not deemed sufficiently clear to negative any of those implied conditions, effectively meaning that the buyers would be entitled to recover any loss or damage caused by the breach of these terms.[187] Flaux J's *obiter* in *The Union Power* has not received support from the legal community.[188] It has also been heavily criticised by Leggatt J in *Hirtenstein v Hill Dickinson LLP*,[189] a case of professional negligence in handling the purchase of a yacht, which has cast serious doubts on the correctness of Flaux J's *obiter* view. [190] In *Hirtenstein*, Leggatt J could not see how the proposed interpretation of the term 'as is', namely to exclude only the right to reject the goods but not the right to claim damages, reflected the expectations of ordinary business people.[191] He further opined that the background understanding of the parties to the sales contract against which the documentation was negotiated was that buying the yacht 'as is, where is' meant that the buyers would take it 'in its existing condition, good or bad, with no subsequent right to sue the seller for any fault which the Yacht might turn out to have'.[192] It should be noted though that the judge addressed the issue of the interpretation of 'as is' in the context of the background understanding of the parties, which may be of assistance to the sellers in an attempt to establish a custom or usage to that effect,[193] and not of the interpretation of the contract of sale *per se*.[194] Thus, until a case on the interpretation of the term 'as is' under the requirement of the 'express agreement' in the Sale of Goods Act 1979 section 55(1) is brought before the courts, there will be ambiguity with respect to the rights of the parties under the BARECON 2001 hire/purchase agreement and specifically whether its terms exclude the Sale of Goods Act 1979 conditions as to the vessel's conformity with the contractual description and the satisfactory quality.

In the light of this uncertainty, it is necessary to explore whether other terms in

183 Ibid., at [84].

184 *Mariola Marine Corp v Lloyd's Register of Shipping (The Morning Watch)* [1990] 1 Lloyd's Rep 547 (QB), 555–556 (Phillips J).

185 *Indigo International Holdings Ltd v Owners &/or Demise Charterers of the Vessel Brave Challenger (The Brave Challenger)* [2003] EWHC 3154 [177] (Steele J).

186 *Dalmare SpA v Union Maritime Ltd (The Union Power)* [2012] EWHC 3537; [2013] 1 Lloyd's Rep 509 [84].

187 Ibid.

188 For a typical attack see Chapter 8 above.

189 [2014] EWHC 2711.

190 Ibid., at [55]–[56].

191 Ibid., at [55].

192 Ibid., at [56].

193 The parties may also negative the implied conditions by pleading the existence of a relevant custom or market meaning, which will be rather difficult to establish as they will have to prove the universality of the alleged custom or usage (See, e.g., *Choil Trading SA v Sahara Energy Resources Ltd* [2010] EWHC 374 [106] (Clarke J)).

194 Ibid.

BARECON 2001 meet the criteria of the 'express agreement' in section 55(1) so as to be effective to exclude such implied terms. To that end, one should consider whether the expressly agreed relief of the sellers from any responsibility for 'possible faults or deficiencies of any description' evinces the charterers' clear intention to waive the implied terms as to both the vessel description and satisfactory quality. A number of cases have considered the term 'all faults' to conclude that such words are apt to negate only the implied condition as to satisfactory quality but not that of the correspondence of the goods with the contractual description.[195] In a similar vein, the words 'deficiencies of any description' are not capable of ousting the implied term as to description. Such a term is not equivalent to 'errors in description' or 'errors of description', for which there are at least some authorities holding that such a term can oust the application of the Sale of Goods Act 1979, section 13.[196] Its literal meaning is 'deficiencies' or defects 'of any kind' or 'whatsoever', a wording which the courts have found to be incapable of excluding the implied condition of the correspondence with the contractual description.[197] Moreover, another difficulty arises from the decision of the House of Lords in *Ashington Piggeries Ltd v Christopher Hill Ltd*,[198] which has read a clause in the contract of sale providing that 'the goods shall be taken with all faults and defects' as intending to exclude the buyer's right to reject goods for faults and defects but not as purporting to oust their right to sue the sellers for compensation for any consequential damage which they may have suffered by reason of the acceptance of defective goods.[199]

Having so far established the potential problems in construing the relevant wording in BARECON 2001 as negating the terms implied into the sales contract by the Sale of Goods Act, it is worth looking at alternative defences to exonerate the owners as sellers from any liability arising out of the breach of any of the implied conditions. In terms of the correspondence of the vessel with the contractual description, such an issue may only arise if the vessel tendered for delivery under the conditional contract of sale does not meet the specifications agreed in the shipbuilding contract. A claim to that effect may be easily disposed of by the sellers by relying on the doctrine of waiver by election (affirmation).[200] Put simply, if the vessel does not satisfy the contractual description, this is because the charterers made an informed decision[201] to accept such a vessel even though they had the right

195 *Shepherd v Kain* (1821) 5 B&Ald 240 (KB), 241; *Munro v Meyer* [1930] 2 KB 312 (KB), 327 (Wright J).

196 *Lloyd del Pacifico v Board of Trade (No. 1)* (1929) 35 LlL Rep 217 (CA), 223 (Lawrence LJ) (*obiter*); *Couchman v Hill* [1947] KB 554 (CA), 557 (Scott LJ). *Contra*, *Nicholson & Venn v Smith-Marriott* (1947) 177 LT 189 (KB) (Hallett J); *D & M Trailers (Halifax) Ltd v Stirling* [1978] RTR 468 (CA), 472 (Browne LJ), 473 (Megaw LJ) (*obiter*).

197 *Shepherd v Kain* (1821) 5 B&Ald 240 (KB), 241; *Taylor v Bullen* (1850) 5 Ex 779 (Exch Ct), 785–786 (Parke B), also holding that the seller's liability for unintentional misdescription may be excluded only by the use of the term 'any error whatever'.

198 [1972] AC 441 (HL).

199 Ibid., at 471 (Hodson LJ) and 511 (Diplock LJ).

200 *Motor Oil Hellas (Corinth) Refineries SA v Shipping Corp of India (The Kanchenjunga)* [1990] 1 Lloyd's Rep 391 (HL), 398 (Goff LJ).

201 The charterers/buyers had the right to send a representative to the shipyard to inspect the construction of the vessel. In addition, they had the final say in rejecting or not the vessel (BARECON 2001, Part III, cll.1(c) and 2(c)). This satisfies the requirement that the party who has to make the choice should either know or have obvious means of knowledge of the facts giving rise to the right, and possibly

to reject her and terminate the combined hire/purchase agreement.[202] Thus, having elected one of the inconsistent legal rights available to them, namely to accept the vessel and treat the financing bareboat charterparty as continuing, the charterers have unequivocally abandoned their inconsistent right to reject the vessel and terminate the hire/purchase contract.[203] In other words, by their conduct, the charterers/buyers have precluded themselves from rejecting the vessel when tendered to them later under the terms of the conditional purchase agreement as not complying with the agreed description.[204] Similarly, the charterers/buyers will be debarred from claiming damages for breach of the condition as to the description from the sellers (owners/lessors). Should they have wanted to recover such damages, they should have exercised their right to instruct the owners/lessors to claim such damages from the shipyard when renegotiating the terms of the delivery of the vessel.[205] If not, they will be deemed to have also foregone their rights to claim damages for the tendered vessel's failure to correspond with its description.

The above discussion has demonstrated the difficulties arising out of the construction of the BARECON 2001 terms on the delivery of the vessel under the conditional sale agreement. There is a chance that owners/lessors may be held liable for breach of the implied terms under the Sale of Goods Act, even though they have not been involved in either the construction or the operation of the vessel. To avoid uncertainty and future litigation, BIMCO should consider revising the pertinent clause in BARECON 2001 with a view to employing words which will be sufficiently clear to exclude the implied terms of satisfactory quality/description. To that effect, a redrafted clause shall for instance, make express reference to the terms such as 'conditions' or 'all obligations or liabilities express or implied arising by law', as such a wording has already been construed by the courts as negating the implied conditions of the Sale of Goods Act 1979.[206]

13.5 Conclusions

Reverting to the question posed at the title of this chapter as to whether the division of the parties' rights and obligations under BARECON 2001 is fairly balanced, the answer is, it is suggested, no. The above analysis has clearly demonstrated that BARECON 2001, which is drafted on back-to-back terms with the shipbuilding contract, promotes to a great extent the owners/lessors' interests but not at all times that of the charterers/buyers. In particular, taking into account the owners/lessors' role as financiers of the shipbuilding contract, it expressly shields them from all

of the existence of the right (*Panchaud Frères SA v Établissements General Grain Co* [1970] 1 Lloyd's Rep 53 (CA), 57–58 (Winn LJ)).

202 BARECON 2001, Part III, cl. 2(c).

203 Ibid., at cl. 2(c)(ii); Sale of Goods Act 1979, s. 11(2). See further *Motor Oil Hellas (Corinth) Refineries SA v Shipping Corp of India (The Kanchenjunga)* [1990] 1 Lloyd's Rep 391 (HL), 399 (Goff LJ).

204 *Toepfer v Warinco AG* [1978] 2 Lloyd's Rep 569 (QB), 576 (Brandon J). *Motor Oil Hellas (Corinth) Refineries SA v Shipping Corp of India (The Kanchenjunga)*, ibid., at 397–398 (Goff LJ).

205 BARECON 2001, Part III, cl. 2(c)(ii).

206 *Bominflot Bunkergesellschaft für Mineraloele mbH & Co KG v Petroplus Marketing AG (The Mercini Lady)* [2010] EWCA 1145; [2011] 1 Lloyd's Rep 442 [62] (Rix LJ); *Air Transworld Ltd v Bombardier Inc* [2012] EWHC 243; [2012] 1 Lloyd's Rep 349 [31] (Cooke J).

claims they cannot pass on to the shipyard, such as claims for unseaworthiness, the vessel's performance or specifications or defects other than those falling into the guarantee clause, as well delay in her delivery falling into the grace period.[207]

On the contrary, the charterers/buyers are not always afforded the same level of protection. There is no doubt that their involvement in the construction of the vessel protects their interests as at the end of the day they are the ones who will operate the vessel during the hire period and beyond when they acquire her ownership.[208] However, there will be instances where the terms of BARECON 2001 and/or of the shipbuilding contract may jeopardise their interest in the newbuilding vessel. As the above discussion has showed, one of such instances relates to their duty to accept a newbuilding with minor defects, if such delivery is deemed valid under the shipbuilding contract, as BARECON 2001 prohibits the rejection of the newbuilding vessel for reasons other than those provided in the shipbuilding contract.[209] Such a prohibition in this case certainly places the charterers/buyers in a worse position than they would have been if they were parties to the shipbuilding contract. Put simply, when accepting such a vessel, the charterers/buyers have neither a right of direct recourse against the shipyard for the necessary repairs and compensation for any time lost while remedying such minor defects nor the means of forcing the owners/lessors to sue the shipbuilders under the pertinent 'minor defects' clause. Unfortunately, this is not the only difficulty they will face because of the restrictions imposed on them by the privity of contract. In a similar vein, if they discover any guarantee defects, they have no other option but to rely on the owners/lessors' endeavours to enforce the guarantee clause against the shipyard.[210] What also goes against them in such a case is that the owners/lessors' obligation is to be judged by the use of reasonable (not best) endeavours only to that effect and not by the achievement of the performance of the works covered by the guarantee clause or the recovery of their costs. It is thus evident that, as it stands alone, BARECON 2001 may jeopardise the charterers/buyers' rights in the newbuilding vessel.

All these difficulties may, however, be rather easily ironed out though the mechanisms of circumventing the privity of contract, such as the Contracts (Rights of Third Parties) Act 1999 or assignment, as discussed above. What is therefore required is careful drafting of the novated shipbuilding contract to include clauses that either grant the charterers/buyers the right to sue the shipyard in their own right or confer the benefits of the shipbuilding contract to them under the terms of the Contracts (Rights of Third Parties) Act 1999. Another solution, albeit possibly more complicated, is the transfer of the benefit of the minor defects clause and/or guarantee clause to the charterers/buyers via assignment provided though that the owners/lessors will obtain the shipyard's prior consent.[211] Either of the above solutions will restore the balance of the parties' rights by giving the charterers/lessors, who have an operational interest in the vessel, the right of direct action against the shipyard.

Another problematic provision, which this time may be detrimental to the

207 BARECON 2001, Part III, cll 1(d) and 2(a).
208 Ibid., at cll. 1(b)–(c)and 2(c).
209 Ibid., at cll. 2(c)(iii), NEWBUILDCON, art. 27(d)(iv).
210 BARECON 2001, Part III, cl. 1(d), lines 29–34.
211 NEWBUILDCON, art. 45(b)(ii). SAJ, art. XIV.

owners/lessors' interests, is the clause for the delivery of the vessel 'as she is' under the BARECON 2001 hire/purchase agreement.[212] The above analysis has emphasised the uncertainty this provision yields to, namely whether the BARECON 2001 terminology is capable of excluding the implied terms under the Sale of Goods Act 1979.[213] To avoid such ambiguities, it is proposed that BIMCO should consider redrafting the relevant provision to incorporate language which will clearly point to the charterers/buyers' waiver of any rights under the implied conditions as to description and satisfactory quality by, for example, making express reference to 'conditions' or 'all obligations or liabilities express or implied arising by law'.[214]

All in all, there is no doubt that there is room for improvement of the BARECON 2001 terms to restore the balance between the parties' rights. The suggested amendments to the BARECON 2001 will not solve the issues raised unless they are also combined with the proposed additional clauses to be inserted in the novated shipbuilding contract in order to ensure that the charterers/buyers will be granted the same rights they would have had if they were parties to the shipbuilding contract.

212 BARECON 2001, Part IV, lines 45–48.

213 Sale of Goods Act 1979, ss 13–14.

214 *Bominflot Bunkergesellschaft für Mineraloele mbH & Co KG v Petroplus Marketing AG (The Mercini Lady)* [2010] EWCA 1145; [2011] 1 Lloyd's Rep 442 [62] (Rix LJ); *Air Transworld Ltd v Bombardier Inc* [2012] EWHC 243; [2012] 1 Lloyd's Rep 349 [31] (Cooke J).

CHAPTER 14

Shipping finance and sanctions

David Osborne★

14.1 Introduction

The use of sanctions in international politics has expanded during the last 20 years and there is no sign of this changing. The recent temporary and limited relaxation of European Union (EU) and United States (US) sanctions against Iran is an indication of the perceived success of sanctions, rather than the reverse.[1] The escalating Russian sanctions have attracted much wider attention. The EU and US sanctions against Iran are not identical, nor are the respective temporary and limited relaxations identical, but they have many similar features. A common theme of both EU and US sanctions against Russia is the targeting of oil and gas projects. 'Phase three' sanctions imposed by the EU are directed at the sale, supply, transfer or export of specified technologies to Russian parties or for use in Russia – with particular reference to deep water oil exploration and production, Arctic oil exploration and production, and shale oil projects.[2] The US Department of Commerce regulations also impose similar restrictions through the requirement for export licences for such technologies.[3] In addition, US 'sectoral' sanctions introduced in July 2014 are directed at new debt and equity raising by a number of Russian banks and energy companies.[4] EU phase three sanctions also target some Russian banks and energy companies in this way.[5] The lists of banks and energy companies respectively targeted by US sectoral sanctions and EU phase three sanctions have some names in common but are not identical. Russia has retaliated with sanctions of its own in the form of import restrictions.[6] Iranian sanctions have been specifically directed at shipping and marine insurance; this has given rise to a number

★ Partner, Watson, Farley & Williams LLP. The author wishes to thank Jane Freeburg–Sarma of the New York office of Watson Farley & Williams LLP' for assistance on issues relating to US sanctions. This chapter was completed on 25 November 2014 and reflects the legal position with regard to sanctions by that day. Readers should note that this area of law changes rapidly.

1 The initial six-month relaxation was extended to 24 November 2014 and has been further extended to 30 June 2015. For comment, see D. E. Sanger, 'Staying the Course on Iran by Threatening Pain and Offering Relief', *The New York Times*, 20 July 2014, A6.

2 Council Regulation (EU) No. 833/2014, art. 3, as supplemented and amended by Council Regulation (EU) No. 960/2014.

3 FR Doc 2014-18579, filed 1 August, 2014.

4 Executive Order 13662; Sectoral Sanctions Identifications List, 29 July 2014.

5 Council Regulation (EU) No. 833/2014, as supplemented and amended by Council Regulation (EU) No. 960/2014.

6 On 6 August 2014, Russia banned the import of certain foodstuffs from the EU, the US, Australia, Canada and Norway.

of cases before the English courts. The Russian sanctions so far are not specifically directed at shipping but have the potential directly or indirectly to affect shipping activity, especially that which is ancillary to the targeted Russian oil and gas industries.

The current and developing use of sanctions is the modern equivalent of the 'pacific blockade' used by naval powers in the nineteenth century[7] outside a state of war. Instead of naval forces being deployed to impose a physical blockade, laws are used to apply economic pressure. Sanctions are a more far-reaching concept than an export ban or trade embargo – although the latter are often included within a package of sanctions.[8] Sanctions are a legal weapon not involving the use of physical force (or the threat of physical force).[9] There are three features to note: (i) complexity; (ii) jurisdictional and conflict of laws issues; and (iii) justiciability.

First, complexity. Many sanctions provisions are detailed, lengthy and specific as to scope. Some sanctions are directed generally at any activity involving a particular country.[10] Some are directed at specific activities within, or trades to, a particular country.[11] Others are directed at particular companies and individuals within a country.[12] Sometimes, the second and third elements are combined.[13] Sanctions compliance by companies whose businesses are affected has become a major user of time and resources.

Second, jurisdictional and conflict of laws issues. The starting point is that sanctions imposed by law operate on those who are bound by the particular law. This generally means the individual citizens of the relevant state, companies incorporated in the relevant state and, by whatever test is applied, others carrying on activity within the territory of the relevant state. This is the approach taken by the European Union to sanctions.[14] Increasingly, however, it is not the approach taken by the United States. Most US sanctions are binding only on 'US persons'[15] but there has been a recent tendency for US sanctions to be extraterritorial and purport to bind anyone in the world. This has been seen most notably in many of

7 L. Davis and S. Engerman, 'Sanctions: Neither War nor Peace' (2003) 17 *Journal of Economic Perspectives* at 187–197.

8 See, e.g., the restrictions imposed by the EU and the US on, amongst other things, export of specified technology to Russia referred to above. The United Kingdom Export Control Organisation (ECO) of the Department of Business, Innovation and Skills has added Russia to the list of prohibited destinations in respect of certain export licences (Notice to Exporters 2014/14, 14 August 2014).

9 'Sanctions are a necessary middle ground between war and words' (Kofi Annan).

10 See, e.g., the US Cuban Assets Control Regulations, 31 CFR Part 515.

11 See, e.g., the US Iranian Financial Sanctions Regulations, 31 CFR Part 561.

12 See, e.g., the US Somalia Sanctions Regulations, 31 CFR Part 551; Executive Orders 13660, 13661 and 13662 (Ukraine-related sanctions); see also Regulation (EU) No. 269/2014, as amended (Ukraine-related sanctions).

13 See, e.g., Council Regulation (EU) No. 267/2012 (Iran sanctions).

14 EU sanctions typically apply (a) within the territory of the Union; (b) on board any aircraft or any vessel under the jurisdiction of a Member State; (c) to any person inside or outside the territory of the Union who is a national of a Member State; (d) to any legal person, entity or body inside or outside the territory of the Union, which is incorporated or constituted under the law of a Member State; or (e) to any legal person, entity or body in respect of any business done in whole or in part within the Union. See, e.g., Council Regulation (EU) No. 833/2014, art. 13.

15 'US persons' and 'persons subject to the jurisdiction of the US' are broadly defined and include: (i) US citizens, regardless of place of residence; (ii) US green card holders, regardless of place of residence; (iii) US residents, regardless of citizenship; (iv) entities organised under the laws of the US or any state or territory of the US; and (v) entities having a place of business in the US and, depending on the sanctions regime, may include entities owned or controlled by US persons.

the US sanctions directed at Iran in recent years.[16] Apart from the express extraterritorial scope of certain sanctions, the US Government and its agencies have been assiduous in bending non-US persons to their will.[17] All major international banks both have a presence within the US and need access to US-based US dollar clearing systems. Another factor is the legal concept of 'facilitation' which is contained in most US sanctions provisions. This means that a US person is guilty of a sanctions breach if it facilitates a transaction by a non-US person which would be a breach of sanctions if carried out by a US person.[18] The long reach of US sanctions has already come before the English courts, as discussed below, and will doubtless do so again. The determination of the US Government to enforce compliance with US sanctions by non-US persons is further shown by the Executive Order of 1 May 2012 which targets non-US persons who have 'violated, attempted to violate, conspired to violate, or caused a violation of [US sanctions against Iran or Syria]' or 'facilitated deceptive transactions for or on behalf of any person subject to United States sanctions covering Iran or Syria'. The penalty for being found to be a 'foreign sanctions evader' is a broad prohibition on US persons trading with such a person.

One aspect of sanctions which is a feature of the complexity of jurisdictional and conflict of law issues is what can be called 'sanctions clash'. This occurs where

16 See the Iran Sanctions Act (formerly known as the Iran and Libya Sanctions Act) and the Comprehensive Iran Sanctions, Accountability, and Divestment Act of 2010 (CISADA), the National Defense Authorisation Act for Fiscal Year 2012 (NDAA), the Iran Threat Reduction and Syria Human Rights Act 2012 (ITRSHRA) and the Iran Freedom and Counter-Proliferation Act of 2012 (IFCA).

17 In February 2013, the SEC introduced rules requiring foreign US listed companies to disclose information on trading with Iran. Iran Threat Reduction and Syria Human Rights Act of 2012, § 219; SEC, Exchange Act Compliance and Disclosure Interpretations Questions 147.01–147.07.

18 For example, the facilitation provision from the Syrian Sanctions Regulations, 31 CFR § 542.210 provides:

> Except as otherwise authorized, United States persons, wherever located, are prohibited from approving, financing, facilitating, or guaranteeing a transaction by a foreign person where the transaction by that foreign person would be prohibited by §§542.206, 542.207, 542.208, or 542.209 of this part if performed by a United States person or within the United States.

The Syrian Sanctions Regulations also contain an interpretative section that provides some guidance on what constitutes facilitation, 31 CFR §542.413:

> With respect to §542.210, a prohibited facilitation or approval of a transaction by a foreign person occurs, among other instances, when a United States person:
>
> (a) alters its operating policies or procedures, or those of a foreign affiliate, to permit a foreign affiliate to accept or perform a specific contract, engagement or transaction involving Syria or the Government of Syria without the approval of the United States person, where such transaction previously required approval by the United States person and such transaction by the foreign affiliate would be prohibited by this part if performed directly by a United States person or from the United States;
>
> (b) refers to a foreign person purchase orders, requests for bids, or similar business opportunities involving Syria or the Government of Syria to which the United States person could not directly respond as a result of the prohibitions contained in this part; or
>
> (c) changes the operating policies and procedures of a particular affiliate with the specific purpose of facilitating transactions that would be prohibited by this part if performed by a United States person or from the United States.

The interpretative guidance contained in 31 CFR §542.413 is not considered to be an exhaustive list of activities that constitute facilitation, but merely an indication of the types of activities that are considered to fall clearly within the scope of facilitation.

there is inconsistency between different sanctions regimes, largely as a result of anti-sanctions or 'blocking' provisions. Two examples are the Regulation which prohibits EU nationals from complying with certain extraterritorial US sanctions, mainly relating to Cuba[19] and the German *Aussenwirt schaftsverordnung* which prohibits German nationals from complying with sanctions other than those imposed by the United Nations or the European Union.

Third, justiciability. EU sanctions imposed by Council Regulation are subject to judicial review, as are sanctions imposed in the UK by delegated legislation. There has been vigorous and successful legal challenge to certain EU sanctions in the General Court of the European Union by various Iranian individuals and companies, most notably in a shipping context, by Islamic Republic of Iran Shipping Lines (IRISL) and National Iranian Tanker Co (NITC).[20] There have been recent signs that some Russian companies might mount a legal challenge to EU sanctions. In the UK, Bank Mellat successfully challenged its designation pursuant to the Counter-Terrorism Act 2008 in a landmark administrative law case which went to the Supreme Court.[21] Judicial review of the lawfulness of sanctions has so far not been seen in a comparable way in the United States. Under US law, determinations by the Department of Treasury's Office of Foreign Assets Control, which administers sanctions programmes, are reviewed by courts under an 'abuse of discretion' standard; as long as the relevant agency did not abuse its discretion in its application of regulations (such as sanctions) within a statutory framework, a court will not overturn the decision of that agency.[22]

The area of English law which is most directly and obviously engaged by sanctions is the effect of illegality on contracts – either illegality under English law or illegality under foreign law and, in relation to each, either initial illegality or supervening illegality.[23] There is a fourth issue to be mentioned – or rather submission to be made. Considerations of public policy have a limited part to play in the legal interpretation of sanctions and their effect because sanctions fall short of, or are an alternative to, a state of war.[24] The application of public policy should not be expanded in a sanctions context outside the relevant areas of law where, and to the

19 Council Regulation (EC) No. 2271/96.

20 Case T-489/10 *Islamic Republic of Iran Shipping Lines v Council of the European Union*; Case T-565/12 *National Iranian Tanker Company v Council of the European Union.*

21 *Bank Mellat v Her Majesty's Treasury* [2013] UKSC 39.

22 5 USC §706(2)(A). An agency decision may be set aside by a court if 'arbitrary, capricious, an abuse of discretion or otherwise not in accordance with law'.

23 See H.G. Beale (ed.), *Chitty on Contracts* 31st edn, (Sweet & Maxwell, 2012), at [16-001]–[16034] and at [30-357]–[30-370]; Edwin Peel (ed.), *Treitel, The Law of Contract* 13th edn, (Sweet & Maxwell, 2011), chaps 11 and 19; Lord Collins of Mapesbury (ed.), *Dicey, Morris & Collins on the Conflict of Laws* 15th edn, (Sweet & Maxwell, 2012), at [32-094]–[35-105]; Sir Guenter Treitel, *Frustration and Force Majeure* 3rd edn, (Sweet & Maxwell, 2013) (hereafter 'Treitel') esp ch 8; A. Berg, 'The Detailed Drafting of a Force Majeure Clause', published as ch. 4 of E. McKendrick (ed.), *Force Majeure and Frustration of Contract*, 2nd edn (Lloyd's of London Press Ltd, 1995); J. Fisher QC and A. Baradon, 'Assessing and managing the impact of economic sanctions on commercial contracts' PLC. Available at: <http://UK.practicallaw.com/8-584-8145> (last accessed 31 December 2014). See also the Law Commission Report, *The Illegality Defence* (Law Com 320), March 2012. Its recommendations will not be followed.

24 *Ertel Bieber & Co v Rio Tinto Co Ltd* [1918] AC 260 is the classic example of public policy being applied to completely dissolve a contract involving trading with the enemy in time of war. See generally Lord McNair and A. D. Watts, *The Legal Effects of War* 4th edn, (Cambridge University Press, 1966), esp chs 4 and 5.

extent that, it currently applies – most notably certain aspects of the law relating to foreign illegality (touched upon below).[25]

14.2 Some background on sanctions and shipping

Many of the sanctions provisions seen in the last decade or so have been expressly directed at shipping and marine insurance. This can be seen most clearly in sanctions directed at Iran;[26] the reasons are not hard to see. Iranian shipping companies have been seen, correctly or otherwise, to be important facilitators of Iran's nuclear programme.[27] The importance of Iran's oil industry and the relative lack of refining capacity has meant that the transportation of oil from Iran and the import of petroleum and petrochemicals into Iran, respectively, have been focused on as activities to target by sanctions. Shipping has been directly in the sanctions firing line. The promoters of sanctions quickly realised that the most effective way to give teeth to sanctions directed at shipping activity is to target insurance, especially P&I insurance. This has meant, especially in the last five or so years, that the shipping and insurance industries have been grappling intensively with the effect of sanctions. The P&I Clubs have given advice to members in the form of 'frequently asked questions' memos. Forms of sanctions cancellation clauses have appeared in policies.[28] The Lloyds Market Association has published the 3100 Sanction Limitation and Exclusion Clause. BIMCO and Intertanko have published sanctions clauses for inclusion in charterparties. Ship finance banks now require sanctions language to be included in finance documents. The penalties for breach of sanctions are severe; many banks have suffered huge fines in the United States. Apart from penalties, there is reputational risk. This, coupled with the perception of shipping as a high-risk activity in sanctions terms, has led to a sharp focus on ship finance from bank compliance departments.

There have already been a number of cases in the English courts which have arisen from Iranian sanctions. The focus here is on commercial cases but this should not detract from the importance of – often successful – legal challenges by Iranian individuals and companies to the imposition of sanctions, both in the English courts[29] and the General Court of the European Union.[30]

25 'Granted that the basis of discharge in cases of supervening illegality is public policy, it is important to bear in mind that the policy considerations raised by different types of illegality can vary considerably in weight.' (Treitel at [8-003]). The US position is that furthering the objective of sanctions programmes is part of public policy – another reason why court challenges to sanctions in the US are usually not successful and often dismissed at an early stage despite the apparent similarity between the US test for setting aside an administrative decision (n. 22) and the position under English law.

26 See, e.g., Council Regulation (EU) No. 267/2012 and the United States Comprehensive Iran Sanction, Accountability, and Divestment Act of 2010 (CISADA).

27 The two main Iranian shipping companies, Islamic Republic of Iran Shipping Lines (IRISL) and National Iranian Tanker Company (NITC), have each successfully challenged their being made subject to sanctions in the General Court of the European Union. *Supra*, n. 20.

28 See *Arash Shipping Enterprises Company Limited v Groupana Transport* [2011] EWCA Civ 620; [2011] I CLC 984.

29 *Bank Mellat v Her Majesty's Treasury* [2013] UKSC 39; [2013] 3 WLR 179.

30 *Supra*, n. 20.

14.2.1 The Iranian insurance cases

Not surprisingly in view of the express targeting of marine insurance by sanctions, three of the cases relate to insurance. The sanctioning of the Islamic Republic of Iran Shipping Lines (IRISL) led to an important case in 2010 which it brought against its P&I club, Steamship Mutual (the Club).[31] On 12 October 2009, HM Treasury issued an Order pursuant to the Counter-Terrorism Act 2008 prohibiting transactions between UK companies in the financial sector and IRISL and Bank Mellat.[32] Licences were granted authorising the Club to continue providing cover until 30 October, when a further licence was issued but on more restrictive terms. The Club took the view that the licence issued on 30 October did not permit it to continue to provide cover and on that date issued a notice terminating cover. On 31 October, one of IRISL's vessels suffered a casualty in China resulting in bunker oil pollution (and also leading to the vessel being a constructive total loss). The two main issues were: whether the terms of the Order sanctioning IRISL and the 30 October licence permitted the Club to continue to provide cover to IRISL to the extent required by the International Convention on Civil Liability for Bunker Oil Pollution Damages (the Bunkers Convention);[33] and whether the effects of the Order and licence were to discharge the insurance cover by frustration.[34] Beatson J held that, as a matter of construction, the effect of the 30 October licence was to permit the Club to continue to provide cover to the extent of risks required to be insured by the Bunkers Convention. The more interesting part of the case relates to the second issue, the pleading by the Club of discharge by frustration. Beatson J held that there was no frustration. The 30 October licence substantially narrowed the type of cover which was permitted but this did not mean that the contract as a whole was frustrated. Beatson J referred to two of the leading cases on frustration: *Davis Contractors Ltd v Fareham Urban District Council*[35] and *Edwinton Commercial Corporation v Tsavliris Russ (Worldwide Salvage and Towage) Ltd (The Sea Angel)*.[36] He cited the test set out by Lord Radcliff in *Davis Contractors* – the rendering of the contractual obligations 'radically different' from that which was undertaken.[37] He also cited the 'multi-factorial approach' espoused by Rix LJ in *The Sea Angel*.[38] Both parts of the *Davis Contractors* test were passed. The effect of the judgment is that, at least in an insurance context, a contract which has reduced scope but still retains its original attributes is not frustrated. However, it can be speculated that this analysis might not, depending on the facts, necessarily apply to other types of contract.

31 *Islamic Republic of Iran Shipping Lines v Steamship Mutual Underwriting Association (Bermuda) Limited* [2010] EWHC 2661 (Comm); [2011] I Lloyd's Rep 195.

32 This was ultimately successfully challenged by Bank Mellat as regards itself; see *Bank Mellat v Her Majesty's Treasury* [2013] UKSC 39.

33 *Islamic Republic of Iran Shipping v Steamship Mutual* [2010] EWHC 2661 (Comm) at [55]–[98].

34 *Ibid.* at [99]–[129].

35 [1956] AC 696.

36 [2007] 2 Lloyd's Rep 517.

37 [1956] AC 696, at 729 (referred to in *Islamic Republic of Iran Shipping v Steamship Mutual* [2010] EWHC 2661 (Comm) at [102]).

38 [2007] 2 Lloyd's Rep 517 at [109]–[133] (referred to in *Islamic Republic of Iran Shipping v Steamship Mutual* [2010] EWHC 2661 (Comm) at [105], [106]).

In *Sea Glory Maritime Co v Al Sagr National Insurance Co (The Nancy)*[39] insurers alleged vitiation of an English law marine insurance policy on five grounds, the first four of which included misrepresentation, non-disclosure and breach of warranty. The fifth ground was that the policy was tainted by illegality arising from breach of US sanctions against Iran.[40] This aspect of the case is worth close attention in the present context. The vessel was voyage chartered by its (non-US) owner to a Hong Kong charterer for the carriage of sulphur from Iran to China. That did not in itself involve any breach of US sanctions. After completion of the voyage to China the vessel suffered a casualty in Russia resulting in it being declared a constructive total loss. The insurers' argument in relation to illegality was ingenious. The owner, through its brokers, had issued US dollar invoices to the charterer. It was found as a matter of fact that the Iranian shipper had indirectly benefited from these payments. This had caused the US bank through which the payments were processed to breach sanctions, including by facilitating a transaction by a non-US person (the owner) that would be unlawful if performed by a US person.

By issuing the invoices the owner had caused a violation of sanctions by the US bank. Blair J held that there had also been a breach of US sanctions by the owner itself, albeit unintentional.[41] It is worth pausing to note the tangential nature of the breach and, arguably, its circularity. A non-US person (the owner) was held to be in breach of US sanctions by causing a US person (the bank) to breach sanctions, that breach of sanctions by the bank being the facilitation of a transaction (trading with Iran) by the same non-US person (the owner) which would have been unlawful if performed by a US person (which the owner was not). The lynch-pin is the express 'facilitation' provision in the US sanctions.[42] A US person can be liable for facilitation of sanctioned activity by a non-US person. In addition a non-US person can be in breach of US law by *causing* facilitation by a US person of sanctioned activity by a non-US person (either the first-named non-US person or a different non-US person). This gives US sanctions effective extraterritorial reach even where, as in this case, they do not have express extraterritorial effect. This result was of course based on expert evidence of US federal law from which it appears, not surprisingly, that what constitutes facilitation as a matter of US law is a difficult area.[43] The issue will be returned to later when looking at the approach taken by ship finance banks to sanctions compliance.

Having found that there was a breach of US law by the owner Blair J then considered what effect that breach had on the insurance policy. The first issue was whether the warranty of legality of the 'adventure' implied by the Marine Insurance Act 1906, section 41 had been breached.[44] He held that the 'adventure' is the exposure of the ship to maritime perils; the payment of freight was not within the insured adventure, so the insurers' claim in relation to section 41 failed. Blair J then considered the issue of illegality under common law.[45] As with the issue under section 41, the

39 [2013] EWHC 2116 (Comm); [2014] I Lloyd's Rep 14.

40 *Ibid.* at [245]–[305].

41 *Ibid.* at [276]–[282].

42 On 'facilitation', *supra* n. 18. Two of the three ways in which the US bank was found to have breached sanctions involved facilitation.

43 *The Nancy* [2013] EWHC 2116 (Comm) at [269]–[275].

44 *Ibid.* at [290]–[295].

45 *Ibid.* at [296]–[305].

unlawful processing of the freight payments was found to be separated from the claim for indemnity under the policy. The insurers' case failed on the illegality issue, as it did on all other issues argued.

The case contains some remarks about the effect on English law contracts of illegality under foreign law. These should be seen in the context of the main finding that the illegality was separated from the claim under the policy, that is, there was no causal connection between the illegality and the claim. As regards foreign illegality and the Marine Insurance Act 1906, section 41, Blair J referred to authority for the proposition that section 41 only relates to illegality under English law, not foreign law.[46] That authority comes from remarks made by Staughton J in *Euro-Diam Ltd v Bathurst*[47] which were referred to by Rix J in *Royal Boskalis Westminster NV v Mountain*;[48] also an equivocal comment by Phillips LJ in the same case in the Court of Appeal.[49] Although not conclusive, and noted by Blair J not to be binding on him, these statements are at the very least strongly persuasive.

Blair J next addressed the question of the effect under common law of foreign illegality. This part of the judgment touches upon, but does not fully address, a complex and difficult area.[50] The treatment is brief and *obiter* because of the finding on lack of causation. Blair J said, 'It is not in dispute however, that the court may treat insurance in respect of an adventure that is illegal by foreign law as unenforceable on similar principles that apply in the case of illegality under domestic law' and referred to the relevant authorities on illegality under English law.[51] The issue of 'taint' by foreign illegality was considered by Staughton J in *Euro-Diam Ltd v Bathurst*[52] and his remarks were approved by Kerr LJ in the Court of Appeal.[53] From this it appears that illegality under foreign law can, in principle, be sufficient to 'taint' an English law contract so as to render it unenforceable – no less than illegality under English law. In *Euro-Diam*, however, it was found that there was insufficient connection between the illegality and the claim under the contract to allow the claim of 'taint' to succeed.[54] The position was similar in *The Nancy*;[55] in that case it was not just that there was insufficient connection but that the plaintiffs did not have to rely on the illegal act at all in making their claim. Blair J went on to say:

> It would be different in my view if at the time of entering into the policy, the assured intended to perform the adventure in a manner which involved a breach of US law. However, it was accepted on behalf of the defendant in oral closing submissions that insofar as unlawful intent refers to knowledge of the illegality, a case based on knowledge of the illegality was not being pursued. This reflected the evidence in the case.[56]

---

46 *Ibid.* at [294], [294]

47 [1990] 1 QB 1, at 11.

48 [1997] Lloyds Reinsurance LR 523, at 589–590.

49 [1999] QB 674, at 736.

50 [2013] EWHC 2116 (Comm) at [296]–[299].

51 *Ibid.* at [298]. The main cases referred to by Blair J were *Bowmakers v Barnet Instruments Ltd* [1945] KB 65 and *Beresford v Royal Insurance Co Ltd* [1938] AC 586.

52 [1990] 1 QB1, at 12–16.

53 *Ibid.* at 25.

54 *Ibid.* at 26–28.

55 [2013] EWHC 2116 (Comm) at [299]–[301].

56 *Ibid.* at [302].

He also referred to the principle of *ex turpi causa non oritur actio* as applied generally and specifically in relation to insurance, noting that the principle did not apply where the illegality arises from an inadvertent breach of law.[57] Finally, he addressed the defendant's invocation of a general principle of public policy even where illegality was inadvertent, in circumstances where sanctions were imposed against Iran by the UN, EU and the UK, in addition to the US, against the background of a perceived nuclear threat. This argument was dismissed on the grounds that a breach of 'international sanctions' would involve a breach of domestic law, that is, English law, the same law as governs the contract, that '. . . inadvertence is very different from a case of deliberate breach' and that 'there is no reason why public policy should be applied so as to give insurers a defence to a claim under an insurance policy which is completely unconnected with a breach of US law'.[58]

There are some difficult issues lurking in this part of Blair J's judgment which space does not permit to examine in detail. A brief reminder of the issues is necessary, however, because some or all of them are likely to arise in future cases where the effect of illegality under foreign law (most likely US sanctions) on insurance policies, charterparties, shipbuilding contracts or finance documents will be more direct than was the case on the facts of *The Nancy*.[59] A starting point as good as any is the 'rule' in *Ralli Brothers v Campania Naviera Sota y Aznar*[60] that an English law contract will not be enforced where its terms require an unlawful act in the place of performance.[61] This was not relevant in *The Nancy* because the insurance contract did not require performance in the United States.

In *The Nancy* no explicit reference was made to the important authorities of *Foster v Driscoll*[62] and *Regazzoni v KC Sethia (1944) Ltd.*[63] However, it must have been these cases which Blair J was alluding to when he made his brief comment about the effect of an intention to breach US law (which was found not to be present).[64] These two cases relate to initial illegality under foreign law where the parties are aware of the foreign illegality and knowingly engage in an activity or venture which is directly contrary to the foreign law. The cases are expressed to give effect to a principle of English public policy. That principle appears to be, or at least to have started as, separate from the rule in *Ralli Brothers* relating to illegality in the place

57 *Ibid.* at [303].

58 *Ibid.* at [305].

59 For a recent case which foreshadows the future see *Soeximex SA v Agrocorp International Pte Ltd.* [2011] EWHC 2743 (Comm); 1 Lloyd's Rep 52. An award of the Grain and Feed Trade Associations (GAFTA) Board of Appeal was submitted for reconsideration by reason of failure to address the effect of US and EU sanctions against Myanmar (Burma) on the obligation of a party to open a letter of credit with a New York bank.

60 [1920] 2 KB 287.

61 See Lord Collins of Mapesbury (ed.), *Dicey, Morris & Collins on the Conflicts of Laws* 15th edn, (Sweet & Maxwell, 2012), at [32-097]–[32-102] for a discussion of the long-standing controversy about whether the 'rule' in *Ralli Brothers* is one of English domestic law relating to supervening illegality or of conflict of laws and hence (if the latter) affected by the Rome Convention or superseded by the Rome I Regulation. The rule is often stated in terms which are broad enough to apply both to initial illegality and to supervening illegality, although the case itself was on supervening illegality. See also J. Gilman *et al* (ed.), *Arnould: Law of Marine Insurance and Average* 18th edn, (Sweet & Maxwell, 2013), at [21-07]–[21-09].

62 [1929] 1 KB 470.

63 [1958] AC 301.

64 [2013] EWHC 2116 (Comm) at [302].

of performance. However, the relationship between *Ralli Brothers* on the one hand and *Foster v Driscoll* and *Regazzoni v Sethia* on the other hand has not always been clear. The basis of Scrutton LJ's dissenting judgment in *Foster v Driscoll* was that the contract could have been performed in a way which did not violate United States law. Scrutton LJ was a member of the Court of Appeal in *Ralli Brothers*. His dissenting judgment in *Foster v Driscoll* might have reflected a reluctance to establish a principle beyond that in *Ralli Brothers*. However, the judgments of the majority in *Foster v Driscoll* (Lawrence LJ and Sankey LJ) clearly indicate such a further principle.[65] The speeches of Lord Reid and Lord Keith of Avonholme in *Regazzoni v Sethia*, at least, also indicate a principle separate from that in *Ralli Brothers*.[66] *Ralli Brothers* and *Foster v Driscoll* were explained at some length by Robert Walker LJ in *Ispahani v Bank Melli Iran*[67] as being separate but from the same 'rootstock'.[68] The carrying out of a prohibited act within the relevant foreign country is a necessary element of *Foster v Driscoll*. 'The contract did, not by its express terms but on a realistic view of the practicalities, necessitate shipping from India'.[69] This explanation was recently followed in *Beijing Jianlong Heavy Industry Group v Golden Ocean Group Ltd*.[70] The apparent initial gap between the *Ralli Brothers* on the one hand and *Foster v Driscoll* and *Regazzoni v Sethia* on the other hand has been narrowed. The former requires that a term of the contract necessitates performance in a jurisdiction where that performance is illegal. The latter requires an intention to commit an illegal act in a friendly foreign country.[71] The principle in *Foster v Driscoll*[72] and *Regazzoni v Sethia*[73] applies to foreign law contracts as much as to English law contracts.[74]

We can expect to see in the future cases arising out of foreign sanctions which bring into play the principles, respectively, in *Ralli Brothers* or *Foster v Driscoll* and *Regazzoni v Sethia*. The long-established practice of US dollar payments being cleared through clearing systems located in the United States means that the requirement for an illegal act in the territory of the United States may well be satisfied so as to engage one or other of the principles.[75] There are areas of uncertainty

65 *Foster v Driscoll* [1929] 2 KB 470, at 513 and 514 (Lawrence LJ) and at 518–522 (Sankey LJ).

66 *Regazzoni v KC Sethia 1 (1944) Ltd* [1958] AC 301, at 324 and 325 (Lord Reid) and at 326 and 327 (Lord Keith of Avonholme).

67 [1998] Lloyd's Rep Bank 133. See also *Toprak Mahsulleri Ofisi v Finagrain Compagnie Commerciale Agricole et Financière SA* [1979] 2 Lloyd's Rep 98, at 107 (Goff J).

68 [1998] Lloyd's Rep Bank 133, at 140.

69 *Ibid.*, Robert Walker LJ, explaining *Foster v Driscoll*.

70 [2013] EWHC 1063 (Comm); 2 Lloyd's Rep 61.

71 An English law contract will be enforceable despite foreign illegality in the absence of facts which engage either the *Ralli Brothers* principle or the *Foster v Driscoll/Regazzoni v Sethia* principle. See *Kleinwort Sons & Co v Ungarische Baumwolle Industrie AG* [1939] 2 KB 678, as followed in *British Nylon Spinners v ICI Ltd* [1955] Ch 37 and *Toprak Mahsulleri Ofisi v Finagrain Compagnie Commerciale Agricole et Financière SA* [1979] 2 Lloyd's Rep 98; also *Libyan Foreign Arab Bank v Bankers Trust Co* [1989] 1 QB 728 and *Deutsche Bank AG v Unitech Limited* [2013] EWHC 2793 (Comm), at [99]–[115].

72 [1929] 1 KB 470.

73 [1958] AC 301.

74 *Royal Boskalis Westminster NV v Mountain* [1999] QB 674, at 692E (Stuart-Smith LJ). Phillips LJ did not agree with the majority (Stewart-Smith LJ and Pill LJ). He drew a distinction between enforcing a contract with requires an illegal act in the place of performance and recognising its effect. He said that English courts would not do the former but would do the latter (the 'contract' in question being a waiver of claims).

75 However, in *Libyan Foreign Arab Bank v Bankers Trust Co* [1989] 1 QB 728, illegality under US

under English law in relation to foreign illegality which might be clarified. One such issue is the precise nature and extent of the law relating to 'taint' by foreign illegality as addressed in *The Nancy*[76] and previously in *Euro-Diam v Bathhurst*.[77] A second issue is what falls within the scope of the domestic law of contract and what falls within the conflict of laws so as to be governed by the Rome Convention or the Rome I Regulation.[78] A third issue is the extent to which English law will give effect to foreign laws having extraterritorial effect.[79] A fourth issue is whether the public policy basis of the respective rules in *Ralli Brothers* and *Foster v Driscoll/Regazzoni v Sethia* will become important in determining if the foreign country in whose territory an illegal act occurs or is intended to occur is regarded as 'friendly'. A fifth issue is whether some foreign sanctions might be regarded as 'political' so as not to be recognised.[80]

The third of the trio of Iranian insurance cases is *Arash Shipping Enterprises Company Limited v Groupana Transport*.[81] In contrast with *The Nancy* this can be mentioned quickly because the result was not surprising. EU insurers of an Iranian controlled shipowner issued a notice of cancellation in reliance on a sanctions cancellation clause in the insurances and EU Regulation 961/3010. This was held to be valid so as to deny cover.

14.2.2 The Iranian banking cases

Turning from cases on the effect of sanctions on insurance to their effect on financing transactions, an obvious point should be made at the outset. The effect of a sanctions provision will depend on exactly how it is worded. Sanctions of the old-fashioned type, which prohibit payments to persons resident in a specified country will make unenforceable a loan obligation arising from a transaction entered into in contravention of the prohibition. This is illustrated by *Al-Kishtaini v Shanshal*.[82] Following the Iraqi invasion of Kuwait in 1990 the UK Treasury made directions pursuant to powers contained in the Emergency Laws (Re-enactments and Repeals) Act 1964, the effect of which was to prohibit payments to persons resident in Iraq.

---

law did not affect the liability of the London branch of a US bank to repay a US dollar deposit because payment could be made in cash (in London). The traditional and long-established systems for clearing US dollar payments are the Clearing House Interbank Payments System (CHIPS) and Fedwire, both based in the United States. Offshore (i.e. outside the United States) clearing systems have been developed, most recently and notably in a number of Asian financial centres. Use of these would seem to allow contract parties to avoid the performance of an illegal act in the United States – as would an agreed change of the currency of payment to a currency other than US dollars.

76 [2013] EWHC 2116 (Comm).

77 [1990] 1 QB 1.

78 *Supra* n. 61. In *Ispahani v Bank Melli Iran*, Robert Walker LJ said that the principles in both *Ralli Brothers* and *Foster v Driscoll/Regazzoni v Sethia* were part of the English law of contract rather than the conflict of laws; *Ispahani v Bank Meli Iran* [1998] Lloyd's Rep Bank 133, at 137.

79 United States law having extraterritorial effect (the Sherman Act anti-trust law) was not recognised in England in *British Nylon Spinners v ICI Ltd* [1995] Ch 37. That case related to English patents, regarded as English property, leaving open the possibility that the case might not be followed in a sanctions context where the English property element is absent.

80 In *Regazzoni v Sethia* [1956] 2 QB 490, at 524 short shrift was given to the argument that the Indian law in question was of a 'penal, revenue or political' nature and so not to be recognised (Parker LJ).

81 [2011] EWHC 2116 (Comm); [2011] 1 CLC 984.

82 [2001] EWCA Civ 264; [2001] Lloyd's Rep Bank 174.

The relevant provision is in the same form as that contained in emergency powers legislation which was brought in at the beginning of the Second World War.[83] A UK resident individual (the plaintiff) paid money to an Iraqi resident individual (the defendant) pursuant to a joint venture for the purpose of funding the purchase by the defendant of certain shares. As a result of currency fluctuations, there was a surplus left over which the plaintiff claimed from the defendant. The plaintiff's claim failed. It was based on an illegal act, that is the transfer of funds from the plaintiff to the defendant in breach of the Treasury Directions.[84] This was a case of a transaction being made unenforceable because of initial illegality rather than a case of supervening illegality giving rise to the possibility of frustration. The plaintiff's ignorance of the illegality was held to be irrelevant.[85]

There are two more recent cases which have arisen from sanctions against Iran where, as in *Al-Kishtaini v Shanshal*, a party has invoked illegality arising from a breach of sanctions to try to avoid an obligation – unlike *Al-Kishtani*, the cases involved supervening rather than initial illegality. *DVB Bank SE v Shere Shipping Co Ltd*,[86] related to a syndicated ship finance loan to subsidiaries of Islamic Republic of Iran Shipping Lines (IRISL). After the financing was entered into, EU sanctions were imposed on the borrowers. These sanctions froze all 'funds or economic resources' belonging to, owned, held or controlled by the borrowers and prohibited funds or economic resources being made available, directly or indirectly, to or for the benefit of the borrowers. The relevant EU Regulations permitted the borrowers as designated persons to make payments under contracts entered into before the date of designation if the competent authority authorised the release of frozen funds for this purpose. The issue was therefore one of supervening illegality – with the possibility of a licence or permit being obtained – rather than initial illegality; contrast this with *Al-Kishtaini v Shanshal*.[87] For a short time the borrowers continued to perform their obligations pursuant to a licence from the Bundesbank, as the competent authority. The borrowers then claimed that they were excused performance because of the effect of the EU Regulations imposing sanctions. The borrowers' first argument was that the loan agreement was 'suspended' because it was an 'economic resource' (defined to mean assets of every kind) and was caught by the prohibition on making economic resources available. This argument was dismissed on the ground that a fully drawn loan was not an asset but a liability.[88] The borrowers' second argument was superficially more promising: that payments under the loan agreement were unenforceable because they would be in breach of the asset freeze. This argument failed because there was every possibility that a (further) licence could be obtained from the competent authority.[89] The next argument was ingenious but flawed or, in the words of the judge 'commercially unreal': that the

83 Defence (Finance) Regulations 1939, para. 2A made pursuant to the Emergency Powers (Defence) Act 1939.

84 The Control of Gold, Securities, Payments and Credits (Republic of Iraq) Directions 1990 SI No. 1616, art. 2.

85 [2001] EWCA Civ 264 at [49].

86 [2013] EWHC 2321 (Ch).

87 [2001] EWCA Civ 264.

88 [2013] EWHC 2321 (Ch) at [36]–[40].

89 *Ibid.* at [42]–[55].

repayment of the loan would have the effect of making funds available in breach of sanctions because it would enhance the borrowers' equity in their ships. This was dismissed on the grounds that although the equity in the ships would increase, the borrowers' cash reserves would reduce by an exactly corresponding amount.[90] The borrowers' final argument[91] was that the effect of the supervening illegality was to excuse performance rather than to frustrate the contract. It was held that even if the repayment obligations were incapable of being performed indefinitely (contrary to the earlier findings) there would not be a suspension of the contract but a frustration. In these circumstances the lenders would be entitled to recover the loan by operation of the Law Reform (Frustrated Contracts) Act 1943, section 1. The case was an application for summary judgment and did not involve the more thorough analysis which would have been made at a full trial. It is also worth asking what the position could be in a case where licences from the competent authority had been applied for and refused. This would have required a much fuller treatment of the law on frustration by supervening illegality.

Melli Bank plc v Holbud Limited[92] was a slightly earlier case also involving an application for summary judgment and was referred to in *DVB Bank v Shere Shipping.*[93] In the Melli Bank case it was the bank rather than the bank's customer which was designated by sanctions. The applicable provisions froze the funds and economic resources of the bank as a designated person but provided for the possibility of HM Treasury granting exemptions by licence. The customer claimed to be discharged from an obligation to pay commitment commission to the bank under a finance facility which had not been used. The customer's first argument was that the contract had been frustrated. After a brief review of the main cases on frustration the judge held that there was no frustration because there was a possibility of obtaining a licence.[94] The reasoning and result on this issue were therefore the same as in *DVB Bank v Shere Shipping.* Again, the case leaves open the question of what the result would be where a licence is applied for and refused.

14.3 The shipping and finance industry's response to sanctions

Shipping and ship finance have been living with the effect of sanctions for a number of years, with the effects being most directly felt as a result of the increasingly tight EU and US sanctions directed at Iran. This is illustrated by two of cases referred to above involving IRISL.[95] It is also noteworthy that IRISL's inability to obtain insurance cover led to financing default and mortgage enforcement against three vessels in 2010, with careful steps taken by the financing banks to obtain licences to ensure that they could receive and retain the proceeds of sale following court sales in

90 *Ibid.* at [53].

91 There were further and separate arguments raised by guarantors of the loan which are not addressed here; they all failed.

92 [2013] EWHC 1506 (Comm).

93 [2013] EWHC 2321 (Ch).

94 *Ibid.* at [15]–[22].

95 *Islamic Republic of Iran Shipping Lines v Steamship Mutual Underwriting Association (Bermuda) Limited* [2010] EWHC 2661 (Comm); *DVB Bank SE v Shere Shipping Co Ltd* [2013] EWHC 2321 (Ch).

Singapore.[96] However, the effect of sanctions has been felt across the shipping and ship finance industries, apart from its effect on Iranian shipping. During the past three or four years ship finance lenders have developed sanctions clauses which are now almost universally included in loan agreements, although the language differs from one lender to another. The traditional and general provisions in loan agreements relating to compliance with laws, illegality and 'know your customer' are now generally regarded as inadequate to address the issues presented by sanctions. The sanctions issues which ship financiers wish to address are broadly as follows: not lending to a sanctioned entity or financing a sanctioned activity at inception; not lending to an entity which becomes sanctioned even though not sanctioned at inception; not financing an initially 'innocent' activity which becomes sanctioned; not financing a vessel which trades in breach of sanctions; and the borrower complying with sanction laws. The jeopardies potentially arising from some or all of these issues are: unenforceability of documents for initial illegality; possible frustration through supervening illegality; inability to receive and retain payments or the proceeds of enforcement of security, at least without obtaining any applicable licence; and – this is the issue which focuses the attention of financiers more than any other – liability for breaching sanctions, in particular liability for facilitating a sanctioned activity. These issues and jeopardies are now typically addressed by representations and undertakings covering the following: that the borrower is not a prohibited/designated person or acting on behalf of such a person at inception; that no sanctioned activity is being financed or facilitated; that the customer will comply with sanctions applicable to it and its business – in particular in respect of the financed vessel; and an exit mechanism for the financier if the borrower becomes sanctioned or engages in a sanctioned activity.

One of the areas of difficulty is the extent to which a ship-owning borrower may be required by its lender to comply with sanctions by which the borrower is not itself legally bound. International financial institutions all have a presence in the EU and the US and require access to US dollar clearing through New York. They will themselves comply with EU and US sanctions – but how far do they need to go to impose such compliance on their ship-owning borrower customers? There are different views and legal analysis in this area, although the increasing trend is for ship financiers to push very hard to require compliance by their borrowers. This approach has been driven largely by the very large fines which have been imposed on US and foreign banks, coupled with a fear of being liable for 'facilitation' of breach of US sanctions. The issues of causing a breach of sanctions and facilitation, respectively, were addressed in *The Nancy*,[97] as mentioned above. There it was held that a non-US person had breached US sanctions, albeit in a way which appears oblique, if not circular. More pertinently, a US person will be guilty of breach of US sanctions if it facilitates a transaction by a non-US person which would involve a breach of US sanctions if engaged in by a non-US person. It is this issue which drives US ship financiers, and also many non-US ship financiers, to seek to require their ship-owning borrowers to comply with US sanctions regardless of whether they are US

96 The '*Sahand*' and other applications [2011] SGHC 27.
97 [2013] EWHC 2116 (Comm) at [254]–[262] and at [269]–[282].

persons – coupled with the trend for many US sanctions laws relating to Iran to be expressed to have extraterritorial effect.[98] There is some logic in distinguishing between the use to which the financed vessel itself will be put and the activities in which the borrower may otherwise be engaged more generally. Finance transactions involving ships are more often than not 'asset specific', with the purpose of the loan being expressly to finance a specific ship, whether a newbuild or second hand. This remains the case despite the consolidation of the shipping industry over the last 20 years and the appearance of a few owners who are able, and choose to, raise finance on the basis of the strength of their balance sheet. In the vast majority of cases a financing will be clearly linked to an identifiable vessel or vessels with traditional ship security being taken by the lender. The risk of liability for facilitation to a ship financier must be greater as regards the use and employment of a specifically financed vessel where the traditional tight undertakings in respect of operation and employment are required. This issue remains a difficult and hotly-negotiated area.[99] Agreement of the sanctions provisions in a facility agreement is now often one of the most difficult issues in a new financing. Sanctions compliance issues also frequently arise in restructurings of financially troubled shipping companies where old documents do not contain sanctions clauses but lenders require their inclusion in the restructuring documents.

A second difficult area is that created by chartering. A vessel on bareboat charter is under the direct control of the charterer; a vessel on time or voyage charter is required to trade at the direction of the charterer, subject to applicable trading limits clauses in each case. Just as it has been found that traditional loan agreement language does not adequately address sanctions compliance, so traditional language in charterparties is not sufficient to address sanction issues – and the issues here are more direct than the arguably more oblique issues which arise under loan agreements. A shipowner which is bound by sanctions restrictions is put in a difficult position to say the least if a chartered vessel is operated (in the case of a bareboat charter) or required to be operated (in the case of a time or voyage charter) by the charterer in breach of sanctions binding on the owner. The charterer might not be bound by the same sanctions regime as the owner. This issue is recognised by drafters of sanctions but without any great sympathy being shown for the plight of the shipowner. Article 37b of Council Regulation (EU) No. 267/1212 in relation to Iranian sanctions prohibits the making available of oil or petrochemical tankers to any Iranian person, entity or body but also the making available 'to any other person, entity or body, unless the providers of vessels have taken appropriate action to prevent the vessel from being used to carry or store oil or petrochemical products that originate in Iran or have been exported to Iran'. It is not clear what is meant by 'appropriate action' in this context. Guidelines issued by the US State Department in 2011[100] in urging non-US shipping and insurance companies to minimise their Iranian business

98 Extra-territorial effect has (so far) not been a feature of the US sanctions relating to Russia.

99 The most contentious issue is requiring borrowers to comply with US sanctions by which they are not bound and which do not have extraterritorial effect. Compliance with extraterritorial US sanctions is becoming universally expected and accepted.

100 'Guidelines about the Provisions of Goods and Services, including Insurance, to Entities that Ship Refined Petroleum to Iran' – Bureau of Economic, Energy and Business Affairs 23 May 2011.

gave the following as one of the examples of potentially sanctionable activity under CISADA[101] (which has extraterritorial effect): 'Charter of a ship to another company that is using the ship to supply Iran with gasoline, diesel, jet fuel or aviation gasoline. The shipowner may still have engaged in sanctionable activity even if it does not have full control of the ship under the charter agreement.' As mentioned above, BIMCO and Intertanko have published examples of charterparty clauses designed to address this problem for the shipowner. However, there are charterparties in existence, many long-term, which do not contain such clauses and it is not evident that such clauses are invariably included in charterparties being written now.[102]

Although the ship finance industry has been wrestling with sanctions for some years it is only the recent imposition of Russian sanctions, through EU 'third-tier' sanctions and US 'sectoral' sanctions, on some significant energy companies and banks that has attracted the attention of finance industry bodies. In response to questions and fears raised by Russian sanctions both the International Swap Dealers' Association (ISDA) and the Loan Market Association (LMA) have issued advice and guidance to their members in relation to their respective documentation.

14.4 The judicial approach so far – and the future

Looking at the increasing use of sanctions in world affairs and then focusing on the way sanctions issues have so far been addressed before the English courts, two related features can be seen. First, the recent cases so far have mostly involved disputes where the effect of sanctions has been indirect or tangential; it is very likely that future cases will involve facts where the effect of sanctions is more direct. Second, the English courts have shown a willingness to give effect to contractual rights and obligations to the fullest extent possible despite the effect of sanctions, with reluctance to find that contracts have been frustrated and an antipathy to parties who opportunistically plead the effect of sanctions as a means of avoiding obligations. The tendency of the English courts to uphold contracts despite the effect of sanctions can be seen in one form or another in cases going back over nearly 30 years and beginning with *Libyan Arab Foreign Bank v Bankers Trust Co.*[103] Other cases which have been touched upon here and which evidence the same tendency in relation, variously, to initial illegality, frustration by supervening illegality and the effect of foreign illegality are (in chronological order): *Euro-Diam Ltd v Bathurst*;[104] *Islamic Republic of Iran Shipping Limited v Steamship Mutual Underwriting Association (Bermuda) Ltd*;[105] *Melli Bank plc v Holbud Limited*;[106] *Sea Glory Maritime Co v Al*

101 The Comprehensive Iran Sanctions, Accountability and Divestment Act of 2010.

102 The extreme concern of international banks about sanctions as a result of heavy fines imposed by the US authorities might have resulted in awareness of sanctions being higher among banks than among shipping companies. There is an element of putting the cart before the horse. Addressing sanctions issues in commercial shipping documents logically comes before addressing them in finance documents. There will be increasing attention focused on the wording of *force majeure* clauses in charterparties and shipbuilding contracts.

103 [1989] 1 QB 728.

104 [1990] 1 QB 1 – not a case on sanctions.

105 [2010] EWHC 2661 (Comm).

106 [2013] EWHC 1506 (Comm).

Sagr National Insurance Co (The Nancy);[107] and *DVB Bank SE v Shere Shipping Co Ltd.*[108] It might well be that the approach shown in these cases will not survive intact in future cases where the impact of sanctions is, on the facts, clearer and more direct than in most of the cases so far. We are likely to see cases with results more like those in *Al-Kishtani v Shansal*,[109] *Soeximex SA v Agrocorp International Pte Ltd*[110] or *Royal Boskalis Westminster NV v Mountain.*[111] As indicated above in the discussion of issues in *The Nancy*[112] – or rather issues which did not arise on the facts of that case – there are a number of uncertain areas in the law relating to foreign illegality which might be clarified in litigation relating to foreign sanctions.[113]

In the next few years there are likely to be major cases in the English courts addressing at least some of the following issues where the results will be different from those in the cases seen so far because of different facts or where the law as it currently stands will be clarified or even modified:

(i) frustration by supervening illegality under English law (a long-term charter being the likely subject, with reference to the interpretation of the *force majeure*/exceptions clause);
(ii) an English law contract (loan, insurance or charter) being alleged to be invalid by reason of initial illegality under English law;
(iii) an English law contract (loan, insurance or charter) being alleged to be:
  (a) invalid by reason of initial foreign (most likely US) law illegality; or
  (b) frustrated by reason of supervening foreign law illegality;
(iv) variations of (iii) above where the sanctions giving rise to the claim of foreign law illegality do or do not have purported extraterritorial effect;
(v) an insurance claim where the owner's insurances are vitiated by illegality arising from sanctions (whether through applicable language in the policy or as a matter of law) and a claim under a mortgagee's interest insurance is contested; and
(vi) a contested loan acceleration by reason of alleged breach of sanctions-related representations or undertakings.

If the sanctions against Russia so far imposed remain in place for any length of time or are extended or expanded, many of these issues are almost certain to give rise to litigation. The cases are likely to be more significant – and give rise to more difficult issues – than the Iranian cases so far in the insurance and banking sectors.

Finally, for completeness, there should be mentioned a consultation paper given limited circulation by the UK Foreign & Commonwealth Office in February 2014 mooting the idea of a new concept of 'contract sanctions'. The thinking behind this

107 [2013] EWHC 2116 (Comm).
108 [2013] EWHC 2321 (Ch).
109 [2001] EWCA Civ 264.
110 [2011] EWHC 2743 (Comm). *Supra*, n. 59.
111 [1999] QB 674 at [61]. *Foster v Driscoll* and *Regazzoni v Sethia* were applied in this case, subject to the dissent of Phillips LJ in distinguishing between enforcement of a contract involving foreign illegality and recognising the effect of such a contract.
112 [2013] EWHC 2116 (Comm).
113 Foreign illegality was not addressed in the Law Commission Report, *The Illegality Defence* (Law Comm 320), March 2012.

is to use the widespread choice of English law and dispute resolution as a sanctions weapon by making any agreement to transact in breach of sanctions unenforceable – irrespective of the identity of the parties. There are three points to be made. First, it would be a major departure from the position of the EU and the UK up to now not to try to give sanctions extraterritorial effect – in contrast with the position which has been seen in many recent US sanctions. Second, it would cut across and/or re-write the current legal position relating to both initial and supervening illegality. Third, as many commentators have been quick to point out, it would promote a flight from English law and dispute resolution by many foreign parties.[114] The United States has been successful in the express and *de facto* imposition of sanctions extraterritorially because of the primacy of the US dollar as an international currency and the globalisation of the banking system. English law and dispute resolution for international transaction are matters of choice by the parties to a greater extent than the need to transact in US dollars.[115]

114 The Bar Council of the United Kingdom gave a cogently reasoned response to the consultation paper in March 2014, which made this point among a number of others. (It is not clear whether the FCO consultation paper will be taken further.)

115 The primacy of the US dollar as an international currency in now being challenged by initiatives taken by the BRIC countries (Brazil, Russia, India and China). This trend will expand and continue if US sanctions against Russia continue.

CHAPTER 15

Using derivatives to finance ship sales and shipbuilding: a civil law perspective

*Professor Dr Olivier Cachard**

15.1 Introduction

A scarcity of sources of funding for new vessels has resulted from the introduction of the Basel III Rules, which have further reinforced the risk-aversion of European banks, whether as individual players or as contributors to syndicated loans. Consequently, shipowners have been constrained to raise a higher proportion of the required funding themselves, either in the capital markets or through active management of their own commercial risks. Shipping derivatives were designed by financiers as a tool of risk management, whose function was precisely to hedge some of the risks suffered by shipowners.

One highly significant risk factor is generated by the volatility of freight rates which provide shipowners with their main, if not their only, revenue. Owners can hedge this volatility risk through options, futures or freight forward agreements known collectively as 'shipping freight derivatives'. Furthermore, through a kind of domino effect, the volatility of freight rates creates further volatility elsewhere: in particular, in the prices of newbuilds, second-hand vessels, and ships being disposed of for scrap. As for newbuilds, shipowners may hedge their exposure to changes in price by financial activities rather than operations on the physical market: they will, for example, buy or sell *forward ship value agreements* (FOSVAs). As for the prices of ships sent for scrap, shipowners have to decide the right moment to end exploitation of the vessel but may also use derivatives based on the *Baltic Exchange Demolition Assessments* (BDAs).

All this risk management is made possible by the standardisation of market indicators. Not only is there increasing standardisation of vessels, especially bulkers and container ships, depending on the capacity of locks and harbours and the usual needs of cargo buyers, but there also is a standardisation of maritime routes.[1] Data is available as regards commodity prices and bunkers but also as regards freight rates through indexes.[2] There are emerging electronic marketplaces specifically designed

* Honorary Dean of the Faculty of Law of Nancy, Institut François Gény, Avocat à la Cour d'appel de Metz.

1 With numbers assigned to regular chartering routes, see, e.g., C8-03 Gibraltar/Hamburg Trans-Atlantic Round Voyage, Delivery Gibraltar-Hamburg range, 5–15 days ahead of the index date, Trans-Atlantic round voyage duration 30–45 days, redelivery Gibraltar-Hamburg range.

2 Examples include the Baltic Capesize Index, the Baltic Panamax Index, the Baltic Supramax Index, the Baltic Handysize Index, the Baltic Dirty Tanker Index, the Baltic Clean Tanker Index, the Shanghai Containerised Freight Index, and so on.

to match offer and demand in the field. As a result, the use of derivatives has become more and more popular in the industry.

At this stage, it is necessary to explain the functioning of derivatives. Derivatives are based upon the personal subjective appreciation of the level of risk attached to an underlying transaction. The underlying transaction could be the sale of a commodity, a contract of affreightment or the provision of unit spaces on a vessel, the parties then speculating on the rates. The appreciation of the level of risk is carried out by the buyer and the seller in the light of indexes and econometric analysis for a given period which ends with a named 'date of maturity'. On that date, it is accepted that there will be no physical settlement but a cash settlement between the buyer and the seller. The appraisal of the risk requires, to say the least, a good understanding of the market and sophisticated analysis. For instance, in order to anticipate freight rates, market players have to take into account not only the underlying asset market but also other, related, derivative contracts.[3] For instance, it is now clear that the price of commodity futures has a positive knock-on impact on the returns from forward freight agreements.

In the derivatives market, pure investors with no physical position at all operate side-by-side with those genuinely in the shipping sector, such as shipowners and charterers. The expectations of these maritime players can be summarised as follows. When there is a sale of a FOSVA, buyers expect the price of the relevant newbuild to increase before the maturity date, whereas, by contrast, the sellers expect that it will fall. When there is a transaction over a freight forward agreement, shipowners try to minimise the risk of declining freight rates while charterers aim to minimise the risk of booming rates: according to the derivative contract, the charter will be settled at a certain spread level (from the spot level). When there is a transaction over a freight option, the party benefiting from the option pays a premium for the entitlement to require the other to conclude a charterparty at the maturity date at the rate fixed in the agreement. If the value of the underlying asset at the maturity date is more than the option price, the buyer is said to be 'in the money'; if not, he is referred to as being 'out of the money'.

Indeed, the arrangement can be even more sophisticated, as when the asset underlying the derivative is not a commodity, rate or a spread, but instead is itself another derivative. At this second level, a derivative could be built out of another derivative, which is known by financiers as the process of structuration. The result is that a future can be elaborated out of an option, or an option out of a forward agreement. For the layman, this result can look like some kind of devilish chemistry. This opinion may be shared by more experienced people as well. 'Derivatives,' Warren Buffett memorably said in 2003, 'are financial weapons of mass destruction, carrying dangers that while now latent, are potentially lethal'.[4] This prediction turned out to be well-founded: witness the credit crunch of 2009.

To look at the legal framework of this exposure to risks, we will begin with a civil law analysis of, first, the legal nature and a categorisation of shipping derivatives, and secondly the risks created by such derivatives.

3 M. Kavussanos, I. Visvikis and D. Dimitrakopoulos, 'Economic spillovers between related derivatives markets: the case of commodity and freight markets', *Transportation Research* Part E 68 (2014), pp. 79–102.

4 See the annual report for 2002 of Berkshire Hathaway Inc, Warren Buffett's investment company.

15.2 A civil law analysis of the legal nature and a categorisation of shipping derivatives

Looking at the 2004 EU Directive on markets in financial instruments,[5] one can see that shipping derivatives are financial instruments to a large extent governed by special legislation. According to Article 4(2) of that Directive, derivatives belong to the category of 'investment services and activities', some being cleared and settled through recognised clearing houses, and others traded on regulated markets. Section C of Annex I to the Directive provides a list of derivatives belonging to the category of financial instruments, including 'options, futures, swaps and any other derivative contract relating to commodities that can be physically settled provided that they are traded on a regulated market and/or a MTF', 'derivative instruments for the transfer of risk credit', 'financial contracts for differences' and 'options, futures, swaps, forward rate agreements, and any other derivative contracts relating to climatic variables, freight rates, climatic allowances or inflation rates or other official statistics that must be settled in cash or may be settled in cash at the option of one of the parties'. Beyond this poetic inventory, the civil lawyer's task is to ascertain the legal nature of shipping derivatives and to suggest a classification depending on the characteristics of the different types of derivatives.

15.2.1 The legal nature of shipping derivatives

One way to approach the legal nature of shipping derivatives is to start with a distinction between the *purpose* of the operation and the *object* of the obligations undertaken by the parties to it. The *purpose* of shipping derivatives is to hedge a risk: each of the parties, shipowner or charterer, hopes to transfer it to the other, depending on the circumstances. But this *purpose* of transfer of risk is not to be confused with the *object* of the obligations undertaken. The transfer of a risk is not legally the object of these obligations: on the contrary, the parties have other duties that merely *indirectly* transfer the risk. In Roman law there was not, and in much modern civil law there is not, room for a mere sale of an expectation (*venditio spei*). For example, in an option contract over freight, the promisee has the power to require the promisor to sell (in the case of a *freight call* option) or to buy (in a *freight put* option) the underlying asset at the contractually agreed price. Does this type of derivative fall under the prohibition of gaming contracts in jurisdictions such as France, where the Code Civil precludes 'any action for a gaming debt or for the payment of a bet'?[6] The answer is no: this provision has not been declared applicable to derivatives or other financial instruments that are regulated and thus authorised by special legal regimes.

A second possible approach to the legal nature of shipping derivatives lies in the principle of abstraction[7] as understood in commercial law. A shipping derivative is based on an underlying asset whose legal nature has no consequence: it may be a commodity, a currency, a debt, an interest rate, a spread or whatever.

5 Directive 2004/39/EC on markets in financial instruments.

6 See art. 1965: '*La loi n'accorde aucune action pour une dette du jeu ou pour le paiement d'un pari*'.

7 Note that this is an entirely separate concept from the so-called *Abstraktionsprinzip* in the German law governing sales and transfers of ownership.

Two conclusions may be drawn from this abstraction principle.[8] First, a shipping derivative is autonomous as regards the underlying asset. Thus, when the underlying asset is a right *in rem*, it is not relevant at the moment of the settlement whether or not the promisor or seller is still the owner of it, or indeed whether the underlying asset still exists. So too when the asset is a right *in personam*, there is no direct link between it and the derivative: they are neither related contracts nor do they form a chain of contracts. Second, it also follows from the principle of abstraction that the legal regime which would otherwise govern dealings with the underlying asset is irrelevant. For example, even if the underlying asset is a commodity, the law on the sale of goods would not apply; if the underlying asset was the performance of a charterparty, the contractual regime established by the charterparty would not be applicable; and so on.

15.2.2 The classification of shipping derivatives

Different classifications of shipping derivatives can be suggested either on the basis of market criteria or on purely legal criteria. The most obvious market classification of a derivative is one based on the nature of the underlying asset – goods, commodities, spreads between two values, other derivatives, and so on. In the shipping sector in particular, the assets could be the ship as a whole, or in part through unitisation, a spread, or rates on the freight or bunker markets. However, it follows from the principle of abstraction that this market-based classification cannot be legally relevant. Another possible market classification[9] focuses on the prerogatives of the parties at the moment of settlement. This would yield a distinction between option contracts[10] on the one hand and forward agreements and futures[11] on the other. In an option contract, the party benefiting from the option is offered the unilateral possibility of buying or selling the underlying asset at the price initially defined, named the *strike price*. By contrast, in forward agreements both parties are irrevocably committed to buy and sell the underlying asset at the maturity date.[12] 'A freight forward agreement,' it has been said, 'is a contract between two parties to financially settle a freight rate for a specified quantity of cargo or type of vessel, and one of the major shipping routes at a certain date in the future'.[13] This second market classification has more legal relevance as it refers to the obligations of the parties. Yet a third market classification is based on the nature of the market in which trading takes place, distinguishing between over-the-counter operations and operation on exchange. This third market classification refers to the existence of standard terms, the market rules, and limiting risks. These three market

8 A. Gaudemet, *Les dérivés*, (Economica, 2010), pp. 105 ff.

9 On which see O. Cachard, 'Les dérivés maritimes: risques financiers et risques juridiques associés', DMF 2009, 45.

10 J.C. Hull, *Options, Futures and Other Derivatives*, 5th edn, (Prentice Hall, 2003). Option contracts were originally invented in the Netherlands in the seventeenth century.

11 R. J. Teweles, F. J. Jones, *The Futures Game*, 2nd edn, (McGraw-Hill Book Company, 1992).

12 J.B. Mojuyé, *Le droit des produits dérivés financiers (Swaps, Options, Futures . . .) en France et aux Etats-Unis*, Paris, LGDJ, § 250, p. 80.

13 S. Albertjin, W. Bessler and W Drobetz, 'Finance Shipping Companies and Shipping Operations, A Risk Management Perspective' (2011) *Journal of Applied Corporate Finance* 76.

classifications may favour the understanding of the transactions but they have little legal density.

Some other classifications are based on purely legal criteria, invoking general principles of contract law[14] rather than simply looking at the financial technique involved. One such legally based classification is bottomed on the issuance of a document of title embedding the rights of the promisee. On this basis, we have to distinguish plain derivative contracts, as mentioned in Article 4(2) of the EU Directive on Markets in Financial Instruments, and derivative documents of title, a distinction that has consequences in particular as regards assignment. A second possible legal classification is based on the nature of the rights of the promisee. The promisee may either be vested with a simple option to exercise a right if he so wishes (known as a potestative right or *droit potestatif*), or there may exist fixed rights in both parties defined *ex ante* (known as *droits fermes*). This second legal classification matches our second market classification[15] opposing option contracts and futures or forward agreements. A third legal classification depends on the question whether the derivative allows the transfer of the underlying asset or whether the settlement is just over the difference of value. This three legal tests may be combined for the sake of the analysis of a given shipping derivative.

15.3 A civil law analysis of the risks created by shipping derivatives

The trade of shipping derivatives creates different types of risks which may affect the parties thereto. One legal risk is related to the validity of the derivative contract, for example in a case where legislation in the jurisdiction concerned prohibits speculative contracts. A second legal risk can arise when a broker or agent does not properly transmit the promisee's order to execute a sale or whatever. A third type of risk is counterparty risk, namely the risk that the buyer of a derivative product may default, thus depriving the seller of payment in the event of a failure to perform. A fourth type of risk, global in nature, results from the snowball effect of a succession of defaults, contributing to a systemic risk. To increase confidence and liquidity in the markets, confidence has to be restored by mechanisms aimed at countering risks of these types, such mechanisms being intended to protect either the parties themselves or third parties who might otherwise lose out. Finally, the specific risk created by conflict of laws will briefly be examined.

15.3.1 Framing the risk between the parties to derivative contracts

Several strategies are available to counter the risk of the parties to shipping derivatives. Requiring precontractual information is a first element that modestly contributes to the limitation of risks. Among the provisions of the EU Directive on markets in financial instruments is Article 19, requiring investment firms to provide fair, clear and non-misleading information. The provision of collateral is a second element which increases security. The collateral could be a financial collateral arrangement

14 A. Gaudemet, *Les dérivés, op. cit.*
15 See above.

within the meaning of the 2002 Directive[16] on financial collateral arrangements. For instance, in Luxembourg, parties to a shipping derivative can set up a pledge over bank accounts or shares belonging to the debtor, with full protection of the pledgee in case of insolvency procedures.[17] The collateral could also be a first demand guarantee or a trust by way of security (known as a *fiducie sûreté* in French law). A third possibility could be to adjust the assets encumbered by the security right: some of the rules of clearing houses where shipping freight derivatives are traded require the parties to adjust the coverage to the level of risk.

15.3.2 Limiting the risks incurred by third parties

The parties to shipping derivative contracts may well frame protection of their own interests either through the terms of the derivatives or through the rules governing the marketplace. But problems of privity of contract clearly set limits to the protection of third parties. Who are these third parties in the ambit of shipping derivatives? They are mainly the creditors of the immediate parties, that is, the shipowner and charterer (or whatever). For third parties in this position, many civil law jurisdictions have created general remedies exemplified in France by the *action paulienne* and the *action oblique*. Under the former, creditors are given the right to avail themselves of 'all the rights and causes of actions which belonged to their debtor, with the exception of those which are purely personal to the debtor'.[18] Under the latter, effectively a derivative action, creditors are given the right to 'attack in their own name the acts performed by their debtor, whereby he has defrauded them of their rights'.[19] But the category of third parties also encompasses other operators such as brokers, persons providing investment services and banks and other systemic services; and furthermore, the protections designed by the civil law may lack efficacity when default is due to insolvency. The question therefore arises whether third parties can be protected either through the actions of the immediate parties, or through regulations and clearing house rules.

To begin with, third parties may be indirectly protected through provisions protecting their own debtor who appears to have concluded a shipping derivative contract. Derivative contracts, and sometimes clearing house rules, may well provide for automatic termination or at least the right to terminate in the event of default, with close-out provisions based on an indemnity. A set-off (compensation) may also be established through a 'netting clause' that is useful when there is no default statutory rule allowing set-off.[20] In case of several derivative agreements

16 Directive 2002/47/EC.

17 See *Loi du 5 août 2005 sur les contrats de garantie financière* (Luxembourg), art. 20, §4; O. Cachard, Y. Hilpert and C. Raths, *Luxembourg Maritime Code*, Larcier 2014.

18 '*Néanmoins, les créanciers peuvent exercer tous les droits et actions de leur débiteur, à l'exception de ceux qui sont exclusivement attachés à la personne.*' (*Code Civil*, art. 1166).

19 '*Ils peuvent aussi, en leur nom personnel, attaquer les actes faits par leur débiteur en fraude de leurs droits.*' (*Code Civil*, art. 1167).

20 Such as exists in France, where it is provided in general that: 'When two persons are indebted to each other, set-off applies between them which wipes out both debts in the manner and instances appearing below' ('*Lorsque deux personnes se trouvent débitrices l'une envers l'autre, il s'opère entre elles une compensation qui éteint les deux dettes, de la manière et dans les cas ci-après exprimés.*') – *Code Civil*, art. 1289.

with different entities, the drafters will include a cross-agreement bridge to extend the effects of set-off.

Second, third parties may be indirectly protected through specific regulations and clearing house rules. Both on-exchange and over-the-counter derivative contracts must be concluded through an investment service provider within the meaning of the EU Directive on markets in financial instruments, which offers prudential guarantees. Although shipping freight derivatives were initially traded over-the-counter through brokers, the current trend favours on-exchange transactions. When a derivative contract is concluded through a clearing house, the clearing house provides a performance guarantee which guarantees the performance of all obligations by its members to the benefit of all the other members. Therefore, the clearing house absorbs the counterparty risk and turns it into a clearing house risk, which is deemed to be smaller.[21]

15.3.3 Conflict of laws

To the four types of risks already suggested, it is possible to add the determination of the applicable law to shipping derivatives. What is the law applicable to these abstract contracts or documents of title? A straightforward answer can normally be found in the choice-of-law clause very frequently stipulated in derivative contracts (which forms part of the International Swaps and Derivatives Association's model terms).[22] In the absence of such a clause, the applicable law to the derivative contract must be determined according to the conflict rules laid down in the Rome I Regulation.[23] As shipping derivatives are governed by the abstraction principle, it seems improper to refer to the underlying asset to identify the proper special rule specifying the governing law: for instance, if the underlying asset is the sale of bunkers, it seems wrong to refer to the conflicts rules governing sales. In fact, the courts in such a case have to determine which of the parties to the derivative contract must render the 'characteristic performance' under it, and then presumptively apply the law of that person's habitual residence.[24] Some other elements are equally complex, such as the application of the mandatory rules of the state where the clearing house is located.[25] Another complexity stems out of the neutralisation of the impact of insolvency on the conflict of laws by EU financial directives. For example, the 2002 EU Directive on Financial Collateral Arrangements expressly disapplies certain insolvency provisions of the applicable *lex concursus* in case of winding up. Article 8 of that Directive states that: 'Member States shall ensure that a financial collateral arrangement, as well as the provision of financial collateral

21 I. Riassetto, *Les garanties des contrats à termes sur le MATIF*, PhD Strasbourg, 1999.

22 See the ISDA Master Agreement, Cross-border, Multi-currency.

23 More fully, Regulation (EC) No. 593/2008 of the European Parliament and of the Council of 17 June 2008 on the law applicable to contractual obligations.

24 Rome I, art. 4.2.

25 *Ibid.* art. 9.3. ('Effect may be given to the overriding mandatory provisions of the law of the country where the obligations arising out of the contract have to be or have been performed, in so far as those overriding mandatory provisions render the performance of the contract unlawful. In considering whether to give effect to those provisions, regard shall be had to their nature and purpose and to the consequences of their application or non-application.')

under such arrangement, may not be declared invalid or void or be reversed on the sole basis that the financial collateral arrangement has come into existence, or the financial collateral has been provided on the day of commencement of winding up proceedings or reorganisation measures but prior to the order or decree making that commencement.'

APPENDICES

APPENDIX 1

BIMCO NEWBUILDCON

Explanatory Notes for NEWBUILDCON are available from BIMCO at www.bimco.org

BIMCO

NEWBUILDCON

STANDARD NEWBUILDING CONTRACT

PART I

1. **Place and date of Contract** (Cl. 3, Cl. 44(b), Cl. 47)

2. **Builder's name, full style address and contact details** (Definitions)

Name:
Address:

Country:
Phone/Fax:
E-mail:

Company registration No.

Additional names, addresses and contact numbers:

Name:
Address:

Country:
Phone/Fax:
E-mail:

Company registration No.

3. **Buyer's name, full style address and contact details** (Definitions)

Name:
Address:

Country:
Phone/Fax:
E-mail:

Company registration No.

Additional names, addresses and contact numbers:

Name:
Address:

Country:
Phone/Fax:
E-mail:

Company registration No.

4. **Vessel description/type** (Definitions, Cl. 2(b))

State vessel type:
(i) Dry bulk carrier:
(ii) Tanker:
(iii) Container vessel:
(iv) Other (state type):

A. **Main dimensions** (Cl. 2(b)):

(i) LOA (m):
(ii) Length between perpendiculars (m):
(iii) Deadweight capacity DWT (mts):
(iv) Mean draft in salt water (m):

B1. **Cargo capacity** (Cl. 2(b)(v) and Cl. 11):

(i) Cubic capacity:
(ii) Bale capacity:
(iii) Grain capacity:

B2. **TEU carrying capacity** (only if applicable) (state number of containers):
(i) Total on deck
a. 20'/40'/45" TEU:
b. No. of reefers:
(ii) Total under deck
a. 20'/40'/45"TEU:
b. No. of reefers:
(iii) No. of TEU homogenous loaded at 14 mts.

C. **Main engine(s):** (Cl. 2(b)(ii), Cl. 2(b)(iv) and Cl. 9)

(i) Maker/Type:
(ii) Max. Continuous Rating (MCR) (kilowatts at MCR):
(iii) RPM at MCR:
(iv) Specific Fuel Oil Consumption at MCR:
(v) Normal Continuous Rating (NCR):
(vi) RPM at NCR:
(vii) Type of fuel and specification (including Calorific Value (kcal/kg)):

D. **Average speed** (Cl. 2(b)(i) and Cl. 8)

(i) Service speed at design draft (m):
(ii) Min. number of knots:
(iii) Engine output (kilowatts at MCR):
(iv) Percentage of engine's max. continuous power/sea margin:
(v) RPM:

E. **Other matters** (optional)(state any other technical requirements for the particular vessel type)(Cl. 2(b)(vi) and Cl. 12)

1

Explanatory Notes for NEWBUILDCON are available from BIMCO at www.bimco.org

5. **Shipyard(s)** (if different from Box 2) (Full style address and contact details) (Definitions)

Name:
Address:

Country:
Phone/Fax:
E-mail:

Name:
Address:

Country:
Phone/Fax:
E-mail:

Additional names, addresses and contact numbers:

Name:
Address:

Country:
Phone/Fax:
E-mail:

Additional names, addresses and contact numbers:

Name:
Address:

Country:
Phone/Fax:
E-mail:

6. **Builder's Hull Number** (Definitions, Cl. 2(a))

7. **Flag State** (Definitions)

8. **Classification Society/Class Notation** (Definitions, Cl. 3)

9. **Contract Price and Currency** (Definitions, Cl. 7)

(a) Price
(b) Currency

10. **Contractual Date of Delivery** (Definitions, Cl. 14(c)(iii)(1))

12. **Builder's Bank Account Details** (Cl. 15(d))

Name:
Address:

Country:
Phone/Fax:
E-mail:
Sort code:
Account number:
Account name:

11. **Payment Amounts and Time Due** (Definitions, Cl.7, Cl.15, Cl.39(c))

1st Instalment (see Cl. 15(a)(i))

2nd Instalment and time due

3rd Instalment and time due

4th Instalment and time due

Final Instalment (see Cl. 15(a)(iv))

13. **Speed Deficiency** (Cl. 8, Cl. 39(a)(iv))

(i) Contract Price reduction amount:
(ii) Maximum amount: (state monetary limit):

14. **Excessive Fuel Consumption** (Cl. 9, Cl. 39(a)(v))

(i) Contract Price reduction amount:
(ii) Maximum amount: (state monetary limit):

15. **Deadweight Deficiency** (Cl. 10, Cl. 39(a)(vi))

(i) Deadweight tolerance:
(ii) Contract Price reduction amount:
(iii) Maximum amount: (state monetary limit):

16. **Cubic Deficiency** (Cl. 11, Cl. 39(a)(vii))

(i) Cubic capacity tolerance:
(ii) Contract Price reduction amount:
(iii) Maximum amount: (state monetary limit):

17. **Other Deficiencies** (Optional)(Cl. 12, Cl. 39(a)(viii))

Explanatory Notes for NEWBUILDCON are available from BIMCO at www.bimco.org

18. **Late Delivery Compensation** (Cl. 13) (i) Amount per day: (ii) Maximum amount: (state monetary limit):	19. **Guarantees** (Cl. 14(a) and (b)) (a) Buyer's guarantees (i) Number of days after signing Contract: (ii) Buyer's Instalment/Performance Guarantee: (b) Builder's guarantees (i) Number of days after signing Contract:
20. **Guarantee Period** (state number of months. If left blank 12 months shall apply) (Cl. 35(a)(i)) 21. **Additional Guarantee Period** (state number of months) (Cl. 35(e))	22. **Suspension and Termination** (Cl. 39) (i) Running period (state number of days): (ii) Notice period (state number of days):
23. **Governing law and Dispute Resolution** (Cl. 41 and Cl. 42) (a) Governing law (b) Place of dispute resolution **N/A**	24. **Guarantee Engineer** (state monthly lump sum) (Cl. 36(b))
25. **Effective Date of Contract** (state conditions to be fulfilled) (Cl. 44(a))	26. **State number of days within which conditions have to be satisfied** (Cl. 44(b))
27. **Optional additional vessels** (state number) (Cl. 46)	28. **Optional additional vessels contract price and delivery dates** (Cl. 46)
29. **Declaration of Options** (state number of months after Effective date) (Cl. 46)	30. **Interest** (state rate of interest) (Cl. 18, Cl. 38(b)(ii)(2)(i), Cl. 39(e) and 39(f) (iv))
31. **Buyer's Guarantor** (state name of bank or party as appropriate, full style address and contact details((Cl. 14(a)) Name: Address: Country: Phone/Fax: E-mail:	32. **Builder's Guarantor** (state name of bank or party as appropriate, full style address and contact details((Cl. 14(b), Cl. 27(d)(iv)(3)) Name: Address: Country: Phone/Fax: E-mail:
33. **Additional Annexes**	34. **Numbers of Additional Clauses**

This Contract consists of PART I including additional clauses, if any agreed and stated in Box 34, and PART II as well as any Annexes agreed and attached hereto and shall be performed subject to the conditions contained herein. In the event of a conflict of conditions the provisions of PART I shall prevail over those of PART II to the extent of such conflict, but no further.

The Specification, Maker's List, Plans, and/or Drawings hereafter approved by the Buyer shall form part of this Contract, but in the event of conflict between the provisions of this Contract and the Specification, Maker's List, Plans and/or drawings, the provisions of this Contract shall prevail. In the event of inconsistency between the Specification and Maker's List, on the one hand and the Plans and/or Drawings on the other, the Specifications/Maker's List shall prevail. In the case of inconsistency between any of the Plans and/or Drawings, the later in date shall prevail.

Signature (Builder)	**Signature** (Buyer)

NEWBUILDCON - Standard Newbuilding Contract - PART II

LIST OF CLAUSES

v1.31

NEWBUILDCON - Standard Newbuilding Contract - PART II

v1.31

DEFINITIONS
In this Contract:

"Banking Day" means a day on which banks are open in the places stated in Box 2 and Box 3 and, where a remittance is in US dollars, in New York.

"Builder" means the company or companies stated in Box 2, organised and existing under the laws of the country or countries stated in Box 2 having their principal office at the address stated in Box 2 and including their personnel. If more than one company is stated in Box 2 then they shall be jointly and severally liable.

"Buyer" means the company or companies stated in Box 3, organised and existing under the laws of the country or countries stated in Box 3 having their principal office at the address stated in Box 3 and including their personnel.

"Buyer's Representative" means the named representative of the Buyer who may be present at the Shipyard throughout the construction of the Vessel.

"Buyer's Supplies" means all of the items to be provided by the Buyer in accordance with the Specification at its own risk, cost and expense.

"Classification Society" means the classification society stated in Box 8.

"Contract Price" means the amount stated in Box 9 as may be adjusted in accordance with the terms of this Contract.

"Contract" means this BIMCO Standard Newbuilding Contract consisting of Part I including additional Clauses, if any agreed, and Part II as well as any Annexes (including the Specification and Maker's List) and Plans and Drawings attached hereto.

"Contractual Date of Delivery" means the contractual date of delivery stated in Box 10.

"Defects" means any deficiencies or defects in the design, construction, material and/or workmanship on the part of the Builder or its Sub-contractors.

"Delivery Date" means the Contractual Date of Delivery as may be adjusted in accordance with the terms of this Contract.

"Final Instalment" means the last instalment payable at delivery calculated in accordance with Clause 15 (Payments).

"Flag State" means the State of the flag which the Vessel will fly when registered, as stated in Box 7.

"Instalments" means the amounts payable in accordance with Box 11.

"In writing" means any method of legible communication.

"Maker's List" means the list of suppliers for equipment, machinery and services approved by the Parties and stated in Annex C.

"Parties" means the Builder and the Buyer.

"Party" means the Builder or the Buyer, as the case may be.

"Permissible Delays" means delays to the construction and/or delivery of the Vessel and which

v1.31

entitle the Builder to extend the Delivery Date in accordance with Clause 34 (Permissible Delays).

"Personnel" means the employees, agents, servants, suppliers and independent contractors engaged by either Party in order to perform work or duties under this Contract for which that Party is responsible.

"Plans and Drawings" means the plans and drawings attached hereto or listed and/or described in the Specification.

"Regulatory Authorities" means the regulatory authorities whose rules and regulations must be complied with in the construction and delivery of the Vessel. Such bodies shall include the Flag State together with the other bodies listed in the Specification.

"Shipyard" means the place or places stated in Box 5 where the Vessel will be assembled and/or constructed.

"Specification" means the technical details contained in Annex B.

"Sub-contractor" means any person (not being a servant or employee of the Builder) or company with whom the Builder has entered into a contract for the design, construction, manufacture or supply of any item, equipment, work or service for the Vessel.

"Vessel" means the vessel described in Box 4 (including its machinery, equipment and appurtenances described in the Specification) with hull number as per Box 6, built in accordance with this Contract.

INTERPRETATION

Singular/Plural

In this Contract the singular includes the plural and vice versa as the context admits or requires.

Headings

The index and headings to the Clauses and Annexes to this Contract are for convenience only and will not affect its construction or interpretation.

Jointly and severally

All covenants, agreements, undertakings, indemnities, representations and warranties by more than one person are entered into, given or made by such persons jointly and severally.

NEWBUILDCON - Standard Newbuilding Contract - PART II

v1.31

SECTION 1 – VESSEL

1. Builder's and Buyer's obligations

It is mutually agreed between the Builder and the Buyer that:

(a) the Builder shall design, construct, test and survey, launch, equip, complete, sell and deliver the Vessel to the Buyer all in accordance with good international shipbuilding and marine engineering practice; and

(b) the Buyer shall purchase, take delivery of and pay for the Vessel.

2. Description

(a) The Vessel shall be constructed at the Shipyard and shall have the Builder's Hull Number stated in Box 6.

(b) The Vessel shall have the dimensions and characteristics as stated in Box 4 and the Specification. These shall be defined, measured and calculated in accordance with the Specification or, if omitted from the Specification, in accordance with the following:

(i) Speed – The Vessel's average speed on a sea trial undertaken in both directions over a measured distance of one (1) nautical mile, with clean hull, in weather with wind speed and sea state not exceeding Beaufort Wind Force Scale 3 and Douglas Sea State Scale 2 respectively on a draft as stated in Box 4D(i) shall be at least the number of knots stated in Box 4D(ii). During such a sea trial the engine's output in kilowatts shall be as stated in Box 4D(iii) corresponding to the percentage of the engine's maximum continuous power output stated in Box 4D(iv) at the approximate revolutions per minute stated in Box 4D(v).

(ii) Fuel Consumption - The fuel consumption of the main engine on the test bed using fuel of the type and specification stated in Box 4C(vii) shall not exceed the number of grams per kilowatt/hour stated in Box 4C(iv) when the engine develops the number of kilowatts with an effective calorific value of the number of kilocalories per kilogram stated in Box 4C(ii) and Box 4C(vii) respectively.

(iii) Deadweight - The Vessel's deadweight shall be the number of metric tons stated in Box 4A(iii) on international summer freeboard, corresponding to a mean draft in saltwater (specific gravity 1.025) as stated in Box 4A(iv). The specified deadweight shall include fuel, provisions, stores, freshwater, crew and passengers in addition to spare parts not less than the requirements of the Classification Society.

(iv) Propulsion - The Vessel's propulsion machinery shall be of the type and with maximum continuous power in kilowatts at the number of revolutions per minute as stated in Box 4C(i), 4C(ii) and 4C(iii).

(v) Cargo Capacity – The Vessel's cargo capacity shall be the capacities stated in Box 4B1 and 4B2.

(vi) Other matters – The Vessel shall meet the technical requirements stated in Box 4E.

3. Classification, Rules and Regulations

(a) The Vessel shall be designed, constructed, surveyed, tested and delivered in compliance with the applicable laws, rules, regulations and requirements of the

Classification Society stated in Box 8, and the Regulatory Authorities:

(i) in force as of the date of this Contract stated in Box 1, or

(ii) if not in force as of the date of this Contract, which are ratified and promulgated on or before the date of this Contract and which will be compulsory for the Vessel on or before the delivery of the Vessel in accordance with Clause 28 (Delivery).

All such laws, rules, regulations and requirements of the Classification Society and the Regulatory Authorities shall be complied with without qualification (see Clause 26 (Changes in Rules and Regulations)).

(b) The final decisions of the Classification Society or Regulatory Authorities shall be binding on the Parties as to the Vessel's compliance with their respective applicable laws, rules, regulations and requirements.

(c) All costs, fees and charges incidental to and in respect of compliance with the applicable laws, rules, regulations and requirements of the Classification Society or Regulatory Authorities referred to above shall be for the Builder's cost and expense.

4. IMO Hazardous Materials Inventory
The Builder shall, in accordance with the IMO Guidelines on Ship Recycling, Resolution A.962 (23), with amendments in force as of the date of this Contract:

(a) Endeavour to take due account of the Vessel's ultimate disposal when designing and constructing the Vessel by:

(i) using materials which can be recycled safely and in an environmentally sound manner; and

(ii) by minimising the use of materials known to be potentially hazardous to health and the environment.

(b) Inconsultation with equipment manufacturers provide the Buyer with a Green Passport Statement of Compliance issued by the Classification Society containing information including the Vessel's hull number and main particulars and listing any and all materials known to be potentially hazardous utilised in the construction of the Vessel, its equipment and systems.

The list of the materials known to be potentially hazardous shall contain the location and the approximate quantity/volume of each identified material on board the Vessel.

5. Protective Coatings
The Vessel's double-side skin spaces and dedicated seawater ballast tanks shall be coated in accordance with the Specification. In any event the minimum coating standard shall be in accordance with the requirements of the IMO Performance Standard for Protective Coatings for dedicated seawater ballast tanks in all types of ships and double-side skin spaces of bulk carriers (IMO PSPC, Resolution MSC. 215(82)) and, where applicable, in accordance with the IACS Common Structural Rules for Bulk Carriers and for Oil Tankers or subsequent modifications or replacement applicable in accordance with Clause 3(a) (Classification, Rules and Regulations).

6. Source of Origin
If so requested by the Buyer, the Builder shall identify the country of origin of all the main

components listed in the Maker's List and Specification.

NEWBUILDCON - Standard Newbuilding Contract - PART II

v1.31

SECTION 2 – FINANCIAL

7. Contract Price

The Contract Price shall be the amount stated in Box 9(a) as may be adjusted in accordance with the terms of this Contract.

The Contract Price and any other payments to be made to the Builder pursuant to this Contract shall be paid in the currency stated in Box 9(b) and in accordance with the payment terms stated in Box 11 and Clause 15 (Payments).

8. Speed Deficiency

If the speed of the Vessel as stated in Box 4D(ii) is not achieved in the manner stated in the Specification or Clause 2(b)(i) the following shall apply:

(a) There shall be no adjustment of the Contract Price except to the extent provided in Sub-clause 8(b).

(b) If the reduction in speed is greater than 2/10ths of a knot, the Contract Price shall be reduced by the amount stated in Box 13(i) for each whole 1/10th of a knot reduction in speed in excess of 2/10ths of a knot as liquidated damages up to the maximum amount stated in Box 13(ii).

(c) If the reduction in speed would entitle the Buyer to a reduction in the Contract Price greater than the maximum amount stated in Box 13(ii), the Buyer shall have the option to terminate this Contract in accordance with Clause 39(a)(iv) (Suspension and Termination).

9. Excessive Fuel Consumption

If the fuel consumption of the Vessel's main engine on the test bed using the fuel specified in Box 4C(vii) exceeds the figure stated in Box 4C(iv) the following shall apply:

(a) There shall be no adjustment of the Contract Price except to the extent provided in Sub-clause 9(b).

(b) If the excess fuel consumption is greater than 2% (two percent) of the specified fuel consumption the Contract Price shall be reduced by the amount stated in Box 14(i) for each whole percentage in excess of 2% as liquidated damages up to a maximum amount as stated in Box 14(ii).

(c) If the excess fuel consumption would entitle the Buyer to a reduction in the Contract Price greater than the maximum amount stated in Box 14(ii), the Buyer shall have the option to:

(i) accept the main engine at a reduction in the Contract Price corresponding to the maximum amount stated in Box 14(ii) or

(ii) reject the main engine and either

(1) require the Builder to rectify the deficiency and repeat the trial or replace the main engine with one that conforms to the requirements of the Contract. (The time taken to rectify the deficiency and repeat the trial or replace the main engine in accordance with this Sub-Clause shall not be a Permissible Delay) or

(2) terminate this Contract forthwith in accordance with Clause 39(a)(v) (Suspension

and Termination).

10. Deadweight Deficiency

If the actual deadweight of the Vessel determined in accordance with the Specification is less than the deadweight stated in Box 4A(iii) the following shall apply:

(a) There shall be no adjustment of the Contract Price except to the extent provided in Sub-clause 10(b)

(b) If the reduction in deadweight is greater than the number of metric tonnes stated in Box 15(i) then for each whole metric tonne in excess of the figure in Box 15(i) below the deadweight stated in Box 4A(iii) the Contract Price shall be reduced by the amount stated in Box 15(ii) as liquidated damages up to a maximum amount as stated in Box 15(iii).

(c) If the reduction in deadweight would entitle the Buyer to a reduction in the Contract Price greater than the maximum amount stated in Box 15(iii) the Buyer shall have the option to terminate this Contract in accordance with Clause 39(a)(vi) (Suspension and Termination).

11. Cubic Capacity Deficiency

If the actual cubic capacity of the Vessel determined in accordance with the Specification is less than the cubic capacity specified in Box 4B1(i) the following shall apply:

(a) There shall be no adjustment of the Contract Price except to the extent provided in Sub-clause 11(b)

(b) If the reduction in cubic capacity is greater than the number of cubic metres stated in Box 16(i) then for each full cubic metre in excess of the figure in Box 16(i) below the cubic capacity stated in Box 4B1(i) the Contract Price shall be reduced by the amount stated in Box 16(ii) as liquidated damages up to a maximum amount as stated in Box 16(iii).

(c) If the reduction in cubic capacity would entitle the Buyer to a reduction in the Contract Price greater than the maximum amount stated in Box 16(iii) the Buyer shall have the option to terminate this Contract in accordance with Clause 39(a)(vii) (Suspension and Termination).

12. Other Deficiencies (optional Clause)

NOTE: This Clause has been left blank intentionally to allow the parties to insert, where appropriate, a liquidated damages provision covering deficiencies in any special technical requirements for a particular vessel type and specified by the parties in Box 4E and Box 17. (See also Clause 39(a)(viii)).

13. Late Delivery for non-permissible delays

If delivery takes place more than 30 days after the Delivery Date then for each day thereafter the Contract Price shall be reduced by the amount stated in Box 18 per day as liquidated damages up to a maximum delay of 180 days (comprising a 30 day grace period plus 150 days).

If the delay exceeds 180 days the Buyer shall have the option to terminate this Contract in accordance with Clause 39(a)(iii) (Suspension and Termination).

14. Guarantees

(a) Buyer's Instalment/Performance Guarantee

To secure the Buyer's obligation to pay the instalments of the Contract Price prior to delivery the Buyer shall, within the number of days stated in Box 19(a)(i) after the signing of this Contract, deliver to the Builder an irrevocable and unconditional guarantee issued by the bank or party stated in Box 31 substantially in the form and substance set out in Annexes A(i) (Instalments) or A(ii) (Performance) as stated in Box 19(a)(ii), failing which the Builder shall have the option to terminate this Contract in accordance with Clause 39(b)(iv) (Suspension and Termination).

(b) Builder's Refund Guarantee
To secure the Builder's obligation to refund the Buyer's pre-delivery instalments pursuant to this Contract the Builder shall, within the number of days stated in Box 19(b)(i) after the signing of this Contract and before the date for payment of the first instalment in accordance with Clause 15(a)(i) (Payments – Instalments), provide the Buyer with a Refund Guarantee issued by the bank or party named in Box 32 substantially in the form and substance set out in ANNEX A(iii)(Refund Guarantee), failing which the Buyer shall have the option to terminate this Contract in accordance with Clause 39(a)(ix) (Suspension and Termination).

(c) Guarantee Compliance and Expiry
The Parties shall ensure that any guarantee issued on their behalf shall:

(i) comply with the laws, regulations, constitution and procedures of the guarantor and its country of issue, including but not limited to, its registration with any necessary authorities; and

(ii) on expiry be returned to the guarantor; and

(iii) in the case of the Refund Guarantee (sub-Clause 14(b)), remain in force until either

(1) a date at least 300 days after the Contractual Date of Delivery stated in Box 10 or 30 days after the final resolution of any dispute under Clause 42 (Dispute Resolution), whichever is the later; or

(2) delivery of the Vessel to, and acceptance of the Vessel by, the Buyer whichever is the sooner.

15. Payments
(a) Instalments
The Contract Price shall be paid by the Buyer to the Builder by Instalments, when due and payable in accordance with Box 11 and this Clause, the pre-delivery Instalments being paid as advances and not deposits as follows:

(i) Unless otherwise stated in Box 11 the first Instalment shall be due and payable by the Buyer five (5) Banking Days after the Refund Guarantee has been provided in accordance with Clause 14(b) (Builder's Refund Guarantee).

(ii) The Builder shall give the Buyer invoices for each Instalment under this Contract. With the exception of the first and Final Instalment the Builder shall give the Buyer an invoice to cover the sum due to it not less than ten (10) Banking Days prior to the due date of each Instalment.

(iii) All Instalments other than the first and Final Instalment shall be payable within four (4) Banking Days of the due date thereof.

(iv) The Final Instalment shall be due and payable upon delivery of the Vessel in accordance with Box 11 and Clause 28 (Delivery).

(b) Payment for Modifications and other items
(i) The sums due or refundable as a result of modifications and changes, and changes in Rules and Regulations under Clause 24 (Modifications and Changes) and Clause 26 (Changes in Rules and Regulations) shall be added to or deducted from the Final Instalment.

(ii) All expenses payable in accordance with Clause 27(c)(iii) (Conduct of the Sea Trial) and Clause 22(b) shall be paid together with the Final Instalment.

(iii) Sums due for other items shall be paid within fifteen (15) Banking Days after receipt by the Buyer of the Builder's invoice.

(c) Payment of Liquidated Damages
Any amounts for liquidated damages under Clause 8 (Speed Deficiency), Clause 9 (Excessive Fuel Consumption), Clause 10 (Deadweight Deficiency), Clause 11 (Cubic Capacity Deficiency), Clause 12 (Other Deficiencies) and Clause 13 (Late Delivery for non-permissible delays) shall be calculated and determined before delivery and may be deducted from the Final Instalment.

(d) Payment Procedures
(i) If the date on which any payment is due in accordance with the provisions of this Contract does not fall on a Banking Day, payment shall be made on the next Banking Day.

(ii) Payment of sums due in accordance with the provisions of this Contract shall be made, in the case of payments to the Builder, by electronic transfer to the Builder's account stipulated in Box 12 and, in the case of payments to the Buyer by electronic transfer to such bank as the Buyer by notice to the Builder nominates to receive payments on its behalf.

(iii) The cost of remitting payments shall be for the account of the payer.

(iv) Payments by either Party to the other under this Contract, and their receipt, shall not be deemed a waiver of any right or claim either Party may have against the other.

(v) In the event of late payment of Instalments by the Buyer, the Builder shall have the right to suspend work under this Contract in accordance with Clause 39(c) (Suspension of Work).

16. Taxes, duties, stamps, dues and fees
(a) The Builder shall bear and pay all taxes, duties, stamps, dues and fees imposed in the place stated in Box 2 in connection with the execution and/or performance of this Contract, excluding any taxes, duties, stamps, dues and fees imposed in the place stated in Box 2 upon the Buyer's Supplies which shall be for the Buyer's cost and expense.

(b) The Buyer shall bear and pay all taxes, duties, stamps, dues and fees imposed outside the place stated in Box 2 in connection with the execution and/or performance of this Contract, except for taxes, duties, stamps, dues and fees imposed upon those items and services procured by the Builder for construction of the Vessel.

(c) If either Party pays any taxes, duties, stamps, dues and fees for which the other Party is responsible under this Clause, the other Party shall reimburse the paying Party within fifteen (15) Banking Days of receipt of notice to that effect, together with evidence of the

v1.31

amount paid.

17. Right to set-off

The Buyer shall not have the right to retain or set-off any amount against any payment due to the Builder under this Contract except in relation to the Final Instalment as specifically provided in this Contract (see Clause 15 (Payments) and Clause 30 (Final Instalment)).

18. Interest

If either Party fails to pay any sum due in accordance with the terms of this Contract, the other Party shall have the right to charge interest from the due date at the rate stated in Box 30 on such outstanding sums (see also Clause 39 (Suspension and Termination)).

SECTION 3 – PRODUCTION

19. Sub-contracting

The Builder shall employ the Sub-contractors as set out in the Specification or Maker's list. Except for minor work, the Builder shall not employ other sub-contractors without the Buyer's approval, which shall not be unreasonably withheld.

Notwithstanding any sub-contracting, the Builder shall remain fully responsible for the due performance of such work as if undertaken by the Builder at the Shipyard.

20. Approvals

The times and numbers specified in this Clause shall apply unless otherwise stated in the Specification.

(a) As soon as possible and not later than sixty (60) running days after the effective date of the Contract (see Clause 44 (Effective date of Contract)) the Builder shall provide the Buyer with proposed detailed building and testing schedules. The Buyer shall comment on the schedules as soon as possible and at the latest within fourteen (14) running days after receipt of the proposals. The Builder shall thereafter prepare and issue in writing amended building and testing schedules incorporating the Buyer's comments within fourteen (14) running days thereafter.

(b) The Builder shall despatch to the Buyer a total of three (3) full sets of the Plans and Drawings for the Buyer's approval and shall also submit such other technical information as the Buyer may reasonably require, not less than thirty (30) running days before any construction works commence. The Builder shall give notice to the Buyer advising the date of despatch of the Plans and Drawings and the Buyer shall give notice to the Builder confirming receipt thereof. The Buyer shall within fourteen (14) running days of receipt send to the Builder one (1) set of the Plans and Drawings with the Buyer's approval or approval with comments, amendments or reservations.

In the event that the Buyer needs additional time to consider the Plans and Drawings submitted pursuant to this Clause, it shall request the same in writing of the Builder whose agreement shall not be unreasonably withheld. In the event that the Buyer's comments, amendments or reservations are unclear, unspecified or illegible, the Builder may give notice requesting clarification. If the Buyer fails to respond to the request to provide clarification within five (5) running days of receipt of the Builder's notice, the Builder shall determine whether and to what extent it can adopt the comments, amendments or reservations.

If requested by the Buyer in writing, the Plans and Drawings shall also be sent in an agreed electronic format.

(c) The Builder shall take due note of the Buyer's comments, amendments or reservations (if any) on Plans and Drawings submitted pursuant to this Clause and, if such comments, amendments or reservations are not of such a nature or extent as to constitute a modification or change of the Specification within the meaning of Clause 24 (Modifications and Changes), then the Builder shall commence or continue construction of the Vessel in accordance with the corrected or amended Plans and Drawings.

(d) If the Builder considers the comments, amendments or reservations to the Plans and Drawings are of a nature or extent that constitutes a modification or change under Clause 24 (Modifications and Changes), the Builder shall notify the Buyer accordingly and proceed in accordance with Clause 24 (Modifications and Changes). If the Buyer disagrees the matter shall be resolved in accordance with Clause 24(e).

(e) In the event that the Buyer fails to return any Plans and Drawings to the Builder with approval or approval with comments, amendments or reservations, if any, within the time limit stated above, such Plans and Drawings shall be deemed to have been approved by the Buyer.

(f) The Buyer's approval or deemed approval of any Plans and Drawings shall not affect the obligations of the Builder to design, construct and deliver, or the obligations of the Buyer to take delivery of, and pay for, the Vessel in accordance with the other provisions of this Contract; nor shall it diminish the Builder's responsibility in respect of its obligations under this Contract nor shall it constitute any acceptance by the Buyer of any responsibility for any defect in the Vessel.

(g) The Builder shall give the Buyer, as soon as practicable, copies of all relevant correspondence relating to the Vessel to and from the Classification Society and the Regulatory Authorities, together with all plans approved by the Classification Society.

21. Buyer's Supplies

(a) Buyer

(i) The Buyer shall, at its own risk, cost and expense, supply and deliver to the Builder all of the Buyer's Supplies. Such items shall be delivered at a warehouse or other storage facility at the Shipyard or as otherwise directed by the Builder in a proper condition ready for installation by the Builder or Sub-contractor in or on the Vessel (hereinafter "Installation"), in accordance with the building and testing schedules in Clause 20(a) (Approvals) or as may reasonably be required by the Builder.

(ii) To assist Installation, the Buyer shall provide the Builder with the necessary documentation including specifications, plans, drawings, instruction books, manuals, test reports and certificates required to comply with all applicable rules and regulations. If so requested by the Builder, the Buyer shall, if reasonably possible and at no cost to the Builder, arrange for the representatives of the manufacturers of the Buyer's Supplies to assist the Builder in Installation and/or to carry out the Installation of the Buyer's Supplies by themselves or to make necessary adjustments at the Shipyard in accordance with the manufacturer's instructions, including commissioning.

(iii) The Builder may reject any and all of the Buyer's Supplies when and if found on reasonable grounds to be unsuitable or in improper condition for Installation or not in compliance with the Classification Society or Regulatory Authorities' requirements.

(iv) If delay in delivery of any of the Buyer's Supplies in accordance with Sub-Clause (a)(i) exceeds thirty (30) days and will cause actual delay to the delivery of the Vessel, the Builder shall have the right to proceed with the construction of the Vessel without Installation of the delayed items. The Buyer shall accept and take delivery of the Vessel so constructed.

(b) Builder

(i) The Builder shall safely store and handle the Buyer's Supplies after delivery thereof at a warehouse or other storage facility at the Shipyard or elsewhere as determined by the Builder and shall, at its own cost, expense and responsibility, install them in or on the Vessel in accordance with the Specification, provided that the Builder shall not be responsible for the quality, efficiency and/or performance of any of the Buyer's Supplies.

(ii) The Buyer's Supplies shall be at all times the property of the Buyer but shall be at the Builder's risk from the time of their delivery to the Shipyard until the time of their redelivery to the Buyer whether or not as part of the Vessel.

22. Buyer's Representative, Assistants, Officers and Crew

(a) The Buyer may, at its own cost and expense, have one representative present at the Shipyard throughout the construction together with a reasonable number of assistants and, as appropriate, officers and crew. The Buyer shall notify the Builder in advance in writing of:

(i) the names of the Buyer's Representative, assistants and, as appropriate, officers and crew; and

(ii) the scope of the Buyer's Representative's authority which, in particular, shall include the extent to which the Buyer's Representative has authority to approve plans, drawings and calculations, agree modifications and invoices and attendance at and approval of tests, trials and inspections relating to the Vessel at the Shipyard and/or premises of Sub-contractors; and

(iii) any other information reasonably required by the Builder to facilitate access to the Shipyard and/or premises of Sub-contractors.

(b) The Builder shall, at its own cost and expense, provide the Buyer's Representative and assistants with reasonable office accommodation and facilities (including communication equipment, such as telephone, fax and appropriate internet access, and printers or a connection to the Builder's printers) as the Buyer may reasonably require. The Buyer shall bear the costs of all communication expenses arising from the use by the Buyer's Representative and assistants of the communications equipment provided by the Builder. Such expenses shall be payable by the Buyer on receipt of an invoice from the Builder in accordance with Clause 15(b) (Payments - Payment for Modifications and other items).

(c) The Builder shall have the right to request the Buyer to replace the Buyer's Representative or any assistants but only if the Builder shows that they are carrying out their duties in an unreasonable manner detrimental to the proper progress of the construction of the Vessel, in which case the Buyer shall make proper replacement as soon as possible.

(d) The Buyer's Representative shall have the right to communicate directly with the Classification Society, provided such communication does not unreasonably interfere with the Builder's communication with the Classification Society.

(e) The Builder shall render reasonable assistance to the Buyer in helping to provide suitable accommodation, obtain necessary visas, residence and work permits and any other administrative assistance as the case may be for the Buyer's Representative, assistants and, as appropriate, officers and crew.

23. Inspections, Tests and Trials

(a) To enable the Buyer's Representative and assistants to carry out their duties and inspect the work being done, the Buyer's Representative and/or assistants shall have the right to inspect the Vessel throughout the period of the construction of the Vessel and until its delivery and acceptance.

(b) The Buyer's Representative and/or assistants shall have the right to attend all tests, trials and inspections, including those supervised by the Classification Society and Regulatory Authorities, on any parts of the Vessel whether or not installed. The Builder shall give the Buyer reasonable notice in advance of all such tests, trials and inspections to enable the

Buyer's Representative and/or assistants to attend. If the Buyer's Representative and/or assistants becomes aware of non-conformity of any aspect of the design, construction, material or workmanship arising out of such tests, trials and inspections he/they shall notify the Builder as soon as possible.

(c) For the purposes of attending such inspections, tests and trials the Builder shall, at any time during working hours or at any other time when work is being performed, provide the Buyer's Representative and/or assistants with unimpeded access to the Shipyard, Vessel, workshops, and anywhere else where work on or storage of items connected with the construction of the Vessel is being performed. The Builder shall use its best efforts to arrange similar access for inspection purposes to Sub-contractor's premises during working hours or at any time when work is being performed.

(d) Neither the Buyer's Representative's and/or assistants' inspection and/or attendance at any inspection, test or trial, nor the Buyer's Representative's and/or assistants' failure to notify the Builder of any non-conformity shall relieve the Builder from its obligations under this Contract or be deemed to be or construed as a waiver of any objection to, or any acceptance of, faulty design, construction, material and/or workmanship, or any admission that any materials or workmanship are of the standard required for due performance of this Contract.

24. Modifications and Changes

(a) The Buyer shall have the right at any time to request reasonable modifications or changes in the Specification and/or Plans and Drawings. The Buyer shall request such modifications and/or changes in writing, giving sufficient particulars, documentation and details fully to describe the modifications and/or changes requested.

(b) The Builder shall, as soon as possible after receipt of the written request for modifications or changes, give the Buyer a written proposal of the consequences of implementing such modifications and/or changes. These consequences may include changes in the Contract Price, Delivery Date, capacity, draft, speed, fuel consumption, or any other provisions of this Contract. If in the Builder's reasonable judgement, such modifications and/or changes will adversely affect the Builder's planning or programme in relation to the Builder's other commitments, the Builder shall notify the Buyer that it declines to give such a proposal for the requested modifications and/or changes or part thereof.

(c) The Builder shall use reasonable efforts to minimise the extra costs, delay or other negative impact on the Vessel's capacity, performance or other factors caused by the Buyer's request. The Builder's proposal shall be reasonable for such work.

(d) On the basis of the Builder's proposal the Buyer may elect in writing to agree to the necessary amendments to this Contract, in which case the Builder shall build the Vessel in accordance with this Contract so amended.

(e) If the Buyer does not accept the Builder's notice as provided in Clause 20(d) (Approvals) or if in the Buyer's opinion the Builder's proposal for modifications and/or changes under this Clause is unreasonable, the Buyer may, by giving notice to the Builder, order the Builder to proceed with the requested modifications and/or changes but the consequences of implementing such modifications and/or changes shall be decided in accordance with Clause 42 (Dispute Resolution).

(f) If the Buyer elects not to continue with the request for modifications and/or changes, the Buyer shall notify the Builder accordingly.

(g) If the Buyer does not respond within seven (7) running days after receipt of the Builder's notice in Sub-Clause (b), the Buyer shall be deemed to have withdrawn the request for modifications and/or changes.

25. Builder's Modifications and Substitution of Materials

The Builder shall have the right to make minor modifications and/or changes to the Specification and/or plans if so required by virtue of changes to the Builder's local conditions or facilities, the availability of materials and equipment, the introduction of improved methods or for any other reason of a similar nature provided that the Builder shall first obtain the Buyer's written approval, which shall not be unreasonably withheld or delayed.

Such modifications and/or changes shall satisfy the requirements of the Classification Society and the Regulatory Authorities and shall not relieve the Builder from its obligation to otherwise deliver the Vessel in accordance with this Contract. Any savings obtained shall be credited to the Buyer and the Buyer shall not be obliged to pay any extra for, or suffer any delay in delivery or other adverse consequences of, such modifications and/or changes.

26. Changes in Rules and Regulations

If, after the date of Contract, there are any changes in applicable laws, rules, regulations or requirements (or their application) of the Classification Society or Regulatory Authorities, the following shall apply:

(a) Upon receipt of notice of such changes either Party shall promptly notify the other Party thereof.

(b) If such changes will be compulsory for the Vessel at the time of delivery, the Builder shall, unless the Buyer at its sole discretion seeks and obtains a waiver from the Classification Society or Regulatory Authorities (as appropriate), incorporate such modifications and/or changes into the construction of the Vessel. The Parties shall endeavour to agree on such adjustments to the Contract Price, Delivery Date or other Contract terms as are a direct consequence of the change in applicable laws, rules, regulations or requirements. If the Parties fail to agree on the adjustments, the Builder shall proceed with the required changes and the matter shall be decided in accordance with Clause 42 (Dispute Resolution).

(c) If such changes are not compulsory but the Buyer requires the changes to be incorporated, Clause 24 (Modifications and Changes) shall apply.

27. Sea Trials

The times and numbers specified in this Clause shall apply unless otherwise stated in the Specification.

(a) Notice

The Buyer's Representative, together with a suitable number of assistants, officers and crew, shall have the right to be present at sea trials. The Builder shall give the Buyer at least fourteen (14) running days notice of the time and place and expected duration of sea trials and the Buyer shall promptly acknowledge receipt of such notice.

If neither the Buyer's Representative nor any authorised assistants attend the sea trials for any reason after such notice to the Buyer, such absence shall be deemed to be a waiver by the Buyer of its right to be present. The Builder may then conduct the sea trials without the Buyer's Representative being on board, provided that a representative of the Classification Society and Regulatory Authorities is present. In such circumstances, the results and conditions of the sea trials shall be as confirmed in writing by the Classification Society and/or Regulatory Authorities.

(b) Weather Conditions
The sea trials shall be conducted in weather conditions as described in this Contract and/or Specification. If the sea trials are interrupted or prevented by weather conditions in excess of the stated conditions, any resulting delay in delivery of the Vessel shall be deemed a Permissible Delay in accordance with <u>Clause 34</u> (Permissible Delays). In such an event, the sea trials shall be discontinued or postponed until the first favourable day thereafter when weather conditions permit.

(c) Conduct of the Sea Trials
(i) The sea trials shall be conducted in the presence of representatives from the Classification Society and Regulatory Authorities and in the manner described in this Contract. The sea trials shall be of sufficient scope and duration to enable the Parties to verify and establish that the Vessel conforms in all respects with the performance requirements of this Contract. The Builder shall have the right to repeat any sea trials, subject to appropriate notice to the Buyer.

(ii) The Builder shall provide sufficient crew necessary for the safe navigation of the Vessel.

(iii) All expenses in connection with the sea trials, including the provision of bunkers, lubricating oil, grease, fresh water and stores needed to undertake the sea trials shall be for the Builder's cost and expense. Together with the Final Instalment, the Buyer shall reimburse the Builder at cost price for any quantities of bunkers and unbroached lubricating oil, grease, fresh water and stores remaining on board at delivery.

(d) Method of Acceptance or Rejection
(i) Upon completion of the sea trials the Builder shall give the Buyer the results of the sea trials in writing. If the Builder considers that the results thereof demonstrate that the Vessel conforms to the requirements of this Contract, the Builder shall give the Buyer notice of when delivery will take place. Such notice shall state where and when the Vessel will be ready for delivery, which will be at least fifteen (15) running days after the notice is given. Within five (5) running days after receipt of this notice and the trial results, the Buyer shall notify the Builder in writing of its acceptance for delivery or rejection of the Vessel.

(ii) If the results of the sea trials demonstrate that the Vessel or any part or equipment thereof does not conform to the requirements of this Contract, or if the Buyer rejects the Vessel for other reasons which the Builder accepts as valid, the Builder shall take all necessary steps to rectify such non-conformity. If necessary the Builder shall for its own cost and expense carry out a further sea trial in accordance with this Clause to ascertain that the Vessel complies with the terms of this Contract. Upon demonstration by the Builder that the deficiencies have been corrected, the procedure set out in this <u>Sub-Clause (d)</u> shall apply.

(iii) If the Buyer gives notice of rejection under (i) above or rejects the Vessel under (ii) above, the Buyer shall state in which respects the Vessel does not conform to the requirements of this Contract (hereinafter "Delivery Defects").

(iv) If the Delivery Defects are of minor importance and do not affect Class or the operation

v1.31

of the Vessel in its intended trade but the Builder is unable to rectify the matter within a reasonable time and in any event before the accrual of the Buyer's right to terminate in accordance with Clause 39 (Suspension and Termination), the Builder may nevertheless require the Buyer to take delivery of the Vessel, on condition that the Builder first:

(1) undertakes to remedy the Delivery Defects for its own cost and expense as soon as possible; and

(2) agrees in writing to indemnify the Buyer for any loss incurred as a consequence thereof, including loss of time; and

(3) provides the Buyer with a guarantee issued by the party named in Box 32 (or if Box 32 is not filled in, a bank guarantee from a first class bank) substantially in the form and substance set out in Annex A(iv) for a sum which the Buyer reasonably requests to cover (1) and (2) above, failing agreement such sum to be resolved in accordance with Clause 42 (Dispute Resolution);

whereupon the Buyer shall accept delivery of the Vessel.

(v) If the Builder disputes the rejection of the Vessel by the Buyer, the dispute shall be resolved in accordance with Clause 42 (Dispute Resolution).

NEWBUILDCON - Standard Newbuilding Contract - PART II

v1.31

SECTION 4 – DELIVERY

28. Delivery

Subject to Clause 27(d) (Sea Trials – Method of Acceptance or Rejection) the Vessel shall be delivered to the Buyer on or after the Delivery Date at the Shipyard or at a safe place in the immediate vicinity thereof in a clean and orderly condition, ready for service, upon:

(a) exchange and acceptance by the Parties hereto of a Protocol of Delivery and Acceptance signed by each Party acknowledging delivery of the Vessel by the Builder and acceptance thereof by the Buyer; and

(b) the provision by the Builder of the other documents listed in Clause 29 (Documents on delivery); and

(c) payment by the Buyer of the Final Instalment in accordance with Clause 30 (Final Instalment).

29. Documents on Delivery

Upon exchange of the Protocols of Delivery and Acceptance the Builder shall provide at no cost to the Buyer the following additional documents:

(a) Protocol of Trials made pursuant to the Specification.

(b) Protocol of Inventory and Equipment of the Vessel, including spare parts, as detailed in the Specification.

(c) Protocol of Surplus Consumable Stores which are payable by the Buyer to the Builder.

(d) Plans and Drawings pertaining to the Vessel together with all necessary instruction manuals, as detailed in the Specification.

(e) All certificates including the documents required to be furnished on delivery pursuant to this Contract. All certificates shall be issued without qualification.

If, however, the Classification certificate and/or other required certificates are not available at the time of delivery, the Buyer shall accept interim certificates provided that the Builder, at its cost and expense, provides the Buyer with final certificates as promptly as possible.

The Builder warrants that:

(i) such interim certificates shall enable the Vessel to be registered and trade and operate without restriction; and

(ii) final certificates shall be provided as above.

If the Builder fails to perform (i) and/or (ii) above, the Builder shall compensate the Buyer for any loss incurred as a consequence thereof, including loss of time.

(f) Declaration of Warranty by the Builder that the Vessel is free and clear of any liens, claims, charges, mortgages and other encumbrances.

(g) Builder's Certificate.

NEWBUILDCON - Standard Newbuilding Contract - PART II

(h) Certificate of Non-Registration.

(i) Commercial invoices covering Final Instalment and modifications.

(j) Bill of Sale or other document that certifies that the title of the Vessel passes to the Buyer.

(k) IMO Hazardous Material Inventory Statement of Compliance in accordance with IMO Resolution A.962(23) (as referred to in Clause 4 (IMO Hazardous Materials Inventory)).

(l) Any other documents reasonably required by the Buyer.

The Buyer may require the Builder by giving reasonable notice, prior to delivery, to arrange for any documents listed above to be duly notarised and, if required, legalised at the Buyer's cost and expense.

30. Final Instalment

(a) The Final Instalment shall be adjusted in accordance with this Contract and notified by the Builder to the Buyer not later than seven (7) Banking Days prior to the notified date of delivery (see Clause 27(d) (Sea Trials – Method of Acceptance or Rejection)). Not later than two (2) Banking Days prior to the notified date of delivery the amount of the Final Instalment, as adjusted, shall be deposited with the Builder's Bank as set out in Box 12, with irrevocable instructions that, subject to Sub-Clause (c) below, the amount shall be released to the Builder against presentation by the Builder of a copy of the Protocol of Delivery and Acceptance of the Vessel signed by the Builder and the Buyer. Interest, if any, accruing on such deposit shall be for the benefit of the Buyer.

(b) If the Buyer does not agree the amount of the Final Instalment as adjusted and notified by the Builder, the Buyer shall notify the Builder within five (5) running days. Thereafter the Buyer may take delivery of the Vessel on payment of the Final Instalment as adjusted (or such other amount as the Builder may agree) but without prejudice to the Buyer's rights and remedies under this Contract and the dispute shall be resolved in accordance with Clause 42 (Dispute Resolution).

(c) If the Protocol of Delivery and Acceptance is not so presented within seven (7) days following the date for delivery of the Vessel as notified by the Builder in accordance with Clause 27(d) (Sea Trials – Method of Acceptance or Rejection), the Buyer shall have the right to withdraw the said deposit plus accrued interest. However, if and when a new date for delivery of the Vessel is notified to the Buyer by the Builder in accordance with Clause 27(d) (Sea Trials – Method of Acceptance or Rejection), the Buyer shall make a further cash deposit for the Final Instalment in accordance with the same terms and conditions as set out above.

31. Title and Risk

Title and risk of loss of or damage to the Vessel shall rest with the Builder until exchange of the Protocols of Delivery and Acceptance is effected, immediately upon which title and risk shall pass to the Buyer.

At the time of delivery the Vessel shall be free of all liens, claims, charges, mortgages and other encumbrances.

32. Possession and Removal of the Vessel

(a) The Buyer shall take physical possession of the Vessel immediately upon Delivery and Acceptance thereof.

(b) The Buyer shall remove the Vessel from the place of delivery within five (5) running days after Delivery and Acceptance as aforesaid. If the Buyer does not so remove the Vessel within the said period, the Buyer shall pay to the Builder reasonable mooring charges for the Vessel. The Builder shall also have the right to move the Vessel from the place of delivery to another safe place at its convenience at any time after the five (5) running days' period has expired provided the Buyer is notified accordingly.

33. Vessel Registration
The Buyer shall register the Vessel at its own cost and expense.

v1.31

SECTION 5 – LEGAL

34. Permissible Delays

(a) The Delivery Date shall be extended if any of the following events cause actual delay to the delivery of the Vessel:

(i) Force majeure events

(1) acts of God;

(2) any government requisition, control, intervention, requirement or interference;

(3) threat or act of war, warlike operations, terrorism or the consequences thereof;

(4) riots, civil commotions, blockades or embargoes;

(5) epidemics;

(6) earthquakes, landslides, floods, tidal waves or extraordinary weather conditions;

(7) strikes, lockouts or other industrial action, but only if of a general nature and not limited solely to the Builder and/or the Sub-contractors or their employees;

(8) fire, accident, explosion (whether in the Shipyard or elsewhere);

(9) any interruption to the supply of public utilities to the Builder;

(10) any other cause of a similar nature to the above beyond the control of the Builder or its Sub-contractors;

(11) delays to sea trials in accordance with Clause 27(b) (Sea Trial – Weather Conditions).

(ii) Other events

(1) Late delivery of, or delivery of, any defective Buyer's Supplies in accordance with Clause 21(a)(iv) (Buyer's Supplies);

(2) Delays due to modifications and changes in accordance with Clause 24(b) or (e) (Modifications and Changes);

(3) Delays due to changes in rules and regulations in accordance with Clause 26 (Changes in Rules and Regulations);

(4) An actual or constructive total loss in accordance with Clause 38(b)(ii) (Insurances – Allocation of Insurance Proceeds);

(5) Suspension of work pursuant to Clause 39(c) (Suspension and Termination – Suspension of Work);

(iii) Provided that in respect of (i) and (ii) above:

(1) such events were not caused by the error, neglect, act or omission of the Builder or its Sub-contractors; and

(2) were not, or could not reasonably have been, foreseen by the Builder at the date of the Contract; and

(3) the Builder shall have complied with Sub-Clause (b) hereunder; and

(4) the Builder shall have made all reasonable efforts to avoid and minimise the effects such events have on the delivery of the Vessel.

(b) The Builder shall notify the Buyer within ten (10) running days of when the Builder becomes aware of the occurrence of any event of delay on account of which the Builder asserts that it may have the right to claim an extension of the Delivery Date. A failure to so notify shall bar the Builder from claiming an extension to the Delivery Date. The Builder shall also advise the Buyer in writing (A) within two (2) running days of the ending of any event notified under this Clause that the event has ended, and (B) as soon as reasonably possible after (A), the length of extension of the Delivery Date claimed by the Builder.

35. Builder's Guarantee

(a) The Builder shall guarantee the Vessel against any Defects (see Definitions) provided such Defects are:

(i) discovered within the number of months stated in Box 20 (hereinafter "the Guarantee Period") after delivery of the Vessel in accordance with Clause 28 (Delivery); and

(ii) notice thereof is given to the Builder as soon as reasonably possible after the discovery thereof and latest thirty (30) running days after the expiry of the Guarantee Period describing such Defects so far as reasonably practical

(hereinafter called "Guarantee Defects").

(b) The Builder shall make any necessary repairs or replacements to rectify any Guarantee Defects or damage to the Vessel caused as a direct and immediate consequence of such Guarantee Defects. Such repairs and replacements shall be made at the Shipyard at the Builder's cost and expense.

(c) The Buyer shall have the right to arrange for the necessary repairs to rectify any Guarantee Defects or damage to the Vessel caused as a direct and immediate consequence of such Guarantee Defects to be made elsewhere or obtain any necessary replacement parts and materials:

(i) if it is impractical to bring the Vessel to the Shipyard; or

(ii) if the Builder cannot supply necessary replacement parts and materials without impairing or delaying the operation or working of the Vessel.

(d) In the event that the Buyer makes the necessary repairs or replacements at any other shipyard or works other than the Shipyard, the Buyer shall first, but as soon as possible, give the Builder notice of the time and place such repairs will be made. The Builder shall have the right, without prejudice, to inspect through its own representative the nature and extent of the Guarantee Defects to be replaced or repaired. The Builder shall, in such case, promptly advise the Buyer in writing, after such examination has been completed, of its acceptance or rejection of such Guarantee Defects as ones that are covered by the guarantee.

(i) The Builder shall pay the Buyer in the currency stated in Box 9 the reasonable cost and expenses of such repairs or replacements.

(ii) Where applicable, the Buyer shall return replaced parts to the Builder at the Builder's request and cost and expense provided the Builder makes such request at the time of the replacement. In the event that they are the subject of a dispute under Clause 42 (Dispute Resolution), the Builder shall hold the replaced parts available for inspection by the Buyer. Upon their replacement, the ownership of replaced parts shall revert to the Builder.

(e) The Builder guarantees repairs or replacements to the Vessel made under sub-Clause (b) above for an additional Guarantee Period of the number of months stated in Box 21 from the date of completion of such repairs or replacements provided such work has been performed by the Builder or its Sub-contractors. The additional Guarantee Period shall, however, not end on a date earlier than the end of the original Guarantee Period for any such item.

(f) If, as a result of the guarantee works, the Vessel has been lying idle continuously for a period in excess of thirty (30) days, the Guarantee Period shall be extended by the total number of such days (counting from the first day the Vessel is idle) that fall within the Guarantee Period, whether or not other work was carried out during such period.

(g) Without prejudice to any other rights the Buyer may have under this Contract, following the expiry of the Guarantee Period or in the event that the Builder is in breach of its obligation to rectify Guarantee Defects in accordance with this Clause, the Builder shall at the Buyer's request assign (to the extent to which it may validly do so) to the Buyer, or as the Buyer may direct, the right, title and interest of the Builder in and to all guarantees or warranties given by the Sub-contractors or suppliers of any of the materials or equipment used in the construction of the Vessel.

36. Guarantee Engineer

(a) The Buyer shall have the right to require the Builder to, or the Builder may, appoint a Guarantee Engineer to attend onboard the Vessel for such portion of the Guarantee Period as the Buyer may reasonably require. The Buyer and its employees shall provide the Guarantee Engineer with full co-operation in carrying out his duties. The Guarantee Engineer shall act as the Builder's representative on board and shall give the Buyer full co-operation to enable the Buyer to obtain the most efficient use of the Vessel's machinery and equipment.

(b) The Buyer shall provide the Guarantee Engineer with accommodation and provisions to a standard comparable to the Vessel's Chief Engineer, at no cost to the Builder. The Buyer shall pay the Builder the monthly lump sum stated in Box 24, or pro rata thereof for part of a month, as compensation for part of the cost and charges to be borne by the Builder in connection with the Guarantee Engineer. The Buyer shall also arrange and pay for the transportation of the Guarantee Engineer between the Vessel and his home country.

(c) The Guarantee Engineer shall, at all times and in all respects, be deemed to be the employee of the Builder.

(d) If the Buyer decides not to exercise its right to require the Builder to provide a Guarantee Engineer on board the Vessel, this shall not prejudice the Buyer's rights under the provisions of Clause 35 (Builder's Guarantee).

37. Responsibilities and exclusions from liabilities

NEWBUILDCON - Standard Newbuilding Contract - PART II

v1.31

Builder's exclusion Clauses

(a) Liability for Defects discovered before or at the time of delivery

The Buyer's remedy for delay in delivery of the Vessel, or for Defects discovered before or at the time of such delivery, are set out in Clauses 8 to 13 inclusive and Clause 27(d) (Sea Trials – Method of Acceptance or Rejection).

(b) Liability for Defects discovered after delivery
Except to the extent expressly provided in Clause 35 (Builder's Guarantee), the Builder shall have no liability in contract, tort (including negligence), breach of statutory duty or otherwise for:

(i) any Defect discovered after delivery of the Vessel or

(ii) any loss, damage or expenses caused as a consequence of such Defect (which shall include, but not be limited to, loss of time, loss of profit or earnings or demurrage directly or indirectly incurred by the Buyer).

(c) Liability for third party replacement or repair
The Builder shall not be responsible for any Defects in any part of the Vessel which may, subsequent to delivery of the Vessel, have been replaced or in any way repaired by any contractor, other than the Builder or its Sub-contractors, or for any such Defects which have been caused in whole or part by omission or improper use or maintenance of the Vessel on the part of the Buyer or by ordinary wear and tear.

(d) Implied terms
The guarantee contained in Clause 35 (Builder's Guarantee) replaces and excludes any other liability, guarantee, warranty and/or condition and/or innominate term imposed or implied by the law, customary, statutory or otherwise, by reason of the construction and Sale of the Vessel by the Builder for and to the Buyer.

Mutual exclusion Clauses

(e) Liability following termination
In the event of termination in accordance with the provisions of Clause 39 (Suspension and Termination), neither Party shall have any liability to the other whatsoever or howsoever arising, except as expressly provided in that Clause.

In the event, however, that a Party fails to perform the Contract, or unequivocally indicates its intention not to perform it, in a way which thereby permits the other Party to treat the Contract as at an end other than under the terms of the Contract, any such claim that the other party may have shall not be limited or excluded by the terms of this Contract.

Responsibility Clauses

(f) Responsibility for death and personal injury
Each Party to this Contract shall accept responsibility and liability for the death and personal injury of its Personnel, unless the death or personal injury was inflicted by the other Party or its Sub-contractors with the intent to cause such death or injury, or recklessly and with knowledge that such death or injury would probably result.

Each Party further agrees to indemnify and hold harmless the other Party, as regards both liability and legal costs, in the event of claims relating to or resulting from death or personal injury of its Personnel against the Party who is not responsible for them under this Sub-Clause 37(f).

(g) Responsibility for damage to or loss of property

v1.31

Unless otherwise provided in this Contract, each Party shall accept responsibility and liability for damage to or loss of its property and the property belonging to its Personnel unless such damage or loss was caused by the other Party or its Sub-contractors with the intent to cause such damage or loss, or recklessly and with knowledge that such damage or loss would probably result.

Each Party further agrees to indemnify and hold harmless the other Party, as regards both liability and legal costs, in the event of claims relating to or resulting from damage to or loss of property against the Party who is not responsible for them under this Sub-Clause 37(g).

38. Insurances

(a) Builder's Insurances

From the time of first steel cutting or equivalent (or delivery of the Buyer's Supplies, whichever is earlier) until the Vessel is completed, delivered to and accepted by the Buyer, the Builder shall (in the joint names (as assureds) of the Builder and the Buyer) effect and maintain at no cost to the Buyer, Builder's Risk Insurance for the Vessel and Buyer's Supplies. Such Builder's Risk Insurance shall:

(i) be provided by insurers reasonably acceptable to the Buyer; and

(ii) be on terms no less wide than Institute Clauses for Builder's Risk terms (1/6/88) including Institute War and Institute Strike Clauses ; and

(iii) be in an amount not less than the aggregate of the payments made by the Buyer to The Builder plus the value of the Buyer's Supplies at the Shipyard.

If specifically requested by the Buyer, the Builder shall increase the amount insured under the policy to cover the rebuilding costs of the Vessel or such other amount as the Buyer may request. Any additional premium charged for this shall be paid by the Buyer.

The Builder shall provide the Buyer with copies of the insurance policy as placed.

The Buyer shall notify the Builder of the value of any subsequent changes in the value of the Buyer's Supplies for insurance purposes. Upon receipt of notice of change in value the Builder shall amend the insured value for the Buyer's Supplies accordingly.

(b) Allocation of Insurance Proceeds

(i) In the event that the Vessel is at any time prior to or at delivery damaged by any insured cause and provided such damage does not constitute an actual or constructive total loss of the Vessel, the Builder shall make good such damage and shall apply any amounts recovered under the insurance referred to in Sub-Clause (a) to the costs of any repair or replacement, including repair or replacement of lost or damaged Buyer's Supplies. Such damage shall be made good so as to comply with this Contract and all repairs shall be carried out in compliance with the requirements of the Classification Society and Regulatory Authorities as appropriate without qualification.

(ii) Should the Vessel become an actual or constructive total loss from any insured cause:

(1) the Builder and the Buyer may agree that a new vessel is built or the Vessel reconstructed in accordance with the terms of this Contract provided agreement is reached in writing to an extension of the Delivery Date and/or any other necessary amendment to the Contract, in which case any amounts recovered under the insurance referred to in Sub-Clause (a) will be applied to the construction or reconstruction of

v1.31

the Vessel if appropriate; or

(2) If the Builder and Buyer are unable to agree within a reasonable time on an extension to the Delivery Date and/or any other necessary amendment to the Contract as provided for in <u>Sub-Clause (b)(ii)(1)</u> the Builder shall:

(i) promptly refund to the Buyer the full amount of sums paid by the Buyer to the Builder together with interest thereon at a rate per annum as stated in <u>Box 30</u> from the date of payment to the date of refund; and

(ii) make payment to the Buyer of the insured value of the Buyer's Supplies or alternatively, at the Builder's cost, deliver the Buyer's Supplies to the Buyer in undamaged condition.

Once all payments have been made by the Builder to the Buyer in accordance with <u>Sub-Clause (b)(ii)(2)</u> this Contract shall be deemed terminated and all future rights and obligations of each of the Parties to the other shall cease whereupon the guarantees provided under this Contract shall be returned.

39. Suspension and Termination

(a) Buyer's Termination

The Buyer shall have the right to terminate this Contract forthwith upon giving notice in the event that:

(i) the guarantor providing the Refund Guarantee on behalf of the Builder in accordance With <u>Clause 14(b)</u> (Guarantees – Builder's Refund Guarantee) is deemed insolvent pursuant to <u>Sub-Clause (d)</u> below, unless the Builder provides a replacement Refund Guarantee acceptable to the Buyer within 30 days of the Buyer's notice requiring A replacement Refund Guarantee to be provided, during which period no further payments shall be made to the Builder by the Buyer and provided that notice of termination is given before an acceptable replacement Refund Guarantee is received by the Buyer, or

(ii) the Builder fails to perform any work relating to the construction of the Vessel for a running period of at least the number of days stated in <u>Box 22(i)</u>, excluding Permissible Delays, provided that thereafter the Buyer gives the Builder at least the number of days' written notice stated in <u>Box 22(ii)</u> of its intention to terminate this Contract under this Clause and within that period the Builder fails to remedy its breach and provided further that the notice of termination is given before the Builder has remedied its breach; or

(iii) (1) the delivery of the Vessel is delayed by more than 180 days by virtue of events that fall within <u>Clause 34(a)(i)</u> (Permissible Delays – Force Majeure events); or
(2) the delivery of the Vessel is delayed by more than 180 days by virtue of events which do not fall within <u>Clause 34(a)(i)</u> or <u>34(a)(ii)</u> (Permissible Delays); or
(3) the aggregate of delays to the delivery of the Vessel in (1) and (2) above is more than 270 days.

The Builder may at any time after the right to terminate has occurred give notice requesting that the Buyer either agrees to a new delivery date or terminates this Contract. Such new delivery date shall be a reasonable estimate by the Builder of the date when the Vessel will be ready for delivery. Within fifteen (15) days of the Builder's request, the Buyer shall notify the Builder of its decision. If the Buyer does not terminate this Contract then the new delivery date shall be deemed to be the Delivery Date provided it does not occur later than thirty (30) days prior to the expiry of the Refund Guarantee (Clause 14(b) (Guarantees – Builder's Refund Guarantee)). Notwithstanding Clause 34(a)(i) (Permissible Delays – Force majeure events) and this Clause 39(a)(iii)(1), (2) or (3) but subject to Clause 34(a)(ii) (Permissible Delays - Other events), if the Vessel is not delivered by that date, the Buyer shall have the right to terminate this Contract. The Builder's right to request the Buyer to agree a new delivery date shall operate on each and every occasion the events stated in this Sub-Clause (a)(iii) give rise to the Buyer's option to terminate.

(iv) The reduction in speed would entitle the Buyer to a reduction in the Contract Price greater than the amount stated in Box 13(ii); or

(v) The Buyer rejects the main engine and terminates the Contract in accordance with Clause 9(c)(ii)(2); or

(vi) The reduction in deadweight would entitle the Buyer to a reduction in the Contract Price greater than the amount stated in Box 15(iii); or

(vii) The reduction in cubic capacity would entitle the Buyer to a reduction in the Contract Price greater than the amount stated in Box 16(iii); or

(viii) The condition of the Vessel is deficient in the manner stated in Clause 12 (Other Deficiencies) and Box 17; or

(ix) The Builder is in breach of Clause 14 (Guarantees).

(b) Builder's Termination
The Builder shall have the right to terminate this Contract forthwith upon giving notice to the Buyer in the event that:

(i) The guarantor providing the Instalment Guarantee or Performance Guarantee on behalf of the Buyer under Clause 14(a) (Buyer's Instalment/Performance Guarantee) Is deemed insolvent pursuant to Sub-Clause (d) below, unless the Buyer can provide A replacement Performance Guarantee acceptable to the Builder within 30 days and provided that notice of termination is given before an acceptable Buyer's Instalment or Performance Guarantee is received by the Builder, or

(ii) The Buyer fails to pay any sums due under this Contract for a period of twenty-one (21) Banking Days provided that the Builder thereafter gives the Buyer at least 5 Banking Days notice of its intention to terminate under this Clause, and within that period the Buyer fails to remedy the breach and provided that notice of termination Is given before the Buyer pays the outstanding sums due, or

(iii) The Buyer fails to take delivery of the Vessel tendered in accordance with this Contract, or

(iv) The Buyer is in breach of Clause 14 (Guarantees).

(c) Suspension of Work
Without prejudice to Sub-Clause (b) above the Builder shall have the right to suspend Work under this Contract if the Buyer fails to pay any instalment stated in Box 11 due for a period of fifteen (15) Banking Days until payment of such outstanding sums.

(d) Deemed Insolvency
A Party or the guarantor providing the Refund Guarantee shall be deemed insolvent if proceedings are commenced against the insolvent Party or the guarantor for winding

NEWBUILDCON - Standard Newbuilding Contract - PART II

v1.31

up, dissolution or reorganisation (otherwise than for the purpose of amalgamation or reconstruction), liquidation, the appointment of a receiver, trustee or similar officer, bankruptcy, suspension of payments or similar events.
A Party shall have the right to terminate this Contract forthwith upon giving notice if the Other Party or the guarantor is deemed insolvent.

(e) Effect of Buyer's Termination
If this Contract is terminated by the Buyer, the Builder shall refund all sums paid by the Buyer to the Builder under Clause 7 (Contract Price) and Clause 15 (Payments) hereof plus interest thereon at the rate stated in Box 30 per annum from the date of payment to the date of refund. The Builder shall also return the Buyer's Supplies, or if they cannot be returned, the Builder shall pay to the Buyer an amount equal to the Buyer's cost for Such Buyer's Supplies.

(f) Effect of Builder's Termination
If this Contract is terminated by the Builder, the Builder shall have the right to retain the Buyer's Supplies together with any instalments paid by the Buyer and shall have the right and power either to complete or not to complete the Vessel as it deems fit but in any event shall sell the Vessel (either in its complete or incomplete form), including those Buyer's Supplies which are installed or have been utilised on board the Vessel, at the best price reasonably obtainable at a public or private sale on reasonable terms and conditions.

(i) In the event of the sale of the Vessel in its complete form the proceeds of the sale received by the Builder shall be applied in the following order:

(1) to payment of all expenses incurred by the Builder in respect of the sale and otherwise incurred by the Builder as a result of the Buyer's default;

(2) to payment of all unpaid instalments of the Contract Price including any which would have been payable after the date of termination and interest on such instalments at the rate of interest stated in Box 30 from the respective due dates thereof to the date of application.

(ii) In the event of the sale of the Vessel in its incomplete form the proceeds of sale received by the Builder shall be applied in the following order:

(1) to payment of all expenses incurred by the Builder in respect of the sale and otherwise incurred by the Builder as a result of the Buyer's default;

(2) to payment of all unpaid instalments of the Contract Price to the extent due but not yet paid at the date of termination and interest on such instalments at the rate of interest stated in Box 30 from the respective due dates thereof to the date of application;

(3) to payment of all costs of part construction of the Vessel less any paid instalments and less any sums credited under (2) above;

(4) to payment of the Builder's reasonable net loss of profit caused by the Buyer's default.

(iii) In either of the above events if the proceeds of sale exceed the sums to which such proceeds are to be applied as aforesaid the Builder shall promptly pay any such excess to the Buyer without interest thereon, provided that the amount of such payment to the Buyer shall in no event exceed the total amount of instalments paid by the Buyer. The Builder shall at the same time either permit the Buyer to remove the Buyer's Supplies which are not installed or utilised onboard the Vessel (if any) from the Shipyard for the cost and expense of the Buyer, or give credit to the Buyer for the full value thereof.

(iv) If the proceeds of sale are insufficient to pay the Builder the total amounts due from the Buyer as aforesaid, the Builder may sell the Buyer's Supplies which are not installed or utilised onboard the Vessel (if any) at the best price reasonably obtainable at a public auction or private sale on reasonable terms and conditions, applying the proceeds of such sale toward the unsatisfied amounts due from the Buyer, and giving credit to the Buyer for any excess.

(v) If the proceeds of sale are still insufficient to pay the Builder the total amounts due from the Buyer as aforesaid, the Buyer shall pay to the Builder the amount of such deficiency, plus interest at the rate stated in Box 30 to cover periods whenever payments from the Buyer became overdue.

40. Copyrights, Trade Marks and Patents

(a) Where they are owned and supplied by a Party, that Party shall retain all copyright, Trade mark, patent or similar rights (hereinafter called "Intellectual Property Rights") with respect to the Specification, Plans and Drawings, technical descriptions, calculations, Test results and other data, and information and documents concerning the design and construction of the Vessel. The other Party undertakes not to disclose the same or divulge any information contained therein to any third parties without the prior written consent of the first Party, except where it is necessary for usual operation, repair and maintenance of the Vessel and to subsequent owners.

(b) Each Party shall ensure that any manufacture and/or supply according to specifications, drawings, models or other instructions supplied by it shall not infringe any Intellectual Property Rights of third parties. Should claims nevertheless be made against the other Party in respect of Intellectual Property Rights arising out of or in any way related to the performance of the Contract, the first Party shall keep the other Party indemnified against the cost of such claims, including any legal costs in connection therewith.

(c) For the purpose of this Sub-Clause (c), "Information" means technical information relating to the Vessel designated by one Party as confidential, except information which corresponds in substance to information which:

(i) was developed by and in possession of the other Party prior to first receipt from the first Party; and/or

(ii) at the date hereof or hereafter, through no wrongful act or failure to act on the part of the other Party, enters the public domain.

Where it is necessary during the performance of this Contract for the first Party to make Information available to the other Party, the other Party shall hold all such Information In confidence and not disclose it to any third parties or use it for any purpose other than as provided herein without the prior written consent of the first Party, which shall not be unreasonably withheld.

41. Governing law

This Contract shall be subject to English law unless another law is stated in Box 23(a) In which case the law stated in Box 23(a) shall apply.

42. Dispute Resolution

(a) Classification/Regulatory Authorities

Any dispute concerning the Vessel's compliance or non compliance with the rules,

NEWBUILDCON - Standard Newbuilding Contract - PART II

v1.31

regulations and requirements of the Classification Society or other Regulatory Authorities shall be referred to the Classification Society or other Regulatory Authorities, as the case may be, the final decision of which shall be final and binding upon the Parties hereto. All other disputes shall be referred to expert determination or arbitration in accordance with Sub-Clauses (b) through (e).

(b) Expert determination
Unless Sub-Clause (a) applies or Sub-Clauses (c) to (e) apply, in the event that a dispute arises under this Contract either Party may require by notice in writing to the other Party that such dispute be referred to an independent third party (an "Expert") as the Parties jointly nominate in writing, subject to the following procedure:

(i) if the Parties fail to nominate an Expert within seven (7) days of the date of the notice referred to in this Sub-Clause (b), the dispute shall be resolved in accordance with Sub-Clauses (c) to (e) below;

(ii) the Expert shall act as an expert and not as an arbitrator and his decision shall be final and binding upon the Parties;

(iii) the Expert's determination shall be conducted in accordance with the following rules, unless otherwise agreed by the Parties:

(1) the Parties may make written representations within seven (7) days of the Expert's appointment and shall copy in full such written representations to the other Party within such time period;

(2) the Parties shall have a further seven (7) days to make written comments on each other's representations and shall copy in full such written comments to the other Party within such time period;

(3) the Expert may call for such other documents and written evidence from the Parties as the Expert may reasonably require and the Parties shall provide such documents and written evidence within the period specified by the Expert. The Parties shall copy, in full, such documents and written evidence to the other Party within such time period provided that if either Party claims any such information is confidential to it then, provided in the reasonable opinion of the Expert that Party has properly claimed the same as confidential, the Expert shall not disclose the same to the other Party or to any third party;

(4) the Expert shall decide whether or not to take oral representations from or on behalf of either Party, but if he does so he shall give the other Party the opportunity to be present;

(5) the Expert shall have regard to all representations and evidence before him when making his decision, which shall be in writing, and give full reasons for his decision; and

(6) the Expert shall use all reasonable endeavours to publish his decision within twenty-eight (28) days of his appointment.

(iv) Unless the Parties agree otherwise, each Party shall bear its own costs of a reference to the Expert, and fees and expenses of the Expert shall be borne equally between the Parties.

NEWBUILDCON - Standard Newbuilding Contract - PART II

v1.31

(v) Without prejudice to the rest of this Sub-Clause (b) the Parties shall consider on an ongoing basis whether or not it would be suitable to refer any dispute to an Expert or to enter into mediation in accordance with Sub-Clause (e).

(c) * Arbitration and Mediation
Unless Sub-Clause (a) or (b) applies and unless Box 23(b) states a place other than London, any dispute arising out of or in connection with this Contract shall be referred to arbitration in London in accordance with the Arbitration Act 1996 or any statutory modification or re-enactment thereof save to the extent necessary to give effect to the provisions of this Clause.

The arbitration shall be conducted in accordance with the London Maritime Arbitrators Association (LMAA) Terms current at the time when the arbitration proceedings are commenced.

The reference shall be to three arbitrators. A Party wishing to refer a dispute to arbitration shall appoint its arbitrator and send notice of such appointment in writing to the other Party requiring the other Party to appoint its own arbitrator within fourteen (14) calendar days of that notice and stating that it will appoint its arbitrator as sole arbitrator unless the other Party appoints its own arbitrator and gives notice that it has done so within the fourteen (14) days specified. If the other Party does not appoint its own arbitrator and give notice that it has done so within the fourteen (14) days specified, the Party referring a dispute to arbitration may, without the requirement of any further prior notice to the other Party, appoint its arbitrator as sole arbitrator and shall advise the other Party accordingly. The award of a sole arbitrator shall be binding on both Parties as if he had been appointed by agreement.

Nothing herein shall prevent the Parties agreeing in writing to vary these provisions to provide for the appointment of a sole arbitrator.

In cases where neither the claim nor any counterclaim exceeds the sum of US$100,000 (or such other sum as the Parties may agree) the arbitration shall be conducted in accordance with the LMAA Small Claims Procedure current at the time when the arbitration proceedings are commenced.

(d) *Unless Sub-Clauses (a), (b) or (c) apply, any dispute arising out of or in connection with this Contract shall be referred to arbitration at the place stated in Box 23(b), subject to the procedures applicable there.

(e) Notwithstanding Sub-Clauses (c) and (d) above, the Parties may agree at any time to refer to mediation any difference and/or dispute arising out of or in connection with this Contract.

In the case of a dispute in respect of which arbitration has been commenced under (c) or (d) above, the following shall apply:

(i) Either Party may at any time and from time to time elect to refer the dispute or part of the dispute to mediation by service on the other Party of a written notice (the "Mediation Notice") calling on the other Party to agree to mediation.

(ii) The other Party shall thereupon within fourteen (14) calendar days of receipt of the Mediation Notice confirm that they agree to mediation, in which case the Parties shall thereafter agree a mediator within a further fourteen (14) calendar days, failing which on the application of either Party a mediator will be appointed promptly by the Arbitration Tribunal ("the Tribunal") or such person as the Tribunal may designate for that purpose. The mediation shall be conducted in such place and in accordance

NEWBUILDCON - Standard Newbuilding Contract - PART II

with such procedure and on such terms as the Parties may agree or, in the event of disagreement, as may be set by the mediator.

(iii) If the other Party does not agree to mediate, that fact may be brought to the attention of the Tribunal and may be taken into account by the Tribunal when allocating the costs of the arbitration as between the Parties.

(iv) The mediation shall not affect the right of either Party to seek such relief or take such steps as it considers necessary to protect its interest.

(v) Either Party may advise the Tribunal that they have agreed to mediation. The arbitration procedure shall continue during the conduct of the mediation but the Tribunal may take the mediation timetable into account when setting the timetable for steps in the arbitration.

(vi) Unless otherwise agreed or specified in the mediation terms, each Party shall bear its own costs incurred in the mediation and the Parties shall share equally the mediator's costs and expenses.

(vii) The mediation process shall be without prejudice and confidential and no information or documents disclosed during it shall be revealed to the Tribunal except to the extent that they are disclosable under the law and procedure governing the arbitration.

(Note: The Parties should be aware that the mediation process may not necessarily interrupt time limits.)

* Sub-Clauses (c) and (d) are alternatives; state place of dispute resolution in Box 23(b).

If Box 23(b) is not appropriately filled in, Sub-Clause (c) of this Clause shall apply.

NEWBUILDCON - Standard Newbuilding Contract - PART II

v1.31

SECTION 6 – SUNDRY

43. Notices

(a) All notices given by either Party or their agents to the other Party or their agents in accordance with the provisions of this Contract shall be in writing and shall, unless specifically provided in this Contract to the contrary, be sent to the address for that other Party as set out in Box 2 or Box 3 as appropriate or to such other address as the other Party may designate in writing.

(b) A notice may be sent by post, facsimile, electronically or delivered by hand in accordance with Sub-Clause (a).

(c) Any notice given under this Contract shall take effect on receipt by the other party And shall be deemed to have been received:

(i) if posted, on the seventh (7th) day after posting;

(ii) if sent by facsimile or electronically, on the day of transmission;

(iii) if delivered by hand, on the day of delivery.

And in each case proof of posting, transmission or handing in shall be proof that notice Has been given.

44. Effective date of Contract

(a) This Contract shall become effective when the conditions stated in Box 25 have been satisfied. If no conditions are stated in Box 25 then the effective date of the Contract shall be the date stated in Box 1. The Parties shall immediately notify each other when the conditions stated in Box 25 relevant to that Party have been satisfied.

(b) If any of the conditions referred to above have not been satisfied within the number of days stated in Box 26 after the date of this Contract stated in Box 1, this Contract shall be deemed null and void and both Parties shall immediately be relieved of any obligations or liabilities to the other Party under this Contract.

45. Assignment

(a) Builder's assignment
The Builder shall have the right to assign the benefits of this Contract to the Builder's financiers for the purpose of securing the Builder's financing.

(b) Buyer's assignment

(i) The Buyer shall have the right to assign the benefits of this Contract to the Buyer's financiers for the purpose of securing the Buyer's financing.

(ii) The Buyer shall have the right, subject to the Builder's consent which shall not be unreasonably withheld, to assign, transfer or novate this Contract to any other third party.

46. Options

The Buyer shall have the option for the construction by the Builder of additional vessels as stated in Box 27 at the contract price and delivery dates stated in Box 28, but otherwise on the same terms and conditions as this Contract with logical amendments. Such option must be declared by the Buyer to the Builder within the number of months stated in Box 29 following the Effective date of this Contract referred to in Clause 44 (Effective date of

Contract).

47. Entire Agreement
This Contract constitutes the entire agreement between the Parties and no promise, undertaking, representation, warranty or statement by either Party prior to the date of this Contract stated in Box 1 shall affect this Contract. Any modification of this Contract shall not be of any effect unless in writing signed by or on behalf of the Parties.

48. Third party rights
Unless expressly identified in this Contract, no third parties shall have the right to enforce any term of this Contract.

ANNEX "A" – (GUARANTEES)
BIMCO STANDARD NEWBUILDING CONTRACT
CODE NAME: NEWBUILDCON

ANNEX A(i)

BUYER'S IRREVOCABLE LETTER OF GUARANTEE FOR THE 2ND & 3RD INSTALMENTS

To: [here insert name and address of the Builder]

1. In this Guarantee, the following terms have the following meanings:

'Award Interest' means any interest which may be awarded against the Buyer in connection with the final determination of any dispute notified to us in accordance with paragraph 5 below.

'Contract' means the contract dated [here insert date] made between the Buyer and you for the construction of the Vessel, as the same is amended at any time.

'Contractual Interest' means the sum payable on an Instalment at the Interest Rate in accordance with the terms of the Contract.

'Demand' means a written demand for payment under this Guarantee.

'Instalment' means the amount of each of the [here identify the instalments to be guaranteed] payments in respect of the contract price under the Contract (to the extent that it has not been paid) which is made on, before or after the date of this Guarantee to you by the Buyer.

'Interest Rate' means the rate of interest prescribed by the Contract as applicable to any part of an Instalment which the Buyer is obliged to pay under the terms of the Contract calculated from such date as is prescribed by the Contract to the date of your receipt of the payment.

'Maximum Liability' means our maximum liability under this Guarantee, including Contractual Interest which shall be [here insert amount] plus any Award Interest.

'Buyer' means [here insert name and address of Buyer]

'Vessel' means [here insert technical description and/or name]

2. In consideration of you entering into the Contract, agreeing to construct the Vessel in accordance with the terms of the Contract, and agreeing to accept this Guarantee pursuant to the Contract, at the request of the Buyer we irrevocably and unconditionally guarantee (but as primary obligor and not by way of secondary liability only) that in the event that the Buyer fails punctually to pay to you any Instalment we shall, upon receipt by us from you of a Demand for the same (together with a copy of a demand made by you against the Buyer for payment), pay to you or to your order upon the expiry of thirty (30) days from receipt of such Demand the sum demanded by you by way of the payment of any Instalment together with Contractual Interest and Award Interest (if any) provided that our total liability shall not exceed the Maximum Liability.

This Guarantee shall become effective in the amount corresponding to the amount of each Instalment with effect from the date when it becomes due to you under the Contract together with Contractual Interest, if any.

3. This Guarantee shall not be affected by any indulgence or delay allowed to the Buyer nor by any amendment to, or variation of, the Contract whether as to time or otherwise that may be agreed between you and the Buyer nor by any circumstances that would otherwise discharge our liability as guarantor.

4. Subject to paragraph 5 below, this Guarantee shall remain in force until the first to occur of (a) due delivery of the Vessel to, and acceptance of the Vessel by, the Buyer, (b) the payment to you by the Buyer or by us of all sums secured by this Guarantee, and (c) the Buyer's valid and lawful cancellation and/or rescission of the Contract pursuant to the terms of the Contract. However, notwithstanding the foregoing, if within twenty-eight (28) days of our receipt of a Demand we receive a written notice from you or the Buyer that your claim for payment of the sums referred to in the Demand has been disputed and that such dispute will be resolved in accordance with the Contract, the period of validity of this Guarantee shall be extended until thirty (30) days after the dispute has been finally determined in accordance with paragraph 5 below.

NEWBUILDCON - Standard Newbuilding Contract – PART II

5. Notwithstanding the other terms of this Guarantee, if within twenty-eight (28) days of our receipt of a Demand we receive written notice from you or from the Buyer stating that your claim for payment of the sums referred to in the Demand has been disputed and that such dispute will be resolved in accordance with the Contract (including, as may be, by Class, an Expert, a Mediator or in arbitration), then we shall not be obliged to make any payment to you under this Guarantee until thirty (30) days after the dispute has been finally determined or in the event of an appeal from an arbitration award, until thirty (30) days after delivery of the final unappealable judgment; or in the event that the court remits the matter to the arbitrator, until thirty (30) days after the publication of the revised final award or in the event of an appeal from the award, until thirty (30) days after delivery of the final unappealable judgment.

6. All payments to be made under this Guarantee shall be made without any set off or counterclaim and without deduction or withholding for or on account of any taxes, duties or charges whatsoever unless we are compelled by law to deduct or withhold the same in which case we shall make the minimum deduction or withholding permitted and will pay such additional amounts as may be necessary in order that the amount received by you after such deductions or withholdings shall be equal to the amount which would have been received had no such deduction or withholding been made.

7. All payments to be made under this Guarantee shall be made in [here insert currency].

8. Notwithstanding any provision in the Contract, this Guarantee shall be freely assignable by you and by any assignee. Upon assignment, all references in this Guarantee to "you" shall be read as references to the assignee or subsequent assignees.

9. This Guarantee is governed by the laws of [here state applicable law. In the absence of any statement the law of England and Wales shall apply] and we hereby submit to the exclusive jurisdiction of [here state place of dispute resolution. In the absence of any statement the High Court of England and Wales shall apply]. [We hereby authorise and nominate the following agent to accept service of any court proceedings on our behalf: [here insert name and address.]]

10. Any notice, claim or Demand to be given or made by you under this Guarantee shall be in writing signed by one of your officers and may be served on us either by post or by tested telex/authorised SWIFT or equivalent, and if sent by post to [here insert address] (or such other address as we may notify to you in writing) and if by tested telex at [here insert number] or if by SWIFT or equivalent at [here insert number] via your bank and shall be effective only upon actual receipt.

11. To the extent that we may be or may hereafter become entitled, in any jurisdiction, to claim for ourselves or our property, assets or revenue immunity (whether by reason of sovereignty or otherwise) in respect of our obligations under this Guarantee from service of process, suit, jurisdiction, judgment, order, award, attachment (before or after judgment or award), set off, execution of a judgment or other legal process and to the extent that in any such jurisdiction there may be attributed to us or any of our property, assets or revenue such an immunity (whether or not claimed) we hereby irrevocably agree not to claim and hereby irrevocably waive such immunity to the fullest extent permitted by the laws of such jurisdiction.

12. We hereby warrant that we are permitted by any relevant law to which we are subject (including, where relevant, the laws of the place or places of each of our incorporation, establishment, regulation, registration and residence) to:
12.1 issue a guarantee in this form,
12.2 make payment under this Guarantee in a currency other than that of the place of (where relevant) each of our incorporation, establishment, regulation, registration and residence in case of a Demand for payment under this Guarantee, and
12.3 designate the place stated in paragraph 9 above as the forum and the place of jurisdiction to which we irrevocably submit.

13. We hereby warrant that this Guarantee has been, or will be, duly registered with the relevant State authority in any legal jurisdiction in which such registration is required for any reason.

14. We hereby warrant that we have obtained all necessary approvals and authorisations to issue this Guarantee.

Dated the day of 2

.... (signature)

for and on behalf of [here insert name of Guarantor]

ANNEX A(ii)

IRREVOCABLE LETTER OF GUARANTEE FOR PERFORMANCE OF BUYER'S OBLIGATIONS

To: [here insert name and address of the Builder]

1. In this Guarantee, the following terms have the following meanings:

   'Contract' means the contract dated [here insert date] made between the Buyer and you for the construction of the Vessel, as the same may be amended at any time.

   'Buyer' means [here insert name and address of Buyer]

   'Vessel' means [here insert technical description and/or name]

2. In consideration of you entering into the Contract, agreeing to construct the Vessel in accordance with the terms of the Contract, and agreeing to accept this Guarantee pursuant to the Contract, at the request of the Buyer we irrevocably and unconditionally guarantee (but as primary obligor and not by way of secondary liability only) performance by the Buyer of all its liabilities and responsibilities under the Contract, including but not limited to due and punctual payment of any instalment of the contract price by the Buyer to you under the Contract, and we shall, upon receipt by us from you of a written demand for the same (together with a copy of a demand made by you against the Buyer for payment), pay to you or to your order upon the expiry of thirty (30) days from receipt of such written demand the sum demanded by you together with interest due under the Contract.

3. This Guarantee is governed by the laws of [here state applicable law. In the absence of any statement the law of England and Wales shall apply] and we hereby submit to the exclusive jurisdiction of [here state place of dispute resolution. In the absence of any statement the High Court of England and Wales shall apply]. [We hereby authorise and nominate the following agent to accept service of any court proceedings on our behalf: [here insert name and address.]]

Dated the day of 2

……………………………………………………………………. (signature)

for and on behalf of [here insert name of Guarantor]

NEWBUILDCON - Standard Newbuilding Contract – PART II

ANNEX A(iii)

REFUND GUARANTEE

To: [here insert name and address of the Buyer]

1. In this Guarantee, the following terms have the following meanings:

'Award Interest' means any interest which may be awarded against the Builder in connection with the final determination of any dispute notified to us in accordance with paragraph 5 below.

'Contract' means the contract dated [here insert date] made between the Builder and you for the construction of the Vessel, as the same is amended at any time.

'Contractual Interest' means the sum payable on an Instalment at the Interest Rate in accordance with the terms of the Contract.

'Demand' means a written demand for payment under this Guarantee.

'Instalment' means the amount of each payment in respect of the contract price under the Contract (to the extent that it has not been refunded) which is made on, before or after the date of this Guarantee to the Builder (or at the Builder's direction) by you or on your behalf.

'Interest Rate' means the rate of interest prescribed by the Contract as applicable to any part of an Instalment which the Builder is obliged to repay under the terms of the Contract calculated from the date on which the Builder received the Instalment to the date of your receipt of the repayment.

'Maximum Liability' means our maximum liability under this Guarantee, including Contractual Interest which shall be [here insert amount] plus any Award Interest.

'Builder' means [here insert name and address of shipbuilder]

'Vessel' means [here insert technical description and/or name]

2. In consideration of you entering into the Contract, agreeing to pay an Instalment or Instalments to the Builder, and agreeing to accept this Guarantee pursuant to the Contract, at the request of the Builder we irrevocably and unconditionally guarantee (but as primary obligor and not by way of secondary liability only) that if the Builder becomes liable under the Contract to repay any part of any Instalment we shall, upon receipt by us from you of a Demand for the same (together with a copy of a demand made by you against the Builder for repayment), pay to you or to your order upon the expiry of thirty (30) days from receipt of such Demand the sum demanded by you by way of the repayment of any Instalment together with Contractual Interest and Award Interest (if any) provided that our total liability shall not exceed the Maximum Liability.

This Guarantee shall become effective in the amount corresponding to the amount of each and every Instalment paid to the Builder under the Contract together with Contractual Interest as and when each such Instalment has been received by the Builder.

3. This Guarantee shall not be affected by any indulgence or delay allowed to the Builder nor by any amendment to, or variation of, the Contract whether as to time or otherwise that may be agreed between you and the Builder nor by any circumstances that would otherwise discharge our liability as guarantor.

4. Subject to paragraph 5 below, this Guarantee shall remain in force until the first to occur of (a) due delivery of the Vessel to, and acceptance of the Vessel by, you (b) the payment to you by the Builder or by us of all sums secured by this Guarantee[, and (c) three-hundred (300) days after the Contractual Date of Delivery]. However, notwithstanding the foregoing, if within twenty-eight (28) days after our receipt of a Demand we receive a written notice from you or the Builder that your claim for the repayment of any sums referred to in the Demand has been disputed and that such dispute will be resolved in accordance with the Contract, the period of validity of this Guarantee shall be extended until thirty (30) days after the dispute has been finally determined in accordance with paragraph 5 below.

5. Notwithstanding the other terms of this Guarantee, if within twenty-eight (28) days after our receipt of a Demand we receive a written notice from you or from the Builder stating that your claim for repayment of any sums referred to in the Demand has been disputed and that such dispute will be resolved in accordance with the Contract (including, as may be, by Class, an Expert, a Mediator or in arbitration), then we shall not be obliged to make any payment to you

NEWBUILDCON - Standard Newbuilding Contract – PART II

under this Guarantee until thirty (30) days after the dispute has been finally determined or in the event of an appeal from an arbitration award, until thirty (30) days after delivery of the final unappealable judgment; or in the event that the court remits the matter to the arbitrator, until thirty (30) days after the publication of the revised final award or in the event of an appeal from the award, until thirty (30) days after delivery of the final unappealable judgment.

6. All payments to be made under this Guarantee shall be made without any set off or counterclaim and without deduction or withholding for or on account of any taxes, duties or charges whatsoever unless we are compelled by law to deduct or withhold the same in which case we shall make the minimum deduction or withholding permitted and will pay such additional amounts as may be necessary in order that the amount received by you after such deductions or withholdings shall be equal to the amount which would have been received had no such deduction or withholding been made.

7. All payments to be made under this Guarantee shall be made in [here insert currency].

8. Notwithstanding any provision in the Contract, this Guarantee shall be freely assignable by you and by any assignee. Upon assignment, all references in this Guarantee to "you" shall be read as references to the assignee or subsequent assignees.

9. This Guarantee is governed by the laws of [here state applicable law. In the absence of any statement the laws of England and Wales shall apply] and we hereby submit to the exclusive jurisdiction of [here state place of dispute resolution. In the absence of any statement the High Court of England and Wales shall apply]. [We hereby authorise and nominate the following agent to accept service of any court proceedings on our behalf: [here insert name and address.]]

10. Any notice, claim or Demand to be given or made by you under this Guarantee shall be in writing signed by one of your officers and may be served on us either by post or by tested telex/authorised SWIFT or equivalent, and if sent by post to [here insert address] (or such other address as we may notify to you in writing) and if by tested telex at [here insert number] or if by SWIFT or equivalent at [here insert number] via your bank and shall be effective only upon actual receipt.

11. To the extent that we may be or may hereafter become entitled, in any jurisdiction, to claim for ourselves or our property, assets or revenue immunity (whether by reason of sovereignty or otherwise) in respect of our obligations under this Guarantee from service of process, suit, jurisdiction, judgment, order, award, attachment (before or after judgment or award), set off, execution of a judgment or other legal process and to the extent that in any such jurisdiction there may be attributed to us or any of our property, assets or revenue such an immunity (whether or not claimed) we hereby irrevocably agree not to claim and hereby irrevocably waive such immunity to the fullest extent permitted by the laws of such jurisdiction.

12. We hereby warrant that we are permitted by any relevant law to which we are subject (including, where relevant, the laws of the place or places of each of our incorporation, establishment, regulation, registration and residence) to:
12.1 issue a guarantee in this form,
12.2 make payment under this Guarantee in a currency other than that of the place of (where relevant) each of our incorporation, establishment, regulation, registration and residence in case of a Demand for payment under this Guarantee, and
12.3 designate the place stated in paragraph 9 above as the forum and the place of jurisdiction to which we irrevocably submit.

13. We hereby warrant that this Guarantee has been, or will be, duly registered with the relevant State authority in any legal jurisdiction in which such registration is required for any reason.

14. We hereby warrant that we have obtained all necessary approvals and authorisations to issue this Guarantee.

Dated the day of 2

…………………………………………………………………… (signature)

for and on behalf of [here insert name of Guarantor]

NEWBUILDCON - Standard Newbuilding Contract – PART II

ANNEX A(iv)

IRREVOCABLE LETTER OF GUARANTEE FOR BUILDER'S OBLIGATIONS UNDER CLAUSE 27

To: [here insert name and address of the Buyer]

1. In this Guarantee, the following terms have the following meanings:

   'Award Interest' means any interest which may be awarded against the Builder in connection with the final determination of any dispute notified to us in accordance with paragraph 5 below.

   'Contract' means the contract dated [here insert date] made between the Builder and you for the construction of the Vessel, as the same may be amended at any time.

   'Demand' means a written demand for payment under this Guarantee.

   'Maximum Liability' means our maximum liability under this Guarantee, including Award Interest which shall be [here insert amount].

   'Builder' means [here insert name and address of shipbuilder]

   'Vessel' means [here insert technical description and/or name]

2. In consideration of you entering into the Contract, agreeing to take delivery of the Vessel, and agreeing to accept this Guarantee pursuant to the Contract, at the request of the Builder we irrevocably and unconditionally guarantee (but as primary obligor and not by way of secondary liability only) that if the Builder becomes liable to pay any sum to you in accordance with the terms of Clause 27(d)(iv) of the Contract in respect of Delivery Defects, as defined in the Contract, we shall, upon receipt by us from you of a Demand for the same (together with a copy of a demand made by you against the Builder for payment), pay to you or to your order upon the expiry of thirty (30) days from receipt of such Demand the sum demanded by you and Award Interest (if any) provided that our total liability shall not exceed the Maximum Liability.

3. This Guarantee shall not be affected by any indulgence or delay allowed to the Builder nor by any amendment to, or variation of, the Contract whether as to time or otherwise that may be agreed between you and the Builder nor by any circumstances that would otherwise discharge our liability as guarantor.

4. Subject to paragraph 5 below, this Guarantee shall remain in force until the first to occur of (a) the repair of the Delivery Defects and (b) the payment to you by the Builder or by us of all sums secured by this Guarantee.

5. Notwithstanding the other terms of this Guarantee, if within twenty-eight (28) days after our receipt of a Demand we receive written notice from you or from the Builder stating that your claim to payment of the sums referred to in the Demand has been disputed and that such dispute will be resolved in accordance with the Contract (including, as may be, by Class, an Expert, a Mediator or in arbitration), then we shall not be obliged to make any payment to you under this Guarantee until thirty (30) days after the dispute has been finally determined or in the event of an appeal from an arbitration award, until thirty (30) days after delivery of the final unappealable judgment; or in the event that the court remits the matter to the arbitrator, until thirty (30) days after the publication of the revised final award.

6. All payments to be made under this our Guarantee shall be made without any set off or counterclaim and without deduction or withholding for or on account of any taxes, duties or charges whatsoever unless we are compelled by law to deduct or withhold the same in which case we shall make the minimum deduction or withholding permitted and will pay such additional amounts as may be necessary in order that the amount received by you after such deductions or withholdings shall be equal to the amount which would have been received had no such deduction or withholding been made.

7. All payments to be made under this Guarantee shall be made in [here insert currency].

8. Notwithstanding any provision in the Contract, this Guarantee shall be freely assignable by you and by any assignee. Upon assignment, all references in this Guarantee to "you" shall be read as references to the assignee or subsequent assignees.

9. This Guarantee is governed by the laws of [here state applicable law. In the absence of any statement the laws of England and Wales shall apply] and we hereby submit to the exclusive jurisdiction of [here state place of dispute resolution. In the absence of any statement the High Court of England and Wales shall apply]. [We hereby authorise and nominate the following agent to accept service of any court proceedings on our behalf: [here insert name and

NEWBUILDCON - Standard Newbuilding Contract – PART II

address.]]

10. Any notice, claim or Demand to be given or made by you under this Guarantee shall be in writing signed by one of your officers and may be served on us either by post or by tested telex/authorised SWIFT or equivalent, and if sent by post to [here insert address] (or such other address as we may notify to you in writing) and if by tested telex at [here insert number] or if by SWIFT or equivalent at [here insert number] via your bank and shall be effective only upon actual receipt.

11. To the extent that we may be or may hereafter become entitled, in any jurisdiction, to claim for ourselves or our property, assets or revenue immunity (whether by reason of sovereignty or otherwise) in respect of our obligations under this Guarantee from service of process, suit, jurisdiction, judgment, order, award, attachment (before or after judgment or award), set off, execution of a judgment or other legal process and to the extent that in any such jurisdiction there may be attributed to us or any of our property, assets or revenue such an immunity (whether or not claimed) we hereby irrevocably agree not to claim and hereby irrevocably waive such immunity to the fullest extent permitted by the laws of such jurisdiction.

12. We hereby warrant that we are permitted by any relevant law to which we are subject (including, where relevant, the laws of the place or places of each of our incorporation, establishment, regulation, registration and residence) to:
12.1 issue a guarantee in this form,
12.2 make payment under this Guarantee in a currency other than that of the place of (where relevant) each of our incorporation, establishment, regulation, registration and residence in case of a Demand for payment under this Guarantee, and
12.3 designate the place stated in paragraph 9 above as the forum and the place of jurisdiction to which we irrevocably submit.

13. We hereby warrant that this Guarantee has been, or will be, duly registered with the relevant State authority in any legal jurisdiction in which such registration is required for any reason.

14. We hereby warrant that we have obtained all necessary approvals and authorisations to issue this Guarantee.

Dated the day of 2

.... (signature)

for and on behalf of [here insert name of Guarantor]

NEWBUILDCON - Standard Newbuilding Contract – PART II

ANNEX "B" - (SPECIFICATION)
BIMCO STANDARD NEWBUILDING CONTRACT
CODE NAME: NEWBUILDCON

ANNEX "C" - (MAKER'S LIST)
BIMCO STANDARD NEWBUILDING CONTRACT
CODE NAME: NEWBUILDCON

APPENDIX 2

CHINA MARITIME ARBITRATION COMMISSION STANDARD NEWBUILDING CONTRACT (SHANGHAI FORM)

CHINA MARITIME ARBITRATION COMMISSION STANDARD NEWBUILDING CONTRACT (SHANGHAI FORM)(ENGLISH VERSION)

CMAC STANDARD NEWBUILDING CONTRACT

(HULL NO: ________)

THIS CONTRACT, is made this (date) day of (month), (year), by and between, (full name of the BUYER), a corporation organized and existing under laws of (country or area of the BUYER), having its principal office at (country or city), as one party (hereinafter called the "BUYER" or "Principal"), and (full name of the shipbuilding trading company) a corporation organised and existing under laws of P.R. China, having its principal office at P.R. China as well as (full name of the shipbuilder) a corporation organized and existing under laws of P.R. China, having its principal office at P.R. China, together as the other party (hereinafter called the "SELLER" or "BUILDER").

WITNESSTH:

In consideration of the mutual covenants contained herein, the SELLER agrees to design, build ,equip, launch and complete at the SELLER's shipyard and to sell to the BUYER after accomplishment and successful trial of (type, name and number of the vessels), The technical Specifications will be described in Article I of the Contract. The VESSEL will be registered under the flag of (name of flag country). The BUYER agrees to purchase and take delivery of the aforesaid VESSEL from the SELLER and to pay for the same in accordance with the terms and conditions hereinafter set forth.

SECTION 1 VESSEL

ARTICLE I DESCRIPTION

1. HULL NUMBER

The Vessel shall be constructed at SELLER's shipyard or the shipyard specified by the SELLERS and shall have the BUILDER's Hull No: _________.

2. PRINCIPAL TECHNICAL DOCUMENTATION

The VESSEL shall be constructed, equipped and completed in accordance with the following principal technical documentation:

(1) Specification (Drawing No.)

(2) General Arrangement (Drawing No.)

(3) Midship Section (Drawing No.)

(4) Makers list (Drawing No.)

3. PRINCIPAL TECHNICAL PARAMETERS OF THE VESSEL

The principal technical parameters of the VESSEL as per the Specifications are as below:

(1) Principal Paticulars and Dimensions of the VESSEL

Length overall	abt.	m
Length between perpendiculars		m
Breadth, moulded		m
Depth, moulded		m
Designed Draft		m
Speed		knot
Deadweight		t
Hold Capacity		m^3

(2) Propelling Machinery:

The VESSEL shall be equipped, in accordance with the Specifications, with set (s)
.__________of Main Engine of ______type with MCR_____kW and rated revolution rpm.

The technical parameters of the VESSEL shall be defined, measured, and approved in accordance with the "Specifications":

4. GUARANTEES

(1) Speed

The Builder guarantees that the trial speed, after correction, is to be not less than nautical miles per hour on the trial condition of the weather and clean hull as stipulated in the Specifications. The speed shall be corrected for wind speed and shallow water effect with the method as specified in the Specifications.

(2) Deadweight

The term, "Deadweight", as used in the Contract, shall be as defined in the Specifications. The Builder guarantees that the VESSEL is to have a deadweight of not less than metric tons at the designed loaded draft of________ meters in sea water of 1.025t/ m^3 specific gravity.

(3) Hold Capacity

The Builder guarantees that the VESSEL's actual hold capacity is to be not less than m^3 as specified in the Specifications.

(4) Fuel Oil Consumption

The Builder guarantees that the fuel oil consumption of the Main Engine at shop test is not to exceed_________grams/brake horse power/hour at normal continuous output at shop trial based on diesel fuel oil having a lower calorific value of ________kilocalories per kilogram.

(5) The actual speed, deadweight , hold capacity and fuel oil consumption of the VESSEL shall be measured by the BUILDER in the presence of the BUYER's supervisor(s) or the Representative authorized by the BUYER for purpose of confirmation or approval.

Should there be any dispute between the BUILDER and the BUYER in such measurements, Classification Society shall be appointed to make remeasurements subject to the agreement by both parties. The result made by the Classification Society shall be final and binding to both parties.

ARTICLE II CLASSIFICATION,RULES AND REGULATIONS

1. The VESSEL, including its design construction inspection test and delivery shall conform to the rules , regulations and the conventions of ________(herein insert the "Classification Society") , and shall obtain the class notation of _________and shall also

meet the requirements of the regulations and the rules of the flag authorities as below:
(1) The regulations and the rules having becoming effective on the date of the signing date of signing of the Contract, or
(2) The regulations and the rules which have not been effective while have been officially approved and published on the date of signing of the Contract and are to come into force pirot to the delivery of the VESSEL.
All the regulations, the rules and requirements of the flag authorities and the Classification Society shall be unconditionally met.
2. All the laws rules and regulations of the authorities shall be final and binding upon the parties hereto.
3. All fees and charges incidental to Classification and to comply with the rules, regulations and requirements of the Contract as described in the Specifications as well as royalties, if any, payable on account of the construction of the VESSEL shall be for the account of the BUILDER, except as otherwise provided and agreed herein.
The key plans, materials and workmanship entering into the construction of the VESSEL shall at all times be subject to inspections and tests in accordance with the rules and regulations of the Classification Society.

ARTICLE III DESIGN—LIABILITY IN THE DESIGN CONTRACT

The design of the VESSEL shall be carried out by a design company in accordance with the Design Specifications based on the requirements of the BUYER. The design shall determine the performance and technical parameters of the VESSEL (including navigation areas and shipping course, ship type, trading purpose, nationality and class, propelling equipment, speed and reserved power, endurance, deadweight, hold capacity and accommodation facilities), as well as compliance with the rules and regulations of CLASSIFICATION SOCIETY and relevant authorities. A design contract shall be executed by the PARTIES for quality assurance and clarification of the obligations.
If the BUYER assigns ______ design company from outside of China to design the VESSEL the BUYER shall enter into a design contract with this design company. Thereafter, the BUILDER enters into a subcontract with _______ (the design company) for the detailed design and workshop design. In such event, the BUILDER shall undertake responsibility neither for any defect of the VESSEL and mistake arising from the design, or any deficiency of speed, deadweight, fuel oil consumption and hold capacity due to non-compliance with the relevant regulations and rules, nor for any extension of the Delivery Time of the VESSEL due to the delay in the delivery of the drawings.
If the design company is assigned by the BUILDER, the design contract shall be signed between the BUILDER and this design company. In such event, the BUILDER shall undertake responsibility for any defect of the VESSEL and mistake arising from the design, or any deficiency of speed, deadweight, fuel oil consumption and hold capacity due to non-compliance with the relevant regulations and rules, and for any extension of the Delivery Time of the VESSEL due to the delay in the delivery of the drawings.

ARTICLE IV ENVIRONMENTAL PROTECTION

1. IMO HAZARDOUS MATERIALS INVENTORY

The BUILDER shall, in accordance with the IMO Guidelines on Ship Recycling, with amendments in force as of the date of the Contract:

(1) Endeavour to take due account of the Vessel's Recycling disposal when designing and constructing the Vessel;

(2) Maximizing the use of the materials of green, low carbon and being recycled safely and environmental friendly;

(3) Minimizing the use of the materials which have been known to be potentially hazardous to human health and the environment.

In consultation with equipment manufactures, the BUILDER will provide the Buyer with a Green Passport Statement of Compliance issued by the Classification Society containing information including the Vessel's hull number and main particulars and listing any and all equipment and systems having been known to be potentially hazardous utilized in the construction of the Vessel.

The list of the materials which are potentially hazardous shall contain the location and the approximate quantity/volume of each material on board of the Vessel.

2. Protective Coating Standard

The Vessel's double-side skin spaces and dedicated seawater ballast tanks shall be coated in accordance with the Specifications. The Protective Coating Standard should be in compliance with the applicable laws, regulations, rules and requirements of the Classification Society and the Regulatory Authorities.

3. Source of Origin

If so requested by the Buyer, the BUILDER shall identify the country of origin of the main components listed in the Maker's List and Specifications.

SECTION 2 FINANCIAL

ARTICLE V CONTRACT PRICE & TERMS OF PAYMENT

1. CONTRACT PRICE:

The purchase price of the Vessel is ____________________________ (Amount and Currency, hereinafter called the "Contract Price"), net receivable by the BUILDER which is exclusive of the BUYER's Supplies as provided in Article XI hereof, and shall be adjusted subject to Article VI under the Contract.

2. CURRENCY:

Any and all payments by the BUYER to the BUILDER under the Contract shall be made in ___________________ (Currency and Symbol).

3. TERMS OF PAYMENT:

The Contract Price shall be paid by the BUYER to the BUILDER in installments as follows:

(1) The 1st Installment:

The sum of United States Dollars_____________________ (USD______________), representing

.__________ percent (___________%) of the Contract Price, shall become due and payable and be paid by the BUYER concurrently within three (3) working days after the

BUILDER issued the Letter of Refund Guarantee.

(2) The 2nd Installment:

The sun of United States Dollars____________________ (USD______________), representing

.__________ percent (__________%) of the Contract Price, shall become due and payable and be paid within three (3) business days after the cutting of the first steel plate of the Vessel.

(3) The 3rd Installment:

The sun of United States Dollars____________________ (USD______________), representing

.__________ percent (__________%) of the Contract Price, shall become due and payable and be paid within three (3) business days after keel-laying of the first section of the Vessel.

(4) The 4th Installment:

The sun of United States Dollars____________________ (USD______________), representing

.__________ percent (__________%) of the Contract Price, shall become due and payable and be paid within three (3) business days after launching of the Vessel.

(5) The 5th Installment:

The sun of United States Dollars____________________ (USD______________), representing

.__________ percent (__________%) of the Contract Price plus any increase or minus any decrease due to modifications and/or adjustments to the Contract Price in accordance with provisions of the relevant Article hereof, shall become due and payable and be paid by the BUYER to the BUILDER on date of delivery of the Vessel. The BUILDER shall send to the BUYER an email or facsimile demand for this installment ten (10) days prior to the scheduled date of delivery of the Vessel.

(6) The 6th Installment:

The sun of United States Dollars____________________ (USD______________), representing

.__________ percent (__________%) of the Contract Price shall be paid within three (3) working days after the expiration of the guarantee period of the Vessel.

4. METHOD OF PAYMENT

(1) 1st Installment:

The BUYER shall remit the amount of this installment to the receiving bank nominated by the BUILDER (to name the bank here), for credit to the account of the BUILDER, within three (3) business days after receiving notice in writing from the BUILDER which should also be confirmed and signed by the BUYER's authorized representative (same as below) regarding effectiveness of the Contract.

(2) 2nd Installment:

The BUYER shall remit the amount of this installment to the receiving bank nominated by the BUILDER, for credit to the account of the BUILDER, within three (3) business days after receiving notice in writing from the BUILDER regarding cutting steel of the Vessel.

(3) 3rd Installment:

The BUYER shall remit the amount of this installment to the receiving bank nominated by the BUILDER, for credit to the account of the BUILDER, within three (3) business days after receiving notice in writing from the BUILDER regarding keel-laying of the Vessel.

(4) 4th Installment:

The BUYER shall remit the amount of this installment to the receiving bank nominated by the BUILDER, for credit to the account of the BUILDER, within three (3) business days after receiving notice in writing from the BUILDER regarding launching of the Vessel.

(5) 5th Installment (Payable upon delivery of the Vessel):

The BUYER shall, at least three (3) bank business days prior to the scheduled date of delivery of the Vessel, make a cash deposit with the bank nominated by the BUILDER, covering the amount of this installment as adjusted in accordance with the provisions of the Contract as agreed by the Parties hereto, with an irrevocable instruction that the said amount shall be released to the BUILDER against presentation by the BUILDER to the said Bank, of a copy of the Protocol of Delivery and Acceptance signed by the BUYER's authorized representative and the SELLER. Interest, if any, accrued from such deposit, shall be for the benefit on the BUYER.

Unless otherwise provided in the Contract, the BUYER shall not defer or refuse payment of any installment for the reason of any disputes or arguments of whatsoever nature between the Parties.

(6) The 6th Installment:

The BUYER shall remit the amount of this installment to the receiving bank nominated by the BUILDER, for credit to the account of the BUILDER within three (3) working days after receipt of the notice from the BUILDER stating the expiration of the guarantee period of the Vessel.

5. PREPAYMENT:

The BUYER shall have the obligation to make prepayment of any and all installments before delivery of the Vessel, by giving to the SELLER at least thirty (30) calendar days prior notice in writing, without any price adjustment of the Vessel for such prepayment.

6.SECURITY FOR PAYMENT OF INSTALLMENTS BEFORE DELIVERY:

The BUYER shall, concurrently when the Contract being signed, deliver to the BUILDER an irrevocable and unconditional Letter of Guarantee in the from annexed hereto as Exhibit "B" in favor of the BUILDER issued by a first class international bank (hereinafter called the "Guarantor") acceptable to BUILDER's bank and the BUILDER. This guarantee shall secure the BUYER's obligation for the payment of all 2nd, 3rd and 4th installments of the Contract Price.

7. REFUNDS

All payments made by the BUYER prior to delivery of the Vessel shall be in the nature of advance to the BUILDER, and in the event the Contract is rescinded or canceled by the BUILDER, all in accordance with the specific terms of the Contract permitting such rescission or cancellation, the BUILDER shall refund to the BUYER the full amount of all sums already paid by the BUYER to the BUILDER under the Contract, together with interest (at the rate set out in respective provision thereof) from the respective payment date(s) to the date of remittance by telegraphic transfer of such refund to the account specified by the BUYER.

As security to the BUYER, the BUILDER shall deliver to the BUYER, concurrently with the

Contract being signed, a Refund Guarantee to be issued by (to name bank here) in the form as per ANNEX "A" annexed hereto.

ARTICLE VI ADJUSTMENT OF THE CONTRACT PRICE

The Contract Price of the VESSEL shall be adjusted subject to (1) the actual technical particulars different from the Specification and/or (2) the external changes in the market different from the situation when signing the Contract . It is mutually agreed by both parties that the said technical particulars refer to the speed, the deadweight, the hold capacity, the fuel consumption of the main engine, delivery time and other items, and the said external changes in the market refer to the price of major materials and exchange rate. It is understood by both parties that the adjustment of the Contract Price refer to the increase or reduction of the Contract Price and any adjustment of the Contract Price is by way of liquidated damages and not by way of penalty.

This Article specifies the amount of liquidated damages for the deficiency of abovementioned technical particulars and changes in the market as below.

1. SPEED

(1) If the reduction of the speed (as determined by the sea trial after correction according to the Specifications, same is to apply to below) is less than three-tenths (3/10) of one knot below the guaranteed speed as specified in Paragraph 4 (1) of Article I, there shall be no adjustment of the Contract Price.

(2) However, if the reduction of the speed is greater than or equal to three-tenths (3/10) of one knot below the guaranteed speed as specified in Paragraph 3 (1) of Article I, the Contract Price shall be reduced as follows:

In case of reduction of less than 0.3 knot USD 0

In case of reduction

at or above 0.30 but below 0.40 knot USD______

at or above 0.40 but below 0.50 knot USD______

at or above 0.50 but below 0.60 knot USD______

at or above 0.60 but below 0.70 knot USD______

at or above 0.70 but below 0.80 knot USD______

at or above 0.90 but below 1.00 knot USD______

(3) If the reduction in speed is greater than 1 knot below the guaranteed speed, the BUYER shall at his own discretion either reject the VESSEL or cancel the Contract in accordance with Article XXVII, or may accept the VESSEL at a reduction in the Contract Price as above provided, by United States Dollars____________ only (USD_________) being the maximum.

(4) The BUILDER shall be entitled to carry out further sea trials after correcting the defects which result in the speed reduction to achieve the guaranteed speed as specified in the Specification.

(5) If the actual speed is more than the guaranteed speed as specified in the Specification, the BUYER shall provide bonus to the BUILDER as below method:

If the increase of the speed is less than three-tenths (3/10) of one knot more than the guaranteed speed, there shall be no adjustment of the Contract Price.

Thereafter the Contract Price shall be increased by USD _______ for each one-tenths (1/10) of one knot more than the guaranteed speed, but the maximum amount shall not be more than United States Dollars ________ (USD_____).

2. DEADWEIGHT

(1) If the actual deadweight of the VESSEL at design draft determined in accordance with the Specifications is not less than 98% of the guaranteed deadweight (_______metric ton), there shall be no adjustment of the Contract Price.

(2) The Contract Price shall be decreased by the sum of United States Dollars (USD_____) for each full metric ton of such reduction being less than 98% of the guaranteed deadweight (________) metric tons.

(3) If the reduction in the VESSEL'S actual deadweight is greater than ______metric tons below the guaranteed deadweight, the BUYER may, at its option, reject the VESSEL or cancel the Contract in accordance with the provisions of Article XVII of the Contract, or may accept the VESSEL with reduction in the Contract Price up to the maximum amount of United States Dollars______(USD_______).

(4) If the increase of actual deadweight of the VESSEL at design draft determined in accordance with the Specifications is less than 102% of the guaranteed deadweight (______ metric ton), there shall be no adjustment of the Contract Price, while the Contract Price shall be increased by the sum of United States Dollars (USD_____) for each full metric ton of such increase being more than 102% of the guaranteed deadweight (_______) metric tons. But the maximum amount increased shall not be more than United States Dollars_______(USD____).

3. HOLD CAPACITY

(1) If the actual hold capacity of the VESSEL is not less than 98% of the guaranteed hold capacity as specified in the Specification, there shall be no adjustment of the Contract Price.

(2) If the actual hold capacity of the VESSEL is less than 98% of the hold capacity as specified in the Specification, then for each full cubic meter thereafter below the hold capacity, the Contract Price shall be reduced by the sum of USD ______for each full cubic meter as liquidated damages up to a maximum amount United States Dollars (USD______)

(fractions of one percent to be prorated).

(3) If the reduction of the VESSEL'S actual hold capacity is greater than______below the hold capacity as specified in the Specification, the BUYER may reject the VESSEL or cancel the Contract.

(4) If the actual hold capacity of the VESSEL is more than 102% of the hold capacity as specified in the Specification, then for each full cubic meter thereafter above the hold capacity, the Contract Price shall be increased by the sum of United States Dollars (USD_____) for each cubic meter. But the maximum amount increased shall not be more than United States Dollars _______ (USD_____) (proportionally the amount less than 1 cubic meter can be exempted).

4. EXCESSIVE FUEL CONSUMPTION OF THE MAIN ENGINE

(1) If the actual fuel consumption of the Main Engine, as determined on the test bed is equal to or less than five percent (5%) of the guaranteed fuel consumption as specified under the provision of the Contract and the Specifications, there shall be no adjustment of the Contract Price.

(2) However, if the actual fuel consumption as determined by the test bed is greater than five percent (5%) above the guaranteed fuel consumption then, the Contract Price shall be reduced by the sum of United States Dollars_________(USD) for each whole percentage (1%) increase in fuel consumption in excess of the above said five (5%) (fractions of one percent to be prorated).

(3) If as determined by test bed such actual fuel consumption of the Main Engines is greater than ten percent (10%). i.e.______ gram/bhp/h in excess of the guaranteed fuel consumption, the BUYER may reject the main engine or reject the VESSEL or cancel the Contract or may accept the VESSEL at a reduction in the Contract Price but the maximum reduction shall not be more than United States Dollars________(USD_______).

5. DELIVERY

(1) There shall be no adjustment of the Contract Price for the first thirty (30) days of delay in delivery of the VESSEL beyond the Delivery Date as defined in Article XIV hereof ending as of twelve o'clock midnight of the thirtieth (30th) day of delay.

(2) If the delivery of the VESSEL is delayed more than thirty (30) days after the Delivery Date as defined in Article XIV hereof, then, in such event, beginning at twelve o'clock midnight of the thirtieth (30th) day after the date on which delivery is required under the Contract, the Contract Price of the VESSEL shall be reduced by deducting the sum of United States Dollars ________

(USD______) per day.

Unless the parties hereto agree otherwise, the total reduction in the Contract Price shall be deducted from the fifth installment of the Contract Price and in any event (including the event that the BUYER consents to take the VESSEL at the later delivery date after the expiration of

.___________days delay of delivery shall not be more than _________ days at the above specified rate of reduction, that is United States Dollars ________(USD______) being the maximum.

(3) If the delay in the delivery exceeds_________(______) days (being the total of permissible and non-permissible Delays) after the Delivery Date as defined in Article XIV, in such event, the BUYER may, at its option, terminate or cancel the Contract in accordance with the provisions of Article XIII. The BUILDER may at any time after the expiration of the aforementioned ________

(______) days, if the BUYER has not served notice of cancellation pursuant to Article X, notify the BUYER of the date upon which the SELLER estimates the VESSEL will be ready for delivery and demand in writing that the BUYER make an election, in which case the BUYER shall, within thirty (30) days after such demand is received by the BUYER, either notify the SELLER of its decision to cancel the Contract, or consent to take delivery of the VESSEL at an agreed future date, it being understood and agreed by the parties hereto that, if the VESSEL is not delivered by such future date, the BUYER shall have the same right of cancellation upon the same terms, as herein provided.

(4) The delivery of the VESSEL shall not be deemed delayed and the Contract Price shall not be reduced for any period when the Delivery Date of the VESSEL is extended by reason of causes and provisions of Articles XI, XIII. The Contract Price shall not be adjusted or reduced if the delivery of the VESSEL is delayed by reason of permissible delays as defined in Article XV hereof.

(5) If the delivery of the VESSEL shall be made earlier than the Delivery Date as

stipulated in Article XIV, the BUILDER shall notify the BUYER and that such notification shall be given not less than_____(_____) days prior to the newly planned delivery date. A certain amount of bonus shall be given by the BUYER to the BUILDER as follows:

In the event that the delivery shall be made within fifteen (15) days earlier than the Delivery Date, the Contract Price shall remain unchanged. In the event that the delivery shall be made more than fifteen (15) days earlier than the specified Delivery Date, then a bonus shall be added to the Contract Price at a rate of United States Dollars________Only (USD_____) per day for each full day earlier than the fifteenth (15th) day prior to the Delivery Date. The total increase of the Contract Price for the earlier delivery shall be added to the fifth instalment of the Contract Price. While the maximum amount increased shall not be more than United States Dollars_______Only (USD______)

If the actual delivery time is after the Delivery Date as specified in Article XIV and before the permissible extended delivery date, the BUILDER shall not be entitled to be given bonus as specified in above paragraph 5 (5).

(6) In the event that the BUILDER is unable to deliver the VESSEL on the newly planned delivery date as declared, the VESSEL can, nevertheless, be delivered by the BUILDER at a date after such declared newly planned date.

In such circumstances, and for the purpose of determining the liquidated damages to the BUYER (according to the provisions of Paragraph 1 (1) of this Article) and the BUYER's right to cancel the Contract (according to the provisions of Paragraph 1 (3) of this Article), the newly planned delivery date declared by the BUILDER shall not be in any way be treated or be taken as having substituted the original Delivery Date as defined in Article XIV. The BUYER's aforesaid right for liquidated damages and to cancel the Contract shall be exercised to the extent as described in Paragraph 4 (1), 4 (2) and/or 4 (3) of Article XIII. In whatever circumstances, the Delivery Date as defined in Article VII (not the newly planned delivery date as declared by the BUILDER) shall be used to exercise the BUYER's right for liquidated damages and to rescind the Contract and the BUILDER's liability to pay the aforesaid liquidated damages resulting from the delay in delivery of the VESSEL as specified in paragraph 5 (1), 5 (2), and 5 (3).

If the actual delivery time is after the declared newly planned date and before the Delivery Date as specified in Article XIV, the BUILDER shall be entitled to be given bonus as specified in above paragraph 5 (5).

6. MAJOR MATERIALS AND EQUIPMENT

The major materials and equipment referred in this paragraph refer to steels and main propulsion equipment.

The Contract Price of this VESSEL is determined by the market price (basic price) of steels and main propulsion equipment when signing the Contract.

If the price difference between the actual purchase price contracted between the BUILDER and the suppliers is _____% higher than the basic price, the price difference shall be shared by both parties with the BUILDER undertaking ______% and the BUYER undertaking ______%.

7.EXCHANGE RATE

As the construction will last for a long time and the fluctuation of the exchange rate between USD and RMB may cause relevant financial changes of the parties, it is agreed that the exchange rate published by Bank of China on the date when the Contract signed will be deemed as a reference, any price difference due to the fluctuation when paying each installment should be shared by 50% by each party. The maximum of the exchange rate fluctuation shall not exceed 2% of the Contract Price, the part more than 2% should be borne by the BUYER.

8. OTHERS (OPTIONAL CLAUSE)
The purpose of setting aside this clause is to allow the parties to insert special technical requirements for a particular type of vessel and adjustment of the Contract Price arising from the changes (failure to meet or exceeding) in the values of the said special technical requirements.
Any adjustment of the Contract Price made as per Paragraph 1. 2. 3. 4. 5 and 6 of this article shall be specified in writing before the payment of the 5th installment.

9. EFFECTIVENESS OF CONTRACT RESCISSION
It is expressly understood and agreed by the parties hereto that in any case as stated above, if the BUYER rescinds or cancel the Contract pursuant to any provision under this Article, the BUYER shall be entitled to have its rights and compensation as set out in Article XXVII hereof shall not be entitled to any liquidated damage or compensation whether described above or otherwise.

ARTICLE VII TAXES AND DUTIES

1. TAXES
All taxes, if any, including stamp duties, incurred in connection with this Contract in the People's Republic of China shall be borne by the BUILDER. The BUILDER shall bear all taxes and/or duties imposed upon the equipment and facilities procured by the BUILDER in the People's Republic of China for the construction of the VESSEL.

2. DUTIES
The BUILDER shall indemnify the BUYER for, and hold it harmless against, any duties imposed in the People's Republic of China upon material and equipment which under the terms of the Contract and/or the Specifications will, or may be, supplied by the BUYER from the abroad for installation in the VESSEL as well as any duties imposed in the People's Republic of China upon running stores, provisions and supplies furnished by the BUYER from abroad to be stocked on board the VESSEL and also the duties or taxed if any, to be imposed upon the export of the VESSEL as a whole or upon any of its parts or equipment.
Any tax of duty other than those described hereinabove, if any, shall be borne by the BUYER.

SECTION 3 PRODUCTION

ARTICLE VIII APPROVAL AND ACCEPTANCE OF PLANS AND DRAWINGS

1. APPROVAL OF PLANS AND DRAWINGS
The BUILDER shall send to the BUYER (or its representatives) three (3) copies of plans drawings and technical documents for the BUYER's approval as required in the Specification, Within fourteen (14) running days upon receipt by the BUYER or twenty-one

(21) running days upon dispatch by the BUILDER, whichever is earlier, the BUYER shall return to the BUILDER one copy of such documentation with the BUYER's approval or approval with remarks. In the event that the BUYER fails to return such documentation to the BUILDER within the above-designated period of time the said documentation shall be deemed to have been completely approved by the BUYER, unless an extension is requested in writing by the BUYER with the written consent of the BUILDER.

If the BUYER returns such documentation with remarks to the BUILDER within the above –designated period of time and such remarks are of a nature or substance constitute necessary modifications, the BUILDER shall commence or continue construction of the VESSEL in accordance with the modified drawings and technical documents. In the event that the BUYER's remarks are unclear or unspecified, the BUILDER shall have the right to make construction in its own interpretation.

A Makers List is attached with the Specifications, which is a list of the suppliers recommended by the BUILDER and to be known by the BUYER. Within ten (10) running days upon receipt by the BUYER or fifteen (15) running days upon notification by the BUILDER, whichever is shorter, the BUYER shall give a prior written request for alternative suppliers with sufficient reasons.

The BUILDER may, at its sole discretion, choose any of the suppliers who are approved by the BUYER. In the event that the BUYER prefer to a supplier other than the one preferred by the BUILDER (whether recommended by the BUILDER or not) and such preference affects the cost, the BUILDER shall notify the BUIDER of the cost changed. Within ten (10) running days upon receipt by the BUYER or fifteen (15) running days upon notification by the BUILDER, whichever is shorter, the BUYER shall have the right to notify the BUILDER of its insistence on the decision. In such event, the cost to be adjusted shall be added into or deducted from the CONTRACT PRICE. Should the BUYER take no measure or give up its decision, the BUILDER shall, at its sole discretion and under its guarantee obligation, choose any one of the recommended suppliers.

It is agreed by the parties that all communications relating to the VESSEL with the BUILDER's suppliers shall be carried out by the BUILDER. The BUYER shall use its best efforts to approve the documentation related soonest within the above-designated time Such approval shall not diminish the BUILDR's responsibility for the VESSEL.

A prior written consent by the BUYER shall be obtained if the BUILDER turns to other makers not included in the Maker List.

2. ACCEPTANCE OF PLANS AND DRAWINGS

The parties shall make a list of all the plans and drawings, which are to be sent to the BUYER for approval (hereinafter called "the LIST") within________running days after signing of the Contract. Before arrival of the Supervisor at the BUILDER's Shipyard, the plans and the drawings specified in the LIST shall be sent to the BUYER and the BUYER shall, within _____( ) days after receipt thereof (excluding mailing time), return such plans and drawings submitted by the BUILDER with approval or remarks.

Concurrently with the arrival of the Supervisor at the BUILDER's shipyard, the BUYER shall notify the BUILDER in writing, stating the authority which the said Supervisor shall have on behalf of the BUYER, to approve or disapprove those of the plans and drawings specified in the LIST which have not yet been sent to the BUYER. The Supervisor shall, within_____( ) days after receipt thereof, return those plans and drawings with approval or remarks.

Unless notification is given to the BUILDER by the Supervisor or the BUYER of approval or disapproval of any plans and drawings within the above-designated period of time for each case, the said plans and drawings shall be deemed to have been automatically approved.
The plans and drawings approved by the BUYER Supervisor shall be final, and any alteration thereof shall be regarded as modification specified in Article XII of the Contract.

3. NOTES
(1) The drawings and the Specifications have been signed by the parties on or before the effective date of the Contract and shall form an integral part of the Contract and have the same effect as the Contract.
(2) In the event of any inconsistency or conflict between the provisions of the Contract and the Drawings and the Specifications, the provisions of the Contract shall prevail, without prejudice to the full effect of the drawings and the Specifications under other circumstances. In the event of any inconsistency or conflict between the drawings and the Specifications, the Specifications shall prevail. Any work required by the Specifications but not shown in the drawings, or shown in the drawings but not required by the Specifications, shall be accomplished by the BUILDER as if is a part of the Contract. Once finding any inconsistency or conflict between the provisions of the Contract and the drawings and the Specifications or between the drawings and the Specifications, one party shall promptly give the other party a written notice.
(3) In the event of any modification or change on the Specifications agreed by the parties after the effective date of the Contract, which may lead to the inconsistency or conflict between the Specifications and the drawings, such modification or change shall prevail.

ARTICLE IX SUPERVISION AND INSPECTION

1. APPOINTMENT OF THE BUYER'S SUPERVISOR
The BUYER shall send in good time to and maintain at the BUILDER's Shipyard, at the BUYER's own cost and expense, one or more representative (s) (hereinafter called the "Supervisor"), who shall be duly appointed in writing by the BUYER to supervise and inspect the construction of the VESSEL, her engines and accessories. The BUILDER shall be liable for obtaining of the necessary visa for the Supervisor to enter China or the BUILDER's Shipyard. The Supervisor shall ensure his observance of the rules, regulations and laws of the People's Republic of China. The BUYER undertakes to give the SELLER an adequate notice for the application of the visa.

2. SUPERVISION AND INSPECTION BY SUPERVISOR
The necessary inspection of the VESSEL, her machinery, equipment and out fittings shall be carried out by the Classification Society and inspection team of the BUYER's throughout the entire period of construction in order to ensure that the construction of the VESSEL is duly performed in accordance with the Contract and the Specifications.
The Supervisor shall have, at all time until delivery of the VESSEL, the right to attend tests according to the mutually agreed test list and inspect the VESSEL, her engines, accessories and materials at the BUILDER's Shipyard, its subcontractors or any other place where work is done or materials stored in connection with the VESSEL. In the event that the Supervisor discovers that any construction or material or workmanship does not

or will not conform to the requirements of the Contract and the Specifications, the Supervisor shall promptly give the BUILDER a notice in writing as to such non-conformity, upon receipt of which the BUILDER shall correct such nonconformity if the BUILDER agrees with the Supervisor. Should the BUILDER disagree to the Supervisor's comments, the BUILDER shall provide with reasons and explanations or discuss with the Supervisor for solution. Should there be no agreement reached, such dispute shall be submitted to Classification Society for solution, such solution made by the Classification Society shall be final binding to both parties. In such event, the BUILDER may proceed with construction. The BUYER undertakes and ensures the BUILDER that the Supervisor shall carry and usual shipbuilding practice and in a way as to minimize any increase in building costs and delays in the construction of the VESSEL.

The BUILDER shall provide, free of charge, the Supervisor with office accommodation, and other necessary facilities at or near the BUILDER's Shipyard. At all times, during the construction of the VESSEL until delivery thereof, the Supervisor shall be given unimpeded access to the VESSEL, her engines and accessories, and anywhere else where the work on, or the materials are being processed or stored, in connection with the construction of the VESSEL, including the BUILDER's yards, workshops, warehouses and its subcontractor's premises for working or storing materials in connection with the VESSEL's construction. The travel expenses to the subcontractor's premises outside of the city shall be at the BUYER's account. The transportation for the Supervisor in the city shall be offered by the BUILDER.

3. LIABILITY OF THE BUILDER

The Supervisor engaged by the BUYER under the Contract shall at all times be deemed to be the employee of the BUYER. The BUILDER shall be under no liability to the BUYER, or to the Supervisor or to the BUYER's employees or agents for personal injuries (including death), which incur on the VESSEL or/and within the premises of the BUILDER or its subcontractors while they or one of them is engaging in the construction of the VESSEL, unless such personal injuries (including death) were caused by gross negligence of the BUILDER, or of any of the BUILDRS's employees or agents or subcontractors. Nor shall the BUILDER be under any liability for damage, loss or destruction on the properties in China of the Supervisor, the BUYER's employees or agents, unless such damage, loss or destruction was under gross negligence of the BUILDER, or of any of the BUILDER's employees or agents or sub-contractors.

4. SALARIES AND EXPENSES

All salaries and daily expenses of the Supervisor or any other employees appointed by the BUYER shall be for the BUYER's account in accordance with this article.

5. REPORT OF PROGRESS

The BUYER has the right to request the BUILDER for a monthly report of the construction progress during the construction of the VESSEL.

6. REPLACEMENT OF SUPERVISOR

The BUILDER has the right to request the BUYER in writing to replace the Supervisor who is deemed to be unsuitable and unsatisfactory for inspection on the progress of the construction of the VESSEL. If necessary, the BUYER may send its representative to the BUILDER's yard for investigation. Once such request of the BUILDER is justified by the

BUYER, the BUYER shall make the replacement as soon as practicable.

ARTICLE X SUBCONTRACTING

The BUILDER may, at its sole discretion and responsibility, subcontract any portion of the construction work of the VESSEL to qualified and experienced subcontractors as set out in the Specifications of the Maker's List, but delivery and final assembly into the VESSEL of any such work subcontracted shall be at the BUILDER's Shipyard. The BUILDER shall remain fully responsible for due performance of such subcontracted work...

ARTCLE XI BUYER'S SUPPLIES

1. ARTICLE X BUYER'S LIABILITY

(1) The BUYER shall, at its own risk, cost and expenses, supply to the BUILDER the items as specified in the Specifications (hereinafter called the "BUYER'S SUPPLIES") in appropriate time during the construction of the VESSEL. Such items shall be delivered at a warehouse or other storage facilities at the Shipyard as set out in the Specifications in a proper condition ready for installation on the VESSEL.

(2) To assist installation, the BUYER shall provide the BUILDER with the necessary documentation including specifications, drawings, instruction books, manuals, test reports and others required to comply with all applicable rules and regulations. If so requested by the BUILDER, the BUYER shall, at no cost to the BUILDER, arrange for the representatives of the manufacturers of the BUYER'S SUPPLIES to assist the BUILDER in installation or to carry out the installation of the BUYER'S SUPPLIES by themselves, or to carry out commission at the Shipyard.

(3) The BUILDER may reject any and all of the BUYER'S SUPPLIES if found to be unsuitable or in improper condition for installation. But if required by the BUYER, the BUILDER may make repair of adjustment of the BUYER'S SUPPLIES, which shall neither affect any other rights of the BUILDER under the Contract, nor shall constitute any bearing by the BUILDER of the consequences raised thereafter. In such event, the BUYER shall pay to the BUILDER all costs arising from the repair or adjustment and the delivery of the VESSEL shall automatically be extended for a period of time taken for such repair or adjustment.

(4) Should the BUYER fail to deliver to the BUILDER the BUYER'S SUPPLIES within the specified time, the delivery of the VESSEL shall automatically be extended for a period of such delay, which the BUYER shall be liable for. In such event, the BUYER shall pay to the BUILDER all losses sustained by the BUILDER due to such delay and such payment shall be made upon delivery of the VESSEL.

If delay in delivery of the BUYER'S SUPPLIES exceeds thirty (30) days, the BUILDER shall have the right to proceed with the construction of the VESSEL without the installation of such delayed items. The BUYER shall accept and take delivery of the VESSEL so constructed, without prejudice to any other rights of the BUILDER.

2. BUILDER's LIABILITY

The BUILDER shall store and properly handle the BUYER'S SUPPLIES after delivery to the Shipyard and install them on the VESSEL. Unless otherwise provided in the Contract, the BUILDER shall not be responsible for the quality and performance of the BUYER's

SUPPLIES.

ATRICLE XII MODIFICATIONS, CHANGES AND EXTRAS

1. HOW EFFECTED

The Specification and/or drawings to which the VESSEL is built can be modified or changed at any time provided that the parties reach agreement and if in the BUILDER's reasonable judgement, such modifications or changes or accumulated modifications will not adversely affect the BUILDER's other commitments, and the BUYER shall agree as below provided to the adjustment of the Contract Price, extension of the Delivery Time and other terms of the Contract (if any). Subject to above conditions, the BUILDER shall use the best efforts to accommodate the BUYER's reasonable request so that the said changes and/or modifications shall be made at reasonable costs and within the period of time as short as possible.

Such agreements on the modifications or changes shall refer to an agreement on the increase or decrease, if any, in the Contract Price of the VESSEL and an agreement as to any extension or reduction in the time for delivery, and furthermore to provide an additional guarantee to the BUILDER to the BUILDER's satisfaction, or an agreement as to any other alternations in the Contract or an agreement on the modification of the Specification arising from the changes. The agreements as to the changes and modifications to the Specifications pursuant to the foregoing shall be effective by exchanges of such documents or other written documents and shall constitute an amendment to Specification and shall be deemed to be a part of the Contract and Specification. Upon agreements as to the changes and/or modifications to the Specifications have been reached, the BUILDER shall alter the construction of the VESSEL in accordance therewith, including any additions to, or reductions from the work to be performed in connection with such construction. If the parties fail to agree on the adjustment to the Contract Price, or extension of the delivery time or providing additional guarantee to the BUILDER or modifications to the Contract, the BUILDER is not obliged to undertake any request from the BUYER for modifications.

2. CHANGES IN RULES AND REGULATIONS

(1) If, after the date of signing the Contract, any requirements as to the rules and regulations as specified in the Contract and the Specifications to which the construction of the VESSEL is required to conform, are modified or changed by the Classification Society or the other regulatory authorities to make such modifications or changes, the BUYER or the BUILDER, upon receipt of the notice thereof, shall transmit such information in full to each other in writing, whereupon within(   ) days after receipt of the said notice by the BUYER form the BUILDER or vice versa, the BUYER shall instruct the BUILDER in writing as to the modifications or changes, if any, to be made in the VESSEL which the BUYER, in its sole discretion, shall decide. The SELLER shall promptly comply with such modifications or changes, if any in the construction of the VESSEL, provided that the BUYER shall first agree:

(a) to any increase or decrease in the Contract Price of the VESSEL occasioned by the cost for such modifications and changes; and/or

(b) to any extension in the delivery time of the VESSEL due to such modifications and changes; and/or

(c) to any increase or decrease in the guaranteed deadweight and speed of the VESSEL, if modifications and changes results in increased or reduced deadweight and speed; and/or

(d) to any other modifications in the terms of the Contract or of Specifications or both

(e) to the additional guarantee provided by the BUYER to the BUILDER's satisfaction due to the increase of the Contract Price.

Agreement as to such modifications or changes under this Paragraph shall be made in the same manner as provided above for modifications and/or changes of the Specifications and/or Plans.

(2) If, due to whatever reasons, the parties shall fail to agree on the adjustment of the Contract Price or extension of the delivery time or increase or decrease of the guaranteed speed, deadweight, or providing additional security to the BUILDER or any modifications of the terms of the Contract, the BUILDER shall be entitled to proceed with the construction of the VESSEL in accordance with the terms of the Contract and Specifications without making any such modifications or changes and the BUYER shall abide by the Contract and the Specifications without such modifications or changes.

(3) Even if the BUILDER and the BUYER fail to agree on the adjustment of the Contract Price or extension of the delivery time due to the modifications to meet the compulsory rules and regulations, the BUILDER shall proceed with such modifications as required by the said rules and regulations while the BUILDER is entitled to reserve the right by stating in writing the adjustment of the Contract Price or extension of the delivery time. Such dispute shall be settled as per Article XXVI of the Contract.

3. SUBSTITUTION OF MATERIALS AND/OR EQUIPMENT

In the event that any of the materials and/or equipment required by the Specifications and the Contract for the construction of the VESSEL cannot be procured in time to meet the delivery time of the VESSEL, the BUILDER may, with providing adequate evidence and subject to the approval by the BUYER in writing, supply other materials and/or equipment of the equivalent quality, and meeting the requirements of the Classification Society and of the rules and the regulations with which the construction of the VESSEL must comply.

ARTICLE XIII SEA TRIALS

1. NOTICE

The BUILDER shall, after obtaining the certificate for sea trails, give the BUYER and the supervisors at least thirty (30) calendar days prior notice of the estimated date and seven (7) definite days prior notice in writing of the time and place of the sea trial to be carried out as per the specification and the BUYER and the supervisors shall promptly acknowledge the receipt of such notice. The representative of the BUYER and/or the supervisor shall be present to witness such trials and approve the performance thereof during the sea trials. Failure of the BUYER's representative to be present in time at the trials after such notice to the BUYER, the Delivery Date of the VESSEL shall be extended by the period of delay caused by such failure of the BUYER's representative delays to attend such trails. If the absence of the BUYER's representative causes the delay in the sea trials more than seven (7) days, the BUYER shall be deemed to have waived the right to be present at the trials, and the BUILDER may conduct the sea trials without the BUYER's representative being onboard, and in such circumstances, the BUYER shall be obliged to accept the VESSEL on the basis of a certificate jointly signed by the BUILDER and the Classification after the sea trails and minor corrections and remedies as mentioned in this Article certifying that the VESSEL conforms to the requirements of the Contract and the Specification with satisfaction in all respects. The BUILDER shall make all necessary arrangements for the visa of the BUYER's representative, otherwise the sea trials shall be postponed until the BUYER's representative arrives in the shipyard, and any resulting delay in delivery of the VESSEL shall not be deemed as a Permissible Delay. But in case the nationality and the personal particulars of the BUYER's representative are not acceptable to the BUILDER as per current regulation and/or law of the People's Republic of China, the BUYER shall immediately replace this representative or other representative as per the BUILDER's request. The parties understand that the weather conditions of the area where the sea trials are to be carried out is momentarily changeable without advance warning, therefore if the sea trials are interrupted or prevented by the sudden unfavorable weather conditions, the parties shall agree to stop the sea trials and postpone the sea trials until the first favorable day thereafter when weather conditions permit. Unless otherwise the parties agree in writing that the BUYER accepts the VESSEL on the basis of the results achieved in the sea trials prior to sudden change in the weather conditions. The resulting delay due to the bad weather conditions shall be deemed as Permissible Delay as described by the Contract.

2. CONDUCT OF THE SEA TRIALS

(a) All expenses in connection with the sea trials be for the BUILDER's costs and expenses. The BUILDER shall arrange necessary crews onboard to meet safety requirements during sea trials. The sea trials shall be conducted as per the requirements of the Specification and comply with the requirements of the Specification.

The course of the sea trials shall be determined by the BUILDER and the sea trials shall be conducted in the area equipped with speed measuring facilities.

(b) To meet the conditions for sea trial as stipulated in the Specification, the BUILDER shall furnish water and fuel oil with exception of lubricating oil and hydraulic oil which shall be supplied by the BUYER. The said provision by the parties shall comply with the requirements of the applicable engine Specification. All the costs of the water, bunkers, lubricating oil, hydraulic oil and grease consumed during the sea trials shall be for the BUILDER's account.

3. SEA TRIALS AT LOAD DRAFT

In addition to the provisions as required by this Article, the BUILDER shall for the BUILDER's account provide fresh water and the stores for the sea trials to have the VESSEL achieve the ballast conditions (with fresh water, sea water of other stores as may be required) as specified by the Specifications for reaching the load draft.

4. METHOD OF ACCEPTANCE OR REJECTION

(a) The BUYER or the BUYER's representative, within six (6) working days after receiving the notice from the BUILDER of the completion of the sea trials, shall notify the BUILER in writing of its acceptance for delivery or rejection of the VESSEL together with reasons therefore.

(b) However, should the results of the sea trials demonstrate that the VESSEL or any part or equipment therefore does not conform to the requirements of the Specification, the

BUILDER shall together with the BUYER's representative inspect the cause of such non-conformity and the BUILDER shall take all necessary steps to rectify such non-conformity. If necessary the BUILDER shall for its own cost and expense carry out further sea trial. If the BUILDER notifies the BUYER of the completion of the deficiencies and the new sea trial, the BUYER, within six (6) working days after receiving the said notice from the BUILDER, shall notify the BUILDER in writing of its acceptance for delivery or rejection of the VESSEL together with reasons therefore.

(c) If the BUYER fails to notify the BUILDER in writing of its acceptance or rejection of the VESSEL together with reasons therefore with reasons within six (6) working days as stipulated in above item (a) and (b), the BUYER shall be deemed to have accepted the VESSEL.

(d) If the parties disputes any results of the sea trials and tests, the disputes shall be resolved in accordance with Article XXVI for arbitration.

(e) The BUYER shall not reject to take delivery of the VESSEL due to its remarks made after the sea trials and/or further sea trial. While the BUILDER shall remove the remarks (if such remarks are acceptable to the BUILDER) before effecting delivery of the VESSEL to the BUYER under the Contract.

5. DISPOSITION LF SURPLUS CONSUMABLE STORES

Upon taking delivery of the VESSEL, the BUYER shall reimburse the BUILDER in the manner as stipulated in 3 (5) and 4 (5) under Article V at current price same as the one in the delivery port for any quantities of the bunkers, fresh water and other consumable stores remaining onboard after sea trial.

The BUILDER shall reimburse the BUYER in the manner as stipulated in 3 (5) and 4 (5) under Article V for any quantities of lubricating oil and hydraulic oil actually consumed during sea trials at same price as the BUYER purchased.

6. EFFECT OF THE ACCEPTANCE

The BUYER's written notification of acceptance of the VESSEL delivered to the BUILDER as above stipulated, shall be final and binding provided the VESSEL conforms to the Specifications, and the BUYER is not entitled to refuse the BUILDER's request to deliver the VESSEL provided that the BUILDER complies with all other conditions for delivery as above set forth.

SECTION 4 DELIVERY

ARTICLE XIV DELIVERY AND DELIVERY DOCUMENTS

1. TIME AND PLACE

Subject to Article XIII after the sea trials (or possible new sea trials) and the acceptance by the BUYER, the VESSEL shall be delivered on or before __________safely afloat by the BUILDER to the BUYER at the BUILDER's shipyard in accordance with the Specifications and with all Classification and Statutory Certificates provided that. In the even of delays in the construction of the VESSEL or any performance required under the Contract due to causes which under the terms of the Contract permit extension of the delivery time, the aforementioned delivery time of the VESSEL shall be extended accordingly.

Abovementioned date or such later date to which delivery is extended pursuant to the terms of the Contract is hereinafter called the "Delivery Date".

2. WHEN AND HOW EFFECT
Provided that the BUYER and the BUILDER have fulfilled all of their respective obligations as stipulated in the Contract, delivery of the VESSEL shall be effected forthwith by the concurrent delivery by each of the parties hereto, one to the other, of the Protocol of Delivery and Acceptance, acknowledging delivery of the VESSEL by the BUILDER and acceptance thereof by the BUYER, and the said Protocol of Delivery and Acceptance shall be prepared in quadruplicate and executed by each of the parties hereto after signing.

3. DOCUMENTS TO BE DELIVERED TO THE BUYER
Upon delivery and acceptance of the VESSEL by the BUYER, the BUILDER shall provide to the BUYER the following documents together with the aforementioned Protocol of Delivery and Acceptance:

(1) Protocol of Trials made by the BUILDER pursuant to the Specifications.

(2) Protocol of Inventory and Equipment of the VESSEL including spare parts and the like, all as specified in the Specifications, made by the BUILDER.

(3) Protocol of Surplus Consumable Stores made by the BUILDER.

(4) Finished Drawings and Plans pertaining to the VESSEL as stipulated in the Specifications, made by the BUILDER.

(5) Protocol of Deadweight and Inclining Experiment, made by the BUILDER.

(6) All below certificates required to be furnished upon delivery of the VESSEL pursuant to the Contract and as provided in the Specifications.

...

Certificates shall be issued by relevant Authorities or Classification Society. The VESSEL shall comply with the above rules and regulations which are in force at the time of signing the Contract. All the certificates shall be delivered in one (1) original to the VESSEL and two (2) copies to the BUYER.

If the complete certificate or certificates are unable to be issued at the time of delivery by the Classification Society or any third party other than the BUILDER, then the BUYER shall accept the interim certificate or certificates as issued by the Classification Society or the third party other than the BUILDER, provided that the full final certificates shall be furnished by the BUILDER before the expiry of the interim certificate or certificates.

(7) Declaration of Warranty issued by the BUILDER that the VESSEL is delivered to the BUYER free and clear of any liens, charges, claims, mortgages, or other encumbrances upon the BUYES's title thereto, and in particular, that the VESSEL is absolutely free of all burdens in the nature If imposts, duties, taxes or charges imposed by the province or country of the port of delivery, as well as of all liabilities of the BUILDER to its sub-contractors, employees and crews and/or all liabilities arising from the operation of the VESSEL in trial run or trial runs, or otherwise, prior to delivery.

(8) Commercial Invoice made by the BUILDER.

(9) Bill of Sale made by the BUILDER.

4. OUTSTANDING ITEMS WHEN DELIVERY OF THE VESSEL
If there are few outstanding items which don't affect the safety and navigation of the

VESSEL when delivery of the VESSEL, the parties shall list the quantity of the outstanding items and date to complete these items. All outstanding items shall be completed within four (4) days after delivery of the VESSEL, which shall be recorded in the delivery documents. Such outstanding items shall not affect the delivery of the VESSEL and the BUYER shall take delivery of the VESSEL.

ARTICLE XV DELAYS & EXTENSION OF TIME FOR DELIVERY(FORCE MAJEURE)

1.CAUSE OF DELAY

If, at any time before actual delivery, either the construction of the VESSEL, or any performance required hereunder as a prerequisite of delivery of the VESSEL, is delayed due to war. blockade, revolution, insurrection, mobilization, civil commotions, riots, strikes, sabotage, lock-outs, epidemics, epidemics, abnormal weather conditions, typhoon, or other causes beyond the control of the BUILDER or of its sub-contractors for instance the long restriction of electric current from an outside source and the defects of casting and forging components, as the case may be. And other or by force majeure of any description, whether of the nature indicated by the forgoing or not, or by destruction of the BUILDER or works of the BUILDER or its sub-contractors, or of the VESSEL or any part thereof ,by fire, flood, or other causes beyond the control of the SELLER or its sub-con-tractors as the case may be, or due to the bankruptcy of the equipment and/or material supplier or suppliers, then, in the event of delay due to the happening of any of the aforementioned contingencies, the Seller shall not be liable for such delay and the time for delivery of the VESSEL under this Contract shall be extended without any reduction in the Contract Price for a period of time which shall not exceed the total accumulated time of all such delays, subject nevertheless to the BUYER·s right of cancellation and subject however to all relevant provisions of this Contract which authorize and permit extension of the time of delivery of the VESSEL.

2.NOTICE OF DELAY

Within seven (7)days from the date of commencement of any delay on account of which the SELLER claims that it is entitled under this Contract to an extension of the time for delivery of the VESSEL, the SELLER shall advise the BUYER by telex confirmed in writing, of the date such delay commenced, and the reasons therefore.

Likewise within seven(7)days after such delay ends, the SELLER shall advise the BUYER in writing or by telex confirmed in writing of the date such delay ended, and also shall specify the maximum period of the time by which the date for delivery of the VESSEL is extended by reason of such delay. Failure of the BUYER to acknowledge of SELLER·s notification of any claim for extension of the Delivery Date within thirty (30)days after receipt by the BUYER of such notification, shall be deemed to be a waiver by the BUYER of its right to object to such extension,.

3.RIGHT TO CANCEL FOR EXCESSIVE DELAY

If the total accumulated time of all permissible delays and non-permissible delays aggregate to______days or more, excluding delays due to arbitration as provided for in Article XVI here of due to default in performance by the BUYER, of due to delays in delivery of the BUYER·s supplied items, and excluding delays due to cause which, under

Article V, VI, XI, and XII hereof, permit extension or postponement of the time for delivery of the VESSEL, then in such event, the BUYER may in accordance with the provisions set out herein cancel this Contract by serving upon the SELLER telexed notice of cancellation which shall be confirmed in writing and the provisions of Article X of this Contract shall apply. The SELLER may, at any time, after the accumulated time of the aforementioned delays justifying cancellation by the BUYER as above provided for, demand in writing that the BUYER shall make an election, in which case the BUY-ER shall, within thirty(30)days after such demand is received by the BUYER either notify the SELLER of its. Intention to cancel, or consent to an extension of the time of delivery to an agreed future date, it being understood and agreed by the parties hereto that if any further delay occurs on account of causes justifying cancellation as specified in this Contract, the BUYER shall have the same right of cancellation upon the same terms as hereinabove provided.

4.DEFINITON OF PERMISSIBLE DELAY

Delays on account of such causes as provided for in Paragraph 1 of this Article, but excluding any other extensions of a nature which under the terms of this Contract permit postponement of the Delivery Date, shall be understood to be (and are herein referred to as)permissible delays, and are to be distinguished from non-permissible delays on account of which the Contract Price of the VESSEL is subject to adjustment provided for in Article III hereof.

ARTICLE XVI TITLE AND RISK

Title to and risk of the VESSEL shall rest with the BUILDER before delivery and pass to the BUYER immediately upon delivery and acceptance thereof.

At the time of delivery, the VESSEL shall be free of all liens, mortgages, claims, charges, and other encumbrances.

ARTICLE XVII POSSESSION AND REMOVAL OF VESSEL

1. THE POSSESSION OF THE VESSEL

The BUYER shall take physical possession of the VESSEL immediately upon delivery and acceptance thereof.

2. THE REMOVAL OF THE VESSEL

The BUYER shall remove the VESSEL from the port of the BUILDER or delivery place within five (5) calendar days after delivery and acceptance as above. If the BUYER does not remove the VESSEL within the aforesaid five (5) calendar days, in such event, the BUYER shall pay to the BUILDER the reasonable mooring charge of the VESSEL and the BUILDER is entitled to remove the VESSEL at any time thereafter from the delivery place to another safe premise with prior notice to the BUYER.

ARTICLE XVIII VESSEL REGISTRATION

The BUYER shall register the VESSEL at its own cost and expense in accordance with (here to fill in the name of the country where the VESSEL is to be registered) law after delivery and acceptance.

ARTICLE XIX BUILDER'S GURANTEE OF QUALITY

1. GURANTEE OF MATERIAL AND WORKMANSHIP
Within a period of twelve (12) months following the Vessel delivery to the BUYER, the BUILDER shall guarantee the Vessel, her hull and machinery and all parts and equipments thereof that are manufactured or furnished or supplied by the BUILDER and/or its sub-contractors under the Contract including material, equipment (however excluding any parts for the Vessel which have been supplied by or on behalf of the BUYER) against all defects which are due to defective materials, and/or poor workmanship.

2. NOTICE OF DEFECTS
The BUYER shall give the notice to the BUILDER in writing as promptly as possible, after the discovery of any defect or deviation for which a claim may be made under this guarantee. The BUYER's written notice shall describe the nature of the defect and the extent of the damage caused thereby. The BUILDER shall have no obligation under this guarantee for any defects discovered prior to the expiry date of the guarantee period, unless such notice is received by the BUILDER not later than thirty (30) days after the expiry date. The written notice of defects received within thirty (30) days after the expiry date of guarantee period shall explain the nature of such defect and extent of damage, so that the forthcoming claim will be sufficient compliance with the requirements as to time.

3. REMEDY OF DEFECTS
The BUILDER shall remedy any defects against which the Vessel or any parts of the equipment thereof is guaranteed under this Article by making all necessary repairs and/replacement. Such repairs and/or replacement shall be made at the BUILDER's cost and expense.
However, if it is impractical to make the repair by the BUILDER, as if the BUILDER's supply of replacement parts, and materials can not be accomplished without impairing or delaying the operation or working of the Vessel, then in such event that the BUYER makes the necessary repairs or replacements elsewhere at the discretion of the BUYER, the BUYER shall first and in all events, will, as soon as possible, give the BUILDER the notice in writing or orally of the time and place such repairs will be made. Without impairing or delaying the operation or working of the Vessel, the BUILDER shall have the right to inspect through its own representative or that of Classification Society the nature and extent of the Guarantee Defects to be replaced or repaired claimed by the BUYER. The BUILDER shall, in such case, promptly advise the BUYER in writing, after such examination has been completed, of its acceptance or rejection of such Guarantee Defects as ones that covered by the guarantee. In most cases, the Guarantee Engineer, as hereinafter provided for the Vessel, will act for and on behalf of the BUILDER.
In any circumstances as set out below, the BUILDER shall immediately pay to the BUYER in the currency of United States Dollars by telegraphic transfer the actual costs for such repairs or replacements including transportation costs, or at the average lever for making similar repairs or replacements including transportation costs as quoted by the leading shipyards, whichever is lower:

(1) Upon the BUILDER's acceptance of the defects as justifying remedy under this Article, or

(2) If the BUILDER neither accepts nor rejects the notice of defects as above provided, nor refer to the arbitration within thirty (30) days after the receipt of the BUYER's notice of defects.

Any dispute shall be referred to arbitration in accordance with the provisions of Article XXVI hereof.

4. EXTENT OF THE BUILDER'S LIABILITY

The BUILDER shall have no obligation and/or liabilities with respect to any defects discovered after the expiration of the Guarantee Period specified as above.

The BUILDER shall be liable to the BUYER for defects and damages caused by any of the defects specified in Paragraph 1 of this Article provided that such liability of the BUILDER shall be limited to the defects discovered within the Guarantee Period specified in Paragraph 1. The BUILDER shall not be obligated to repair, or be liable for, damages to the Vessel, or to any part of the equipment thereof, due to ordinary wear and tear or caused by the defects other than those specified in Paragraph 1 as above, nor shall there be any BUILDER's liability hereunder for defects in the Vessel, or any part of the equipment thereof, caused by fire or accidents at sea or elsewhere, or management mistake, accidents, negligence, or willful neglect, on the part of the BUYER, its employees or agents including the Vessel's officers, crew and passengers, or any persons on or doing work on the Vessel other than the BUILDER, its employees, agents or sub-contractors. Likewise, the BUILDER shall not be liable for defects in the Vessel, or the equipment or any part thereof, due to repairs or replacement which made by those than the BUILDER and/or their sub-contractors.

Upon the delivery of the Vessel to the BUYER, in accordance with the terms of the Contract, the BUILDER shall thereby and thereupon be released of all responsibility and liability whatsoever and howsoever arising under or by virtue of the Contract (save in respect of those obligations to the BUYER expressly provided for in this Article IX), including but not limit to any responsibility or liability for defective workmanship, materials, or equipment, design or in respect of any other defects whatsoever and any loss or damage resulting from any act, omission or default of the BUILDER. The BUILDER shall, in any circumstances, be liable for any consequential loss or special loss, or expense arising from any cause whatsoever including not limit to, loss of time, profit or earning or demurrage directly from any commitments of the BUYER in connection with the Vessel.

The Guarantee provided in this Article and the obligations and the liabilities of the BUILDER hereunder are exclusive and in lieu of and the BUYER hereby waives all other remedies, warranties, guarantees or liabilities, express or implied, arising by Law or otherwise (including but not limit to any obligations of the BUILDER with respect to fitness, merchantability and consequential damage) or whether or not occasioned by the BUILDER's negligence, The Guarantee shall not be extended, altered or varied except by a written instrument signed by the duly authorized representatives of the BUILDER, and the BUYER.

ARTICLE XX GUARANTEE PERIOD AND GUARANTEE ENGINEER

1. GUARANTEE PERIOD

The SELLER shall guarantee the VESSEL for a period of twelve (12) months from the month of delivery to the last day of the twelfth month.

2. GUARANTEE ENGINEER
The BUILDER shall appoint one or two Guarantee Engineer (s) to serve the VESSEL as the BUILDER's representative (s) for a period of three (3) months from the delivery of the VESSEL. The BUYER, and his employees, shall give such Guarantee Engineer (s) full co-operation in carrying out his/their duties as the representative (s) of the SELLER on board the VESSEL. The BUYER shall accord the Guarantee Engineer (s) the treatment comparable to the VESSEL's Chief Engineer, and shall provide him with accommodation and subsistence at no cost to the SELLER and/or the Guarantee Engineer (s).
The BUYER shall pay to the Guarantee Engineer (s) the sum of United States Dollars per month per person to cover his/their miscellaneous expenses including wages and also the BUYER shall pay the expense of repatriation to the location of the BUILDER by air upon termination of his/their service, the expense of his/their communications with the BUILDER when made in performance of his/their duties as the Guarantee Engineer (s) and the expenses. if any, of his/their medical and hospital care. The BUYER, its successor (s) and/or assign (s), shall be liable to and indemnify the SELLER and/or the Guarantee Engineer (s) and /or the SELLER for personal injuries, including death and damages to, or loss or destruction of property of the Guarantee Engineer (s), if such death, injuries, damages, loss and/or destruction were caused by gross negligence or willful misconduct of the BUYER and/or, its successor (s) and/or assign (s) or its employees and/or agents.
Pertaining to the detailed particulars of this Paragraph, a written agreement shall be made according to this effect between the parties hereto upon delivery of the VESSEL...

SECTION 5 LEGAL

ARTICL XXI LAW APPLICABLE

The parties hereto agree that the Contract should be governed by and interpreted in accordance with the Laws of People's Republic of China or the laws of other state agreed by the two parties.

ARTICLE XXII BUYER'S DEFAULT

1. DEFINITION OF DEFAULT
The BUYER shall be deemed in default of its obligation under the Contract if any of the following events occurs:
(1) The BUYER fails to pay the Second or Third or Fourth installment to the BUILDER when any such installment becomes due and payable under the provisions of Article V hereof and provided the BUYER shall have received the BUILDER's demand for payment in accordance with Article IV hereof, or
(2) The BUYER fails to pay the fifth installment to the BUILDER in accordance with Paragraph 3 (5) and 4 (5) of Article XIII hereof provided the BUYER shall have received the BUILDER's demand for payment in accordance with Article IV hereof; or
(3) The BUYER fails to take delivery of the Vessel, when the Vessel is duly tended for

delivery by the BUILDER under the provisions of Article XIII hereof.

2. NOTICE OF DEFAULT

If the BUYER is in default of payment or in performance of its obligations as provided hereinabove, the BUILDER shall notify the BUYER in writing on the date of occurrence of the default as per Paragraph 1 of this Article and the BUYER shall forthwith acknowledge in writing to the BUILDER that such notification has been received. In case the BUYER does not give the aforesaid acknowledgement in writing to the BUILDER within three (3) calendar days it shall be deemed that such notification has been duly received by the BUYER.

3. INTEREST AND CHARGE

(1) If the BUYER is in default of payment as to any installment as provided in Paragraph 1 (1) and/or 1 (1) of this Article, the BUYER shall pay interest on such installment at the rate of _____

percent (________%) per annum for a period of 15 days from the due date thereof and thereafter at the rate of _______percent (_______%) per annum until the date of the payment of the full amount, including all aforesaid interest.

In case the BUYER shall fail to take delivery of the Vessel as provided in Paragraph 1 (3) of this Article, the BUYER shall be deemed in default of payment of the fifth installment and shall pay interest thereon at the same rate as aforesaid form (and including) the day on which the Vessel is tendered for delivery, as provided in Article XIII Paragraph 6 hereof.

(2) In any event of default by the BUYER under 1 (1) or 1 (2) or 1 (3) above, the BUYER shall not costs, charges and expenses incurred by the BUILDER in consequence of such default unless provided by relevant.

4. DEFAULT BEFORE DELIVERY OF THE VESSEL

(1) If any default by the BUYER occurs as defined in Paragraph 1 (1) or 1 (2) or 1 (3) of this Article, the Delivery Date shall at the BUILDER's option, be postponed for a period of continuance of such default by the BUYER.

(2) If any such default as defined in Paragraph 1 (1) or 1 (2) or 1 (3) of this Article committed by the BUYER continues for a period of fifteen (15) days, then, the BUILDER shall have all following rights and remedies:

(i) The BUILDER may, at its option, cancel or rescind the Contract, provided the BUILDER has notified the BUYER of such default pursuant to Paragraph 2 of this Article, by giving notice of such effect to the BUYER by telex confirmed in writing. Upon receipt by the BUYER of such telex notice of cancellation or rescission, all of the BUYER's Supplies shall forthwith become the sole property of the BUILDER, and the Vessel and all its equipment and machinery shall be at the sole disposal of the BUILDER for sale or otherwise; and

(ii) (Applicable to any BUYER's default defined in 1 (1) of this Article) The BUILDER shall, without prejudice to the BUILDER's right to recover from the BUYER the 5th installment, interest, costs and/or expenses by applying the proceeds to be obtained by sale of the Vessel in accordance with the provisions set out in the Contract, have the right to declare all unpaid 2nd, 3rd and 4th installments to be forthwith due and payable, and upon such declaration, the BUILDER shall have the right to immediately demand the payment of the aggregate amount of all unpaid 2nd, 3rd and 4th installments from the Guarantor in accordance with the terms and conditions of the guarantee issued by the guarantor.

5. SALE OF THE VESSEL

(1) In the event of cancellation or rescission of the Contract as above provided, the BUILDER shall have full right and power either to complete or not to complete the Vessel at it deems fit, and to sell the VESSEL at a public or private sale on such terms and conditions as the BUILDER thinks fit without being answerable for any loss or damage occasioned to the BUYER thereby.

In the case of sale VESSEL, the BUILDER shall give telex or written notice to the BUYER.

(2) In the event of the sale of the VESSEL in its completed state, the proceeds of sale received by the BUYER shall be applied firstly to payment of all expenses attending such sale and otherwise incurred by the BUILDER as a result of the BUYER's default, and then to payment of all unpaid installments and/or unpaid balance of the Contract Price and interest on such installment at the interest rate as specified if the relevant provisions set out above from the respective due dates thereof to the date of application.

(3) In the event of the sale of the VESSEL in its incomplete state, the proceeds of sale received by the BUILDER shall be applied firstly to all expenses attending such sale and otherwise incurred by the BUILDER as a result of the BUYER's default, and then to payment of all costs of construction of the VESSEL (such costs of construction, as herein mentioned, shall include but are not limited to all costs of labor and/or the costs paid by the BUILDER for the equipment and/or materials to be installed on the VESSEL) and/or any fees, charges, expenses and/or royalties incurred and/or to be incurred for the VESSEL less the installments so retained by the BUILDER, and compensation to the BUILDER for a reasonable sum of loss of profit due to the cancellation or rescission of the Contract.

(4) In either of the above events of sale, if the proceed of sale exceeds the total of the amounts to which such proceeds are to be applied as aforesaid, the BUILDER shall promptly pay the excesses to the BUYER without interest, provided, however that the amount of each payment to the BUYER shall in no event exceed the total amount of installments already paid by the BUYER and the cost of the BUYER's supplies, if any.

(5) If the proceed of sale are insufficient to pay such total amounts payable as aforesaid, the BUYER shall promptly pay the deficiency to the BUILDER upon request.

ARTICLE XXIII BUILDER'S DEFAULT

Failure to accomplish the following obligations by the BUILDER shall constitute events of default:

1. The failure of the BUILDER to prosecute the Contract work with such diligence and in such manner as will disable it to complete the VESSEL construction by the Delivery Date stipulated in the Contract, except and to the extent that such failure is due to the causes stated in Article XI of the Contract for which the BUILDER would be entitled to an extension of the Delivery Date, such as the delay of the VESSEL delivery caused by force majeure; provided that the BUYER shall have given the BUILDER notice of such failure and that the BUILDER shall not, within fifteen (15) days of the date of receipt of such notice, have shown to the satisfaction of the BUYER that it has taken steps sufficient to remedy the failure in a manner satisfactory to the BUYER.

2. The failure of the BUILDER in any other respect to use due diligence in the performance of the Contract work or its failure to perform any of the covenants, agreements or undertakings on its part to be performed under the Contract, including but not limited to, the BUILDER's agreement to make prompt payment for all labor, material, services and other charges which are to be paid by the BUILDER; provided that the BUYER shall give notice to the BUILDER as to such failure, and the BUILDER shall not, within fifteen (15) days after being so notified, correct any failure to use due diligence or undertake the performance of said covenants, undertakings or agreements required to cure such failure, and thereafter prosecute in good faith to completion all such work or performance required to cure such failure.

3. The BUILDER being dissolved or adjudicated a bankrupt or making a general assignment for the benefit of its creditors, or the appointment of a receiver or receivers of any kind whatsoever, whether or not appointed in bankruptcy, common law or equity proceedings, whether temporary or permanent for the property of the BUILDER, or the filing by the BUILDER of a petition for reorganization or other proceedings under any of the provisions of the Bankruptcy Code, or the filing of such petition by creditors, stakeholders or any other person whatsoever and approval thereof by the Court.

4. Where there is a cause of delay in the Contract beyond prior expectation (except such causes stated in Article XI of the Contract for which the BUILDER would be entitled to extent the completion), and such delay would disable the BUILDER to accomplish the obligations under the Contract as determine by the BUYER.

ARTICLE XXIV ASSIGNMENT OF THE CONTRACT

1. ASSIGNMENT OF THE BUILDER
To prevent the committed financing from suffering any loss, the BUILDER shall have the right to assign the benefit arising under or by virtue of the Contract to the financing party.

2. ASSIGNMENT OF THE BUYER
(1) To prevent the committed financing from suffering any loss, the BUYER shall have the right to assign the benefit arising under or by virtue of the Contract to the financing party.
(2) After obtaining consent from the BUILDER (such consent not to be unreasonably withheld), the BUYER shall assign or transfer the right and obligation under the Contract or assign the Contract to the third party.
(3) Assigner and assignee shall jointly undertake and responsible for the execution of the provisions set forth under the Contract between the assignee and the BUILDER.

ARTICLE XXV LIEN AND MORTGAGE OF THE VESSEL

1. LIEN OF THE VESSEL
If the BUILDER is under no default, the BUILDER shall have the right to lien the possessing constructing VESSEL to insure the repayment of the construction cost when the contractual payment obligation has not been performed by the BUYER.

2. MORTGAGE OF THE VESSEL
The VESSEL which is under construction shall be eligible for shipping mortgage. The

BUILDER shall have the right to be the Pledger by pledging the VESSEL for the shipping mortgage and should sign a contract with the Pledgee in writing.

The mortgage registration shall be done jointly by the Pledger and the Pledgee in the VESSEL Registry for the shipping mortgage; the third party shall not be confronted for those without registration.

Registration of the shipping mortgage shall include the following main items:

(1) The name, title and address of the Pledgee and Pledger

(2) The name, nationality, the VESSEL ownership certificate issuing authority and the certificate number of the VESSEL to be mortgaged

(3) The loan amount, interest rate and the tenor to be guaranteed

The Pledger shall procure insurance for the mortgaged VESSEL unless the Contract agreed otherwise. The Pledgee shall have the right to procure insurance for the mortgaged VESSEL should it not be insured and the premium shall be borne by the Pledger.

ARTICLE XXVI DISPUTE RESOLUTION AND ARBITRAITON

1. CLASSIFICATION SOCIETY OR OTHER REGULATORY AUTHORITY

Disputes about whether or not the VESSEL is in accordance with the rules, regulations, and requirements of the Classification Society or other Regulatory Authority may submit to the Classification Commission of the Classification Society, or other Regulatory Authority. The determination of the authority is final and binding upon both parties. Any other disputes under the contract should be submitted to conciliation or arbitration for resolution, which is stipulated in below paragraph 2, and 3.

2. CONCILIATION

Before the dispute be submitted to arbitration, both parties agree to submit any of their disputes arising form the Contract or relating to the Contract to Shanghai Maritime Conciliation Center of CMAC for conciliation, which apply to the Center's Conciliation Rules and Panel of conciliators. Both parties agree to submit their settlement agreement by conciliation to Shanghai Sub-Commission of CMAC and jointly appoint or entrust the chairman of CMAC shall appoint one sole arbitrator to render an arbitration award according to the law and the content of conciliation settlement between the parties. The arbitration procedure and time limit shall not be confined by the Arbitration Rules of CMAC. Unless either party opposes, the conciliator appointed in the conciliation procedure may continue to act as the sole arbitrator to render an arbitration award, if the conciliator is also listed among the Panel of Arbitrators of CMAC.

If both parties don't agree to conciliate according to above regulations, Conciliation Rules should be adopted by the parties.

3. ARBITRATION

Any dispute under or in connect with the Contract shall be submit to Shanghai Sub-Commission of CMAC for arbitration which shall be conducted in accordance with the Arbitration Rules of CMAC in effect at the time of applying for arbitration. The arbitration award is final and binding upon both parties.

If both parties don't agree to submit the dispute to the above-mentioned arbitration institute, the both parties agree to submit to ___(location or institute) and apply to ___Arbitration Rules.

ARTICLE XXVII SUSPENSING AND TERMINATION

1. BUYER'S TERMINATION

During the period of performing the contract, the BUYER shall have the right to terminate the Contract upon giving notice in the event that:

(1) The Guarantor providing the refund guarantee on behalf of the BUILDER is deemed insolvent, unless the BUILDER provides a replacement and acceptable refund guarantee to the BUYER within 30 days of the BUYER's notice requiring a replacement refund guarantee to be provided, or

(2) Excluding permissible delays, the BUILDER fails to perform any work relating to the shipbuilding for a running period of at least_____days, provided that thereafter the BUYER gives the BUILDER at least_____days' notice in writing of his intention to terminate the Contract under this clause and within that period the BUILDER fails to remedy its breach, or.

(3) In the event that:

(a) The delivery date is delayed by more than 180 days by fore majeure events,

(b) The delivery of the VESSEL is delayed by more than 180 days, excluding permissible delays,

(c) The aggregate of delays in the delivery of the VESSEL in (a) and (b) above is more than 270 days.

The BUILDER may at any time after the BUYER is entitled to terminates the Contract give notice to the BUYER requesting that the BUYER either agrees to a new delivery date or terminate the Contract. Such new delivery date shall be a reasonable estimated date by the BUILDER, when the VESSEL will be ready for delivery. Within 15 days of the request by the BUILDER, the BUYER shall notify the BUILDER of his determination. If the BUYER doesn't terminate the Contract, then the new delivery date shall be deemed to be the delivery date provided it doesn't occur later than 30 days prior to the expiry of the refund guarantee. If the VESSEL is not delivered by that date, the BUYER shall have the right to terminate this Contract.

(4) The reduction in speed, deadweight, hold capacity and the excessive fuel consumption of the main engine entitles the BUYER to reduce the Contract Price, terminate or cancel the Contract as per clause 1, 2, 3 and 4 in Article VI greater than the reduction amount stated in contract;

(5) There are deficiencies of the VESSEL as stated in the Contract to entitle the BUYER to suspend the performance of the Contract;

(6) The BUILDER's guarantee is in breach of relevant regulations of this Contract.

2. BUILDER'S TERMINATION

During the period of performing the contract, the BUILDER shall have the right to terminate the Contract upon giving notice to the BUYER in the event that:

(1) The Guarantor providing the installment or performance guarantee on behalf of the BUYER is deemed insolvent, unless the BUYER provides a replacement and acceptable refund guarantee to the BUILDER within 30 days; or

(2) The BUYER fails to pay any sums due under the contract for a period of 21 banking days, provided that the BUILDER thereafter gave the BUYER at least 5 banking days notice of his intention to terminate the Contract under this Article and within that period the BUYER failed to remedy the breach;

(3) The BUYER fails to take delivery of the VESSEL tendered in accordance with the Contract;

(4) The BUYER fails to provide installment or performance guarantee according to this Contract.

3. Suspension of BUILDER'S Construction Work

Without prejudice to BUILDER's right of termination, the BUILDER has the right to suspend the construction work if the BUYER fails to pay any installment due according to the contract for 15 banking days until payment of such outstanding sums.

4. DEEMED INSOLVENCY

A party or the Guarantor providing the refund guarantee shall be deemed insolvent if the proceedings are commenced against the insolvent party or the Guarantor for winding up, dissolution or reorganization (otherwise than for the purpose of amalgamation or reconstruction), liquidation, the appointment of a receiver, trustee or similar officer, bankruptcy, suspension of payments or similar events. A party shall have the right to terminate the Contract forthwith upon giving notice if the other party or Guarantor is deemed insolvent.

5. EFFECTIVENESS of BUYER'S TERMINATION

If the Contract is terminated by the BUYER, the BUILDER shall refund all sums paid by the BUYER to the BUILDER according to the contract here of plus interest thereon at the rate stated in contract per annum from the date of payment to the date of refund. The BUILDER shall also return the BUYER's SUPPLIERS, or if they cannot be returned, the BUILDER shall pay to the BUYER an amount equal to the BUYER's cost for such BUYER's SUPPLIERS.

6. EFFECTIVENESS OF THE BUILDER'S TERMINATION

If this Contract is terminated by the BUILDER, the BUILDER shall have the right to retain the BUYER's SUPPLIERS together with any installments paid by the BUYER and shall have the right and power either to complete or not to complete the shipbuilding, but in any event shall sell the VESSEL (either in its complete or incomplete form), including those BUYER's supplies which are installed or have been utilized on board the VESSEL, at the best price reasonably obtainable at a public or private sale on reasonable terms and conditions.

(1) In the event of the sales of the VESSEL in its complete form, the proceeds of the sale received by the BUILDER shall be applied in the following order:

(a) to pay all expenses incurred by the BUILDER in respect of the sale and otherwise incurred by the BUILDER as a result of the BUYER's breach of the contract;

(b) to pay unpaid installments of the contract price including any which would have been payable after the date of termination and interest on such installments at the rate of interest stated in contract from the respective due dates thereof to the date of application.

(2) In the event of the sale of the VESSEL in its incomplete form the proceeds of sale received by BUILDER shall be applied in the following order:

(a) To pay all expenses incurred by the BUILDER in respect of the sale and otherwise incurred by the BUILDER as a result of the BUYER's default;

(b) To pay all unpaid installments of the Contract Price to the extent due but not yet paid at the date of termination and interest on such installments at the rate of interest stated in the

Contract from the respective due date thereof to the date of application;
(c) To pay all costs of part shipbuilding less any paid installments and less any sums credited in paragraph (b) as above;
(d) To compensate the loss of the BUILDER's reasonable net profits caused by the BUYER's breach of the Contract.
(3) In any of the above events, if the proceeds of sale exceed the sums to which such proceeds are to be applied as aforesaid, the BUILDER shall promptly pay any such excess to the BUYER without interest thereon, provided that the amount of such payment to the BUYER shall in no event exceed the total amount of installments paid by the BUYER. The BUILDER shall at the same time either permit the BUYER to remove the BUYER's supplies which are not installed or utilized onboard the VESSEL (if any) from the shipyard for the cost and expense of the BUYER, or give credit to the BUYER for the value thereof.
(4) If the proceeds of sale are insufficient to pay the BUILDER the total amounts due from the BUYER as aforesaid, the BUILDER may sell the BUYER's SUPPLIES which are not installed or utilized onboard the VESSEL (if any) at the best price reasonable obtainable at a public auction or private sale on reasonable terms and conditions, applying the proceeds of such sale toward the unsatisfied amounts due from the BUYER, and giving credit to the BUYER for any excess.
(5) If the proceeds of sale are still insufficient to pay the BUILDER the total amounts due from the BUYER as aforesaid, the BUYER shall pay the BUILDER the amount of such deficiency, plus interest at the rate stated in the Contract calculated from the date when the payments become due.

ARTICLE XXVIII INSURANCE

1. EXTENT OF INSURANCE COVERAGE
From the time of steel cutting of the VESSEL until the same is completed and delivered to the BUYER, the BUILDER shall keep the VESSEL insured with a qualified Chinese insurance company for shipbuilder's insurance. Such insurance shall cover the damages or losses of the VESSEL's materials, hull and equipments which incurred by various marine perils, inland perils or the BUILDER's errors and omissions during the period of construction and trail, as well as any additional costs and liability to a third party including the BUYER's SUPPLIES.
The amount of such insurance coverage shall be in an amount equal to the amount of the Contract Price, and not less than, the aggregate of the payments made by the BUYER to the BUILDER and the value of the BUYER's SUPPLIES. Before the delivery of the VESSEL, if the BUILDER and the BUYER adjust the Contract Price in accordance with the Article VI of the Contract, it is the BUILDER's obligation to adjust the insurance amount accordingly.
All losses under such insurance policy shall be payable to the BUILDER.
The BUYER shall inform the BUILDER of any subsequent changes of the value of the BUYER's SUPPLIES upon receipt of the notice of such changes of value, the BUILDER is obliged to accordingly change the insured value of the materials and equipment which supplied by the BUYER.
The BUYER is obliged to assist the BUILDER in arranging the aforesaid insurance and

providing relevant documents required by the insurance company. The BUILDER is obliged provide the BUYER with the copy of the insurance policy as backup.

2. APPLICATION OF RECOVERED AMOUNT

(1) Partial Loss

In the event the VESSEL shall be damaged by any insured cause whatsoever prior to acceptance and delivery thereof by the BUYER and in the further event that such damage shall not constitute an actual or a constructive total loss of the VESSEL, the BUILDER shall apply the amount recovered under the insurance policy referred to in Paragraph 1 of this Article to the repair of such damage satisfactory to the Classification Society and the Administrative Authority as described in the Specifications. No additional expenses should be collected by the BUILDER and the BUYER shall accept the VESSEL in accordance with this Contract.

(2) Total Loss

In the event that the VESSEL is determined to be an actual or constructive total loss:

(a) By the mutual agreement between the PARTIES hereto, proceed in accordance with terms of the Contract, in which case the amount recovered under said insurance policy shall be applied to the reconstruction and/or repair of the VESSEL's damages and/or reinstallation of BUYER's supplies without additional expenses to BUYER, provided the PARTIES hereto shall have first agreed in writing as to such reasonable extension of the delivery date and adjustment of other terms of other terms of this Contract including the Contract Price as may be necessary for the completion of such reconstruction; or

(b) If due to whatever reasons the parties fail to agree on the above, then refund immediately to the BUYER the amount of all installments paid to the BUILDER under this Contract without interest together with recovered amount for BUYER's SUPPLIES onboard, whereupon this Contract shall be deemed to be canceled and all rights, duties, liabilities and obligations of each of the parties to the other shall terminate forthwith.

Within thirty (30) days after receiving notice of any damage to the VESSEL constituting an actual or a constructive total loss, the BUYER shall notify the BUILDER in writing of its agreement or disagreement under this sub-paragraph. In the event the BUYER fails to so notify the BUILDER, then such failure shall deemed as rescinded and canceled and the BUYER receive the refund as hereinabove provided and this paragraph (b) shall apply.

Concurrently with the refundment to be made by the BUILDER to the BUYER as per the paragraph (b). the Contract is deemed to be terminated. all the right and obligation of one party to another party under the Contract shall be null and invalid, and the guarantees under the Contract shall be returned.

3. TERMINATION OF INSURANCE

The BUILDR's interest under the insurance policy shall cease and terminate forthwith upon delivery of the VESSSEL.

ARTICLE XXIX PATENTS, TRADEMARKS AND COPYRIGHTS

The machinery and equipment of the VESSEL may bear the patent number, trademarks or trade names of the manufacturers. The BUILDER shall defend and save harmless the BUYER from patent liability of claims of patent infringement of any nature or kind, including costs and expenses for, or on account of any patented or patentable invention made or used in the performance of the Contract and also including cost and expense of litigation, if an. If the BUYER involved into a patent infringement because of the BUILDER, it may lodge claims against the BUILDER for its losses arising therefrom.

Nothing contained herein shall be construed a transferring any patent or trademark rights or copyright in equipment covered by the Contract, and all such rights are hereby expressly reserved to the true and lawful owners thereof. Notwithstanding any provisions contained herein to the contrary, the BUILDER's obligation under this Article should not be terminated by the passage of any specified period of time.

The BUILDER's indemnity hereunder does not extend to equipment or parts supplied by the BUYER to the BUILDER if any.

SECTION 6 SUNDRY

ARTICLE XXX NOTICE AND LANGUAGE

1. NOTICE

Any and all notices and communications in connection with the Contract shall be addressed as follows:

P.C.

To the BUYER (the full name of the BUYER)

Address: (the legal address of the BUYER)

Tel No:

Fax No:

Email:

To the SELLER (the full name of the SELLER)

Address: (the legal address of the SELLER)

P.C.

Tel No:

Fax No:

Email:

To the BUILDER: (the full name of the BUILDER)

Address: (the legal address of the BUILDER)

P.C.

Tel No:

Fax No:

Email:

Any notice or communication sent to the BUYER singly from the SELLER or the BUILDER shall be deemed as from the SELLER and the BUILDER together.

Any change in the address shall be communicated in writing by the party adding such change to the other party and in the event of failure to give such notice of change, communications addressed to the other party at their last known address shall be deemed sufficient.

Any and all notices, demands, requests, instructions, advices and communications in connection with the Contract shall be deemed to be given at the time when the same is delivered to the address of the party to be served however, that email and fax shall be deemed to be delivered upon dispatch and receipt.

2. LANGUAGE

Any and all documents as well as notices, communications, specifications and drawings and any other written materials in connection with the Contract shall be written in the Chinese or the English language subject to the situation, each language is legally of equal effect while in case there is any disagreement between the English version and Chinese version, the Chinese one shall prevail. Each party hereto shall have no obligation to translate them into any other language.

ARTICLE XXXI OPTION

The BUYER shall have the right to build the optional vessels, the optional VESSEL should be executed at the same contract price and construction time as the VESSEL Such option must be declared by the BUYER to the BUILDER within the number of months that has been confirmed by the parties following the effective date of the Contract.

The BUILDER also has the right to negotiate with the BUYER on issues of the contract price of the optional vessel the construction time can be properly shortened and the delivery date can be properly put forward.

ARTICLE XXXII EFFECTIVE CONDITIONS AND DATE OF CONTRACT

The Contract shall become effective upon fulfillment of all the following and other mutually agreed conditions:

(1) The Contract and the Specifications have been signed and sealed; and

(2) Receipt by the BUILDER of the first installment in accordance with Article V of this Contract; and

(3) Receipt by the BUILDER of a Letter of Guarantee in the annexed hereto as Annex B issued by a first class international bank of the BUYER acceptable to the SELLER in accordance with Article V hereof; and

(4) Receipt by the BUYER of a Refund Guarantee in the form annexed hereto as Annex A issued by a first class international bank of the SELLER acceptable to the BUYER in accordance with Article V hereof; and

(5) Approval of the boards of the two parties.

Unless other written agreements exist in the Contract, the parties shall be expressly released from their obligations to each other in accordance with this Contract, an instruction should be made to indicate the ineffectiveness of the Contract. Meanwhile, the BUILDER should return the BUYER all its payments which has been paid before plus an interest of the BUYER's payment at the rate of_______percent (______%) per annum.

ARTICLE XXXIII ENTIRE AGREEMENT

The Contract constitutes the entire agreement between the parties and no promise, warranty or statement prior to the date of the Contract shall affect the Contract. Any modification of the Contract shall not be of any effect unless in writing negotiated and signed by or on behalf of the parties.

ANNEXES

CMAC Standard Newbuilding Contract (Shanghai Form)

ANNEX A
BUILDER'S IRREVOCABLE LETTER OF REFUND GUARANTEE

To: [insert the full name of the beneficiary]
Date:
To whom it may concern

IRREVOCABLE LETTLE OF GUARANTEE FOR PAYMENT

At the request of a [here insert the name of shipping trade company] and [here insert the name of shipyard or the shipbuilding company] (hereinafter called as "the Seller"), in consideration of you agreeing to pay the Seller the installments before the delivery of the vessel under the building contract concluded by you and Seller dated [ ] (hereinafter called as "the Contract") for a vessel with construction number [ ] and deadweight [metric ton] (hereinafter called as "the Vessel"), we hereby guarantee that the Seller will repay to you an amount up to a total of United States Dollars [ ] representing the aggregate amount paid by you to the Seller under the Contract before the delivery of the Vessel, of the first installment of the Contract Price amounted USD [ ], the second installment amounted USD [ ], the third installment amounted USD [ ] and the fourth installment amounted USD [ ]. Should the Seller fails to repay to you such any or all installments due as provided by the articles under the Contract, and you suspend to terminate the Contract due to the extension of the delivery date, we will make such payment to you without interest. In the event that the delivery date is delayed for [ ] days and you terminate the Contract in accordance with the clause 3 of Article 8 or clause 1 (3), 2 (3), 3 (3), or 4 (3) of Article 3 of the Contract, we shall pay to you the aforesaid amount of installments together with interest at the rate of [ ] percent (%) per annum, or [ ] percent (%) per annum in other circumstances. Within thirty (30) running days upon receipt by us from you of a Repayment Demand, we shall pay to you the sun as follows.

However, in the event that is disagreement between you and the Seller on the following issues:
(1) Whether the Seller is liable for the repayment of any or all installments to you, and
(2) Whether you have the right to require us for payment Should the Seller or you submit such dispute before the arbitration in accordance with Article 26 of the Contract or raise an appeal in accordance with the laws of other country, we is entitled to terminate the contract and postpone any payment until the arbitration award between the Seller and the Buyer is published. We shall not be obligated to make such payment unless repayment of the Seller is ordered in the arbitration award and the Seller waives or disclaims appeal in accordance with the laws of other country. Should the Seller fails to enforce the arbitration award, we will pay off the amount ordered by the award (or the judgment), which shall not exceed the amount to be secured by the Letter of Guarantee, plus the interest mentioned above.
All payments under the Letter of Guarantee shall be made in United States Dollars. The Letter of Guarantee comes to effect upon the receipt of the first installment by the Seller

from you prior to the delivery under the Contract. The sum available under the Letter of Guarantee shall be in line with the amount which has paid by you prior to the delivery. The amount to be paid under the Letter of Guarantee shall not exceed the aggregate amount of the installments paid by you to the Seller before the delivery of the Vessel, plus the interest at the rate of [ ] percent (%) per annum calculated on the amount of each installment from the date of receipt of the same by the Seller until the date on which the same is repaid under the Letter of Guarantee.

The Letter of Guarantee shall remain in force until (a) due delivery of the Vessel to, and acceptance of the Vessel by, the Buyer, or (b) the payment to you by the Buyer or by us of all sums guaranteed by the Letter of Guarantee, or (c) the [ ] day (the thirtieth day after the delivery date under the Article 14 of the Contract), whichever is earlier. Hereafter, you shall return the Letter of Guarantee to us by airmail for the purpose of revocation, unless:

(1) In the event that the Vessel is under construction, the period of validity of the Letter of Guarantee shall be extended until [ ] day (the [ ] day after the delivery date under the Article 13 of the Contract), or the date of due delivery of the Vessel to and acceptance of the Vessel by the Buyer, if first incurs. Or

(2) In the event that you or the Seller submit before the arbitration authority or the court the matter under the Contract, the period of validity of the Letter of Guarantee will be automatically extended until the thirtieth day after the publication of the arbitration award of the judgment.

The Letter of Guarantee is governed by the laws of the Peoples' Republic of China.

Yours truly,

[ ] Bank

CMAC Standard Newbuilding Contract (Shanghai Form)

ANNEX B
BUYER'S IRREVOCABLE LETTER OF GUARANTEE ON THE SECOND, THIRD AND FOURTH INSTALLMENTS
BANK

To: [insert the full name of the beneficiary]
[ ] Shipyard (or Shipbuilding Company)
Date:
To whom it may concern

(1) In consideration of you and [ ] (hereinafter called as "the BUYER") entering into a sales contract (building contract) dated [ ] for the construction of a VESSEL with hull number [ ] and deadweight ton of [metric ton] (hereinafter called as "the VESSEL") we, as the primary obligor and not a guarantor only, hereby irrevocably, absolutely and unconditionally guarantee that the BUYER will punctually pay to you the sum of United States Dollars [ ] representing the aggregate amount of the second, third and fourth installments of the Contract Price as provided by the Section (2) herein.
(2)In accordance with the Contract, the sum to be secured in the Letter of Guarantee includes the second installment of USD [ ], for which the BUYER is liable within three (3) New York Banking Day after the commencement of the construction of the VESSEL, the third installment of USD [ ], for which the BUYER is liable when the first paragraph of the VESSEL is put on the slipway, and the fourth installment of USD [ ], for which the BUYER is liable within three (3) New York Banking Day after launching of the VESSEL.
(3) As the primary obligor and not a guarantor only, we hereby irrevocably, absolutely and unconditionally guarantee that we will punctually pay the interest calculated on the amount of each installment, which the BUYER is obliged to pay, at the rate of [ ] percent (%) per annum from the expiry date of such installment as provided by the Contract until the date on which the same is fully paid.
(4) Should the BUYER fails to punctually pay any installment or interest and such failure is last for fifteen (15) days, we will, upon receipt by us from you of the first written demand for the same, pay to you or to your order the amount of the second, third and fourth installments and relevant interest as provided by Section (3) above, without request to you for a further action, measure or steps on the mortgages holding by you or the SELLER.
(5) The Letter of Guarantee and obligations hereunder shall be freely assignable by you at your own discretion. If assigned, it is guaranteed that any third party consigned by you or the Business Department of the Head Office of Bank of China is your agent. Such third party or Business Department of Head Office of Bank of China is deemed to be authorized as the agent herein.
(6) Any payment under the Letter of Guarantee shall be made in United States Dollars to [here insert the name of bank] and transfer to your account in the Business Department of the Head Office of Bank of China, with you or your agent as the beneficiary and the receiving bank appointed or temporarily appointed by the SELLER.
(7) Our obligations under the letter of guarantee shall not be affected or impaired by the

following factors: any dispute between the BUYER and you, as the SELLER, under the Contract, or delay on delivery or construction of the VESSEL for any reason, change or extension of the period of construction or delivery of the VESSEL, any mortgage or compensations incurs or to incur, any absolution from you or other relevant parties, non-performance or invalidity of the relevant provisions or other factor which may diminish the obligations under the Letter of Guarantee such as any action, omission, facts or conditions.

(8) Any representation or demand in writing made by you shall be signed by one of your duly authorized officers and may be served on us by hand or by post to [ ] (or such other address as we may notify to you in writing), or by authorized communication number with written confirm of Bank of China.

(9) The Letter of Guarantee come to effect upon your receipt and remain in force until (a) due delivery of the VESSEL to, and acceptance of the VESSEL with full performance of his obligation, by the BUYER, or (b) the payment to you by the BUYER or by us of all amount of the second, third and fourth installments and interest, whichever is earlier.

(10) Our liability under the Letter of Obligation shall not exceed the maximum sum of USD [ ], which includes:

(a) The aggregate amount of the second, third and fourth installments to be secured in the Letter of Guarantee, which is USD [ ], and

(b) Interest on the installment for sixty (60) days at the rate of [ ] percent (%) per annum, which is USD [ ].

(11) All payments to be made under the Letter of Guarantee shall be made without any set off or counterclaim and without deduction or withholding for or on account of any taxes, duties or charges whatsoever unless we are compelled by law to deduct or withhold the same in which case we shall make the minimum deduction or withholding permitted and will pay such additional amounts as may be necessary in order that amount received by you after such deduction or withholding shall be equal to the amount which would have been received had no such deduction or withholding been made.

(12) The Letter of Guarantee is to be construed in accordance with the laws of [here insert the name of country]. We hereby submit it to the courts without judicial privileges for any legal action or litigation in the [here insert the name of country].

(13) The Letter of Guarantee will be ended as of the date above. After that, you shall return the Letter of Guarantee without any condition.

(14) In witness whereof, the Letter of Guarantee is effective and submitted by our duly authorized representatives on the date above.

CMAC Standard Newbuilding Contract (Shanghai Form)

ANNEX C
SPECIFICATION

CMAC Standard Newbuilding Contract (Shanghai Form)

ANNEX D

GENERAL ARRANGMENT AND MIDSHIP DRAWING

CMAC Standard Newbuilding Contract (Shanghai Form)

ANNEX E
MAKER'S LIST

CMAC Standard Newbuilding Contract (Shanghai Form)

ANNEX F

Comprehensive Table for the Contract's Element

(B-C Standard Shipbuilding Contract Facility Edition)

<table>
<tr><td colspan="3">1.Place and Date of the Contract</td></tr>
<tr><td>2. BUILDER's name, full style address, and
contract details
ship trading company (if any)
Name:
Address:
Country:
Tel:
Fax:
Email:

BUILDER
Name:
Address:
Country:
Tel:
Fax:
Email:</td><td colspan="2">3. BUYER's name, full style address, and contract details
Name:
Address:
Country:
Tel:
Fax:
Email:</td></tr>
<tr><td>4. VESSEL description/type</td><td colspan="2"></td></tr>
<tr><td>A. Main Dimensions
(i) LOA (m)
(ii) Length between perpendiculars (m)
(iii) Deadweight capacity (DWT) (mts)
(iv) Design draft (m)</td><td>B. cargo capacity
(i) Cubic capacity
(ii)Rold capacity</td><td>B2. TEU carrying capacity(state number of containers)
(i) Total on deck
a. 20'/40'/45'TEU
b. No. of reefers
(ii) Total under deck
a. 20'/40'/45'TEU
b. No. of reefers</td></tr>
<tr><td>Main engine
(i) Maker/Type
(ii) Max. continuous rating (kilowatts at MCR)
(iii) RPM at MCR
(iv) Specific Fuel Oil Consumption at MCR</td><td colspan="2">D. Average speed
(i) Service speed (knot)at design draft

E. Other matters (optional)</td></tr>
<tr><td>5. BUILDER's Hull Number</td><td colspan="2">6. Flag State</td></tr>
<tr><td>7. Classification Society/Class character</td><td colspan="2">8. Contract Price and Currency
(a) Price</td></tr>
</table>

	(b) Currency
9. Contractual Date of Delivery	10. Payments 1st installment 2nd installment 3rd installment 4th installment Final installment
11. BUILDER's Bank Account Data Name: Address: Country: Phone/Fax: Email: Sort code: Account number: Account name:	
12. Adjustment of Contract Price A. Speed (a) Contract Price reduction amount (b) Maximum reduction amount (c) Contract Price addition amount (d) Maximum addition amount	B. Deadweight: (a) Contract Price reduction amount (b) Maximum reduction amount (c) Contract Price addition amount (d) Maximum addition amount
C. Cubic (a) Contract Price reduction amount (b) Maximum reduction amount (c) Contract Price addition amount (d) Maximum addition amount	D. Fuel Consumption (a) Contract Price reduction amount (b) Maximum reduction amount (c) Contract Price addition amount (d) Maximum addition amount
E. Late or Advanced Delivery compensation (a) Amount per day for delay (b) Maximum reduction amount (c) Amount per day for advanced delivery (d) Maximum addition amount	F. Others (optional)
13.Guarantees (a) BUYER's guarantees (i) Number of days after signing contract (ii) BUYER's installment/Performance Guarantee (b) BUILDER's guarantees (i) Numbers of days after signing contract	14. Guarantee Period(state number of months, if left blank 12 months shall apply)
15. Additional Guarantee Period (state number of months)	16. Suspension and Termination (a) Running period (state number of days) (b) Notice period (state number of days)
17. Applicable Law and Dispute Resolution (a) Governing law: (i) Law of P.R.C.,or (ii) Law of other country (b) Institution or place of dispute resolution:(i) CMAC-Shanghai, or (ii) other arbitration institution or place	18. Guarantee Engineer (state total number of months)
19. Effective Date of Contract (state conditions to be satisfied)	20. State number of days within which conditions have to be satisfied

21. Optional additional vessels (state number)	22. Optional additional vessels contract price and delivery dates
23. Declaration of options (state number of months after Effective Date)	24. Interest (state rate of interest)
25. BUYER's Guarantor (state name of bank or party as appropriate, full style address, and contract details) Name: Address: Country: Phone: Fax: Email:	26. BUILDER's Guarantor (state name of bank or party as appropriate, full style address, and contract details) Name: Address: Country: Phone: Fax: Email:
27. Number of Annexes	28. Number of Additional Clauses
Signature (BUILDER)	Signature (BUYER)

APPENDIX 3

SALEFORM 2012

Norwegian Shipbrokers' Association's Memorandum of Agreement for sale and purchase of ships. Adopted by BIMCO in 1956.
Code-name
SALEFORM 2012
Revised 1966, 1983 and 1986/87, 1993 and 2012

MEMORANDUM OF AGREEMENT

Dated:

(*Name of sellers*), hereinafter called the "Sellers", have agreed to sell, and

(*Name of buyers*), hereinafter called the "Buyers", have agreed to buy:

Name of vessel:

IMO Number:

Classification Society:

Class Notation:

Year of Build: Builder/Yard:

Flag: Place of Registration: GT/NT: /

hereinafter called the "Vessel", on the following terms and conditions:

Definitions

"Banking Days" are days on which banks are open both in the country of the currency stipulated for the Purchase Price in Clause 1 (Purchase Price) and in the place of closing stipulated in Clause 8 (Documentation) and (*add additional jurisdictions as appropriate*).

"Buyers' Nominated Flag State" means (*state flag state*).

"Class" means the class notation referred to above.

"Classification Society" means the Society referred to above.

"Deposit" shall have the meaning given in Clause 2 (Deposit)

"Deposit Holder" means (*state name and location of Deposit Holder*) or, if left blank, the Sellers' Bank, which shall hold and release the Deposit in accordance with this Agreement.

"In writing" or "written" means a letter handed over from the Sellers to the Buyers or vice versa, a registered letter, e-mail or telefax.

"Parties" means the Sellers and the Buyers.

"Purchase Price" means the price for the Vessel as stated in Clause 1 (Purchase Price).

"Sellers' Account" means (*state details of bank account*) at the Sellers' Bank.

"Sellers' Bank" means (*state name of bank, branch and details*) or, if left blank, the bank notified by the Sellers to the Buyers for receipt of the balance of the Purchase Price.

1. Purchase Price

The Purchase Price is (*state currency and amount both in words and figures*).

2. Deposit

As security for the correct fulfilment of this Agreement the Buyers shall lodge a deposit of % (per cent) or, if left blank, 10% (ten per cent), of the Purchase Price (the "Deposit") in an interest bearing account for the Parties with the Deposit Holder within three (3) Banking Days after the date that:

(i) this Agreement has been signed by the Parties and exchanged in original or by e-mail or telefax; and

(ii) the Deposit Holder has confirmed in writing to the Parties that the account has been opened.

The Deposit shall be released in accordance with joint written instructions of the Parties. Interest, if any, shall be credited to the Buyers. Any fee charged for holding and releasing the Deposit shall be borne equally by the Parties. The Parties shall provide to the Deposit Holder all necessary documentation to open and maintain the account without delay.

3. Payment

On delivery of the Vessel, but not later than three (3) Banking Days after the date that Notice of Readiness has been given in accordance with Clause 5 (Time and place of delivery and notices):

(i) the Deposit shall be released to the Sellers; and

(ii) the balance of the Purchase Price and all other sums payable on delivery by the Buyers to the Sellers under this Agreement shall be paid in full free of bank charges to the Sellers' Account.

4. Inspection

(a) *The Buyers have inspected and accepted the Vessel's classification records. The Buyers have also inspected the Vessel at/in *(state place)* on *(state date)* and have accepted the Vessel following this inspection and the sale is outright and definite, subject only to the terms and conditions of this Agreement.

(b) *The Buyers shall have the right to inspect the Vessel's classification records and declare whether same are accepted or not within *(state date/period)*.

The Sellers shall make the Vessel available for inspection at/in *(state place/range)* within *(state date/period)*.

The Buyers shall undertake the inspection without undue delay to the Vessel. Should the Buyers cause undue delay they shall compensate the Sellers for the losses thereby incurred.

The Buyers shall inspect the Vessel without opening up and without cost to the Sellers.

During the inspection, the Vessel's deck and engine log books shall be made available for examination by the Buyers.

The sale shall become outright and definite, subject only to the terms and conditions of this Agreement, provided that the Sellers receive written notice of acceptance of the Vessel from the Buyers within seventy-two (72) hours after completion of such inspection or after the date/last day of the period stated in Line 59, whichever is earlier.

Should the Buyers fail to undertake the inspection as scheduled and/or notice of acceptance of the Vessel's classification records and/or of the Vessel not be received by the Sellers as aforesaid, the Deposit together with interest earned, if any, shall be released immediately to the Buyers, whereafter this Agreement shall be null and void.

**4(a) and 4(b) are alternatives; delete whichever is not applicable. In the absence of deletions, alternative 4(a) shall apply.*

5. Time and place of delivery and notices

(a) The Vessel shall be delivered and taken over safely afloat at a safe and accessible berth or anchorage at/in *(state place/range)* in the Sellers' option.

Notice of Readiness shall not be tendered before: *(date)*

Cancelling Date (see Clauses 5(c), 6 (a)(i), 6(a) (iii) and 14): [redacted]

(b) The Sellers shall keep the Buyers well informed of the Vessel's itinerary and shall provide the Buyers with twenty (20), ten (10), five (5) and three (3) days' notice of the date the Sellers intend to tender Notice of Readiness and of the intended place of delivery.

When the Vessel is at the place of delivery and physically ready for delivery in accordance with this Agreement, the Sellers shall give the Buyers a written Notice of Readiness for delivery.

(c) If the Sellers anticipate that, notwithstanding the exercise of due diligence by them, the Vessel will not be ready for delivery by the Cancelling Date they may notify the Buyers in writing stating the date when they anticipate that the Vessel will be ready for delivery and proposing a new Cancelling Date. Upon receipt of such notification the Buyers shall have the option of either cancelling this Agreement in accordance with Clause 14 (Sellers' Default) within three (3) Banking Days of receipt of the notice or of accepting the new date as the new Cancelling Date. If the Buyers have not declared their option within three (3) Banking Days of receipt of the Sellers' notification or if the Buyers accept the new date, the date proposed in the Sellers' notification shall be deemed to be the new Cancelling Date and shall be substituted for the Cancelling Date stipulated in line 79.

If this Agreement is maintained with the new Cancelling Date all other terms and conditions hereof including those contained in Clauses 5(b) and 5(d) shall remain unaltered and in full force and effect.

(d) Cancellation, failure to cancel or acceptance of the new Cancelling Date shall be entirely without prejudice to any claim for damages the Buyers may have under Clause 14 (Sellers' Default) for the Vessel not being ready by the original Cancelling Date.

(e) Should the Vessel become an actual, constructive or compromised total loss before delivery the Deposit together with interest earned, if any, shall be released immediately to the Buyers whereafter this Agreement shall be null and void.

6. Divers Inspection / Drydocking

(a) *

(i) The Buyers shall have the option at their cost and expense to arrange for an underwater inspection by a diver approved by the Classification Society prior to the delivery of the Vessel. Such option shall be declared latest nine (9) days prior to the Vessel's intended date of readiness for delivery as notified by the Sellers pursuant to Clause 5(b) of this Agreement. The Sellers shall at their cost and expense make the Vessel available for such inspection. This inspection shall be carried out without undue delay and in the presence of a Classification Society surveyor arranged for by the Sellers and paid for by the Buyers. The Buyers' representative(s) shall have the right to be present at the diver's inspection as observer(s) only without interfering with the work or decisions of the Classification Society surveyor. The extent of the inspection and the conditions under which it is performed shall be to the satisfaction of the Classification Society. If the conditions at the place of delivery are unsuitable for such inspection, the Sellers shall at their cost and expense make the Vessel available at a suitable alternative place near to the delivery port, in which event the Cancelling Date shall be extended by the additional time required for such positioning and the subsequent re-positioning. The Sellers may not tender Notice of Readiness prior to completion of the underwater inspection.

(ii) If the rudder, propeller, bottom or other underwater parts below the deepest load line are found broken, damaged or defective so as to affect the Vessel's class, then (1) unless repairs can be carried out afloat to the satisfaction of the Classification Society, the Sellers shall arrange for the Vessel to be drydocked at their expense for inspection by the Classification Society of the Vessel's underwater parts below the deepest load line, the extent of the inspection being in accordance with the Classification Society's rules (2) such defects shall be made good by the Sellers at their cost and expense to the satisfaction of the Classification Society without condition/recommendation** and (3) the

Sellers shall pay for the underwater inspection and the Classification Society's attendance.

Notwithstanding anything to the contrary in this Agreement, if the Classification Society do not require the aforementioned defects to be rectified before the next class drydocking survey, the Sellers shall be entitled to deliver the Vessel with these defects against a deduction from the Purchase Price of the estimated direct cost (of labour and materials) of carrying out the repairs to the satisfaction of the Classification Society, whereafter the Buyers shall have no further rights whatsoever in respect of the defects and/or repairs. The estimated direct cost of the repairs shall be the average of quotes for the repair work obtained from two reputable independent shipyards at or in the vicinity of the port of delivery, one to be obtained by each of the Parties within two (2) Banking Days from the date of the imposition of the condition/recommendation, unless the Parties agree otherwise. Should either of the Parties fail to obtain such a quote within the stipulated time then the quote duly obtained by the other Party shall be the sole basis for the estimate of the direct repair costs. The Sellers may not tender Notice of Readiness prior to such estimate having been established.

(iii) If the Vessel is to be drydocked pursuant to Clause 6(a) (ii) and no suitable dry-docking facilities are available at the port of delivery, the Sellers shall take the Vessel to a port where suitable drydocking facilities are available, whether within or outside the delivery range as per Clause 5(a). Once drydocking has taken place the Sellers shall deliver the Vessel at a port within the delivery range as per Clause 5(a) which shall, for the purpose of this Clause, become the new port of delivery. In such event the Cancelling Date shall be extended by the additional time required for the drydocking and extra steaming, but limited to a maximum of fourteen (14) days.

(b) *The Sellers shall place the Vessel in drydock at the port of delivery for inspection by the Classification Society of the Vessel's underwater parts below the deepest load line, the extent of the inspection being in accordance with the Classification Society's rules. If the rudder, propeller, bottom or other underwater parts below the deepest load line are found broken, damaged or defective so as to affect the Vessel's class, such defects shall be made good at the Sellers' cost and expense to the satisfaction of the Classification Society without condition/recommendation**. In such event the Sellers are also to pay for the costs and expenses in connection with putting the Vessel in and taking her out of drydock, including the drydock dues and the Classification Society's fees. The Sellers shall also pay for these costs and expenses if parts of the tailshaft system are condemned or found defective or broken so as to affect the Vessel's class. In all other cases, the Buyers shall pay the aforesaid costs and expenses, dues and fees.

(c) If the Vessel is drydocked pursuant to Clause 6(a) (ii) or 6(b) above:

(i) The Classification Society may require survey of the tailshaft system, the extent of the survey being to the satisfaction of the Classification Society surveyor. If such survey is not required by the Classification Society, the Buyers shall have the option to require the tailshaft to be drawn and surveyed by the Classification Society, the extent of the survey being in accordance with the Classification Society's rules for tailshaft survey and consistent with the current stage of the Vessel's survey cycle. The Buyers shall declare whether they require the tailshaft to be drawn and surveyed not later than by the completion of the inspection by the Classification Society. The drawing and refitting of the tailshaft shall be arranged by the Sellers. Should any parts of the tailshaft system be condemned or found defective so as to affect the Vessel's class, those parts shall be renewed or made good at the Sellers' cost and expense to the satisfaction of the Classification Society without condition/recommendation**.

(ii) The costs and expenses relating to the survey of the tailshaft system shall be borne by the Buyers unless the Classification Society requires such survey to be carried out or if parts of the system are condemned or found defective or broken so as to affect the Vessel's class, in which case the Sellers shall pay these costs and expenses.

(iii) The Buyers' representative(s) shall have the right to be present in the drydock, as observer(s) only without interfering with the work or decisions of the Classification Society surveyor.

(iv) The Buyers shall have the right to have the underwater parts of the Vessel cleaned and painted at their risk, cost and expense without interfering with the Sellers' or the Classification Society surveyor's work, if any, and without affecting the Vessel's timely delivery. If, however, the Buyers' work in drydock is still in progress when the Sellers have completed the work which the Sellers are required to do, the additional docking time needed to complete the Buyers' work shall be for the Buyers' risk, cost and expense. In the event that the Buyers' work requires such additional time, the Sellers may upon completion of the Sellers' work tender Notice of Readiness for delivery whilst the Vessel is still in drydock and, notwithstanding Clause 5(a), the Buyers shall be obliged to take delivery in accordance with Clause 3 (Payment), whether the Vessel is in drydock or not.

**6(a) and 6(b) are alternatives; delete whichever is not applicable. In the absence of deletions, alternative 6(a) shall apply.*

***Notes or memoranda, if any, in the surveyor's report which are accepted by the Classification Society without condition/recommendation are not to be taken into account.*

7. Spares, bunkers and other items

The Sellers shall deliver the Vessel to the Buyers with everything belonging to her on board and on shore. All spare parts and spare equipment including spare tail-end shaft(s) and/or spare propeller(s)/propeller blade(s), if any, belonging to the Vessel at the time of inspection used or unused, whether on board or not shall become the Buyers' property, but spares on order are excluded. Forwarding charges, if any, shall be for the Buyers' account. The Sellers are not required to replace spare parts including spare tail-end shaft(s) and spare propeller(s)/propeller blade(s) which are taken out of spare and used as replacement prior to delivery, but the replaced items shall be the property of the Buyers. Unused stores and provisions shall be included in the sale and be taken over by the Buyers without extra payment.

Library and forms exclusively for use in the Sellers' vessel(s) and captain's, officers' and crew's personal belongings including the slop chest are excluded from the sale without compensation, as well as the following additional items: *(include list)*

Items on board which are on hire or owned by third parties, listed as follows, are excluded from the sale without compensation: *(include list)*

Items on board at the time of inspection which are on hire or owned by third parties, not listed above, shall be replaced or procured by the Sellers prior to delivery at their cost and expense.

The Buyers shall take over remaining bunkers and unused lubricating and hydraulic oils and greases in storage tanks and unopened drums and pay either:

(a) *the actual net price (excluding barging expenses) as evidenced by invoices or vouchers; or

(b) *the current net market price (excluding barging expenses) at the port and date of delivery of the Vessel or, if unavailable, at the nearest bunkering port,

for the quantities taken over.

Payment under this Clause shall be made at the same time and place and in the same currency as the Purchase Price.

"inspection" in this Clause 7, shall mean the Buyers' inspection according to Clause 4(a) or 4(b) (Inspection), if applicable. If the Vessel is taken over without inspection, the date of this Agreement shall be the relevant date.

**(a) and (b) are alternatives, delete whichever is not applicable. In the absence of deletions alternative (a) shall apply.*

8. Documentation

The place of closing: [redacted]

(a) In exchange for payment of the Purchase Price the Sellers shall provide the Buyers with the following delivery documents:

(i) Legal Bill(s) of Sale in a form recordable in the Buyers' Nominated Flag State, transferring title of the Vessel and stating that the Vessel is free from all mortgages, encumbrances and maritime liens or any other debts whatsoever, duly notarially attested and legalised or apostilled, as required by the Buyers' Nominated Flag State;

(ii) Evidence that all necessary corporate, shareholder and other action has been taken by the Sellers to authorise the execution, delivery and performance of this Agreement;

(iii) Power of Attorney of the Sellers appointing one or more representatives to act on behalf of the Sellers in the performance of this Agreement, duly notarially attested and legalised or apostilled (as appropriate);

(iv) Certificate or Transcript of Registry issued by the competent authorities of the flag state on the date of delivery evidencing the Sellers' ownership of the Vessel and that the Vessel is free from registered encumbrances and mortgages, to be faxed or e-mailed by such authority to the closing meeting with the original to be sent to the Buyers as soon as possible after delivery of the Vessel;

(v) Declaration of Class or (depending on the Classification Society) a Class Maintenance Certificate issued within three (3) Banking Days prior to delivery confirming that the Vessel is in Class free of condition/recommendation;

(vi) Certificate of Deletion of the Vessel from the Vessel's registry or other official evidence of deletion appropriate to the Vessel's registry at the time of delivery, or, in the event that the registry does not as a matter of practice issue such documentation immediately, a written undertaking by the Sellers to effect deletion from the Vessel's registry forthwith and provide a certificate or other official evidence of deletion to the Buyers promptly and latest within four (4) weeks after the Purchase Price has been paid and the Vessel has been delivered;

(vii) A copy of the Vessel's Continuous Synopsis Record certifying the date on which the Vessel ceased to be registered with the Vessel's registry, or, in the event that the registry does not as a matter of practice issue such certificate immediately, a written undertaking from the Sellers to provide the copy of this certificate promptly upon it being issued together with evidence of submission by the Sellers of a duly executed Form 2 stating the date on which the Vessel shall cease to be registered with the Vessel's registry;

(viii) Commercial Invoice for the Vessel;

(ix) Commercial Invoice(s) for bunkers, lubricating and hydraulic oils and greases;

(x) A copy of the Sellers' letter to their satellite communication provider cancelling the Vessel's communications contract which is to be sent immediately after delivery of the Vessel;

(xi) Any additional documents as may reasonably be required by the competent authorities of the Buyers' Nominated Flag State for the purpose of registering the Vessel, provided the Buyers notify the Sellers of any such documents as soon as possible after the date of this Agreement; and

(xii) The Sellers' letter of confirmation that to the best of their knowledge, the Vessel is not black listed by any nation or international organisation.

(b) At the time of delivery the Buyers shall provide the Sellers with:

(i) Evidence that all necessary corporate, shareholder and other action has been taken by the Buyers to authorise the execution, delivery and performance of this Agreement; and

(ii) Power of Attorney of the Buyers appointing one or more representatives to act on behalf of the Buyers in the performance of this Agreement, duly notarially attested and legalised or apostilled (as appropriate).

(c) If any of the documents listed in Sub-clauses (a) and (b) above are not in the English language they shall be accompanied by an English translation by an authorised translator or certified by a lawyer qualified to practice in the country of the translated language.

(d) The Parties shall to the extent possible exchange copies, drafts or samples of the documents listed in Sub-clause (a) and Sub-clause (b) above for review and comment by the other party not later than *(state number of days)*, or if left blank, nine (9) days prior to the Vessel's intended date of readiness for delivery as notified by the Sellers pursuant to Clause 5(b) of this Agreement.

(e) Concurrent with the exchange of documents in Sub-clause (a) and Sub-clause (b) above, the Sellers shall also hand to the Buyers the classification certificate(s) as well as all plans, drawings and manuals, (excluding ISM/ISPS manuals), which are on board the Vessel. Other certificates which are on board the Vessel shall also be handed over to the Buyers unless the Sellers are required to retain same, in which case the Buyers have the right to take copies.

(f) Other technical documentation which may be in the Sellers' possession shall promptly after delivery be forwarded to the Buyers at their expense, if they so request. The Sellers may keep the Vessel's log books but the Buyers have the right to take copies of same.

(g) The Parties shall sign and deliver to each other a Protocol of Delivery and Acceptance confirming the date and time of delivery of the Vessel from the Sellers to the Buyers.

9. Encumbrances

The Sellers warrant that the Vessel, at the time of delivery, is free from all charters, encumbrances, mortgages and maritime liens or any other debts whatsoever, and is not subject to Port State or other administrative detentions. The Sellers hereby undertake to indemnify the Buyers against all consequences of claims made against the Vessel which have been incurred prior to the time of delivery.

10. Taxes, fees and expenses

Any taxes, fees and expenses in connection with the purchase and registration in the Buyers' Nominated Flag State shall be for the Buyers' account, whereas similar charges in connection with the closing of the Sellers' register shall be for the Sellers' account.

11. Condition on delivery

The Vessel with everything belonging to her shall be at the Sellers' risk and expense until she is delivered to the Buyers, but subject to the terms and conditions of this Agreement she shall be delivered and taken over as she was at the time of inspection, fair wear and tear excepted.

However, the Vessel shall be delivered free of cargo and free of stowaways with her Class maintained without condition/recommendation*, free of average damage affecting the Vessel's class, and with her classification certificates and national certificates, as well as all other certificates the Vessel had at the time of inspection, valid and unextended without condition/recommendation* by the Classification Society or the relevant authorities at the time of delivery.

"inspection" in this Clause 11, shall mean the Buyers' inspection according to Clause 4(a) or 4(b) (Inspection), if applicable. If the Vessel is taken over without inspection, the date of this Agreement shall be the relevant date.

**Notes and memoranda, if any, in the surveyor's report which are accepted by the Classification Society without condition/recommendation are not to be taken into account.*

12. Name/markings
Upon delivery the Buyers undertake to change the name of the Vessel and alter funnel markings.

13. Buyers' default
Should the Deposit not be lodged in accordance with Clause 2 (Deposit), the Sellers have the right to cancel this Agreement, and they shall be entitled to claim compensation for their losses and for all expenses incurred together with interest.

Should the Purchase Price not be paid in accordance with Clause 3 (Payment), the Sellers have the right to cancel this Agreement, in which case the Deposit together with interest earned, if any, shall be released to the Sellers. If the Deposit does not cover their loss, the Sellers shall be entitled to claim further compensation for their losses and for all expenses incurred together with interest.

14. Sellers' default
Should the Sellers fail to give Notice of Readiness in accordance with Clause 5(b) or fail to be ready to validly complete a legal transfer by the Cancelling Date the Buyers shall have the option of cancelling this Agreement. If after Notice of Readiness has been given but before the Buyers have taken delivery, the Vessel ceases to be physically ready for delivery and is not made physically ready again by the Cancelling Date and new Notice of Readiness given, the Buyers shall retain their option to cancel. In the event that the Buyers elect to cancel this Agreement, the Deposit together with interest earned, if any, shall be released to them immediately.

Should the Sellers fail to give Notice of Readiness by the Cancelling Date or fail to be ready to validly complete a legal transfer as aforesaid they shall make due compensation to the Buyers for their loss and for all expenses together with interest if their failure is due to proven negligence and whether or not the Buyers cancel this Agreement.

15. Buyers' representatives
After this Agreement has been signed by the Parties and the Deposit has been lodged, the Buyers have the right to place two (2) representatives on board the Vessel at their sole risk and expense.

These representatives are on board for the purpose of familiarisation and in the capacity of observers only, and they shall not interfere in any respect with the operation of the Vessel. The Buyers and the Buyers' representatives shall sign the Sellers' P&I Club's standard letter of indemnity prior to their embarkation.

16. Law and Arbitration
(a) *This Agreement shall be governed by and construed in accordance with English law and any dispute arising out of or in connection with this Agreement shall be referred to arbitration in London in accordance with the Arbitration Act 1996 or any statutory modification or re-enactment thereof save to the extent necessary to give effect to the provisions of this Clause.

The arbitration shall be conducted in accordance with the London Maritime Arbitrators Association (LMAA) Terms current at the time when the arbitration proceedings are commenced.

The reference shall be to three arbitrators. A party wishing to refer a dispute to arbitration shall appoint its arbitrator and send notice of such appointment in writing to the other party requiring the other party to appoint its own arbitrator within fourteen (14) calendar days of that notice and stating that it will appoint its arbitrator as sole arbitrator unless the other party appoints its own arbitrator and gives notice that it has done so within the fourteen (14) days specified. If the other party does not appoint its own arbitrator and give notice that it has done so within the fourteen (14) days specified, the party referring a dispute to arbitration may, without the requirement of any further prior notice to the other party, appoint its arbitrator as sole arbitrator and shall advise the other party accordingly. The award of a sole arbitrator shall be binding on both Parties as if the sole arbitrator had been appointed by agreement.

In cases where neither the claim nor any counterclaim exceeds the sum of US$100,000 the arbitration shall be conducted in accordance with the LMAA Small Claims Procedure current at the time when the arbitration proceedings are commenced.

(b) *This Agreement shall be governed by and construed in accordance with Title 9 of the United States Code and the substantive law (not including the choice of law rules) of the State of New York and any dispute arising out of or in connection with this Agreement shall be referred to three (3) persons at New York, one to be appointed by each of the parties hereto, and the third by the two so chosen; their decision or that of any two of them shall be final, and for the purposes of enforcing any award, judgment may be entered on an award by any court of competent jurisdiction. The proceedings shall be conducted in accordance with the rules of the Society of Maritime Arbitrators, Inc.

In cases where neither the claim nor any counterclaim exceeds the sum of US$ 100,000 the arbitration shall be conducted in accordance with the Shortened Arbitration Procedure of the Society of Maritime Arbitrators, Inc.

(c) This Agreement shall be governed by and construed in accordance with the laws of *(state place)* and any dispute arising out of or in connection with this Agreement shall be referred to arbitration at *(state place)*, subject to the procedures applicable there.

**16(a), 16(b) and 16(c) are alternatives; delete whichever is not applicable. In the absence of deletions, alternative 16(a) shall apply.*

17. Notices

All notices to be provided under this Agreement shall be in writing.

Contact details for recipients of notices are as follows:

For the Buyers:

For the Sellers:

18. Entire Agreement

The written terms of this Agreement comprise the entire agreement between the Buyers and the Sellers in relation to the sale and purchase of the Vessel and supersede all previous agreements whether oral or written between the Parties in relation thereto.

Each of the Parties acknowledges that in entering into this Agreement it has not relied on and shall have no right or remedy in respect of any statement, representation, assurance or warranty (whether or not made negligently) other than as is expressly set out in this Agreement.

Any terms implied into this Agreement by any applicable statute or law are hereby excluded to the extent that such exclusion can legally be made. Nothing in this Clause shall limit or exclude any liability for fraud.

For and on behalf of the Sellers

Name: [redacted]

Title: [redacted]

For and on behalf of the Buyers

Name: [redacted]

Title: [redacted]

APPENDIX 4

SINGAPORE SHIP SALE FORM

APPENDIX 4

MEMORANDUM OF AGREEMENT

Singapore Ship Sale Form [SSF2011]

Date of Agreement:

1. The Sellers:

1(a) Guarantor (optional)*:

2. The Buyers or Nominee**:

2 (a) Guarantor (optional)*:

3. Name of the Vessel:	4. IMO No./Official No./Call Sign:
5. Type, Built Yard, Built Year & GT:	6. (a) Flag/Port of Registry:
7. Classification Society ("Class"):	(b) Bare-boat Registry (if any):

8. Purchase Price:

(i) Deposit (10 % of Purchase Price):

(a) Payee Bank:

(b) Value Date:

(ii) Balance Purchase Price (Purchase Price less Deposit): + any extras under Clause 7

(a) Payee Bank:

(iii) Place of Closing:	(iv) Daily Cost of Delayed Delivery:

9. (i) Physical Inspection (Port and Date):

(ii) Pre-Delivery Divers Inspection (Port):

10. Delivery Place (at safe anchorage or berth in):

Delivery Date (Range): Cancelling Date:

Declaration: It is hereby mutually agreed that this Agreement shall be performed according to the terms and conditions set out herein. Additional clauses, if any, shall be deemed to be fully incorporated into this Agreement.

11. Signatures - For and on behalf of:

The SELLERS: (Name/Title)	The BUYERS: (Name/Title)
GUARANTOR, if any: (Name/Title)	GUARANTOR, if any: (Name/Title)

* *This is an optional clause applicable in instances where either both parties or one of the parties requires to have a guarantor to guarantee the performance of this Agreement. The Guarantor by signing this Agreement irrevocably and unconditionally guarantees the due performance of the relevant party. In such cases, default by a party shall vest the other party with the immediate right to start a single arbitration against both the named party and its guarantor as co-respondents (in accordance with Clause 15 of this Agreement) and thereby to recover damages from the guarantor, who shall be jointly and severally liable with the defaulter.*

** *The Buyers shall have more than one right of nomination provided that the Nominee is nominated latest upon receipt of the 15 days notice to be given under Clause 5 (a) of this Agreement or by such date as may be agreed to by the Sellers and the Buyers, failing which the right to nominate shall be lost. A three-party addendum to this Agreement recording the novation in favour of the Nominee Buyers shall be entered into by the Buyers, Sellers and Nominee Buyers.*

Whereas it is hereby agreed on this day that the Sellers identified in Box 1 have agreed to sell and the Buyers identified in Box 2 have agreed to buy, the Vessel with specifications stated in Box 3, 4, 5, 6, 7, for the Purchase Price stated in Box 8, subject to the following terms and conditions:

1. Deposit

The Buyers shall pay a deposit of 10 per cent of the Purchase Price specified in Box 8 (i) as security for the fulfillment of this Agreement to the bank nominated by the Sellers in Box 8 (i) (a), with a value date no later than that specified upon in Box 8 (i) (b) of this Agreement. Notwithstanding that the amount received may be lesser due to bank remittance charges imposed during the normal course of transfer, such amount shall stand as due fulfillment of the Buyers obligation to pay the deposit and be held in a joint escrow account of both the Sellers and the Buyers, which shall be released to the Sellers as part of the Purchase Price in accordance with joint written instructions of the Sellers and the Buyers. The Sellers are to arrange the opening of the joint escrow account latest by 2 banking days prior to the Value Date. The Buyers, latest together with their remittance of the Deposit, are to arrange bank-to-bank confirmation from the remitting bank to the bank specified in Box 8 (i) (a) that the Buyers, and the remitting party if different, are a known customer of the bank and should it be required by the bank in Box 8 (i) (a), the Buyers will also arrange for the bank-to-bank confirmation to include the confirmation by the remitting bank that they know the source of funds. Both Sellers and Buyers shall comply with the anti-money laundering laws and regulations of the country in which the bank(s) specified in Box 8 are located.

Any interest earned on the deposit shall accrue to the Buyers whereas any closing fee/fees charged for holding the deposit shall be borne equally by the Sellers and the Buyers.

2. Payment

(a) The Buyers shall pay the Balance Purchase Price specified in Box 8 (ii) in full including any extras under Clause 7 free of bank/transfer charges to Sellers' nominated bank account at Sellers' bank stated in Box 8 (ii) (a) upon delivery of the Vessel. The agreed Purchase Price shall be paid for same day value within 3 full banking days, (being banking days in the place of closing and in the country of the Purchase Price currency) after the Sellers tender the written notice* of actual readiness of the Vessel for delivery in accordance with Clause 5 (b).

(b) The Buyers may delay to take delivery of the Vessel for up to a maximum of further seven (7) consecutive days paying to the Sellers the sum specified in Box 8 (iv) per day, or part thereof, as compensation for such delay provided that the Buyers have declared their intention to take late delivery prior to the expiry of the specified 3 full banking days. Any such amount due shall be paid at the time and place and in the same currency as the Purchase Price and any additional amounts due under Clause 7. If such delay exceeds seven (7) consecutive days then the Sellers shall have the right to cancel this Agreement and claim damages for their losses incurred.

**Throughout this Agreement, a written notice is to mean a registered letter, telex, tele-fax, e-mail or other modern form of written communication between the Sellers and the Buyers.*

3. Inspections*

(a) The Buyers have physically inspected the Vessel at the place and on the date specified in Box 9 (i) as well as the Classification records and have accepted the Vessel making the sale outright, subject only to the terms and conditions of this Agreement.

(b) The Sellers shall make the Vessel available for Physical Inspection as per Box 9 (i) hereof.

The Buyers shall undertake the Physical Inspection** without undue delay to the Vessel. Should the Buyers cause undue delay, the Sellers shall be compensated for the losses incurred by them. The sale shall become definite and outright, subject only to the terms and conditions of this Agreement, if the Vessel is accepted by the Buyers after the inspection and a written notice of acceptance from the Buyers is received by the Sellers within 72 hours after completion of Physical Inspection of the Vessel. If the Buyers decline the Vessel or if such notice of acceptance is not received by the Sellers within the afore-mentioned time, the deposit together with any interest earned shall be immediately released to the Buyers, whereafter this Agreement shall be null and void.

** 3 (a) and 3 (b) are alternatives; delete whichever is not applicable. In the absence of deletions, alternative 3 (a) shall apply.*

*** In the context of this Agreement, Physical Inspection of the Vessel is to mean only inspection of the Vessel physically including taking photographs without opening up of the Vessel and without cost to the Sellers. The Physical Inspection to include inspection of Vessel's Classification records, continuous synopsis record, maintenance records, deck and engine log books and available ballast spaces.*

4. Condition on Delivery

Until the Vessel is delivered and taken over by the Buyers, the Vessel and everything belonging to her shall be at Sellers' risk and expense, subject to the terms of this Agreement. The Sellers shall deliver the Vessel to the Buyers in substantially the same condition as the Vessel was at the time of inspection*, with the exception of fair wear and tear, with present Class maintained free from any outstanding Class conditions and/or recommendations**, free from damage affecting Vessel's Class, with all Class and trading certificates (both national and international) clean and valid at the time of delivery. All cargo spaces shall be clean and free of any cargo, subject only to immovable residues. If the Vessel is not in the same condition as the Vessel was at the time of inspection, the Buyers may reject the Vessel but only if the difference in condition has a substantial impact upon the Buyers' ability to trade the Vessel. Otherwise, the Buyers' remedy for differences in condition shall lie in damages.

The burden of proof as to the condition of the Vessel at the time of inspection shall be on the Buyers.

** Inspection shall mean the Buyers' inspection according to 3(a) or 3(b) as applicable. If the Vessel is taken over without inspection, the date of this Agreement shall be the relevant date.*

*** Any notes in Class reports which are accepted by Class without imposing a condition/recommendation are not to be taken into account and shall not constitute outstanding Class conditions and/ or recommendation within the meaning of this Clause.*

5. Notices and Notice of Actual Readiness

(a) Prior to the arrival of the Vessel at the Delivery Place specified in Box 10, the Sellers shall provide the Buyers with 30, 15, 7, and 3 days advance written notices to keep the Buyers advised of the estimated date and port of delivery and of the Vessel's itinerary. Following the tender of any notice, Sellers are to take reasonable steps not to hinder delivery by the date set out in the notice.

(b) Upon the arrival of the Vessel at the Delivery Place and when the Vessel is physically ready in accordance with Clause 4 for delivery and Sellers have ready all of the Sellers' documents required by Clause 8 (save for the Certificate of Ownership or equivalent, Class Maintained Certificate, Invoice for Bunkers and Lubricants and the Protocol of Delivery and Acceptance), the Sellers shall tender a written Notice of Actual Readiness of the Vessel to the Buyers. Subject only to Clause 2 (b), the Buyers shall take delivery of the Vessel within 3 full banking days after the Sellers tender such notice.

(c) However, if the Vessel becomes an actual, constructive or compromised total loss before delivery, the Sellers incur no liability under this Agreement, the Buyers are entitled to the immediate return of the deposit and any interest earned thereon and thereafter this Agreement shall be null and void.

6. Pre-Delivery Divers Inspection

Prior to delivery, the Sellers shall make the Vessel available to the Buyers for underwater inspection. The Sellers shall be responsible for ensuring that the port, anchorage or berth chosen for underwater inspection of the Vessel is suitable and permitting such inspection.

(a) The Buyers shall have the right to appoint, at their own expense, a Class approved diver to inspect the Vessel's underwater parts below the deepest load line including rudder and propeller upon the Vessel's arrival at the port specified in Box 9(ii). The Sellers shall grant Buyers sufficient daylight hours within which to conduct the said inspection and Sellers shall be obliged to ensure attendance of the Class surveyor to monitor the said inspection which may be attended by Buyers' and Sellers' representatives without interference to Class and/or the divers. However, should the Buyers fail to arrange for such inspection then they shall lose the right of such divers inspection.

(i) If any defects are found during underwater inspection including rudder and propeller that shall affect the Vessel's present Class and the repair of which Class agrees can be deferred to the Vessel's next scheduled dry-docking, the Buyers' sole remedy shall be the payment by the Sellers of the estimated cost of repair of such defects only excluding any dry-dock costs, as per the average of the quotations of two reputable repair yards independent of the Sellers and the Buyers in the delivery area, one to be selected by each party. The said average amount in respect of the cost of repair shall be deducted from the Purchase Price to be paid to the Sellers at the time of delivery of the Vessel. The costs of Class attendance and divers fees incurred for the underwater inspection shall be borne by the Buyers unless damage is found and the Class imposes a recommendation in which case both costs shall be borne by the Sellers.

(ii) If damage is found for which Class requires immediate repair, then Sellers shall repair such damage without delay prior to delivery. Should the Sellers be required to dry-dock the Vessel to repair such damage, then Clause 6 (b) shall apply.

(b) Where the Sellers are required to dry-dock the Vessel under Clause 6 (a) (ii), the Sellers shall also enable the inspection of the Vessel's bottom, rudder, propeller, tail shaft and other underwater parts by a surveyor of the Classification Society to the satisfaction of the Classification Society standards. The Sellers shall be obliged to rectify any defects found that affect the present Class of the Vessel within the agreed time or if no agreement is reached then latest within 14 days of such damage being found (and, insofar as necessary, the Cancelling Date shall be extended to allow the full agreed or 14 days' repair period), failure of which shall enable the Buyers to cancel the Agreement and recover the deposit together with interest.

(i) The Buyers shall bear the cost of the survey of the tail shaft system unless the Classification Society requires such survey to be carried out, in which case the Sellers shall bear the cost. The expenses in connection with putting the Vessel in and taking her out of dry-dock including dry-dock dues and Classification Society's fees shall be paid by the Sellers if any condition or recommendation, excluding surveyor notes is issued as a result of the survey. In all other cases Buyers shall pay the aforesaid expenses, dues and fees.

(ii) The Buyer shall have the right to place a representative for observation whilst the Vessel is in dry-dock without interfering with the Classification surveyor's work or decisions, during the Classification Society inspections.

(iii) The Buyers shall have the right to have the underwater parts of the Vessel cleaned and painted at their risk and expense without interfering with the Classification Society's or the Sellers' work, and without affecting timely delivery of the Vessel. Upon the completion of the Sellers' work, the Sellers may tender Notice of Actual Readiness of the Vessel for delivery notwithstanding the non-completion of Buyers' work and notwithstanding that the Vessel is not at the Delivery Place, upon which the Buyers shall be obliged to take delivery of the Vessel in accordance with the aforesaid notice. All dry-docking expenses incurred after such delivery except undocking expenses under Clause 6 (a) (ii) shall be borne by the Buyers.

The Classification Society shall be the only entity to determine whether any underwater damage constitutes a condition of Class, and such determination shall be final and binding on both parties.

7. Spares/ Bunkers & Others

The Sellers shall deliver the Vessel to the Buyers with everything belonging to her including all spare parts and spare equipment on board and on shore except those spare parts that are on order. Any forwarding charges shall be the Buyers' expense. However, the Sellers are not required to replace any spare parts that are taken out of spare and used as replacement in the Vessel prior to delivery unless required by Class. The radio installation and navigational equipment shall be included in the sale, along with all unused stores and provisions without extra payment. Any crockery, plates, cutlery, linen and other items bearing the Sellers' name, if taken by the Sellers, shall be replaced with unmarked items. However, the following items shall be excluded without compensation;

(a) Items that are on use exclusively in Sellers' Vessel like library, forms etc;
(b) Personal belongings including slop chest of the Vessel captain, officers and crew;
(c) Items on hire; ....
(d) Others, if any................................
....

The remaining bunkers, unused lubricants in designated storage tanks (not header tanks) and unopened drums shall be taken over by the Buyers, on payment of the net price excluding barging expenses paid by the Sellers at the date of last supply to the Vessel and evidenced by relevant invoices or vouchers; copies of which shall be made available to the Buyers at the time of delivery. Payment under this Clause shall be made in the same currency and at the same time and place as the Purchase Price.

8. Documentation

(a) As soon as practically possible after the Deposit in Box 8 (i) has been paid in accordance with Clause 1: the Sellers shall forward the Buyers scanned or photocopies of all requested Plans, Registry, Class, Safety/Trading Certificates and other documents reasonably required for preparation of Buyers registration and management documentation.

(b) At the Place of Closing specified in Box 8 (iii) at the time of delivery the Sellers and the Buyers shall sign and deliver to each other a signed Protocol of Delivery and Acceptance stating the date, time and place of delivery of the Vessel from the Sellers to the Buyers.

The Sellers shall furnish the Buyers with the following documents (unless otherwise specified all to be originals in English or with official English translations) in exchange for payment of the full Purchase Price of the Vessel:

(i) Two (2) Bills of Sale to be notarially attested and then legalized by the appropriate authorities as required by the Buyers' incoming flag specifying that the Vessel is free from all encumbrances as set out in Clause 9(a) of this Agreement. The notarial certificate is to confirm the identity of the signatory, his/her ability to bind the Sellers and the authenticity of the signature.

(ii) Resolutions of the Sellers' Board of Directors and Shareholders meetings authorizing the sale and transfer of the Vessel pursuant to this Agreement and appointing persons to represent the Sellers in connection with the sale of the Vessel and the execution of Bills of Sale and any other documents in connection with the sale of the Vessel including the issuance of a Power of Attorney. Such Resolutions are to be notarially attested and then legalized by the appropriate authorities as required by the Buyers' incoming flag. The notarial certificate is to confirm the identity of the signatory, his/her ability to bind the Sellers and the authenticity of the signature.

(iii) Power of Attorney issued by the Sellers authorizing their named representative(s) to effect the sale and transfer of the Vessel to the Buyers, pursuant to this Agreement and carry out any delivery/closing formalities including receiving the Deposit and the Balance Purchase Price or any other amounts pursuant to this Agreement. Such Power of Attorney is to be notarially attested and then legalized by the appropriate authorities as required by the Buyers' incoming flag. The notarial certificate is to confirm the identity of the signatory, his/her ability to bind the Sellers and the authenticity of the signature.

(iv) One (1) Certificate of Ownership or equivalent, dated on the date of Vessel's delivery or such other date as the parties may agree, issued by the competent authorities showing that the Vessel is registered in the ownership of the Sellers and is free from registered mortgages and encumbrances.

(v) A certified true copy of the Sellers' constitutive documents.

(vi) A current Certificate of Good Standing or Equivalent.

(vii) Three (3) Commercial Invoices setting out the main particulars of the Vessel and the Purchase Price of the Vessel.

(viii) One (1) Commercial Invoice setting out the particulars and cost of bunkers and lubricants remaining on board together with copies of the respective vouchers.

(ix) Certificate of Deletion or in lieu thereof, a Letter of Undertaking to provide the Certificate of Deletion and closed CSR from the present Ship Registry within 30 days from the date of delivery.

(x) Letter from the Sellers confirming at the time of delivery that the Vessel is free from all encumbrances, charters, mortgages, maritime liens, writs (save where security has been furnished), port state and other administrative detentions, stowaways, trading commitments and any other debts whatsoever, and undertaking to indemnify fully Buyers against all consequences of any claims against the Buyers that may arise due to claims against the Vessel originating prior to the time of the Vessel's delivery to the Buyers.

(xi) Three (3) Protocols of Delivery and Acceptance. (One each to be retained by the Buyers, the Sellers and the closing Bank)

(xii) Class Maintained Certificate dated not more than 3 working days prior to the date of delivery. However, if the Class Maintained Certificate is issued prior to the underwater inspection, then a copy of the Class report following the divers' inspection also to be included.

(xiii) The Sellers' letter of confirmation that to the best of their knowledge the Vessel:

- has not sustained grounding or any other damage to underwater parts since underwater inspection (or most recent dry-docking in case there is no divers' inspection).
- is not black listed by any government, state, country, political sub division and union.

(xiv) A copy of Sellers or Sellers manager's letter(s) to the respective authorities confirming cancellation of all Inmarsat and other communication contracts for the Vessel effective at the time of delivery.

(c) At the time of delivery of the Vessel the Buyers shall furnish the Sellers with the following documents (unless otherwise specified all to be originals in English or with official English translations):

(i) A certified true copy of the Buyers' constitutive documents.

(ii) A current Certificate of Good Standing or equivalent.

(iii) Resolution of the Board of Directors of the Buyers approving the purchase of the Vessel from the Sellers and granting a power of attorney to authorized representatives of the Buyers. Such Resolution to be notarially attested and then legalized by the appropriate authorities as required by the Sellers. The notarial certificate is to confirm the identity of the signatory, his/her ability to bind the Buyers and the authenticity of the signature.

(iv) Power of Attorney of the Buyers authorizing the Buyers' representatives or their nominees to do all such acts and things which the attorney may consider necessary or desirable on behalf of the Buyers with respect to the purchase and delivery of the Vessel but including specifically, acceptance of the Bill of Sale, signing of the Protocol of Delivery and Acceptance, release/payment of Deposit and Balance Purchase Price or any other amounts pursuant to the Agreement. Such Power of Attorney to be notarially attested and then legalized by the appropriate authorities as required by the Sellers. The notarial certificate is to confirm the identity of the signatory, his/her ability to bind the Buyers and the authenticity of the signature.

(d) As soon as possible but not later than 14 days prior to the Vessel's expected readiness for delivery, the Sellers and the Buyers shall exchange by fax or email (copies to the extent possible) or drafts of the documents listed in sub clauses (b) & (c) above for the other Party's review and comments. Copies of executed versions which are to be in strict conformity with the drafts are also to be circulated latest 3 days prior to delivery.

At the time of delivery onboard the Vessel, the Sellers shall hand over to the Buyers all Classification Certificates, Plans, Drawings, Record Books, Instruction Manuals (excluding ISM or other documents specific only to the Sellers/their Managers). All other certificates and documents onboard and pertinent to the Vessel shall also be handed over to the Buyers unless the Sellers are required to retain same, in which case photocopies are to be left onboard. All other documents which may be in the Sellers'/Sellers manager's possession shall be promptly forwarded to the Buyers after delivery. Forwarding charges, if any, to be for the Buyers' account. The Sellers may keep the Vessel's log books but the Buyers to be so advised and have the right and opportunity to take copies of same.

9. Encumbrances

a) It is a condition of this Agreement, any breach of which will entitle the Buyers to reject the Vessel, that the Vessel, at the time of delivery, is free from all encumbrances, charters, mortgages, maritime liens, writs (save where security has been furnished), port state and other administrative detentions, stowaways, trading commitments and any other debts whatsoever.

b) The Sellers hereby undertake to indemnify fully the Buyers against all consequences of any claims against the Buyers that may arise due to claims against the Vessel originating prior to the time of delivery of the Vessel to the Buyers.

10. Expenses

The Buyers shall bear all expenses including taxes and fees in connection with the purchase and registration of the Vessel under the Buyers' flag, and similarly the Sellers shall bear all expenses in connection with closing of the Sellers' Registry.

11. Vessel Name

The Buyers, upon delivery of the Vessel, shall change the name of the Vessel and alter its funnel markings.

12. Buyers Default

(a) In the event of failure by the Buyers to pay the agreed Deposit or to provide the bank-to-bank confirmation set out in Clause 1 by the Value Date, the Sellers have the right to cancel this Agreement and they shall be entitled to claim compensation for their losses and expenses (but with no automatic right to compensation in the amount of the Deposit).

(b) The failure to pay the agreed Purchase Price, and any additional amounts due under Clause 7 and Clause 2(b), within the deadline provided by Clause 2(a) or, if applicable, Clause 2(b), shall vest the Sellers with the right to cancel this Agreement and the Deposit with any interest earned thereon shall be forfeited to the Sellers (irrespective of the amount of the Sellers' actual losses and expenses). Insofar as the Deposit does not cover the Seller's actual losses and expenses, they shall be entitled to claim further compensation for those losses and expenses not so covered.

(c) The burden of proving any loss and expense shall be on the Sellers.

13. Sellers Default

(a) In the event of failure on the part of the Sellers to give Notice of Actual Readiness in accordance with Clause 5(b) latest within the Cancelling Date specified in Box 10 or, Notice of Actual Readiness for Delivery having been tendered, failure on the part of the Sellers to provide the documents required by Clause 8 and/or to deliver the Vessel as provided in Clause 9, the Buyers shall have the option to cancel this Agreement.

(b) If after Notice of Actual Readiness has been given but before the Buyers have taken delivery, the Vessel ceases to be physically ready for delivery and is not made physically ready again in accordance with Clause 4 and a Notice of Actual Readiness for Delivery re-tendered latest within the Cancelling Date in Box 10, the Buyers shall retain their right to cancel.

(c) In the event the Buyers choose to cancel this Agreement the Deposit together with interest earned shall be released to them immediately.

(d) In addition, save where the failure was caused by matters outside of the Sellers' reasonable control, the Buyers shall be entitled to claim compensation for all their losses and expenses caused by failure of the Sellers to give Notice of Actual Readiness latest within the Cancelling Date in Box 10 or, if Notice of Actual Readiness for Delivery has been tendered, caused by failure by the Sellers to provide the documents required by Clause 8 and/or to deliver the Vessel as provided in Clause 9. The burden of proving any loss and expense, additional or otherwise, shall be on the Buyers. The burden of proving that the failure was caused by matters outside of the Sellers' reasonable control shall be on the Sellers.

14. Buyers Representatives

The Buyers are entitled to place two representatives on board the Vessel after signing a letter of indemnity in Sellers usual form, for the purpose of familiarization and as observers at their expense and risk after this Agreement has been signed by both parties and the Deposit has been lodged. The Buyers' Representatives are to remain onboard until delivery under the Master's control, but are to be allowed access to the Vessel's main spaces, machinery and equipment without interference to the Vessel or her operations.

15. Arbitration & Governing Law

i)* This Agreement and any guarantee contained herein shall be governed by and construed in accordance with Singapore/English* Law and any and all disputes arising out of or in connection with this Agreement, including any question regarding its existence, validity or termination, shall be referred to and finally resolved by arbitration in Singapore in accordance with the Arbitration Rules of the Singapore Chamber of Maritime Arbitration for the time being in force at the commencement of the arbitration.

ii)* This Agreement and any guarantee contained herein shall be governed by and construed in accordance with .... Law and any and all disputes arising out of or in connection with this Agreement, including any question regarding its existence, validity or termination, shall be referred to and finally resolved by arbitration in ....in accordance with the ....Rules for the time being in force at the commencement of the arbitration.

** 15(i) and (ii) are alternatives; delete whichever is not applicable. In the absence of deletions, alternative 15 (i) and Singapore law shall apply to the exclusion of any other law. In the absence of selection by the parties as to the applicable law, seat of arbitration and arbitration rules under alternative 15 (ii); Singapore law shall apply to the exclusion of any other law, Singapore shall be the seat of arbitration and the arbitration rules of the Singapore Chamber of Maritime Arbitration shall apply.*

16. Confidentiality Clause

Both Parties agree in good faith to keep the terms and conditions of this Agreement private and confidential except as required by law. In the event the sale or details thereof become known or reported in the market neither the Sellers nor the Buyers shall have the right to withdraw from the sale or fail to fulfill all their obligations under this Agreement.

17. Entire Agreement Clause

This Agreement and any Addenda thereto contain the entire agreement between the Sellers and the Buyers relating to the transaction which is the subject of this Agreement and all negotiations, understandings and agreements whether in writing or otherwise between the Sellers and the Buyers are superseded and/or replaced by this Agreement.

APPENDIX 5

BIMCO STANDARD BAREBOAT CHARTER 'BARECON 2001'

1. Shipbroker	**BIMCO STANDARD BAREBOAT CHARTER CODE NAME: "BARECON 2001"** PART I
	2. Place and date
3. Owners/Place of business (Cl. 1)	4. Bareboat Charterers/Place of business (Cl. 1)
5. Vessel's name, call sign and flag (Cl. 1 and 3)	
6. Type of Vessel	7. GT/NT
8. When/Where built	9. Total DWT (abt.) in metric tons on summer freeboard
10. Classification Society (Cl. 3)	11. Date of last special survey by the Vessel's classification society
12. Further particulars of Vessel (also indicate minimum number of months' validity of class certificates agreed acc. to Cl. 3)	
13. Port or Place of delivery (Cl. 3)	14. Time for delivery (Cl. 4) / 15. Cancelling date (Cl. 5)
16. Port or Place of redelivery (Cl. 15)	17. No. of months' validity of trading and class certificates upon redelivery (Cl. 15)
18. Running days' notice if other than stated in Cl. 4	19. Frequency of dry-docking (Cl. 10(g))
20. Trading limits (Cl. 6)	
21. Charter period (Cl. 2)	22. Charter hire (Cl. 11)
23. New class and other safety requirements (state percentage of Vessel's insurance value acc. to Box 29)(Cl. 10(a)(ii))	
24. Rate of interest payable acc. to Cl. 11 (f) and, if applicable, acc. to PART IV	25. Currency and method of payment (Cl. 11)

"BARECON 2001" STANDARD BAREBOAT CHARTER PART I

26. Place of payment; also state beneficiary and bank account (Cl. 11)	27. Bank guarantee/bond (sum and place) (Cl. 24) (optional)
28. Mortgage(s), if any (state whether 12(a) or (b) applies; if 12(b) applies state date of Financial Instrument and name of Mortgagee(s)/Place of business) (Cl. 12)	29. Insurance (hull and machinery and war risks) (state value acc. to Cl. 13(f) or, if applicable, acc. to Cl. 14(k)) (also state if Cl. 14 applies)
30. Additional insurance cover, if any, for Owners' account limited to (Cl. 13(b) or, if applicable, Cl. 14(g))	31. Additional insurance cover, if any, for Charterers' account limited to (Cl. 13(b) or, if applicable, Cl. 14(g))
32. Latent defects (only to be filled in if period other than stated in Cl. 3)	33. Brokerage commission and to whom payable (Cl. 27)
34. Grace period (state number of clear banking days) (Cl. 28)	35. Dispute Resolution (state 30(a), 30(b) or 30(c); if 30(c) agreed Place of Arbitration must be stated (Cl. 30)
36. War cancellation (indicate countries agreed) (Cl. 26(f))	
37. Newbuilding Vessel (indicate with "yes" or "no" whether PART III applies) (optional)	38. Name and place of Builders (only to be filled in if PART III applies)
39. Vessel's Yard Building No. (only to be filled in if PART III applies)	40. Date of Building Contract (only to be filled in if PART III applies)
41. Liquidated damages and costs shall accrue to (state party acc. to Cl. 1) a) b) c)	
42. Hire/Purchase agreement (indicate with "yes" or "no" whether PART IV applies) (optional)	43. Bareboat Charter Registry (indicate with "yes" or "no" whether PART V applies) (optional)
44. Flag and Country of the Bareboat Charter Registry (only to be filled in if PART V applies)	45. Country of the Underlying Registry (only to be filled in if PART V applies)
46. Number of additional clauses covering special provisions, if agreed	

PREAMBLE - It is mutually agreed that this Contract shall be performed subject to the conditions contained in this Charter which shall include PART I and PART II. In the event of a conflict of conditions, the provisions of PART I shall prevail over those of PART II to the extent of such conflict but no further. It is further mutually agreed that PART III and/or PART IV and/or PART V shall only apply and only form part of this Charter if expressly agreed and stated in Boxes 37, 42 and 43. If PART III and/or PART IV and/or PART V apply, it is further agreed that in the event of a conflict of conditions, the provisions of PART I and PART II shall prevail over those of PART III and/or PART IV and/or PART V to the extent of such conflict but no further.

Signature (Owners)	Signature (Charterers)

PART II
"BARECON 2001" Standard Bareboat Charter

1. Definitions
In this Charter, the following terms shall have the meanings hereby assigned to them:
"The Owners" shall mean the party identified in Box 3;
"The Charterers" shall mean the party identified in Box 4;
"The Vessel" shall mean the vessel named in Box 5 and with particulars as stated in Boxes 6 to 12.
"Financial Instrument" means the mortgage, deed of covenant or other such financial security instrument as annexed to this Charter and stated in Box 28.

2. Charter Period
In consideration of the hire detailed in Box 22, the Owners have agreed to let and the Charterers have agreed to hire the Vessel for the period stated in Box 21 ("The Charter Period").

3. Delivery
(not applicable when Part III applies, as indicated in Box 37)
(a) The Owners shall before and at the time of delivery exercise due diligence to make the Vessel seaworthy And in every respect ready in hull, machinery and equipment for service under this Charter.
The Vessel shall be delivered by the Owners and taken over by the Charterers at the port or place indicated in Box 13 in such ready safe berth as the Charterers may direct.
(b) The Vessel shall be properly documented on delivery in accordance with the laws of the flag State indicated in Box 5 and the requirements of the classification society stated in Box 10. The Vessel upon delivery shall have her survey cycles up to date and trading and class certificates valid for at least the number of months agreed in Box 12.
(c) The delivery of the Vessel by the Owners and the taking over of the Vessel by the Charterers shall constitute a full performance by the Owners of all the Owners' obligations under this Clause 3, and thereafter the Charterers shall not be entitled to make or assert any claim against the Owners on account of any conditions, representations or warranties expressed or implied with respect to the Vessel but the Owners shall be liable for the cost of but not the time for repairs or renewals occasioned by latent defects in the Vessel, her machinery or appurtenances, existing at the time of delivery under this Charter, provided such defects have manifested themselves within twelve (12) months after delivery unless otherwise provided in Box 32.

4. Time for Delivery
(not applicable when Part III applies, as indicated in Box 37)
The Vessel shall not be delivered before the date indicated in Box 14 without the Charterers' consent and the Owners shall exercise due diligence to deliver the Vessel not later than the date indicated in Box 15.
Unless otherwise agreed in Box 18, the Owners shall give the Charterers not less than thirty (30) running days' preliminary and not less than fourteen (14) running days' definite notice of the date on which the Vessel is expected to be ready for delivery.
The Owners shall keep the Charterers closely advised of possible changes in the Vessel's position.

5. Cancelling
(not applicable when Part III applies, as indicated in Box 37)
(a) Should the Vessel not be delivered latest by the cancelling date indicated in Box 15, the Charterers shall have the option of cancelling this Charter by giving the Owners notice of cancellation within thirty-six (36) running hours after the cancelling date stated in Box 15, failing which this Charter shall remain in full force and effect.
(b) If it appears that the Vessel will be delayed beyond the cancelling date, the Owners may, as soon as they are in a position to state with reasonable certainty the day on which the Vessel should be ready, give notice thereof to the Charterers asking whether they will exercise their option of cancelling, and the option must then be declared within one hundred and sixty-eight (168) running hours of the receipt by the Charterers of such notice or within thirty-six (36) running hours after the cancelling date, whichever is the earlier. If the Charterers do not then exercise their option of cancelling, the seventh day after the readiness date stated in the Owners' notice shall be substituted for the cancelling date indicated in Box 15 for the purpose of this Clause 5.
(c) Cancellation under this Clause 5 shall be without prejudice to any claim the Charterers may otherwise have on the Owners under this Charter.

6. Trading Restrictions
The Vessel shall be employed in lawful trades for the carriage of suitable lawful merchandise within the trading limits indicated in Box 20.
The Charterers undertake not to employ the Vessel or suffer the Vessel to be employed otherwise than in conformity with the terms of the contracts of insurance (including any warranties expressed or implied therein) without first obtaining the consent of the insurers to such employment and complying with such requirements as to extra premium or otherwise as the insurers may prescribe.
The Charterers also undertake not to employ the Vessel or suffer her employment in any trade or business which is forbidden by the law of any country to which the Vessel may sail or is otherwise illicit or in carrying illicit or prohibited goods or in any manner whatsoever which may render her liable to condemnation, destruction, seizure or confiscation.
Notwithstanding any other provisions contained in this Charter it is agreed that nuclear fuels or radioactive products or waste are specifically excluded from the cargo permitted to be loaded or carried under this Charter. This exclusion does not apply to radio-isotopes used or intended to be used for any industrial, commercial, agricultural, medical or scientific purposes provided the Owners' prior approval has been obtained to loading thereof.

7. Surveys on Delivery and Redelivery
(not applicable when Part III applies, as indicated in Box 37)
The Owners and Charterers shall each appoint surveyors for the purpose of determining and agreeing in writing the condition of the Vessel at the time of delivery and redelivery hereunder. The Owners shall bear all expenses of the On-hire Survey including loss of time, if any, and the Charterers shall bear all expenses of the Off-hire Survey including loss of time, if any, at the daily equivalent to the rate of hire or pro rata thereof.

8. Inspection
The Owners shall have the right at any time after giving reasonable notice to the Charterers to inspect or survey the Vessel or instruct a duly authorised surveyor to carry out such survey on their behalf:-
(a) to ascertain the condition of the Vessel and satisfy themselves that the Vessel is being properly repaired and maintained. The costs and fees for such inspection or survey shall be paid by the Owners unless the Vessel is found to require repairs or maintenance in order to achieve the condition so provided;
(b) in dry-dock if the Charterers have not dry-docked Her in accordance with Clause 10(g). The costs and fees for such inspection or survey shall be paid by the Charterers; and
(c) for any other commercial reason they consider necessary (provided it does not unduly interfere with the commercial operation of the Vessel). The costs and fees for such inspection and survey shall be paid by the Owners.
All time used in respect of inspection, survey or repairs shall be for the Charterers' account and form part of the

PART II
"BARECON 2001" Standard Bareboat Charter

Charter Period.
The Charterers shall also permit the Owners to inspect the Vessel's log books whenever requested and shall whenever required by the Owners furnish them with full information regarding any casualties or other accidents or damage to the Vessel.

9. Inventories, Oil and Stores
A complete inventory of the Vessel's entire equipment, outfit including spare parts, appliances and of all consumable stores on board the Vessel shall be made by the Charterers in conjunction with the Owners on delivery and again on redelivery of the Vessel. The Charterers and the Owners, respectively, shall at the time of delivery and redelivery take over and pay for all bunkers, lubricating oil, unbroached provisions, paints, ropes and other consumable stores (excluding spare parts) in the said Vessel at the then current market prices at the ports of delivery and redelivery, respectively. The Charterers shall ensure that all spare parts listed in the inventory and used during the Charter Period are replaced at their expense prior to redelivery of the Vessel.

10. Maintenance and Operation
(a)(i)Maintenance and Repairs - During the Charter Period the Vessel shall be in the full possession and at the absolute disposal for all purposes of the Charterers and under their complete control in every respect. The Charterers shall maintain the Vessel, her machinery, boilers, appurtenances and spare parts in a good state of repair, in efficient operating condition and in accordance with good commercial maintenance practice and, except as provided for in Clause 14(l), if applicable, at their own expense they shall at all times keep the Vessel's Class fully up to date with the Classification Society indicated in Box 10 and maintain all other necessary certificates in force at all times.

(ii) New Class and Other Safety Requirements - In the event of any improvement, structural changes or new equipment becoming necessary for the continued operation of the Vessel by reason of new class requirements or by compulsory legislation costing (excluding the Charterers' loss of time) more than the percentage stated in Box 23, or if Box 23 is left blank, 5 per cent. of the Vessel's insurance value as stated in Box 29, then the extent, if any, to which the rate of hire shall be varied and the ratio in which the cost of compliance shall be shared between the parties concerned in order to achieve a reasonable distribution thereof as between the Owners and the Charterers having regard, inter alia, to the length of the period remaining under this Charter shall, in the absence of agreement, be referred to the dispute resolution method agreed in Clause 30.

(iii) Financial Security - The Charterers shall maintain financial security or responsibility in respect of third party liabilities as required by any government, including federal, state or municipal or other division or authority thereof, to enable the Vessel, without penalty or charge, lawfully to enter, remain at, or leave any port, place, territorial or contiguous waters of any country, state or municipality in performance of this Charter without any delay. This obligation shall apply whether or not such requirements have been lawfully imposed by such government or division or authority thereof.

The Charterers shall make and maintain all arrangements by bond or otherwise as may be necessary to satisfy such requirements at the Charterers' sole expense and the Charterers shall indemnify the Owners against all consequences whatsoever (including loss of time) for any failure or inability to do so.

(b) Operation of the Vessel - The Charterers shall at their own expense and by their own procurement man, victual, navigate, operate, supply, fuel and, whenever required, repair the Vessel during the Charter Period and they shall pay all charges and expenses of every kind and nature whatsoever incidental to their use and operation of the Vessel under this Charter, including annual flag State fees and any foreign general municipality and/or state taxes. The Master, officers and crew of the Vessel shall be the servants of the Charterers for all purposes whatsoever, even if for any reason appointed by the Owners.
Charterers shall comply with the regulations regarding officers and crew in force in the country of the Vessel's flag or any other applicable law.

(c) The Charterers shall keep the Owners and the mortgagee(s) advised of the intended employment, planned dry-docking and major repairs of the Vessel, as reasonably required.

(d) Flag and Name of Vessel – During the Charter Period, the Charterers shall have the liberty to paint the Vessel in their own colours, install and display their funnel insignia and fly their own house flag. The Charterers shall also have the liberty, with the Owners' consent, which shall not be unreasonably withheld, to change the flag and/or the name of the Vessel during the Charter Period. Painting and re-painting, instalment and re-instalment, registration and re-registration, if required by the Owners, shall be at the Charterers' expense and time.

(e) Changes to the Vessel – Subject to Clause 10(a)(ii), the Charterers shall make no structural changes in the Vessel or changes in the machinery, boilers, appurtenances or spare parts thereof without in each instance first securing the Owners' approval thereof. If the Owners so agree, the Charterers shall, if the Owners so require, restore the Vessel to its former condition before the termination of this Charter.

(f) Use of the Vessel's Outfit, Equipment and Appliances - The Charterers shall have the use of all outfit, equipment, and appliances on board the Vessel at the time of delivery, provided the same or their substantial equivalent shall be returned to the Owners on redelivery in the same good order and condition as when received, ordinary wear and tear excepted. The Charterers shall from time to time during the Charter Period replace such items of equipment as shall be so damaged or worn as to be unfit for use. The Charterers are to procure that all repairs to or replacement of any damaged, worn or lost parts or equipment be effected in such manner (both as regards workmanship and quality of materials) as not to diminish the value of the Vessel. The Charterers have the right to fit additional equipment at their expense and risk but the Charterers shall remove such equipment at the end of the period if requested by the Owners. Any equipment including radio equipment on hire on the Vessel at time of delivery shall be kept and maintained by the Charterers and the Charterers shall assume the obligations and liabilities of the Owners under any lease contracts in connection therewith and shall reimburse the Owners for all expenses incurred in connection therewith, also for any new equipment required in order to comply with radio regulations.

(g) Periodical Dry-Docking - The Charterers shall dry-dock the Vessel and clean and paint her underwater parts whenever the same may be necessary, but not less than once during the period stated in Box 19 or, if Box 19 has been left blank, every sixty (60) calendar months after delivery or such other period as may be required by the Classification Society or flag State.

11. Hire
(a) The Charterers shall pay hire due to the Owners punctually in accordance with the terms of this Charter

PART II
"BARECON 2001" Standard Bareboat Charter

in respect of which time shall be of the essence.

(b) The Charterers shall pay to the Owners for the hire of the Vessel a lump sum in the amount indicated in Box 22 which shall be payable not later than every thirty (30) running days in advance, the first lump sum being payable on the date and hour of the Vessel's delivery to the Charterers. Hire shall be paid continuously throughout the Charter Period.

(c) Payment of hire shall be made in cash without discount in the currency and in the manner indicated in Box 25 and at the place mentioned in Box 26.

(d) Final payment of hire, if for a period of less than thirty (30) running days, shall be calculated proportionally according to the number of days and hours remaining before redelivery and advance payment to be effected accordingly.

(e) Should the Vessel be lost or missing, hire shall cease from the date and time when she was lost or last heard of. The date upon which the Vessel is to be treated as lost or missing shall be ten (10) days after the Vessel was last reported or when the Vessel is posted as missing by Lloyd's, whichever occurs first. Any hire paid in advance to be adjusted accordingly.

(f) Any delay in payment of hire shall entitle the Owners to interest at the rate per annum as agreed in Box 24. If Box 24 has not been filled in, the three months Interbank offered rate in London (LIBOR or its successor) for the currency stated in Box 25, as quoted by the British Bankers' Association (BBA) on the date when the hire fell due, increased by 2 per cent., shall apply.

(g) Payment of interest due under sub-clause 11(f) shall be made within seven (7) running days of the date of the Owners' invoice specifying the amount payable or, in the absence of an invoice, at the time of the next hire payment date.

12. Mortgage

(only to apply if Box 28 has been appropriately filled in)

*) **(a)** The Owners warrant that they have not effected any mortgage(s) of the Vessel and that they shall not effect any mortgage(s) without the prior consent of the Charterers, which shall not be unreasonably withheld.

*) **(b)** The Vessel chartered under this Charter is financed by a mortgage according to the Financial Instrument. The Charterers undertake to comply, and provide such information and documents to enable the Owners to comply, with all such instructions or directions in regard to the employment, insurances, operation, repairs and maintenance of the Vessel as laid down in the Financial Instrument or as may be directed from time to time during the currency of the Charter by the mortgagee(s) in conformity with the Financial Instrument. The Charterers confirm that, for this purpose, they have acquainted themselves with all relevant terms, conditions and provisions of the Financial Instrument and agree to acknowledge this in writing in any form that may be required by the mortgagee(s). The Owners warrant that they have not effected any mortgage(s) other than stated in Box 28 and that they shall not agree to any amendment of the mortgage(s) referred to in Box 28 or effect any other mortgage(s) without the prior consent of the Charterers, which shall not be unreasonably withheld.

*) *(Optional, Clauses 12(a) and 12(b) are alternatives; indicate alternative agreed in Box 28).*

13. Insurance and Repairs

(a) During the Charter Period the Vessel shall be kept insured by the Charterers at their expense against hull and machinery, war and Protection and Indemnity risks (and any risks against which it is compulsory to insure for the operation of the Vessel, including maintaining financial security in accordance with sub-clause 10(a)(iii)) in such form as the Owners shall in writing approve, which approval shall not be un-reasonably withheld. Such insurances shall be arranged by the Charterers to protect the interests of both the Owners and the Charterers and the mortgagee(s) (if any), and The Charterers shall be at liberty to protect under such insurances the interests of any managers they may appoint. Insurance policies shall cover the Owners and the Charterers according to their respective interests. Subject to the provisions of the Financial Instrument, if any, and the approval of the Owners and the insurers, the Charterers shall effect all insured repairs and shall undertake settlement and reimbursement from the insurers of all costs in connection with such repairs as well as insured charges, expenses and liabilities to the extent of coverage under the insurances herein provided for.

The Charterers also to remain responsible for and to effect repairs and settlement of costs and expenses incurred thereby in respect of all other repairs not covered by the insurances and/or not exceeding any possible franchise(s) or deductibles provided for in the insurances.

All time used for repairs under the provisions of sub-clause 13(a) and for repairs of latent defects according to Clause 3(c) above, including any deviation, shall be for the Charterers' account.

(b) If the conditions of the above insurances permit additional insurance to be placed by the parties, such cover shall be limited to the amount for each party set out in Box 30 and Box 31, respectively. The Owners or the Charterers as the case may be shall immediately furnish the other party with particulars of any additional insurance effected, including copies of any cover notes or policies and the written consent of the insurers of any such required insurance in any case where the consent of such insurers is necessary.

(c) The Charterers shall upon the request of the Owners, provide information and promptly execute such documents as may be required to enable the Owners to comply with the insurance provisions of the Financial Instrument.

(d) Subject to the provisions of the Financial Instrument, if any, should the Vessel become an actual, constructive, compromised or agreed total loss under the insurances required under sub-clause 13(a), all insurance payments for such loss shall be paid to the Owners who shall distribute the moneys between the Owners and the Charterers according to their respective interests. The Charterers undertake to notify the Owners and the mortgagee(s), if any, of any occurrences in consequence of which the Vessel is likely to become a total loss as defined in this Clause.

(e) The Owners shall upon the request of the Charterers, promptly execute such documents as may be required to enable the Charterers to abandon the Vessel to insurers and claim a constructive total loss.

(f) For the purpose of insurance coverage against hull and machinery and war risks under the provisions of sub-clause 13(a), the value of the Vessel is the sum indicated in Box 29.

14. Insurance, Repairs and Classification

(Optional, only to apply if expressly agreed and stated in Box 29, in which event Clause 13 shall be considered deleted).

(a) During the Charter Period the Vessel shall be kept insured by the Owners at their expense against hull and machinery and war risks under the form of policy or policies attached hereto. The Owners and/or insurers shall not have any right of recovery or subrogation against the Charterers on account of loss of or any damage to the Vessel or her machinery or appurt-enances covered by such insurance, or on account of payments made to discharge claims against or liabilities of the Vessel or the Owners covered by such insurance. Insurance policies shall cover the Owners and the

Charterers according to their respective interests.

(b) During the Charter Period the Vessel shall be kept insured by the Charterers at their expense against Protection and Indemnity risks (and any risks against which it is compulsory to insure for the operation of the Vessel, including maintaining financial security in accordance with sub-clause 10(a)(iii)) in such form as the Owners shall in writing approve which approval shall not be unreasonably withheld.

(c) In the event that any act or negligence of the Charterers shall vitiate any of the insurance herein provided, the Charterers shall pay to the Owners all losses and indemnify the Owners against all claims and demands which would otherwise have been covered by such insurance.

(d) The Charterers shall, subject to the approval of the Owners or Owners' Underwriters, effect all insured repairs, and the Charterers shall undertake settlement of all miscellaneous expenses in connection with such repairs as well as all insured charges, expenses and liabilities, to the extent of coverage under the insurances provided for under the provisions of sub-clause 14(a). The Charterers to be secured reimbursement through the Owners' Underwriters for such expenditures upon presentation of accounts.

(e) The Charterers to remain responsible for and to effect repairs and settlement of costs and expenses incurred thereby in respect of all other repairs not covered by the insurances and/or not exceeding any possible franchise(s) or deductibles provided for in the insurances.

(f) All time used for repairs under the provisions of sub-clauses 14(d) and 14(e) and for repairs of latent defects according to Clause 3 above, including any deviation, shall be for the Charterers' account and shall form part of the Charter Period.
The Owners shall not be responsible for any expenses as are incident to the use and operation of the Vessel for such time as may be required to make such repairs.

(g) If the conditions of the above insurances permit additional insurance to be placed by the parties such cover shall be limited to the amount for each party set out in Box 30 and Box 31, respectively. The Owners or the Charterers as the case may be shall immediately furnish the other party with particulars of any additional insurance effected, including copies of any cover notes or policies and the written consent of the insurers of any such required insurance in any case where the consent of such insurers is necessary.

(h) Should the Vessel become an actual, constructive, compromised or agreed total loss under the insurances required under sub-clause 14(a), all insurance payments for such loss shall be paid to the Owners, who shall distribute the moneys between themselves and the Charterers according to their respective interests.

(i) If the Vessel becomes an actual, constructive, compromised or agreed total loss under the insurances arranged by the Owners in accordance with sub-clause 14(a), this Charter shall terminate as of the date of such loss.

(j) The Charterers shall upon the request of the Owners, promptly execute such documents as may be required to enable the Owners to abandon the Vessel to the insurers and claim a constructive total loss.

(k) For the purpose of insurance coverage against hull and machinery and war risks under the provisions of sub-clause 14(a), the value of the Vessel is the sum indicated in Box 29.

(l) Notwithstanding anything contained in sub-clause 10(a), it is agreed that under the provisions of Clause 14, if applicable, the Owners shall keep the Vessel's Class fully up to date with the Classification Society indicated in Box 10 and maintain all other necessary certificates in force at all times.

15. Redelivery

At the expiration of the Charter Period the Vessel shall be redelivered by the Charterers to the Owners at a safe and ice-free port or place as indicated in Box 16, in such ready safe berth as the Owners may direct. The Charterers shall give the Owners not less than thirty (30) running days' preliminary notice of expected date, range of ports of redelivery or port or place of redelivery and not less than fourteen (14) running days' definite notice of expected date and port or place of redelivery. Any changes thereafter in the Vessel's position shall be notified immediately to the Owners.
The Charterers warrant that they will not permit the Vessel to commence a voyage (including any preceding ballast voyage) which cannot reasonably be expected to be completed in time to allow redelivery of the Vessel within the Charter Period. Notwithstanding the above, should the Charterers fail to redeliver the Vessel within The Charter Period, the Charterers shall pay the daily equivalent to the rate of hire stated in Box 22 plus 10 per cent. or to the market rate, whichever is the higher, for the number of days by which the Charter Period is exceeded. All other terms, conditions and provisions of this Charter shall continue to apply.
Subject to the provisions of Clause 10, the Vessel shall be redelivered to the Owners in the same or as good structure, state, condition and class as that in which she was delivered, fair wear and tear not affecting class excepted.
The Vessel upon redelivery shall have her survey cycles up to date and trading and class certificates valid for at least the number of months agreed in Box 17.

16. Non-Lien

The Charterers will not suffer, nor permit to be continued, any lien or encumbrance incurred by them or their agents, which might have priority over the title and interest of the Owners in the Vessel. The Charterers further agree to fasten to the Vessel in a conspicuous place and to keep so fastened during the Charter Period a notice reading as follows:
"This Vessel is the property of (name of Owners). It is under charter to (name of Charterers) and by the terms of the Charter Party neither the Charterers nor the Master have any right, power or authority to create, incur or permit to be imposed on the Vessel any lien whatsoever."

17. Indemnity

(a) The Charterers shall indemnify the Owners against any loss, damage or expense incurred by the Owners arising out of or in relation to the operation of the Vessel by the Charterers, and against any lien of whatsoever nature arising out of an event occurring during the Charter Period. If the Vessel be arrested or otherwise detained by reason of claims or liens arising out of her operation hereunder by the Charterers, the Charterers shall at their own expense take all reasonable steps to secure that within a reasonable time the Vessel is released, including the provision of bail.
Without prejudice to the generality of the foregoing, the Charterers agree to indemnify the Owners against all consequences or liabilities arising from the Master, officers or agents signing Bills of Lading or other documents.

(b) If the Vessel be arrested or otherwise detained by reason of a claim or claims against the Owners, the Owners shall at their own expense take all reasonable steps to secure that within a reasonable time the Vessel is released, including the provision of bail.
In such circumstances the Owners shall indemnify the Charterers against any loss, damage or expense incurred by the Charterers (including hire paid under this Charter) as a direct consequence of such arrest or detention.

PART II
"BARECON 2001" Standard Bareboat Charter

18. Lien
The Owners to have a lien upon all cargoes, sub-hires and sub-freights belonging or due to the Charterers or any sub-charterers and any Bill of Lading freight for all claims under this Charter, and the Charterers to have a lien on the Vessel for all moneys paid in advance and not earned.

19. Salvage
All salvage and towage performed by the Vessel shall be for the Charterers' benefit and the cost of repairing damage occasioned thereby shall be borne by the Charterers.

20. Wreck Removal
In the event of the Vessel becoming a wreck or obstruction to navigation the Charterers shall indemnify the Owners against any sums whatsoever which the Owners shall become liable to pay and shall pay in consequence of the Vessel becoming a wreck or obstruction to navigation.

21. General Average
The Owners shall not contribute to General Average.

22. Assignment, Sub-Charter and Sale
(a) The Charterers shall not assign this Charter nor sub-charter the Vessel on a bareboat basis except with the prior consent in writing of the Owners, which shall not be unreasonably withheld, and subject to such terms and conditions as the Owners shall approve.
(b) The Owners shall not sell the Vessel during the currency of this Charter except with the prior written consent of the Charterers, which shall not be unreasonably withheld, and subject to the buyer accepting an assignment of this Charter.

23. Contracts of Carriage
*) **(a)** The Charterers are to procure that all documents issued during the Charter Period evidencing the terms and conditions agreed in respect of carriage of goods shall contain a paramount clause incorporating any legislation relating to carrier's liability for cargo compulsorily applicable in the trade; if no such legislation exists, the documents shall incorporate the Hague-Visby Rules. The documents shall also contain the New Jason Clause and the Both-to-Blame Collision Clause.
*) **(b)** The Charterers are to procure that all passenger tickets issued during the Charter Period for the carriage of passengers and their luggage under this Charter shall contain a paramount clause incorporating any legislation relating to carrier's liability for passengers and their luggage compulsorily applicable in the trade; if no such legislation exists, the passenger tickets shall incorporate the Athens Convention Relating to the Carriage of Passengers and their Luggage by Sea, 1974, and any protocol thereto.
*) *Delete as applicable.*

24. Bank Guarantee
(Optional, only to apply if Box 27 filled in)
The Charterers undertake to furnish, before delivery of the Vessel, a first class bank guarantee or bond in the sum and at the place as indicated in Box 27 as guarantee for full performance of their obligations under this Charter.

25. Requisition/Acquisition
(a) In the event of the Requisition for Hire of the Vessel by any governmental or other competent authority (hereinafter referred to as "Requisition for Hire") irrespective of the date during the Charter Period when "Requisition for Hire" may occur and irrespective of the length thereof and whether or not it be for an indefinite or a limited period of time, and irrespective of whether it may or will remain in force for the remainder of the Charter Period, this Charter shall not be deemed thereby or thereupon to be frustrated or otherwise terminated and the Charterers shall continue to pay the stipulated hire in the manner provided by this Charter until the time when the Charter would have terminated pursuant to any of the provisions hereof always provided however that in the event of "Requisition for Hire" any Requisition Hire or compensation received or receivable by the Owners shall be payable to the Charterers during the remainder of the Charter Period or the period of the "Requisition for Hire" whichever be the shorter.
(b) In the event of the Owners being deprived of their ownership in the Vessel by any Compulsory Acquisition of the Vessel or requisition for title by any governmental or other competent authority (hereinafter referred to as "Compulsory Acquisition"), then, irrespective of the date during the Charter Period when "Compulsory Acquisition" may occur, this Charter shall be deemed terminated as of the date of such "Compulsory Acquisition". In such event Charter Hire to be considered as earned and to be paid up to the date and time of such "Compulsory Acquisition".

26. War
(a) For the purpose of this Clause, the words "War Risks" shall include any war (whether actual or threatened), act of war, civil war, hostilities, revolution, rebellion, civil commotion, warlike operations, the laying of mines (whether actual or reported), acts of piracy, acts of terrorists, acts of hostility or malicious damage, blockades (whether imposed against all vessels or imposed selectively against vessels of certain flags or ownership, or against certain cargoes or crews or otherwise howsoever), by any person, body, terrorist or political group, or the Government of any state whatsoever, which may be dangerous or are likely to be or to become dangerous to the Vessel, her cargo, crew or other persons on board the Vessel.
(b) The Vessel, unless the written consent of the Owners be first obtained, shall not continue to or go through any port, place, area or zone (whether of land or sea), or any waterway or canal, where it reasonably appears that the Vessel, her cargo, crew or other persons on board the Vessel, in the reasonable judgement of the Owners, may be, or are likely to be, exposed to War Risks. Should the Vessel be within any such place as aforesaid, which only becomes dangerous, or is likely to be or to become dangerous, after her entry into it, the Owners shall have the right to require the Vessel to leave such area.
(c) The Vessel shall not load contraband cargo, or to pass through any blockade, whether such blockade be imposed on all vessels, or is imposed selectively in any way whatsoever against vessels of certain flags or ownership, or against certain cargoes or crews or otherwise howsoever, or to proceed to an area where she shall be subject, or is likely to be subject to a belligerent's right of search and/or confiscation.
(d) If the insurers of the war risks insurance, when Clause 14 is applicable, should require payment of premiums and/or calls because, pursuant to the Charterers' orders, the Vessel is within, or is due to enter and remain within, any area or areas which are specified by such insurers as being subject to additional premiums because of War Risks, then such premiums and/or calls shall be reimbursed by the Charterers to the Owners at the same time as the next payment of hire is due.
(e) The Charterers shall have the liberty:
(i) to comply with all orders, directions, recommendations or advice as to departure, arrival, routes, sailing in convoy, ports of call, stoppages, destinations, discharge of cargo, delivery, or in any other way whatsoever, which are given by the Government of the Nation under whose flag the Vessel sails, or any other Government, body or group whatsoever acting with the power to compel

PART II
"BARECON 2001" Standard Bareboat Charter

compliance with their orders or directions;

(ii) to comply with the orders, directions or recommendations of any war risks underwriters who have the authority to give the same under the terms of the war risks insurance;

(iii) to comply with the terms of any resolution of the Security Council of the United Nations, any directives of the European Community, the effective orders of any other Supranational body which has the right to issue and give the same, and with national laws aimed at enforcing the same to which the Owners are subject, and to obey the orders and directions of those who are charged with their enforcement.

(f) In the event of outbreak of war (whether there be a declaration of war or not) (i) between any two or more of the following countries: the United States of America; Russia; the United Kingdom; France; and the People's Republic of China, (ii) between any two or more of the countries stated in Box 36, both the Owners and the Charterers shall have the right to cancel this Charter, whereupon the Charterers shall redeliver the Vessel to the Owners in accordance with Clause 15, if the Vessel has cargo on board after discharge thereof at destination, or if debarred under this Clause from reaching or entering it at a near, open and safe port as directed by the Owners, or if the Vessel has no cargo on board, at the port at which the Vessel then is or if at sea at a near, open and safe port as directed by the Owners. In all cases hire shall continue to be paid in accordance with Clause 11 and except as aforesaid all other provisions of this Charter shall apply until redelivery.

27. Commission

The Owners to pay a commission at the rate indicated in Box 33 to the Brokers named in Box 33 on any hire paid under the Charter. If no rate is indicated in Box 33, the commission to be paid by the Owners shall cover the actual expenses of the Brokers and a reasonable fee for their work.

If the full hire is not paid owing to breach of the Charter by either of the parties the party liable therefor shall indemnify the Brokers against their loss of commission. Should the parties agree to cancel the Charter, the Owners shall indemnify the Brokers against any loss of commission but in such case the commission shall not exceed the brokerage on one year's hire.

28. Termination

(a) Charterers' Default

The Owners shall be entitled to withdraw the Vessel from the service of the Charterers and terminate the Charter with immediate effect by written notice to the Charterers if:

(i) the Charterers fail to pay hire in accordance with Clause 11. However, where there is a failure to make punctual payment of hire due to oversight, negligence, errors or omissions on the part of the Charterers or their bankers, the Owners shall give the Charterers written notice of the number of clear banking days stated in Box 34 (as recognised at the agreed place of payment) in which to rectify the failure, and when so rectified within such number of days following the Owners' notice, the payment shall stand as regular and punctual. Failure by the Charterers to pay hire within the number of days stated in Box 34 of their receiving the Owners' notice as provided herein, shall entitle the Owners to withdraw the Vessel from the service of the Charterers and terminate the Charter without further notice;

(ii) the Charterers fail to comply with the requirements of:
(1) Clause 6 (Trading Restrictions)
(2) Clause 13(a) (Insurance and Repairs)
provided that the Owners shall have the option, by written notice to the Charterers, to give the Charterers a specified number of days grace within which to rectify the failure without prejudice to the Owners' right to withdraw and terminate under this Clause if the Charterers fail to comply with such notice;

(iii) the Charterers fail to rectify any failure to comply with the requirements of sub-clause 10(a)(i) (Maintenance and Repairs) as soon as practically possible after the Owners have requested them in writing so to do and in any event so that the Vessel's insurance cover is not prejudiced.

(b) Owners' Default

If the Owners shall by any act or omission be in breach of their obligations under this Charter to the extent that the Charterers are deprived of the use of the Vessel and such breach continues for a period of fourteen (14) running days after written notice thereof has been given by the Charterers to the Owners, the Charterers shall be entitled to terminate this Charter with immediate effect by written notice to the Owners.

(c) Loss of Vessel

This Charter shall be deemed to be terminated if the Vessel becomes a total loss or is declared as a constructive or compromised or arranged total loss. For the purpose of this sub-clause, the Vessel shall not be deemed to be lost unless she has either become an actual total loss or agreement has been reached with her underwriters in respect of her constructive, compromised or arranged total loss or if such agreement with her underwriters is not reached it is adjudged by a competent tribunal that a constructive loss of the Vessel has occurred.

(d) Either party shall be entitled to terminate this Charter with immediate effect by written notice to the other party in the event of an order being made or resolution passed for the winding up, dissolution, liquidation or bankruptcy of the other party (otherwise than for the purpose of reconstruction or amalgamation) or if a receiver is appointed, or if it suspends payment, ceases to carry on business or makes any special arrangement or composition with its creditors.

(e) The termination of this Charter shall be without prejudice to all rights accrued due between the parties prior to the date of termination and to any claim that either party might have.

29. Repossession

In the event of the termination of this Charter in accordance with the applicable provisions of Clause 28, the Owners shall have the right to repossess the Vessel from the Charterers at her current or next port of call, or at a port or place convenient to them without hindrance or interference by the Charterers, courts or local authorities. Pending physical repossession of the Vessel in accordance with this Clause 29, the Charterers shall hold the Vessel as gratuitous bailee only to the Owners. The Owners shall arrange for an authorised representative to board the Vessel as soon as reasonably practicable following the termination of the Charter. The Vessel shall be deemed to be repossessed by the Owners from the Charterers upon the boarding of the Vessel by the Owners' representative. All arrangements and expenses relating to the settling of wages, disembarkation and repatriation of the Charterers' Master, officers and crew shall be the sole responsibility of the Charterers.

30. Dispute Resolution

*) **(a)** This Contract shall be governed by and construed in accordance with English law and any dispute arising out of or in connection with this Contract shall be referred to arbitration in London in accordance with the Arbitration Act 1996 or any statutory modification or re-enactment thereof save to the extent necessary to give effect to

the provisions of this Clause.
The arbitration shall be conducted in accordance with the London Maritime Arbitrators Association (LMAA) Terms current at the time when the arbitration proceedings are commenced.
The reference shall be to three arbitrators. A party wishing to refer a dispute to arbitration shall appoint its arbitrator and send notice of such appointment in writing to the other party requiring the other party to appoint its own arbitrator within 14 calendar days of that notice and stating that it will appoint its arbitrator as sole arbitrator unless the other party appoints its own arbitrator and gives notice that it has done so within the 14 days specified. If the other party does not appoint its own arbitrator and give notice that it has done so within the 14 days specified, the party referring a dispute to arbitration may, without the requirement of any further prior notice to the other party, appoint its arbitrator as sole arbitrator and shall advise the other party accordingly. The award of a sole arbitrator shall be binding on both parties as if he had been appointed by agreement.
Nothing herein shall prevent the parties agreeing in writing to vary these provisions to provide for the appointment of a sole arbitrator.
In cases where neither the claim nor any counterclaim exceeds the sum of US$50,000 (or such other sum as the parties may agree) the arbitration shall be conducted in accordance with the LMAA Small Claims Procedure current at the time when the arbitration proceedings are commenced.

*) **(b)** This Contract shall be governed by and construed in accordance with Title 9 of the United States Code and the Maritime Law of the United States and any dispute arising out of or in connection with this Contract shall be referred to three persons at New York, one to be appointed by each of the parties hereto, and the third by the two so chosen; their decision or that of any two of them shall be final, and for the purposes of enforcing any award, judgement may be entered on an award by any court of competent jurisdiction. The proceedings shall be conducted in accordance with the rules of the Society of Maritime Arbitrators, Inc.
In cases where neither the claim nor any counterclaim exceeds the sum of US$50,000 (or such other sum as the parties may agree) the arbitration shall be conducted in accordance with the Shortened Arbitration Procedure of the Society of Maritime Arbitrators, Inc. current at the time when the arbitration proceedings are commenced.

*) **(c)** This Contract shall be governed by and construed in accordance with the laws of the place mutually agreed by the parties and any dispute arising out of or in connection with this Contract shall be referred to arbitration at a mutually agreed place, subject to the procedures applicable there.

(d) Notwithstanding (a), (b) or (c) above, the parties may agree at any time to refer to mediation any difference and/or dispute arising out of or in connection with this Contract.
In the case of a dispute in respect of which arbitration has been commenced under (a), (b) or (c) above, the following shall apply:-

(i) Either party may at any time and from time to time elect to refer the dispute or part of the dispute to mediation by service on the other party of a written notice (the "Mediation Notice") calling on the other party to agree to mediation.

(ii) The other party shall thereupon within 14 calendar days of receipt of the Mediation Notice confirm that they agree to mediation, in which case the parties shall thereafter agree a mediator within a further 14 calendar days, failing which on the application of either party a mediator will be appointed promptly by the Arbitration Tribunal ("the Tribunal") or such person as the Tribunal may designate for that purpose. The mediation shall be conducted in such place and in accordance with such procedure and on such terms as the parties may agree or, in the event of disagreement, as may be set by the mediator.

(iii) If the other party does not agree to mediate, that fact may be brought to the attention of the Tribunal and may be taken into account by the Tribunal when allocating the costs of the arbitration as between the parties.

(iv) The mediation shall not affect the right of either party to seek such relief or take such steps as it considers necessary to protect its interest.

(v) Either party may advise the Tribunal that they have agreed to mediation. The arbitration procedure shall continue during the conduct of the mediation but the Tribunal may take the mediation timetable into account when setting the timetable for steps in the arbitration.

(vi) Unless otherwise agreed or specified in the mediation terms, each party shall bear its own costs incurred in the mediation and the parties shall share equally the mediator's costs and expenses.

(vii) The mediation process shall be without prejudice and confidential and no information or documents disclosed during it shall be revealed to the Tribunal except to the extent that they are disclosable under the law and procedure governing the arbitration.

(Note: The parties should be aware that the mediation process may not necessarily interrupt time limits.)

(e) If Box 35 in Part I is not appropriately filled in, sub-clause 30(a) of this Clause shall apply. Sub-clause 30(d) shall apply in all cases.

*) *Sub-clauses 30(a), 30(b) and 30(c) are alternatives; indicate alternative agreed in Box 35.*

31. Notices

(a) Any notice to be given by either party to the other party shall be in writing and may be sent by fax, telex, registered or recorded mail or by personal service.

(b) The address of the Parties for service of such communication shall be as stated in Boxes 3 and 4 respectively.

"BARECON 2001" Standard Bareboat Charter

OPTIONAL PART

PART III
PROVISIONS TO APPLY FOR NEWBUILDING VESSELS ONLY

(Optional, only to apply if expressly agreed and stated in Box 37)

1. **Specifications and Building Contract**

(a) The Vessel shall be constructed in accordance with the Building Contract (hereafter called "the Building Contract") as annexed to this Charter, made between the Builders and the Owners and in accordance with the specifications and plans annexed thereto, such Building Contract, specifications and plans having been counter-signed as approved by the Charterers.

(b) No change shall be made in the Building Contract or in the specifications or plans of the Vessel as approved by the Charterers as aforesaid, without the Charterers' consent.

(c) The Charterers shall have the right to send their representative to the Builders' Yard to inspect the Vessel during the course of her construction to satisfy themselves that construction is in accordance with such approved specifications and plans as referred to under sub-clause (a) of this Clause.

(d) The Vessel shall be built in accordance with the Building Contract and shall be of the description set out therein. Subject to the provisions of sub-clause 2(c)(ii) hereunder, the Charterers shall be bound to accept the Vessel from the Owners, completed and constructed in accordance with the Building Contract, on the date of delivery by the Builders. The Charterers undertake that having accepted the Vessel they will not thereafter raise any claims against the Owners in respect of the Vessel's performance or specification or defects, if any. Nevertheless, in respect of any repairs, replacements or defects which appear within the first 12 months from delivery by the Builders, the Owners shall endeavour to compel the Builders to repair, replace or remedy any defects or to recover from the Builders any expenditure incurred in carrying out such repairs, replacements or remedies. However, the Owners' liability to the Charterers shall be limited to the extent the Owners have a valid claim against the Builders under the guarantee clause of the Building Contract (a copy whereof has been supplied to the Charterers). The Charterers shall be bound to accept such sums as the Owners are reasonably able to recover under this Clause and shall make no further claim on the Owners for the difference between the amount(s) so recovered and the actual expenditure on repairs, replacement or remedying defects or for any loss of time incurred. Any liquidated damages for physical defects or deficiencies shall accrue to the account of the party stated in Box 41(a) or if not filled in shall be shared equally between the parties. The costs of pursuing a claim or claims against the Builders under this Clause (including any liability to the Builders) shall be borne by the party stated in Box 41(b) or if not filled in shall be shared equally between the parties.

2. **Time and Place of Delivery**

(a) Subject to the Vessel having completed her acceptance trials including trials of cargo equipment in accordance with the Building Contract and specifications to the satisfaction of the Charterers, the Owners shall give and the Charterers shall take delivery of the Vessel afloat when ready for delivery and properly documented at the Builders' Yard or some other safe and readily accessible dock, wharf or place as may be agreed between the parties hereto and the Builders. Under the Building Contract the Builders have estimated that the Vessel will be ready for delivery to the Owners as therein provided but the delivery date for the purpose of this Charter shall be the date when the Vessel is in fact ready for delivery by the Builders after completion of trials whether that be before or after as indicated in the Building Contract. The Charterers shall not be entitled to refuse acceptance of delivery of the Vessel and upon and after such acceptance, subject to Clause 1(d), the Charterers shall not be entitled to make any claim against the Owners in respect of any conditions, representations or warranties, whether express or implied, as to the seaworthiness of the Vessel or in respect of delay in delivery.

(b) If for any reason other than a default by the Owners under the Building Contract, the Builders become entitled under that Contract not to deliver the Vessel to the Owners, the Owners shall upon giving to the Charterers written notice of Builders becoming so entitled, be excused from giving delivery of the Vessel to the Charterers and upon receipt of such notice by the Charterers this Charter shall cease to have effect.

(c) If for any reason the Owners become entitled under the Building Contract to reject the Vessel the Owners shall, before exercising such right of rejection, consult the Charterers and thereupon

(i) if the Charterers do not wish to take delivery of the Vessel they shall inform the Owners within seven (7) running days by notice in writing and upon receipt by the Owners of such notice this Charter shall cease to have effect; or

(ii) if the Charterers wish to take delivery of the Vessel they may by notice in writing within seven (7) running days require the Owners to negotiate with the Builders as to the terms on which delivery should be taken and/or refrain from exercising their right to rejection and upon receipt of such notice the Owners shall commence such negotiations and/or take delivery of the Vessel from the Builders and deliver her to the Charterers;

(iii) in no circumstances shall the Charterers be entitled to reject the Vessel unless the Owners are able to reject the Vessel from the Builders;

(iv) if this Charter terminates under sub-clause (b) or (c) of this Clause, the Owners shall thereafter not be liable to the Charterers for any claim under or arising out of this Charter or its termination.

(d) Any liquidated damages for delay in delivery under the Building Contract and any costs incurred in pursuing a claim therefor shall accrue to the account of the party stated in Box 41(c) or if not filled in shall be shared equally between the parties.

3. **Guarantee Works**

If not otherwise agreed, the Owners authorise the Charterers to arrange for the guarantee works to be performed in accordance with the building contract terms, and hire to continue during the period of guarantee works. The Charterers have to advise the Owners about the performance to the extent the Owners may request.

4. **Name of Vessel**

The name of the Vessel shall be mutually agreed between the Owners and the Charterers and the Vessel shall be painted in the colours, display the funnel insignia and fly the house flag as required by the Charterers.

5. **Survey on Redelivery**

The Owners and the Charterers shall appoint surveyors for the purpose of determining and agreeing in writing the condition of the Vessel at the time of re-delivery. Without prejudice to Clause 15 (Part II), the Charterers shall bear all survey expenses and all other costs, if any, including the cost of docking and undocking, if required, as well as all repair costs incurred. The Charterers shall also bear all loss of time spent in connection with any docking and undocking as well as repairs, which shall be paid at the rate of hire per day or pro rata.

"BARECON 2001" Standard Bareboat Charter

OPTIONAL PART

PART IV
HIRE/PURCHASE AGREEMENT

(Optional, only to apply if expressly agreed and stated in Box 42)

On expiration of this Charter and provided the Charterers have fulfilled their obligations according to Part I and II as well as Part III, if applicable, it is agreed, that on payment of the final payment of hire as per Clause 11 the Charterers have purchased the Vessel with everything belonging to her and the Vessel is fully paid for.

In the following paragraphs the Owners are referred to as the Sellers and the Charterers as the Buyers.

The Vessel shall be delivered by the Sellers and taken over by the Buyers on expiration of the Charter.

The Sellers guarantee that the Vessel, at the time of delivery, is free from all encumbrances and maritime liens or any debts whatsoever other than those arising from anything done or not done by the Buyers or any existing mortgage agreed not to be paid off by the time of delivery. Should any claims, which have been incurred prior to the time of delivery be made against the Vessel, the Sellers hereby undertake to indemnify the Buyers against all consequences of such claims to the extent it can be proved that the Sellers are responsible for such claims. Any taxes, notarial, consular and other charges and expenses connected with the purchase and registration under Buyers' flag, shall be for Buyers' account. Any taxes, consular and other charges and expenses connected with closing of the Sellers' register, shall be for Sellers' account.

In exchange for payment of the last month's hire instalment the Sellers shall furnish the Buyers with a Bill of Sale duly attested and legalized, together with a certificate setting out the registered encumbrances, if any. On delivery of the Vessel the Sellers shall provide for deletion of the Vessel from the Ship's Register and deliver a certificate of deletion to the Buyers.
The Sellers shall, at the time of delivery, hand to the Buyers all classification certificates (for hull, engines, anchors, chains, etc.), as well as all plans which may be in Sellers' possession.

The Wireless Installation and Nautical Instruments, unless on hire, shall be included in the sale without any extra payment.

The Vessel with everything belonging to her shall be at Sellers' risk and expense until she is delivered to the Buyers, subject to the conditions of this Contract and the Vessel with everything belonging to her shall be delivered and taken over as she is at the time of delivery, after which the Sellers shall have no responsibility for possible faults or deficiencies of any description.

The Buyers undertake to pay for the repatriation of the Master, officers and other personnel if appointed by the Sellers to the port where the Vessel entered the Bareboat Charter as per Clause 3 (Part II) or to pay the equivalent cost for their journey to any other place.

"BARECON 2001" Standard Bareboat Charter

OPTIONAL PART

PART V
PROVISIONS TO APPLY FOR VESSELS REGISTERED IN A BAREBOAT CHARTER REGISTRY

(Optional, only to apply if expressly agreed and stated in Box 43)

1. Definitions

For the purpose of this PART V, the following terms shall have the meanings hereby assigned to them:

"The Bareboat Charter Registry" shall mean the registry of the State whose flag the Vessel will fly and in which the Charterers are registered as the bareboat charterers during the period of the Bareboat Charter.

"The Underlying Registry" shall mean the registry of the state in which the Owners of the Vessel are registered as Owners and to which jurisdiction and control of the Vessel will revert upon termination of the Bareboat Charter Registration.

2. Mortgage

The Vessel chartered under this Charter is financed by a mortgage and the provisions of Clause 12(b) (Part II) shall apply.

3. Termination of Charter by Default

If the Vessel chartered under this Charter is registered in a Bareboat Charter Registry as stated in Box 44, and if the Owners shall default in the payment of any amounts due under the mortgage(s) specified in Box 28, the Charterers shall, if so required by the mortgagee, direct the Owners to re-register the Vessel in the Underlying Registry as shown in Box 45.

In the event of the Vessel being deleted from the Bareboat Charter Registry as stated in Box 44, due to a default by the Owners in the payment of any amounts due under the mortgage(s), the Charterers shall have the right to terminate this Charter forthwith and without prejudice to any other claim they may have against the Owners under this Charter.

APPENDIX 6

INSTITUTE CLAUSES FOR BUILDERS' RISKS 1988

INSTITUTE CLAUSES FOR BUILDERS' RISKS
(This insurance is subject to English law and practice)

VESSEL {Response} **Contract or Yard No.** {Response}

BUILDERS' {Response}

BUILDERS' YARDS {Response}

SUBJECT OF INSURANCE

(Where more than one part of the subject-matter insured is described in Section I(A), Section I(B) or Section II below, then the respective wording of Section I(A), Section I(B) or Section II shall be applied to each part separately.)

SECTION I.

Provisional Period {Response} from {Response}

but this insurance to terminate upon delivery to Owners if prior to expiry of Provisional Period.

A. HULL and MACHINERY etc. under construction at the yard or other premises of the Builders.

Description	Contract or Yard No.	Provisionally valued at	To be built at/by
{Response}	{Response}	{Response}	{Response}

The subject-matter of this sub-section A is covered whilst at Builders' Yard and at Builders' premises elsewhere within the port or place of construction at which the Builders' Yard is situated and whilst in transit between such locations. The Underwriters' liability in respect of each item of this sub-section A which is at such locations shall attach from the time:-

(i) of inception of this Section I if such item has already been allocated to the vessel;

(ii) of delivery to Builders of such item (if allocated) when delivered after inception of this Section I;

(iii) of allocation by Builders if allocated after inception of this Section I.

B. MACHINERY etc. insured hereon whilst under construction by Sub-Contractors.

Description	Contract or Yard No.	Provisionally valued at	To be built at/by
{Response}	{Response}	{Response}	{Response}

The subject-matter of this sub-section B is covered whilst at Sub-Contractors' works and at Sub-Contractors' premises elsewhere within the port or place of construction at which the Sub-Contractors' works are situated and whilst in transit between such locations.

The Underwriters' liability in respect of each item of this sub-section B which is at such locations shall attach from the time:-

(i) of inception of this Section I if such item has already been allocated to the vessel;

(ii) of delivery to the Sub-Contractors of such item (if allocated) when delivered after inception of this Section I;

(iii) of allocation by the Sub-Contractors if allocated after inception of this Section I.

The subject-matter of this sub-section B is also covered whilst:-

(a) in transit to Builders if the transit is within the port or place of construction at which the Builders' Yard is situated;

(b) at Builders' Yard and at Builders' premises elsewhere within the port or place of construction at which the Builders' Yard is situated and whilst in transit between such locations.

SECTION II.

Provisional Period {Response} from {Response}

but this insurance to terminate upon delivery to Owners if prior to expiry of Provisional Period.

MACHINERY etc. insured hereon from delivery to Builders.

Description	Contract or Yard No.	Provisionally valued at	To be built at/by
{Response}	{Response}	{Response}	{Response}

The subject-matter of this Section II is covered whilst at Builders' Yard and at Builders' premises elsewhere within the port or place of construction at which the Builders' Yard is situated and whilst in transit between such locations. The Underwriters' liability in respect of each item of this Section II shall attach from the time of delivery to Builders.

1. INSURED VALUE

1.1 Whereas the value stated herein is provisional, it is agreed that the final contract price, or the total building cost plus {Response}% whichever is the greater, of the subject-matter of this insurance shall be the insured value.

1.2 Should the insured value, determined as above,

1.2.1 exceed the provisional value stated herein, the Assured agree to declare to the Underwriters hereon the amount of such excess and to pay premium thereon at the full policy rates, and the Underwriters agree to accept their proportionate shares of the increase,

or

1.2.2 be less than the provisional value stated herein, the sum insured by this insurance shall be reduced proportionately and the Underwriters agree to return premium at the full policy rates on the amounts by which their respective lines are reduced.

1.3 Nevertheless, should the insured value exceed 125% of the provisional value, then the limits of indemnity under this insurance shall be 125% of the provisional value, any one accident or series of accidents arising out of the same event.

1.4 Notwithstanding the above it is understood and agreed that any variation of the value for insurance on account of a material alteration in the plans or fittings of the vessel or a change in type from that originally contemplated does not come within the scope of this clause and such a variation requires the specific agreement of the Underwriters.

2. TRANSIT

Held covered at a premium to be arranged for transit not provided for in Section I or II above.

3. DELAYED DELIVERY

Held covered at a premium to be arranged in the event of delivery to Owners being delayed beyond the provisional period(s) mentioned above, but in no case shall any additional period of cover extend beyond 30 days from completion of Builders' Trials.

4. DEVIATION OR CHANGE OF VOYAGE

Held covered in case of deviation or change of voyage, provided notice be given to the Underwriters immediately after receipt of advices and any amended terms of cover and any additional premium required by them be agreed.

5. PERILS

5.1 SUBJECT ALWAYS TO ITS TERMS, CONDITIONS AND EXCLUSIONS this insurance is against all risks of loss of or damage to the subject-matter insured caused and discovered during the period of this insurance including the cost of repairing replacing or renewing any defective part condemned solely in consequence of the discovery therein during the period of this insurance of a latent defect. In no case shall this insurance cover the cost of renewing faulty welds.

5.2 In case of failure of launch, the Underwriters to bear all subsequent expenses incurred in completing launch.

6. EARTHQUAKE AND VOLCANIC ERUPTION EXCLUSION

In no case shall this insurance cover loss damage liability or expense caused by earthquake or volcanic eruption. This exclusion applies to all claims including claims under Clauses 13, 17, 19 and 20.

7. POLLUTION HAZARD

This insurance covers loss of or damage to the vessel caused by any governmental authority acting under the powers vested in it to prevent or mitigate a pollution hazard, or threat thereof, resulting directly from damage to the vessel for which the Underwriters are liable under this insurance, provided such act of governmental authority has not resulted from want of due diligence by the Assured, the Owners, or Managers of the vessel or any of them to prevent or mitigate such hazard or threat. Master, Officers, Crew or Pilots not to be considered Owners within the meaning of this Clause 7 should they hold shares in the vessel.

8. FAULTY DESIGN

Notwithstanding anything to the contrary which may be contained in the policy or the clauses attached thereto, this insurance includes loss of or damage to the subject-matter insured caused and discovered during the period of this insurance arising from faulty design of any part or parts thereof but in no case shall this insurance extend to cover the cost or expense of repairing, modifying, replacing or renewing such part or parts, nor any cost or expense incurred by reason of betterment or alteration in design.

9. NAVIGATION

9.1 With leave to proceed to and from any wet or dry docks, harbours, ways, cradles and pontoons within the port or place of construction and to proceed under own power, loaded or in ballast, as often as required, for fitting out, docking, trials or delivery, within a distance by water of 250 nautical miles of the port or place of construction, or held covered at a premium to be arranged in the event of such distance being exceeded.

9.2 Any movement of the vessel in tow outside the port or place of construction held covered at a premium to be arranged, provided previous notice be given to the Underwriters.

10. DEDUCTIBLE

10.1 No claim arising from a peril insured against shall be payable under this insurance unless the aggregate of all such claims arising out of each separate accident or occurrence (including claims under Clauses 13, 17, 19 and 20) exceeds {Response} in which case this sum shall be deducted. Nevertheless the expense of sighting the bottom after stranding, if reasonably incurred specially for that purpose, shall be paid even if no damage be found. This Clause 10.1 shall not apply to a claim for Total or Constructive Total Loss of the vessel or, in the event of such a claim, to any associated claim under Clause 20 arising from the same accident or occurrence.

10.2 Claims for damage by heavy weather occurring during a single sea passage between two successive ports shall be treated as being due to one accident. In the case of such heavy weather extending over a period not wholly covered by this insurance the deductible to be applied to the claim recoverable hereunder shall be the proportion of the above deductible that the number of days of such heavy weather falling within the period of this insurance bears to the number of days of heavy weather during the single sea passage.

The expression "heavy weather" in this Clause 10.2 shall be deemed to include contact with floating ice.

10.3 Excluding any interest comprised therein, recoveries against any claim which is subject to the above deductible shall be credited to the Underwriters in full to the extent of the sum by which the aggregate of the claim unreduced by any recoveries exceeds the above deductible.

10.4 Interest comprised in recoveries shall be apportioned between the Assured and the Underwriters, taking into account the sums paid by the Underwriters and the dates when such payments were made, notwithstanding that by the addition of interest the Underwriters may receive a larger sum than they have paid.

11. UNREPAIRED DAMAGE

11.1 The measure of indemnity in respect of claims for unrepaired damage shall be the reasonable depreciation in the market value of the vessel at the time this insurance terminates arising from such unrepaired damage, but not exceeding the reasonable cost of repairs.

11.2 In no case shall the Underwriters be liable for unrepaired damage in the event of a subsequent Total Loss (whether or not covered under this insurance) sustained during the period covered by this insurance or any extension thereof.

11.3 The Underwriters shall not be liable in respect of unrepaired damage for more than the insured value at the time this insurance terminates.

12. CONSTRUCTIVE TOTAL LOSS

12.1 In ascertaining whether the subject-matter insured is a Constructive Total Loss, the insured value shall be taken as the repaired value and nothing in respect of the damaged or break-up value shall be taken into account.

12.2 No claim for Constructive Total Loss based upon the cost of recovery and/or repair shall be recoverable hereunder unless such cost would exceed the insured value. In making this determination, only the cost relating to a single accident or sequence of damages arising from the same accident shall be taken into account.

13. GENERAL AVERAGE AND SALVAGE

13.1 This insurance covers the vessel's proportion of salvage, salvage charges and/or general average, reduced in respect of any under-insurance, but in case of general average sacrifice of the vessel the Assured may recover in respect of the whole loss without first enforcing their right of contribution from other parties.

13.2 Adjustment to be according to the law and practice obtaining at the place where the adventure ends, as if the contract of affreightment contained no special terms upon the subject; but where the contract of affreightment so provides the adjustment shall be according to the York-Antwerp Rules.

13.3 When the vessel sails in ballast, not under charter, the provisions of the York-Antwerp Rules, 1974 (excluding Rules XX and XXI) shall be applicable, and the voyage for this purpose shall be deemed to continue from the port or place of departure until the arrival of the vessel at the first port or place thereafter other than a port or place of refuge or a port or place of call for bunkering only. If at any such intermediate port or place there is an abandonment of the adventure originally contemplated the voyage shall thereupon be deemed to be terminated.

13.4 No claim under this Clause 13 shall in any case be allowed where the loss was not incurred to avoid or in connection with the avoidance of a peril insured against.

14. NOTICE OF CLAIM

In the event of loss damage liability or expense which may result in a claim under this insurance, prompt notice shall be given to the Underwriters prior to repair and, if the subject-matter is under construction abroad, to the nearest Lloyd's Agent so that a surveyor may be appointed to represent the Underwriters should they so desire.

15. CHANGE OF INTEREST

Any change of interest in the subject-matter insured shall not affect the validity of this insurance.

16. ASSIGNMENT

No assignment of or interest in this insurance or in any moneys which may be or become payable thereunder is to be binding on or recognised by the Underwriters unless a dated notice of such assignment or interest signed by the Assured, and by the assignor in the case of subsequent assignment, is endorsed on the policy and the policy with such endorsement is produced before payment of any claim or return of premium thereunder.

17. COLLISION LIABILITY

17.1 The Underwriters agree to indemnify the Assured for any sum or sums paid by the Assured to any other person or persons by reason of the Assured becoming legally liable by way of damages for

17.1.1 loss of or damage to any other vessel or property on any other vessel

17.1.2 delay to or loss of use of any such other vessel or property thereon

17.1.3 general average of, salvage of, or salvage under contract of, any such other vessel or property thereon,

where such payment by the Assured is in consequence of the vessel hereby insured coming into collision with any other vessel.

17.2 The indemnity provided by this Clause 17 shall be in addition to the indemnity provided by the other terms and conditions of this insurance and shall be subject to the following provisions:

17.2.1 Where the insured vessel is in collision with another vessel and both vessels are to blame then, unless the liability of one or both vessels becomes limited by law, the indemnity under this Clause 17 shall be calculated on the principle of cross-liabilities as if the respective Owners had been compelled to pay to each other such proportion of each other's damages as may have been properly allowed in ascertaining the balance or sum payable by or to the Assured in consequence of the collision.

17.2.2 In no case shall the Underwriters' total liability under clause 17.1 and 17.2 exceed their proportionate part of the insured value of the vessel hereby insured in respect of any one such collision.

17.3 The Underwriters will also pay the legal costs incurred by the Assured or which the Assured may be compelled to pay in contesting liability or taking proceedings to limit liability, with the prior written consent of the Underwriters.

EXCLUSIONS

17.4 Provided always that this Clause 17 shall in no case extend to any sum which the Assured shall pay for or in respect of

17.4.1 removal or disposal of obstructions, wrecks, cargoes or any other thing whatsoever

17.4.2 any real or personal property, or thing whatsoever except other vessels or property on other vessels

17.4.3 the cargo or other property on, or the engagements of, the insured vessel

17.4.4 loss of life, personal injury or illness

17.4.5 pollution or contamination of any real or personal property or thing whatsoever (except other vessels with which the insured vessel is in collision or property on such other vessels).

18. SISTERSHIP

Should the vessel hereby insured come into collision with or receive salvage services from another vessel belonging wholly or in part to the same Owners or under the same management, the Assured shall have the same rights under this insurance as they would have were the other vessel entirely the property of Owners not interested in the vessel hereby insured; but in such cases the liability for the collision or the amount payable for the services rendered shall be referred to a sole arbitrator to be agreed upon between the Underwriters and the Assured.

19. PROTECTION AND INDEMNITY

19.1 The Underwriters agree to indemnify the Assured for any sum or sums paid by the Assured to any other person or persons by reason of the Assured becoming legally liable, as Owner of the vessel, for any claim, demand, damages and/or expenses, where such liability is in consequence of any of the following matters or things and arises from an accident or occurrence during the period of this insurance:

19.1.1 loss of or damage to any fixed or movable object or property or other thing or interest whatsoever, other than the vessel, arising from any cause whatsoever in so far as such loss or damage is not covered by Clause 17

19.1.2 any attempted or actual raising, removal or destruction of any fixed or moveable object or property or other thing, including the wreck of the vessel, or any neglect or failure to raise, remove, or destroy the same

19.1.3 liability assumed by the Assured under contracts of customary towage for the purpose of entering or leaving port or manoeuvring within the port

19.1.4 loss of life, personal injury, illness or payments made for life salvage.

19.2 The Underwriters agree to indemnify the Assured for any of the following arising from an accident or occurrence during the period of this insurance:

19.2.1 the additional cost of fuel, insurance, wages, stores, provisions and port charges reasonably incurred solely for the purpose of landing from the vessel sick or injured persons or stowaways, refugees, or persons saved at sea

19.2.2 additional expenses brought about by the outbreak of infectious disease on board the vessel or ashore

19.2.3 fines imposed on the vessel, on the Assured, or on any Master Officer crew member or agent of the vessel who is reimbursed by the Assured, for any act or neglect or breach of any statute or regulation relating to the operation of the vessel, provided that the Underwriters shall not be liable to indemnify the Assured for any fines which result from any act neglect failure or default of the Assured their agents or servants other than Master Officer or crew member

19.2.4 the expenses of the removal of the wreck of the vessel from any place owned, leased or occupied by the Assured

19.2.5 legal costs incurred by the Assured, or which the Assured may be compelled to pay, in avoiding, minimising or contesting liability with the prior written consent of the Underwriters.

EXCLUSIONS

19.3 Notwithstanding the provisions of Clauses 19.1 and 19.2 this Clause 19 does not cover any liability cost or expense arising in respect of:

19.3.1 any direct or indirect payment of the Assured under workmen's compensation or employers' liability acts and any other statutory or common law, general maritime law or other liability whatsoever in respect of accidents to or illness of workmen or any other persons employed in any capacity whatsoever by the Assured or others in on or about or in connection with the vessel or her cargo materials or repairs

19.3.2 liability assumed by the Assured under agreement expressed or implied in respect of death or illness of or injury to any person employed under a contract of service or apprenticeship by the other party to such agreement

19.3.3 punitive or exemplary damages, however described

19.3.4 cargo or other property carried, to be carried or which has been carried on board the vessel but this Clause 19.3.4 shall not exclude any claim in respect of the extra cost of removing cargo from the wreck of the vessel

19.3.5 loss of or damage to property, owned by builders or repairers or for which they are responsible, which is on board the vessel

19.3.6 liability arising under a contract or indemnity in respect of containers, equipment, fuel or other property on board the vessel and which is owned or leased by the Assured

19.3.7 cash, negotiable instruments, precious metals or stones, valuables or objects of a rare or precious nature, belonging to persons on board the vessel, or non-essential personal effects of any Master, Officer or crew member

19.3.8 fuel, insurance, wages, stores, provisions and port charges arising from delay to the vessel while awaiting a substitute for any Master, Officer or crew member

19.3.9 fines or penalties arising from overloading or illegal fishing

19.3.10 pollution or contamination of any real or personal property or thing whatsoever.

19.4 The indemnity provided by this Clause 19 shall be in addition to the indemnity provided by the other terms and conditions of this insurance.

19.5 Where the Assured or the Underwriters may or could have limited their liability the indemnity under this Clause 19 in respect such liability shall not exceed Underwriters' proportionate part of the amount of such limitation.

19.6 In no case shall the Underwriters' liability under this Clause 19 in respect of each separate accident or occurrence or series of accidents arising out of the same event, exceed their proportionate part of the insured value of the vessel.

19.7 PROVIDED ALWAYS THAT

19.7.1 prompt notice must be given to the Underwriters of every casualty event or claim upon the Assured which may give rise to claim under this Clause 19 and of every event or matter which may cause the Assured to incur liability costs or expense for which he may be insured under this Clause 19

19.7.2 the Assured shall not admit liability for or settle any claim for which he may be insured under this Clause 19 without the prior written consent of the Underwriters.

20. DUTY OF ASSURED (SUE AND LABOUR)

20.1 In case of any loss or misfortune it is the duty of the Assured and their servants and agents to take such measures as may be reasonable for the purpose of averting or minimising a loss which would be recoverable under this insurance.

20.2 Subject to the provisions below and to Clause 10 the Underwriters will contribute to charges properly and reasonably incurred by the Assured their servants or agents for such measures. General average, salvage charges (except as provided for in Clause 20.4) collision defence or attack costs and costs incurred by the Assured in avoiding, minimising or contesting liability covered by Clause 19 are not recoverable under this Clause 20.

20.3 Measures taken by the Assured or the Underwriters with the object of saving, protecting or recovering the subject-matter insured shall not be considered as a waiver or acceptance of abandonment or otherwise prejudice the rights of either party.

20.4 When a claim for Total Loss of the subject-matter insured is admitted under this insurance and expenses have been reasonably incurred in saving or attempting to save the subject-matter insured and other property and there are no proceeds, or the expenses exceed the proceeds, then this insurance shall bear its pro rata share of such proportion of the expenses, or of the expenses in excess of the proceeds, as the case may be, as may reasonably be regarded as having been incurred in respect of the subject-matter insured.

20.5 The sum recoverable under this Clause 20 shall be in addition to the loss otherwise recoverable under this insurance but shall in no circumstances exceed the amount insured under this insurance in respect of the vessel.

The following clauses shall be paramount and shall override anything contained in this insurance inconsistent therewith.

21. WAR EXCLUSION

In no case shall this insurance cover loss damage liability or expense caused by

21.1 war civil war revolution rebellion insurrection, or civil strife arising therefrom, or any hostile act by or against a belligerent power

21.2 capture seizure arrest restraint or detainment (barratry and piracy excepted), and the consequences thereof or any attempt thereat

21.3 derelict mines torpedoes bombs or other derelict weapons of war.

22. STRIKES EXCLUSION

In no case shall this insurance cover loss damage liability or expense caused by

22.1 strikers, locked-out workmen, or persons taking part in labour disturbances, riots or civil commotions

22.2 any terrorist or any person acting from a political motive.

23. MALICIOUS ACTS EXCLUSION

In no case shall this insurance cover loss damage liability or expense from

23.1 the detonation of an explosive

23.2 any weapon of war

and caused by any person, acting maliciously or from a political motive.

24. NUCLEAR EXCLUSION

In no case shall this insurance cover loss damage liability or expense directly or indirectly caused by or contributed to by or arising from

24.1 ionising radiations from or contamination by radioactivity from any nuclear fuel or from any nuclear waste from the combustion of nuclear fuel

24.2 the radioactive, toxic, explosive or other hazardous properties of any explosive nuclear assembly or nuclear component thereof

24.3 any weapon of war employing atomic or nuclear fission and/or fusion or other like reaction or radioactive force or matter.

1/6/88

APPENDIX 7

LONDON MARINE CONSTRUCTION ALL RISKS WORDING 2007

Reproduced with the kind permission of the International Underwriting Association and Lloyd's Market Association.

LONDON MARINE CONSTRUCTION ALL RISKS WORDING (01/09/07)

SCHEDULE

Note: For items marked * the default provision in the relevant clause shall apply in the absence of any other figure agreed in this Schedule.		Clause number [Default provision *]
Assured: Name and address of Builder:	{Missing}	Clause 56.5
Name and address of Yard:	{Missing}	Clause 56.23
Name and address of Buyer:	{Missing}	Clause 56.6
Vessel: Hull number / project name:	{Missing}	Clause 56.21
Period: Commencement date:	{Missing}	Clause 43.2.2 or 56.16.1 as applicable
Termination date:	{Missing}	Clause 43.2.2 or 56.16.2 as applicable
Estimated Launch Date:	{Missing}	Clause 56.13
Sum Insured (insured value):	{Missing}	Clause 43.2.5 or 56.20 as applicable
Escalation percentage *:	{Missing}	Clause 44.2 [110% of the Sum Insured]

Going-in Value (Pre-existing Vessel) - where applicable:	{Missing}	Clause 43.2.1
Limit of liability: Collision liability legal costs *:	{Missing}	Clauses 6.3 and 45.1.3 [25% of the Sum Insured]
Material (not forming part of the Works Value – Actual total loss, Abandonment) *:	{Missing}	Clauses 29.2, 31.4 and 45.1.6 [10% of the Sum Insured]
Trials *:	{Missing}	Clause 45.1.7 [10% of the Sum Insured]
Failure to launch *:	{Missing}	Clause 45.1.8 [10% of the Sum Insured]

1

Allocated Series Cost *:	{Missing}	Clause 56.3 [None]
Deductibles: Section A - Construction	{Missing}	Clause 5
Section B - Liability	{Missing}	Clause 8
Section C - War	{Missing}	Clause 14
Section D - Strikes etc.	{Missing}	Clause 19
Profit percentage *:	{Missing}	Clause 56.17 [10% of the costs]
Navigation distance *:	{Missing}	Clause 49.1.2 [250 nautical miles]
Optional buybacks: Defects:	Insured / Not insured (delete as applicable)	Clause 57 [Not insured]
Profit on partial loss:	Insured / Not insured (delete as applicable)	Clause 58 [Not insured]
Expediting Expenses: Limit of liability *: Trials amendment	Insured / Not insured (delete as applicable) {Missing} Insured / Not insured (delete as applicable)	Clauses 45.1.9 and 59 [Not insured] [10% of the Sum Insured] Clause 60 [Not insured]

Extended cover for Items *:	{Missing}		Clauses 50.1 and 50.3 [None]
Premium: Section C Section D	{Missing} {Missing} {Missing}		 Clause 15.3 Clause 20.2
Premium payments: First instalment *: Subsequent instalments:	Amount {Missing} {Missing}	Due date {Missing} {Missing}	Clause 47.1 [within 45 days] [on due date]
Additional premiums to be paid within *:	{Missing}		Clause 47.2 [60 days]

2

INDEX

3

4

1. General

Application

1.1 This Clause 1 and Parts I to IV shall apply to this insurance. Clauses 57 to 60 of Part V shall only apply where and to the extent that the Underwriters have expressly so agreed in the Schedule.

1.2 Clause 43 (Pre-existing Vessel) shall apply where the Vessel is a Pre-existing Vessel that is to undergo conversion, repair, lengthening or any other work, as more particularly described in Clause 43.

1.3 This insurance is not evidence of financial responsibility under the Oil Pollution Act 1990 or any similar Federal or State laws and may not be shown or tendered to the United States Coast Guard or any Federal or State agency as evidence of financial responsibility or evidence of insurance. The Underwriters do not agree to be guarantors.

Law and Jurisdiction

1.4 This insurance is subject to English law and practice and to the exclusive jurisdiction of the English High Court of Justice

1.4.1 except as provided in Clause 37 (Arbitration); and

1.4.2 except as may be expressly provided herein to the contrary.

1.5 Subject to the overriding provisions of Clause 1.4, disputes between the Assured and the Underwriters may, if not settled amicably by negotiation, be referred by agreement of the Assured and the Underwriters to mediation or other form of alternative dispute resolution.

Invalidity/Unenforceability

1.6 If any provision of this insurance is held to be invalid or unenforceable, such invalidity or unenforceability shall not affect the other provisions of this insurance which shall remain in full force and effect.

Miscellaneous

1.7 Headings or references to headings are for information only and shall not be used in construing these provisions.

1.8 Words commencing with capital letters (save in headings or references to headings) shall have the meanings assigned to them in Clause 56 (Definitions).

PART I - STANDARD COVER

SECTION A - CONSTRUCTION

2. Perils

2.1 Subject always to Clause 3 (Section A exclusions), the insurance under Section A is against all risks of physical loss of or physical damage to the Subject Matter Insured caused and discovered during the Period of the Insurance.

2.2 For the avoidance of doubt, the Subject Matter Insured shall not be regarded as physically lost or physically damaged solely by virtue of the existence therein of any defect in design, plan or specification, defect in workmanship, defect in material or latent defect.

3. Section A exclusions

In no case shall the insurance under Section A cover the cost of replacing, repairing or rectifying any:

3.1 defect in design, plan or specification;

3.2 defect in workmanship;

3.3 defect in material;

3.4 latent defect.

However, these exclusions shall not extend to the cost of repairing physical loss or physical damage caused by such defect and discovered during the Period of the Insurance, to the extent that the cost of repairing such physical loss or physical damage exceeds the cost that would have been incurred to replace, repair or rectify the said defect had it been discovered immediately prior to the occurrence of the physical loss or physical damage caused thereby.

4. General average and salvage

4.1 The insurance under Section A covers the Vessel's proportion of salvage, salvage charges and/or general average, without reduction in respect of any under-insurance, but in the case of general average sacrifice of the Vessel, the Assured may recover in respect of the whole loss without first enforcing their right of contribution from other parties.

4.2 General average shall be adjusted according to the law and practice obtaining at the place where the adventure ends, as if the contract of affreightment contained no special terms upon the subject; but where the contract of affreightment so provides, the adjustment shall be according to the York-Antwerp Rules.

4.3 When the Vessel sails in ballast, not under charter, the provisions of the York-Antwerp Rules 2004 (excluding Rules XX and XXI) shall be applicable, and the voyage for this purpose shall be deemed to continue from the port or place of departure until the arrival of the Vessel at the first port or place thereafter other than a port or place of refuge or a port or place of call for bunkering only. If at any such intermediate port or place there is an abandonment of the adventure originally contemplated, the voyage shall thereupon be deemed to be terminated.

4.4 The Underwriters shall in no case be liable under this Clause 4 where the loss was not incurred to avoid or in connection with the avoidance of a peril insured under this insurance.

4.5 The Underwriters shall in no case be liable under this Clause 4 for or in respect of:

4.5.1 special compensation payable to a salvor under Article 14 of the International Convention on Salvage 1989 or under any other provision in any statute, rule, law or contract which is similar in substance;

6

4.5.2 expenses or liabilities incurred in respect of damage to the environment, or the threat of such damage, or as a consequence of the escape or release of any pollutant substance from the Vessel, or the threat of such escape or release.

4.6 Clause 4.5 shall not however exclude any sum which the Assured shall pay:

4.6.1 to salvors for or in respect of salvage remuneration in which the skill and efforts of the salvors in preventing or minimising damage to the environment as is referred to in Article 13 paragraph 1(b) of the International Convention on Salvage 1989 have been taken into account

4.6.2 as general average expenditure allowable under Rule XI(d) of the York-Antwerp Rules 2004, but only where the contract of affreightment provides for adjustment according to the York-Antwerp Rules 2004.

4.7 Should the Vessel receive salvage services from another vessel belonging wholly or in part to the Assured or under the same management, the Assured shall have the same rights under this insurance as they would have were the other vessel entirely the property of owners not interested in the Vessel; but in such cases the amount payable for the services rendered shall be referred to a sole arbitrator in accordance with Clauses 37.5 to 37.8 (Arbitration).

5. Section A deductible

5.1 No claim arising from a peril insured under Section A shall be payable under Section A unless the aggregate of all such claims arising out of each separate Occurrence (including claims under Clauses 4 (General average and salvage) and 25 (Duty of the Assured (sue and labour))) exceeds the Section A deductible agreed in the Schedule in which case this sum shall be deducted. Nevertheless the expense of sighting the bottom after stranding, if reasonably incurred specially for that purpose, shall be paid even if no physical damage be found.

5.2 This Clause 5 shall not apply to Abandonment or a claim for total or constructive total loss of the Vessel or, in the event of such a claim, to any associated claim under Clause 25 (Duty of the Assured (sue and labour)) arising from the same Occurrence.

SECTION B - LIABILITY

6. **Collision**

6.1 The Underwriters shall indemnify the Assured for any sum or sums paid by the Assured to any other person or persons by reason of the Assured's interest in the Vessel and by reason of the Assured becoming legally liable by way of damages:

6.1.1 for loss of or damage to any other vessel or property on any other vessel;

6.1.2 for delay to or loss of use of any such other vessel or property thereon;

6.1.3 for general average of, salvage of, or salvage under contract of, any such other vessel or property thereon;

6.1.4 in respect of the liability of any such other vessel arising out of the escape or release of any pollutant substance from any such other vessel,

where such payment by the Assured is in consequence of the Vessel coming into collision with any other vessel during the Period of the Insurance.

6.2 The indemnity provided by this Clause 6 shall be subject to the following provisions:

6.2.1 Where the Vessel is in collision with another vessel and both vessels are to blame then, unless the liability of one or both vessels becomes limited by law, the indemnity under this Clause 6 shall be calculated on the principle of cross-liabilities as if the respective owners had been compelled to pay to each other such proportion of each other's damages as may have been properly allowed in ascertaining the balance or sum payable by or to the Assured in consequence of the collision.

6.2.2 In no case shall the Underwriters' total liability under Clauses 6.1 and 6.2 exceed their proportionate shares of the Sum Insured in respect of any one collision.

6.3 The Underwriters shall also indemnify the Assured for legal costs incurred by the Assured or which the Assured may be compelled to pay in contesting liability or taking proceedings to limit liability, provided always that their prior written agreement to the incurring of such legal costs shall have been obtained and that the total liability of the Underwriters under this Clause 6.3 shall in no case (unless the Underwriters' specific written agreement shall have been obtained) exceed 25% of the Sum Insured.

6.4 The Underwriters shall in no case be liable under this Clause 6 for any sum which the Assured shall pay for or in respect of:

6.4.1 removal or disposal of any obstruction, wreck, cargo or any other thing whatsoever;

6.4.2 any real or personal property or thing whatsoever except any other vessel or property on any other vessel;

6.4.3 the cargo or other property on, or the engagements of, the Vessel;

6.4.4 loss of life, personal injury or illness;

6.4.5 pollution or contamination, or threats thereof, of any real or personal property or thing whatsoever (except any other vessel with which the Vessel is in collision or property on such other vessel or the liability arising out of the escape or release of any pollutant substance from such other vessel) or damage to the environment, or threat thereof, save that this exclusion shall not exclude any sum which the Assured shall pay for or in respect of salvage remuneration in which the skill and efforts of the salvors in preventing or minimising damage to the environment as referred to in Article 13 paragraph 1(b) of the International Convention on Salvage 1989 have been taken into account;

8

6.4.6 loss of or damage to property (not being a vessel) owned, leased or occupied by any Assured or any Additional Assured.

6.5 Should the Vessel come into collision with another vessel belonging wholly or in part to the Assured or under the same management, the Assured shall have the same rights under this insurance as they would have were the other vessel entirely the property of owners not interested in the Vessel; but in such cases the liability for the collision shall be referred to a sole arbitrator in accordance with Clauses 37.5 to 37.8 (Arbitration).

7. Protection and indemnity

7.1 The Underwriters shall indemnify the Assured for any sum or sums paid by the Assured to any other person or persons by reason of the Assured becoming legally liable, by reason of the Assured's interest in the Vessel, for any claim, demand, damages and/or expenses, where such liability is in consequence of any of the matters or things described in Clauses 7.1.1 to 7.1.5 and arises from an Occurrence during the Period of the Insurance:

7.1.1 loss of or damage to any fixed or moveable object or property or other thing or interest whatsoever, other than the Vessel, arising from any cause whatsoever in so far as such loss or damage is not covered by Clause 6 (Collision);

7.1.2 any attempted or actual raising, removal or destruction of any fixed or moveable object or property or other thing, including the wreck of the Vessel, or any neglect or failure to raise, remove, or destroy the same;

7.1.3 liability assumed by the Assured under contracts of customary towage for the purpose of entering or leaving port or manoeuvring within the port;

7.1.4 loss of life, personal injury, illness or payments made for life salvage;

7.1.5 the escape or release from the Vessel of any pollutant substance or the threat of such escape or release (subject always to Clause 7.5).

7.2 The Underwriters shall indemnify the Assured for any of the following arising from an Occurrence during the Period of the Insurance:

7.2.1 the additional cost of fuel, insurance, wages, stores, provisions and port charges reasonably incurred solely for the purpose of landing from the Vessel sick or injured persons or stowaways, refugees, or persons saved at sea;

7.2.2 additional expenses brought about by the outbreak of infectious disease on the Vessel or ashore;

7.2.3 fines imposed on the Vessel, on the Assured, or on any master, officer, crew member or agent of the Vessel who is reimbursed by the Assured, for any act or neglect or breach of any statute or regulation relating to the operation of the Vessel, provided that the Underwriters shall in no case be liable to indemnify the Assured for any fines which result from any act, neglect, failure or default of the Assured, their agents or servants other than any master, officer or crew member;

7.2.4 the expenses of the removal of the wreck of the Vessel from any place owned, leased or occupied by the Assured;

7.2.5 legal costs incurred by the Assured, or which the Assured may be compelled to pay, in avoiding, minimising or contesting liability or taking proceedings to limit liability with the prior written agreement of the Underwriters in respect of a liability covered by this Clause 7.

7.3 The Underwriters shall in no case be liable under this Clause 7 for any liability, cost or expense arising in respect of:

7.3.1 any direct or indirect payment by the Assured or any Additional Assured under workmen's compensation or employers' liability acts and any other statutory or common law, general maritime law or other liability whatsoever in respect of accidents to or illness of workmen or any other persons employed in any capacity whatsoever by the Assured or any Additional Assured or others in, on or about or in connection with the Vessel or its cargo, materials or repairs;

7.3.2 liability assumed by the Assured or any Additional Assured under agreement expressed or implied in respect of death or illness of or injury to any person employed under a contract of service or apprenticeship by the other party to such agreement;

7.3.3 punitive or exemplary damages, however described;

7.3.4 cargo or other property carried, to be carried or which has been carried on the Vessel, but this Clause 7.3.4 shall not exclude any claim in respect of the extra cost of removing cargo from the wreck of the Vessel;

7.3.5 loss of or damage to property owned by the Assured or any Additional Assureds or others employed in connection with the Vessel or property for which any of them is responsible, which is on the Vessel;

7.3.6 loss of or damage to the Subject Matter Insured, but this exclusion shall not extend to claims under Clauses 7.1.2 or 7.2.4;

7.3.7 liability arising under a contract or indemnity in respect of containers, equipment, fuel or other property on the Vessel and which is owned or leased by the Assured or any Additional Assured;

7.3.8 loss of or damage to cash, negotiable instruments, precious metals or stones, valuables or objects of a rare or precious nature, belonging to persons on the Vessel, or non-essential personal effects of any master, officer or crew member;

7.3.9 fuel, insurance, wages, stores, provisions and port charges arising from delay to the Vessel while awaiting a substitute for any master, officer or crew member;

7.3.10 fines or penalties arising from overloading or illegal fishing.

10

7.4 The Underwriters shall indemnify the Assured for any of the following in the event of a claim recoverable under Clause 7.1.5 (subject always to Clause 7.5):

7.4.1 the costs of any measures reasonably taken to prevent the escape or release from the Vessel of any pollutant substance;

7.4.2 the costs of any measures reasonably taken for the purpose of preventing or minimising pollution or any resulting loss or damage together with any liability for loss of or damage to property caused by measures so taken;

7.4.3 the costs or liabilities incurred as a result of compliance with any order or direction given by any government or authority, for the purpose of preventing or minimising pollution, provided always that such compliance is not a requirement for the normal operation or salvage or repair of the Vessel.

7.5 It is a condition precedent to the liability of the Underwriters under Clause 7.1.5 and Clause 7.4 that all of the following conditions are met and are shown by the Assured to have been met:

7.5.1 the escape or release of any pollutant substance or the threat of such escape or release was caused by an Occurrence; and

7.5.2 the Occurrence commenced on an identified specific date during the Period of the Insurance; and

7.5.3 the Occurrence was discovered by the Assured within 72 hours of commencement; and

7.5.4 the Occurrence did not result from the Assured's intentional violation of any statute, rule, ordinance or regulation.

7.6 Where the Assured or any Additional Assured or the Underwriters can or could have limited their liability, the indemnity under this Clause 7 in respect of such liability shall in no case exceed the Underwriters' proportionate shares of the amount of such limitation.

7.7 In no case shall the Underwriters' liability under this Clause 7 arising out of each separate Occurrence exceed their proportionate shares of the Sum Insured.

8. Section B deductible

No claim arising from a peril insured under Section B shall be payable under Section B unless the aggregate of all such claims arising out of each separate Occurrence (including claims under Clause 25 (Duty of the Assured (sue and labour))) exceeds the Section B deductible agreed in the Schedule, in which case this sum shall be deducted.

SECTION C - WAR

9. Attachment

9.1 Subject always to Clause 9.2, the insurance under Section C shall attach on the Launch Date (and not before) and then only to the Vessel as at the Launch Date. The insurance under Section C shall only apply to Items as they are placed on the Vessel on or after the Launch Date.

9.2 The insurance under Section C shall not attach if, between the commencement of the Period of the Insurance and the Launch Date, there has occurred any event which would have automatically terminated the insurance under Section C in accordance with Clause 15 (Termination).

10. Perils

Subject always to Clause 13 (Section C exclusions), the insurance under Section C covers:

10.1 physical loss of or physical damage to the Subject Matter Insured caused by war, civil war, revolution, rebellion, insurrection or civil strife arising therefrom, or any hostile act by or against a belligerent power;

10.2 loss of or physical damage to the Subject Matter Insured caused by capture, seizure, arrest, restraint or detainment, arising from perils covered under Clause 10.1, and the consequences thereof or any attempt thereat;

10.3 physical loss of or physical damage to the Subject Matter Insured caused by derelict mines, torpedoes, bombs or other derelict weapons of war.

11. General average and salvage, protection and indemnity/collision liability

Subject always to Clause 13 (Section C exclusions), the insurance under Section C also covers:

11.1 general average and salvage under Clause 4 (General average and salvage) which is excluded by Clause 22 (War exclusion);

11.2 the liability under Section B which is excluded by Clause 22 (War exclusion).

12. Detainment

12.1 In the event that the Vessel shall have been the subject of capture, seizure, arrest, restraint or detainment covered under Clause 10.2 (Perils), and the Assured shall thereby have lost the free use and disposal of the Vessel for a continuous period of 12 months, then for the purpose of ascertaining whether the Vessel is a constructive total loss, the Assured shall be deemed to have been deprived of possession of the Vessel without any likelihood of recovery.

12.2 The inability of the Vessel to sail from any port, canal, waterway or other place to the high seas for a continuous period of 12 months as a result of the closure of the connecting channel to all vessels of such size or draft shall amount to "restraint" in accordance with

Clause 10.2 (Perils), provided that such closure has arisen through the blockage of the port, canal, waterway or other place by a warlike act or act of national defence.

13. Section C exclusions

In no case shall the insurance under Section C cover:

13.1 loss, damage, liability or expense arising from:

13.1.1 the outbreak of war (whether there be a declaration of war or not) between any of the following countries:

United Kingdom, United States of America, France, the Russian Federation, the People's Republic of China;

13.1.2 confiscation, expropriation, requisition or pre-emption;

13.1.3 capture, seizure, arrest, restraint or detainment by or under the order of the government or any public or local authority of the country in which the Vessel is owned or registered;

13.1.4 arrest, restraint or detainment under quarantine regulations or by reason of infringement of any customs or trading regulations;

13.1.5 the operation of ordinary judicial process, failure to provide security or to pay any fine or penalty or any financial cause;

13.1.6 the loss of or frustration of any voyage or contract for sale or other adventure;

13.2 loss, damage, liability or expense covered under Section A, Section B or Section D or which would be recoverable thereunder but for the deductible clauses therein.

14. SECTION C DEDUCTIBLE

14.1 No claim arising from a peril insured under Section C shall be payable under Section C unless the aggregate of all such claims arising out of each separate Occurrence (including claims under Clauses 11 (General average and salvage, P&I and collision liability) and 25 (Duty of the Assured (sue and labour))) exceeds the Section C deductible, if any, agreed in the Schedule in which case this sum shall be deducted.

14.2 This Clause 14 shall not apply to Abandonment or a claim for total or constructive total loss of the Vessel or, in the event of such a claim, to any associated claim under Clause 25 (Duty of the Assured (sue and labour)) arising from the same Occurrence.

15. Termination

15.1 The insurance under Section C may be cancelled by either the Underwriters or the Assured giving 7 days' notice (such cancellation becoming effective on the expiry of 7 days from midnight of the day on which notice of cancellation is issued by or to the Underwriters). The Underwriters shall however reinstate the insurance under Section C subject to agreement between the Underwriters and the Assured prior to the expiry of such notice of cancellation as to any new rate of premium and/or conditions and/or warranties.

15.2 Whether or not such notice of cancellation has been given the insurance under Section C shall TERMINATE AUTOMATICALLY:

15.2.1 upon the outbreak of war (whether there be a declaration of war or not) between any of the following countries:

United Kingdom, United States of America, France, the Russian Federation, the People's Republic of China;

15.2.2 in the event of the Vessel being requisitioned, either for title or use.

13

15.3 In the event either of cancellation by notice or of automatic termination of the insurance under Section C by reason of the operation of this Clause 15, pro rata net return of premium under Section C shall be payable to the Assured, provided always that the premium under this Section C was separately identified from the premium under Section A, Section B and Section D of this insurance, prior to such cancellation by notice or such automatic termination.

SECTION D - STRIKES, TERRORIST, POLITICAL MOTIVE AND MALICIOUS ACTS

16. Perils

Subject always to Clause 18 (Section D exclusions), the insurance under Section D covers physical loss of or physical damage to the Subject Matter Insured caused by the perils listed in Clause 23 (Strikes, terrorist, political motive and malicious acts exclusion).

17. General average and salvage, protection and indemnity/collision liability

Subject always to Clause 18 (Section D exclusions), the insurance under Section D also covers:

17.1 general average and salvage under Clause 4 (General average and salvage) which is excluded by Clause 23 (Strikes, terrorist, political motive and malicious acts exclusion);

17.2 the liability under Section B which is excluded by Clause 23 (Strikes, terrorist, political motive and malicious acts exclusion).

18. Section D exclusions

In no case shall the insurance under Section D cover:

18.1 loss, damage, liability or expense covered under Section A, Section B or Section C or which would be recoverable thereunder but for the deductible clauses therein;

18.2 loss, damage, liability or expense arising from the loss of or frustration of any voyage or contract for sale or other adventure.

19. Section D deductible

19.1 No claim arising from a peril insured under Section D shall be payable under Section D unless the aggregate of all such claims arising out of each separate Occurrence (including claims under Clauses 17 (General average and salvage, P&I/collision liability) and 25 (Duty of the Assured (sue and labour)) exceeds the Section D deductible, if any, agreed in the Schedule in which case this sum shall be deducted.

19.2 This Clause 19 shall not apply to Abandonment or a claim for total or constructive total loss of the Vessel or, in the event of such a claim, to any associated claim under Clause 25 (Duty of the Assured (sue and labour)) arising from the same Occurrence.

20. Termination

20.1 The insurance under Section D may be cancelled by either the Underwriters or the Assured giving 7 days' notice (such cancellation becoming effective on the expiry of 7 days from midnight of the day on which notice of cancellation is issued by or to the Underwriters). The Underwriters shall however reinstate the insurance under Section D subject to agreement between the Underwriters and the Assured prior to the expiry of such notice of cancellation as to any new rate of premium and/or conditions and/or warranties.

20.2 In the event of such cancellation of the insurance under Section D, pro rata net return of premium under Section D shall be payable to the Assured, provided always that the premium under this Section D was separately identified from the premium under Section A, Section B and Section C of this insurance, prior to such cancellation.

PART II - GENERAL EXCLUSIONS

21. General exclusions

In no case shall this insurance cover loss, damage, liability or expense caused:

21.1 by wilful misconduct of the Assured;

21.2 by insolvency;

21.3 solely by ordinary wear and tear;

21.4 by delay, except such expenses as would be recoverable in principle in English law and practice under the York-Antwerp Rules 2004.

22. War exclusion

In no case shall the insurance under Section A, Section B or Section D cover loss, damage, liability or expense caused by:

22.1 war, civil war, revolution, rebellion, insurrection, or civil strife arising therefrom, or any hostile act by or against a belligerent power;

22.2 capture, seizure, arrest, restraint or detainment (violent theft, barratry and piracy excepted), and the consequences thereof or any attempt thereat;

22.3 derelict mines, torpedoes, bombs or other derelict weapons of war.

23. Strikes, terrorist, political motive and malicious acts exclusion

In no case shall the insurance under Section A, Section B or Section C cover loss, damage, liability or expense caused by:

23.1 strikers, locked-out workmen, or persons taking part in labour disturbances, riots or civil commotions;

23.2 violent theft by persons from outside the Vessel;

23.3 piracy;

23.4 barratry of master, officers or crew;

23.5 any terrorist;

23.6 any person acting maliciously or from a political motive;

23.7 the use of any weapon or the detonation of an explosive by any person acting maliciously or from a political motive.

However, the exclusion in Clause 23.7 shall not extend to physical loss of or physical damage to the Vessel caused by any governmental authority acting under the powers vested in it to prevent or mitigate a pollution hazard or damage to the environment or the threat thereof, resulting directly from physical loss or physical damage to the Vessel for which the Underwriters are liable under Section A, Section C or Section D of this insurance, provided such act of governmental authority has not resulted from want of due diligence by the Assured or any of them to prevent or mitigate such hazard or damage or threat thereof. Master, officers, crew or pilots shall not be considered as the Assured within the meaning of this Clause should they hold shares in the Vessel.

24. Radioactive contamination, chemical, biological, bio-chemical and electromagnetic weapons exclusion

In no case shall the insurance under Section A, Section B, Section C or Section D cover loss, damage, liability or expense directly or indirectly caused by or contributed to by or arising from:

24.1 ionising radiations from or contamination by radioactivity from any nuclear fuel or from any nuclear waste or from the combustion of nuclear fuel;

24.2 the radioactive, toxic, explosive or other hazardous or contaminating properties of any nuclear installation, reactor or other nuclear assembly or nuclear component thereof;

24.3 any weapon or device employing atomic or nuclear fission and/or fusion or other like reaction or radioactive force or matter;

24.4 the radioactive, toxic, explosive or other hazardous or contaminating properties of any radioactive matter. The exclusion in this Clause 24.4 shall not extend to radioactive isotopes, other than nuclear fuel, when such isotopes are being prepared, carried, stored, or used for commercial, agricultural, medical, scientific or other similar peaceful purposes;

24.5 any chemical, biological, bio-chemical or electromagnetic weapon.

PART III - CLAIMS

25. Duty of the Assured (sue and labour)

25.1 In case of any loss or misfortune it is the duty of the Assured and their servants and agents to take such measures as may be reasonable for the purpose of averting or minimising a loss which would be recoverable under this insurance.

25.2 Subject to the provisions of this Clause 25 and to Clauses 5 (Section A deductible), 8 (Section B deductible), 14 (Section C deductible) and 19 (Section D deductible), the Underwriters shall contribute to charges properly and reasonably incurred by the Assured, their servants or agents for such measures. General average, salvage charges (except as provided for in Clause 25.4), special compensation and expenses as referred to in Clause 4.5 (General average and salvage), collision defence or attack costs and costs as referred to in Clause 7.4 (Protection and indemnity) are not recoverable under this Clause 25. For the avoidance of doubt, Expediting Expenses are not recoverable under this Clause 25.

16

25.3 Measures taken by the Assured or the Underwriters with the object of saving, protecting or recovering the Subject Matter Insured shall not be considered as a waiver or acceptance of abandonment or otherwise prejudice the rights of either party.

25.4 When the Underwriters have admitted a claim for total loss of the Vessel or there has been an Abandonment under this insurance and expenses have been reasonably incurred in saving or attempting to save the Vessel and other property and there are no proceeds, or the expenses exceed the proceeds, then this insurance shall bear its pro rata share of such proportion of the expenses, or of the expenses in excess of the proceeds, as the case may be, as may reasonably be regarded as having been incurred in respect of the Vessel, excluding all special compensation and expenses as referred to in Clause 4.5 (General average and salvage).

25.5 The sum recoverable under this Clause 25 shall be in addition to the loss otherwise recoverable under this insurance but shall in no case exceed the Sum Insured.

26. Notice of claim

26.1 In the event of an Occurrence whereby loss, damage, liability or expense may result in a claim under this insurance, notice shall be given to the Underwriters:

26.1.1 as soon as possible after the date on which the Assured or the Assured's project management become aware of such loss, damage, liability or expense; and

26.1.2 in case of physical loss of or physical damage to the Subject Matter Insured, prior to the commencement of any repair work.

26.2 If notice is not given to the Underwriters within 180 days of the Assured or the Assured's project management becoming aware of such loss, damage, liability or expense, no claim shall be recoverable under this insurance in respect of such loss, damage, liability or expense, unless the Underwriters agree to the contrary in writing.

26.3 Subject always to Clause 26.2, if notice is not given to the Underwriters prior to the commencement of repair work to physical loss of or physical damage to the Subject Matter Insured, no claim shall be recoverable under this insurance in respect of such physical loss or physical damage, save to the extent that the Assured can establish that the Underwriters have not suffered any prejudice as a result of not having been so notified.

26.4 It is a condition precedent to the liability of the Underwriters under Section B, Clause 11.2 (Section C, P&I and collision liability) and Clause 17.2 (Section D, P&I and collision liability) that:

26.4.1 prompt notice must be given to the Underwriters of every casualty, event or claim upon the Assured which may give rise to a claim under Section B, Clause 11.2 or Clause 17.2 and of every event or matter which may cause the Assured to incur liability, costs or expense for which they may be insured under Section B, Clause 11.2 or Clause 17.2;

26.4.2 the Assured shall not admit liability for or settle any claim for which they may be insured under Section B, Clause 11.2 or Clause 17.2 without the prior written agreement of the Underwriters.

26.5 It is a condition precedent to the liability of the Underwriters under Clauses 7.1.5 and 7.4 (Protection and indemnity) that written notice of the Occurrence is given to the Underwriters within 30 days of the Assured's or the Assured's project management's discovery of the Occurrence.

27. Partial loss

27.1 The measure of indemnity in respect of partial loss shall, where the Subject Matter Insured has been repaired, be the reasonable cost of the repairs, but shall in no case exceed the Sum Insured in respect of any one Occurrence.

27.2 For the avoidance of doubt, the reasonable cost of the repairs shall not include:

27.2.1 the Assured's profit on such cost;

27.2.2 Expediting Expenses.

28. Constructive total loss

28.1 The measure of indemnity in respect of a constructive total loss shall be the Sum Insured.

28.2 In ascertaining whether the Subject Matter Insured is a constructive total loss, the Sum Insured less the Profit Percentage thereon shall be taken as the repaired value and nothing in respect of the damaged or break-up value shall be taken into account.

28.3 No claim for a constructive total loss based upon the cost of recovery and/or the reasonable cost of the repairs (calculated in accordance with Clause 27 (Partial loss)), shall be recoverable under this insurance unless such cost(s) would exceed the Sum Insured less the Profit Percentage thereon.

28.4 In making the determination under Clause 28.3:

28.4.1 only the cost(s) relating to physical loss or physical damage arising from the same Occurrence shall be taken into account; and

28.4.2 no account shall be taken of:

28.4.2.1 Expediting Expenses; or

28.4.2.2 the Assured's profit on the cost of recovery of and/or the reasonable cost of the repairs to the Subject Matter Insured.

29. Actual total loss

29.1 The measure of indemnity in respect of an actual total loss shall be the Works Value plus the Profit Percentage thereon at the time of loss.

29.2 In the event of an actual total loss, the Underwriters shall also indemnify the Assured for the cost of material (not forming part of the Works Value) where the Assured is under a legal obligation to purchase the material, notwithstanding the actual total loss. The liability of the Underwriters shall not exceed the lesser of the contractual liability of the Assured in respect of such material or the cost of such material. In no case shall the Underwriters' liability under this Clause 29.2 exceed 10% of the Sum Insured or such other percentage agreed in the Schedule.

30. Unrepaired damage

30.1 The measure of indemnity in respect of a claim for unrepaired damage shall be the reasonable cost of the repairs calculated in accordance with Clause 27 (Partial loss), but shall in no case exceed the Contract Deduction in respect of such damage.

30.2 In no case shall the Underwriters be liable for unrepaired damage in the event of a subsequent Abandonment or of a subsequent total loss (whether or not covered under this insurance) sustained during the Period of the Insurance or any extension thereof.

30.3 The Underwriters shall in no case be liable for unrepaired damage for more than the Sum Insured at the time this insurance terminates or is cancelled.

30.4 Any dispute between the Assured and the Underwriters as to whether the Underwriters' agreement is or is not reasonably withheld to the amount by which the sum payable by the Buyer to the Builder is reduced by reason of physical damage to the Vessel not repaired by the Builder at the time this insurance terminates or is cancelled shall be referred to arbitration in accordance with Clauses 37.1 to 37.4 (Arbitration).

31. Abandonment

31.1 In the event of physical loss of or physical damage to the Vessel and in the event that:

31.1.1 the cost of recovery and/or the reasonable cost of the repairs (calculated in accordance with Clause 27 (Partial loss)) of the Vessel required to recover and/or repair the Vessel to a stage equivalent to that at the time of loss exceed(s) the Works Value at the time of loss, but do(es) not exceed the Sum Insured; and

31.1.2 there is an Abandonment,

the measure of indemnity shall be the Works Value plus the Profit Percentage thereon at the time of loss.

31.2 Any dispute between the Assured and the Underwriters as to whether the Underwriters' agreement is or is not reasonably withheld to an agreement between the Builder and the Buyer that the Builder is relieved of the obligation to build or deliver the Vessel to the Buyer, by reason of physical loss of or physical damage to the Vessel, shall be referred to arbitration in accordance with Clauses 37.1 to 37.4 (Arbitration).

31.3 In making the determination under Clause 31.1.1:

31.3.1 only the cost(s) relating to physical loss or physical damage arising from the same Occurrence shall be taken into account; and

31.3.2 no account shall be taken of:

31.3.2.1 Expediting Expenses; or

31.3.2.2 the Assured's profit on the cost of recovery and/or the reasonable cost of the repairs to the Vessel.

31.4 In the event of an Abandonment, the Underwriters shall also indemnify the Assured for the cost of material (not forming part of the Works Value) where the Assured is under a legal obligation to purchase the material, notwithstanding the Abandonment. The liability of the Underwriters shall not exceed the lesser of the contractual liability of the Assured in respect of such material or the cost of such material. In no case shall the Underwriters' liability under this Clause 31.4 exceed 10% of the Sum Insured or such other percentage agreed in the Schedule.

32. Trials

In case of failure of trials resulting from physical loss or physical damage to the Vessel due to a peril insured under this insurance, the Underwriters shall pay the expenses reasonably incurred in re-running trials to the extent that such expenses are necessary to effect the repairs, but in no case shall the Underwriters be liable for expenses incurred in re-running trials undertaken solely by reason of contractual provisions between the Builder and the Buyer.

33. Failure of launch

In case of failure of launch of the Vessel, the Underwriters shall pay the expenses reasonably incurred solely in connection with completion of the launch of the Vessel.

34. Deductibles

In the event of a claim under more than one section of this insurance arising out of the same Occurrence, then only the highest applicable deductible shall be deducted.

35. Claims under more than one section exclusion

The insurance under Section B, Section C and Section D excludes any sum recoverable under Section A of this insurance.

36. Payment of claims

Unless otherwise agreed in writing, claims shall be paid to the broker which placed this insurance and such payment, whether in account or otherwise, shall be a complete discharge of the Underwriters' obligations under this insurance in respect of the amount so paid (and whether or not more than one Assured or Additional Assured may have an interest in such payment).

37. Arbitration

Abandonment/unrepaired damage

37.1 Any dispute between the Assured and the Underwriters concerning whether the Underwriters' agreement is or is not reasonably withheld for the purposes of Clause 30.4 (Unrepaired damage) or Clause 31.2 (Abandonment) shall be referred to arbitration in London in accordance with the Arbitration Act 1996 or any statutory modification or re-enactment thereof for the time being in force and in accordance with Clauses 37.2 to 37.4.

37.2 The arbitration shall be conducted in accordance with the London Maritime Arbitrators Association (LMAA) Terms current at the time when the arbitration proceedings are commenced.

37.3 The reference shall be to three arbitrators. A party wishing to refer a dispute to arbitration shall appoint its arbitrator and send notice of such appointment in writing to the other party requiring the other party to appoint its own arbitrator within 14 days of that notice and stating that it will appoint its arbitrator as sole arbitrator unless the other party appoints its own arbitrator and gives notice in writing that it has done so within the 14 days specified. If the other party does not appoint its own arbitrator and give notice in writing that it has done so within the 14 days specified, the party referring a dispute to arbitration may, without the requirement of any further prior notice to the other party, appoint its arbitrator as sole arbitrator and shall advise the other party accordingly. The award of a sole arbitrator shall be binding on both parties as if he or she had been appointed by the agreement of both parties.

37.4 Nothing in this Clause 37 shall prevent the parties agreeing in writing to vary these provisions to provide for the appointment of a sole arbitrator.

Sister ship salvage/collision

37.5 The amount payable in respect of salvage services as described in Clause 4.7 (General average and salvage) or the liability for any collision as described in Clause 6.5 (Collision) shall be referred to arbitration in London in accordance with the Arbitration Act 1996 or any statutory modification or re-enactment thereof for the time being in force and in accordance with Clauses 37.6 to 37.8.

37.6 The matters described in Clause 37.5 shall be referred to a sole arbitrator to be selected either by the Assured or the Underwriters from the persons currently on the panel of Lloyd's Salvage Arbitrators with a right of appeal from an award made by the arbitrator to either party by notice in writing to the other party within 28 days of the date of publication of the arbitrator's award.

37.7 The arbitrator on appeal shall be the person currently acting as the Lloyd's Appeal Arbitrator.

37.8 Both the arbitrator and the arbitrator on appeal shall have the same powers as an arbitrator and an arbitrator on appeal under LOF 2000 or any standard revision thereof.

38. Subrogation

38.1 In the event of loss, damage, liability or expense covered under this insurance, the Underwriters shall be subrogated to all rights and remedies of the Assured and any Additional Assureds:

38.1.1 save against the Builder (subject to Clause 38.2) and the Buyer, against whom such rights and remedies are waived by the Underwriters; and

38.1.2 save against any Additional Assured (subject to Clause 38.2) whose written contract with the Assured provides that the Underwriters shall waive such rights and remedies, but only to the extent of the waiver (if any) provided by such written contract.

38.2 Where an Assured or any Additional Assured is, under a separate insurance, insured against liability for loss of or damage to the Subject Matter Insured, such Assured or Additional Assured agrees on its own behalf and on behalf of the insurers of such liability insurance:

38.2.1 that the Underwriters shall not waive any rights of subrogation against such Assured or Additional Assured; and

38.2.2 to provide a copy of the policy or other document evidencing such liability insurance to the Underwriters upon their written request; and

38.2.3 that the Underwriters may bring proceedings against such Assured or Additional Assured in respect of loss of or damage to the Subject Matter Insured.

However, the Underwriters shall not seek to recover more from such Assured or Additional Assured than the applicable limit or limits of such liability insurance.

39. Recoveries

39.1 The Assured shall, whether or not the Underwriters have paid a claim or agreed to pay a claim or potential claim under this insurance, take reasonable steps to:

39.1.1 assess as soon as possible whether there are any prospects of a recovery from third parties in respect of matters giving rise to a claim or to a potential claim under this insurance;

39.1.2 protect any claims against such third parties if necessary by the commencement of proceedings and the taking of appropriate steps to obtain security for the claim from third parties;

39.1.3 keep the Underwriters and the appointed average adjuster (if any) advised of the recovery prospects and any action taken against third parties;

39.1.4 co-operate with the Underwriters in the taking of such steps as may be reasonably required to pursue any claims against third parties.

39.2 The Underwriters shall pay the reasonable costs incurred by the Assured pursuant to this Clause 39 in the same proportion as the insured losses bear to the total of the insured and uninsured losses (as defined in Clause 39.4.2).

39.3 Where the Assured have incurred reasonable costs pursuant to Clause 39.1.2 and where no claim is recoverable under this insurance, provided always that the Underwriters' written agreement to the reimbursement of such costs shall have been obtained prior to the incurring of such costs, the Underwriters shall reimburse such costs to the extent agreed, notwithstanding that no claim is recoverable under this insurance.

39.4 In the event of recoveries from third parties in respect of claims which have been paid in whole or in part under this insurance, such recoveries shall be distributed between the Underwriters and the Assured as follows:

39.4.1 the reasonable costs and expenses incurred in making such recoveries from the third party shall be deducted first and returned to the paying party;

39.4.2 the balance shall be apportioned between the Underwriters and the Assured in the same proportion that the insured losses and uninsured losses bear to the total of the insured and uninsured losses. For the purposes of Clause 39.2 and this Clause 39.4.2, "uninsured losses" shall mean loss of or damage to the Subject Matter Insured and any liability or expense which would have been recoverable under this insurance, but for the application of deductible(s) and the limits of this insurance.

39.5 For the purposes of this Clause 39, "third parties" shall not include the Builder or the Buyer or any Additional Assured against whom the Underwriters have waived rights of subrogation in accordance with Clause 38 (Subrogation).

40. Residual value

40.1 Where the Underwriters pay for a total loss pursuant to Clause 28 (Constructive total loss) or Clause 29 (Actual total loss) or in respect of an Abandonment pursuant to Clause 31 (Abandonment), the Underwriters shall be entitled to take over the Assured's interest in the Subject Matter Insured.

40.2 Without prejudice to the generality of Clause 40.1, the Underwriters shall in such event be entitled to the residual value of some or all of the Subject Matter Insured and/or may require the transfer of title to them or to their order of some or all of the Subject Matter Insured.

PART IV - GENERAL PROVISIONS AND DEFINITIONS

41. Assured

Where the Assured consists of the Builder and the Buyer:

41.1 any act or any failure to act (whether before or after the Period of the Insurance commences) by or on behalf of the Builder which prevents recovery by the Builder (or would prevent recovery if a claim were made by the Builder) under this insurance shall also prevent any recovery under this insurance by the Buyer;

41.2 any act or any failure to act (whether before or after the Period of the Insurance commences) by or on behalf of the Buyer which prevents recovery by the Buyer (or would prevent recovery if a claim were made by the Buyer) under this insurance shall also prevent recovery under this insurance by the Builder;

41.3 any communication to or from the Underwriters from or to the Builder or their agents shall be deemed to have been made on behalf of or to the Buyer;

41.4 any communication to or from the Underwriters from or to the Buyer or their agents shall be deemed to have been made on behalf of or to the Builder.

42. Additional Assureds

42.1 Additional Assureds shall be insured under this insurance to the extent of their respective rights and interests in the Subject Matter Insured, but in no case shall any Additional Assured be insured under this insurance to any greater extent than provided for in their written contract with the Assured. For the purposes of Section B, "Assured" shall include an "Additional Assured" where such Additional Assured becomes legally liable by reason of their interest in the Vessel.

42.2 Any act or any failure to act (whether before or after the Period of the Insurance commences) by or on behalf of the Assured which prevents recovery by the Assured (or would prevent recovery if a claim were made by the Assured) under this insurance shall also prevent recovery under this insurance by any Additional Assured.

42.3 The rights of any Additional Assured shall only be exercised through the Assured.

42.4 Additional Assureds shall at all times comply with and be subject to the terms and conditions of this insurance applicable to the Assured (save Clause 47 (Premium payment)). However, a failure to comply by an Additional Assured shall not prejudice the rights of other Additional Assureds or the Assured, save where such other Additional Assureds or the Assured are privy to such failure.

42.5 Any agreement between the Underwriters and the Assured relating to this insurance shall be binding on Additional Assureds and neither the Underwriters nor the Builder nor the Buyer shall be obliged to provide notice of such agreement to Additional Assureds.

42.6 Any communication to or from the Underwriters from or to the Assured or their agents shall be deemed to have been made on behalf of or to the Additional Assureds.

43. Pre-existing Vessel

43.1 This Clause 43 shall apply to this insurance if the Vessel is a Pre-existing Vessel that is to undergo conversion, repair, lengthening or any other work.

43.2 The following words shall have the meanings assigned to them by this Clause. Clauses 56.16 ("Period of the Insurance"), 56.20 ("Sum Insured"), 56.21 ("Vessel") and 56.22 ("Works Value") shall be deleted and replaced by Clauses 43.2.2, 43.2.5, 43.2.6 and 43.2.7 respectively.

43.2.1 "Going-in Value" means the sum agreed in the Schedule as the insured value of the Pre-existing Vessel at the commencement of the Period of the Insurance. The Going-in Value shall not be included in Material Cost.

43.2.2 "Period of the Insurance" under Section A, Section B and Section D means the period commencing on:

43.2.2.1 the handover to and acceptance by the Builder of the Pre-existing Vessel at the Yard; or

43.2.2.2 in respect of Preparatory Work, the commencement of the Preparatory Work at the Yard; or

43.2.2.3 such earlier date(s), if any, agreed in the Schedule,

whichever shall first occur.

"Period of the Insurance" under Section C means the period commencing as described in Clause 43.4; and

ceasing, under Sections A, B, C and D, on:

43.2.2.4 re-delivery of the Vessel to the Buyer; or

43.2.2.5 the date agreed in the Schedule; or

43.2.2.6 such date at which this insurance shall be cancelled in accordance with Clause 47 (Premium payment); or

43.2.2.7 such date at which this insurance shall be terminated automatically in accordance with Clause 52 (Automatic termination),

whichever shall first occur.

43.2.3 "Pre-existing Vessel" means the ship, vessel, craft or other structure to undergo conversion, repair, lengthening or any other work.

43.2.4 "Preparatory Work" means fabrication, manufacturing and other work undertaken at the Yard and the storage of Items at the Yard prior to the arrival of the Pre-existing Vessel at the Yard.

43.2.5 "Sum Insured" means the sum agreed in the Schedule as the insured value of the Subject Matter Insured. Should the aggregate of the Works Value and the Going-in Value exceed the Sum Insured, the Sum Insured shall, subject to Clause 44.2 (Sum insured), be increased in the manner set out in Clauses 44.1 (Sum insured) and 43.3.

43.2.6 "Vessel" means the project bearing the hull number or project name identified in the Schedule and includes the Pre-existing Vessel.

43.2.7 "Works Value" means the total of:

43.2.7.1 the Allocated Series Cost spent by the Assured; and

43.2.7.2 the Contractor Cost spent by the Assured; and

43.2.7.3 the Labour Cost spent by the Assured; and

43.2.7.4 the Material Cost spent by the Assured,

but in no case shall the Works Value include the Going-in Value and in no case shall the Works Value exceed the Sum Insured.

43.3 Clause 44.1 (Sum insured) is deleted from the time the Pre-existing Vessel arrives at the Yard. From that time, if the aggregate of:

43.3.1 the Works Value plus the Profit Percentage thereon; and

43.3.2 the Going-in Value,

at any time exceeds the Sum Insured, the Assured shall advise the Underwriters as soon as possible of the amount of such excess and shall pay pro rata additional premium thereon and the Underwriters shall accept their proportionate shares of such increase.

43.4 Clause 9 (Attachment) shall be deleted and replaced by the following:

43.4.1 Subject always to Clause 43.4.2, the insurance under Section C shall attach on the handover to and acceptance by the Builder of the Pre-existing Vessel at the Yard (and not before) and then only to the Vessel. The insurance under Section C shall only apply to Items as they are placed on the Vessel on or after handover to and acceptance by the Builder of the Pre-existing Vessel at the Yard.

43.4.2 The insurance under Section C shall not attach if, between the commencement of the Period of the Insurance and the handover to and acceptance by the Builder of the Pre-existing Vessel at the Yard, there has occurred any event which would have automatically terminated the insurance under Section C in accordance with the provisions of Clause 15 (Termination).

43.5 Clause 27.1 (Partial loss) shall be deleted and replaced by the following:

43.5.1 The measure of indemnity in respect of partial loss of the Subject Matter Insured, but not including the Pre-existing Vessel, shall, where repairs have been undertaken, be the reasonable cost of the repairs, but shall in no case exceed the Sum Insured in respect of any one Occurrence.

The measure of indemnity in respect of partial loss of the Pre-existing Vessel shall, where repairs have been undertaken, be the reasonable cost of the repairs less the customary deductions. Customary deductions shall not be made where the Assured can establish that the repairs were to those areas of the Pre-existing Vessel intended to form part of the Vessel on completion. The measure of indemnity in respect of partial loss of the Pre-existing Vessel shall in no case exceed the Sum Insured in respect of any one Occurrence.

44. Sum Insured

44.1 If the Works Value plus the Profit Percentage thereon at any time exceeds the Sum Insured, the Assured shall advise the Underwriters as soon as possible of the amount of such excess and shall pay pro rata additional premium thereon and the Underwriters shall accept their proportionate shares of such increase.

44.2 The Sum Insured shall in no case exceed 110% or such other percentage (if any) agreed in the Schedule of the sum agreed in the Schedule at the inception of this insurance, unless the Underwriters' prior written agreement has been obtained and any amended terms of cover and any additional premium required by the Underwriters have been agreed.

45. Limit of liability

45.1 In no case shall the Underwriters' liability arising out of each separate Occurrence exceed their proportionate shares of the total of:

45.1.1 the Sum Insured in respect of loss of or damage to the Vessel (whether caused by a peril insured under Section A, Section C or Section D);

45.1.2 the Sum Insured in respect of general average and salvage under Clause 4 (General average and salvage) or Clauses 11.1 or 17.1;

45.1.3 the Sum Insured in respect of liability under Clause 6 (Collision) and 25% (unless the Underwriters' specific written agreement shall have been obtained) of the Sum Insured in respect of legal costs under Clause 6.3;

45.1.4 the Sum Insured in respect of liability under Clause 7 (Protection and indemnity), subject always to the provisions of Clause 7.6;

45.1.5 the Sum Insured in respect of sue and labour expenditure under Clause 25 (Duty of the Assured (sue and labour));

45.1.6 10% (or such other percentage agreed in the Schedule) of the Sum Insured in respect of material under Clauses 29.2 (Actual total loss) and 31.4 (Abandonment);

45.1.7 10% (or such other percentage agreed in the Schedule) of the Sum Insured in respect of trials under Clause 32 (Trials) (whether or not amended by Clause 60 (Trials amendment));

45.1.8 10% (or such other percentage agreed in the Schedule) of the Sum Insured in respect of failure to launch under Clause 33 (Failure to launch);

45.1.9 10% (or such other percentage agreed in the Schedule) of the Sum Insured in respect of Expediting Expenses under Clause 59 (Expediting Expenses).

45.2 In no case shall the Underwriters' liability hereunder be increased by reason of the inclusion of more than one Assured or Additional Assured under this insurance.

46. Separate insurances

If more than one Vessel is insured under this insurance, each Vessel is deemed to be separately insured, as if there were a separate contract of insurance in respect of each Vessel.

47. Premium payment

47.1 The Assured undertakes that the premium shall be paid:

47.1.1 in full to the Underwriters within 45 days (or such other period agreed in the Schedule) of the inception of this insurance; or

47.1.2 where payment by instalment premiums has been agreed:

47.1.2.1 the first instalment premium shall be paid within 45 days (or such other period agreed in the Schedule) of the inception of this insurance;

47.1.2.2 and the second and any subsequent instalment(s) shall be paid by the date(s) they are due.

47.2 The Assured undertakes that any additional premium payable under Clauses 15.1 (War termination), 20.1 (Strikes, terrorist termination), 44 (Sum Insured), 49 (Navigation) and 50 (Items) shall be paid in full to the Underwriters within 60 days of agreement of the additional premium or such other period agreed in the Schedule.

47.3 If the premium (or the first instalment premium) has not been so paid to the Underwriters by the 46th day (or the day after such period agreed in the Schedule) from the inception of this insurance (and, in respect of the second and any subsequent instalment premiums, by the date(s) they are due) or if the premium payable in accordance with Clause 47.2 has not been paid within such period(s) agreed in the Schedule, the Underwriters shall have the right to cancel this insurance by notifying the Assured via the broker in writing.

47.4 The Underwriters shall give not less than 15 days' prior notice of cancellation in writing to the Assured via the broker. If the premium or instalment premium due is paid in full to the Underwriters before the notice period expires, such notice of cancellation shall automatically be revoked. If not, this insurance shall automatically terminate at the end of the notice period.

47.5 In the event of cancellation of this insurance by the Underwriters under this Clause 47, premium shall be due to the Underwriters on a pro rata basis for the period that the Underwriters are on risk but the full premium shall be payable to the Underwriters in the event of loss, damage, liability or expense arising out of an Occurrence commencing or occurring prior to the date of termination which gives rise to a recoverable claim under this insurance.

47.6 Where the premium is to be paid to the Underwriters through a Market Bureau, payment to the Underwriters shall be deemed to occur on the day of delivery of a premium advice note to such Bureau.

48. Due diligence

48.1 It is the duty of the Assured to exercise due diligence in the conduct of all operations relating to the Subject Matter Insured and in the utilisation of all safety practices and equipment considered prudent for such operations, such duty including but not limited to:

48.1.1 the exercise of due diligence in the selection and employment of Contractors, Sub-contractors and others in connection with the Subject Matter Insured; and

48.1.2 the exercise of due diligence in relation to such precautions as may be reasonably required to prevent loss, damage, liability or expense in connection with the Subject Matter Insured.

48.2 In the event of any breach of the duty under Clause 48.1, the Underwriters shall in no case be liable for any loss, damage, liability or expense attributable to such breach.

49. Navigation

49.1 The Vessel shall not at any time:

49.1.1 be towed outside the port or place of construction; or

49.1.2 proceed under its own power to a distance by water of more than 250 nautical miles from the Yard, or such other distance agreed in the Schedule,

unless the Underwriters' prior written agreement shall have been obtained and any amended terms of cover and any additional premium required by the Underwriters have been agreed.

49.2 In the event of a breach of Clause 49.1, the Underwriters shall in no case be liable for any loss, damage, liability or expense arising out of any Occurrence commencing or occurring during the period of such breach.

50. Items

50.1 Items shall not at any time leave the Yard, save when on the Vessel, unless the Underwriters' prior written agreement shall have been obtained and any amended terms of cover and any additional premium required by the Underwriters have been agreed.

50.2 In the event of a breach of Clause 50.1, the Underwriters shall in no case be liable for any loss, damage, liability or expense in respect of Items whilst outside the Yard.

50.3 Any material may be included as Items prior to its arrival at the Yard or prior to its allocation to the Vessel, but only where the Underwriters' prior written agreement shall have been obtained and any amended terms of cover and any additional premium required by the Underwriters have been agreed.

51. Other insurances

51.1 The insurance under Section A, save that provided under Clause 33 (Failure of launch), shall be primary to any other insurances in respect of loss of or damage to the Subject Matter Insured and the Underwriters shall not seek any contribution from the insurers of such other insurances. For the avoidance of doubt, Clause 38.2 (Subrogation) shall not be affected by this Clause 51.

51.2 In no case shall the insurance under Clause 33 (Failure of launch), Section B, Section C or Section D cover loss, damage, liability or expense recoverable under any other insurance or which would have been so recoverable but for the existence of the insurance under Clause 33 (Failure of launch), Section B, Section C or Section D.

52. Automatic termination

Unless the Underwriters agree to the contrary in writing, this insurance shall terminate automatically:

52.1 at the time of any transfer by the Assured of ownership of the Subject Matter Insured, save in respect of a transfer of ownership from the Builder to the Buyer in so far as such change occurs prior to delivery from the Builder to the Buyer; or

52.2 at the time of an Abandonment; or

52.3 when the Underwriters have admitted a claim for total loss of the Vessel.

53. Assignment

No assignment of or interest in this insurance or in any moneys which may be or become payable under this insurance is to be binding on or recognised by the Underwriters unless a dated notice of such assignment or interest signed by the Assured, and by the assignor in the case of subsequent assignment, is endorsed on the policy or other document evidencing the insurance and such policy or other document with such endorsement is produced before payment of any claim or return of premium under this insurance.

54. Several liability

54.1 The Underwriters' obligations are several and not joint and are limited solely to the extent of their individual subscriptions.

54.2 The Underwriters are not responsible for the subscription of any co-subscribing Underwriter who for any reason does not satisfy all or part of its obligations.

55. Contracts (Rights of Third Parties) Act 1999

55.1 No benefit of this insurance is intended to be conferred on or enforceable by any party other than the Assured, the Additional Assureds and loss payee (if any), save as may be expressly provided herein to the contrary.

55.2 This insurance may by agreement between the Assured and the Underwriters be rescinded or varied without the agreement of the loss payee (if any) or any other third party to whom the right of enforcement of any terms has been expressly provided.

56. Definitions

The following words shall have the meanings assigned to them below:

56.1 "Abandonment" means an agreement, with the prior written agreement of the Underwriters (such agreement not to be unreasonably withheld), between the Builder and the Buyer that the Builder is relieved of the obligation to build or deliver the Vessel to the Buyer, by reason of physical loss of or physical damage to the Vessel.

56.2 "Additional Assured" means:

56.2.1 a Contractor to whom benefit of this insurance has been provided for in their written contract with the Assured;

56.2.2 a Sub-contractor to whom benefit of this insurance has been provided for in their written contract with a Contractor.

For the avoidance of doubt, "Additional Assured" shall not include:

56.2.3 a contractor with whom the Assured has not entered or does not during the Period of the Insurance and prior to loss enter into a written contract in relation to the construction of the Vessel;

56.2.4 a sub-contractor who has not entered or does not during the Period of the Insurance and prior to loss enter into a written contract in relation to the construction of the Vessel with a Contractor or where such written contract has not been approved by the Assured.

56.3 "Allocated Series Costs" means the allocation, agreed in the Schedule, by the Assured to the Vessel of:

56.3.1 Contractor Cost as defined in Clause 56.8, but deleting the word "solely";

56.3.2 Labour Cost as defined in Clause 56.12, but deleting the word "solely";

56.3.3 Material Cost as defined in Clause 56.14, but deleting the word "solely",

where such costs relate to the construction of a series of vessels, of which the Vessel is one, but in no case shall the total of the Allocated Series Costs exceed the sum agreed, if any, in the Schedule.

56.4 "Assured" means the Builder or the Buyer named as the Assured in the Schedule. The Assured shall consist of both the Builder and the Buyer, if both are so named in the Schedule.

56.5 "Builder" means the builder (or if Clause 43 applies, the prime contractor) named in the Schedule.

56.6 "Buyer" means the buyer (or in the event Clause 43 applies, the owners and managers of the Pre-existing Vessel) named in the Schedule.

56.7 "Contractor" means those with whom the Assured have entered or during the Period of the Insurance and prior to loss enter into written contracts in relation to the construction of the Vessel, but shall in no case include the Builder or the Buyer.

56.8 "Contractor Cost" means the cost (to include the Assured's overhead) to the Assured of contractors (whether or not the Assured has a written contract with such contractors) solely in relation to the construction of the Vessel and shall not include the Assured's profit on such cost, nor Labour Cost nor Material Cost. Where the Assured consists of the Builder and the Buyer, the "Contractor Cost" shall be the cost to the party which has retained the contractor.

56.9 "Contract Deduction" means the amount by which the sum payable by the Buyer to the Builder is reduced, with the prior written agreement of the Underwriters (such agreement not to be unreasonably withheld), by reason of physical damage to the Vessel not repaired by the Builder at the time this insurance terminates, but shall not include any amount which the Builder shall bear on account of delay which would be caused were such damage to be repaired or on account of any failure to meet performance criteria.

56.10 "Expediting Expenses" means the expenses incurred to expedite the repair of the Vessel following physical loss or physical damage covered under this insurance, including but not limited to the cost of overtime, night work, work on public holidays and express freight (to include non-scheduled airfreight).

56.11 "Items" means any material within the Yard which has been allocated to the Vessel. For the avoidance of doubt, "Items" shall not include any material:

56.11.1 prior to its arrival at the Yard;

56.11.2 prior to its allocation to the Vessel;

save as agreed by the Underwriters in accordance with Clause 50.3 (Items).

31

56.12 "Labour Cost" means the sums payable by the Assured to labourers directly employed by the Assured solely in relation to the construction of the Vessel plus the Assured's overhead on such sums and shall not include the Assured's profit on such sums. Where the Assured consists of the Builder and the Buyer, the "Labour Cost" shall be the sums payable by the party which has employed the labourers.

56.13 "Launch Date" means the first date on which the Assured intends the Vessel's hull to float on water and the Vessel's hull does so.

56.14 "Material Cost" means the cost (to include the Assured's overhead) to the Assured of materials directly supplied by the Assured solely in relation to the construction of the Vessel and shall not include the Assured's profit on such cost. Where the Assured consists of the Builder and the Buyer, the "Material Cost" shall be the cost to the party which has supplied the material.

56.15 "Occurrence" means one loss, accident, disaster or casualty or series of losses, accidents, disasters or casualties arising out of one event. Without prejudice to the generality of the foregoing:

56.15.1 windstorms, tornadoes, cyclones, hurricanes, similar storms and systems of winds of a violent and destructive nature and any tidal surge arising therefrom within any period of 72 consecutive hours and arising out of the same atmospheric disturbance shall constitute one event;

56.15.2 each earthquake, shock, volcanic eruption or tsunami or tidal wave arising therefrom, shall constitute one event. However, if more than one earthquake, shock, volcanic eruption, tsunami or tidal wave shall occur within any period of 72 consecutive hours, such earthquakes, shocks, volcanic eruptions, tsunamis or tidal waves shall constitute one event.

For the purposes of Clauses 56.15.1 and 56.15.2, where the event lasts for more than 72 consecutive hours, the Assured shall be entitled to select the time and date at which the event commences for the purposes of this insurance, provided always that no one 72 hour period shall overlap with another.

Notwithstanding the provisions of this Clause 56.15, damage caused by heavy weather or contact with ice during a single sea passage between two successive ports shall be treated as having been caused by one event.

56.16 "Period of the Insurance" under Section A, Section B and Section D means the period commencing on:

56.16.1 the commencement of the construction of the Vessel at the Yard; or

56.16.2 such earlier date, if any, agreed in the Schedule.

Period of the Insurance under Section C means the period commencing as described in Clause 9.1 (Attachment); and

ceasing, under Section A, Section B, Section C and Section D, on

56.16.3 delivery of the Vessel to the Buyer; or

56.16.4 the date agreed in the Schedule; or

56.16.5 such date at which this insurance shall be cancelled in accordance with Clause 47 (Premium payment); or

56.16.6 such date at which this insurance shall be terminated automatically in accordance with Clause 52 (Automatic termination),

whichever shall first occur.

56.16.7 Termination or cancellation (whether or not such termination or cancellation be automatic) of the insurance under Section C or Section D shall not terminate or cancel the insurance under Section A or Section B.

56.17 "Profit Percentage" means 10% (or such other percentage (if any) agreed in the Schedule) of the Allocated Series Cost, Contractor Cost, Labour Cost and Material Cost in respect of the Assured's profit.

56.18 "Sub-contractor" means those with whom a Contractor has entered or during the Period of the Insurance and prior to loss enters into a written contract in relation to the construction of the Vessel, provided always the written contract has been approved by the Assured, but shall not include the Builder or the Buyer.

56.19 "Subject Matter Insured" means the Items and the Vessel.

56.20 "Sum Insured" means the sum stated in the Schedule as the sum insured in respect of the Subject Matter Insured. Should the Works Value plus the Profit Percentage exceed the Sum Insured, the Sum Insured shall, subject to Clause 44.2 (Sum insured), be increased in the manner set out in Clause 44.1 (Sum insured).

56.21 "Vessel" means the project bearing the hull number or project name identified in the Schedule.

56.22 "Works Value" means the total of:

56.22.1 the Allocated Series Cost spent by the Assured; and

56.22.2 the Contractor Cost spent by the Assured; and

56.22.3 the Labour Cost spent by the Assured; and

56.22.4 the Material Cost spent by the Assured,

but shall in no case exceed the Sum Insured.

56.23 "Yard" means the premises of the Builder named in the Schedule where the Vessel is being constructed or, if Clause 43 applies, where the Pre-existing Vessel is to undergo conversion, repair, lengthening or any other work.

PART V – OPTIONAL BUYBACKS

57. Defects

If the Underwriters have expressly agreed in writing:

57.1 the insurance under Section A shall also cover the cost that would have been incurred to replace, repair or rectify the defect in workmanship, defect in material or latent defect, had such defect been discovered immediately prior to the occurrence of the physical loss or physical damage covered under Clause 2 (Perils), but in no case shall the insurance under Section A cover any costs incurred to replace, repair or rectify a defect in design, plan or specification.

57.2 the insurance under Section A shall also cover 50% of the costs common to both:

57.2.1 the repair of physical loss and physical damage caused by a defect in design, plan or specification, where such physical loss or physical damage is covered under Clause 2 (Perils); and

57.2.2 the replacement, repair and rectification of the said defect,

but in no case shall the insurance under Section A cover any costs incurred solely in the replacement, repair and rectification of the defect in design, plan or specification or incurred by reason of betterment or alteration in design, plan or specification.

58. Profit on partial loss

58.1 If the Underwriters have expressly agreed in writing, Clause 27.2.1 (Partial loss) shall be deleted and the Profit Percentage shall be added to the Labour Cost, the Material Cost and the Contractor Cost in calculating the reasonable cost of the repairs for a partial loss.

58.2 However, notwithstanding the application of this Clause 58, Clause 27.2.1 (Partial loss) shall still apply for the purposes of Clause 28.3 (Constructive total loss) and Clause 31.1.1 (Abandonment).

58.3 For the avoidance of doubt, the reasonable cost of the repairs shall not include the Assured's profit on such cost, save as may be included within the Profit Percentage.

59. Expediting Expenses

If the Underwriters have expressly agreed in writing, Clauses 27.2.2 (Partial loss), 28.4.2.1 (Constructive total loss) and 31.3.2.1 (Abandonment) shall be deleted and Expediting Expenses shall be covered under this insurance, subject to a limit of 10%, or such other percentage agreed in the Schedule, of the Sum Insured in respect of any one Occurrence.

60. Trials amendment

If the Underwriters have expressly agreed in writing, the words "to the extent that such expenses are necessary to effect the repairs, but in no case shall the Underwriters be liable for the expenses incurred in re-running trials undertaken solely by reason of contractual provisions between the Builder and the Buyer." shall be deleted from Clause 32 (Trials).

CL 371

34

APPENDIX 8

INSTITUTE MORTGAGEES' INTEREST CLAUSES 1997

APPENDIX 8

INSTITUTE MORTGAGEES' INTEREST CLAUSES

HULLS

This insurance is subject to English law and practice

A. RECITAL

Whereas the Assured has entered into a loan agreement commensurate with which the Assured holds certain collateral security including a first mortgage on the Mortgaged Vessel and endorsements of its interests on the Owners' Policies and Club Entries.

Now it is agreed as follows:

1. INSURING CLAUSE

1.1 This insurance will indemnify the Assured for loss resulting from loss of or damage to or liability of the Mortgaged Vessel which, in the absence of an insured peril set out in Clause 2.1 below, would prima facie be covered by the Owners' Policies and Club Entries, and not excluded therein, but in respect of which there is subsequent non-payment (or reduced payment which is approved in advance by the Underwriters hereon) by any of the Underwriters of Owners' Policies and Club Entries as a result of any insured peril, **provided always that such insured peril occurs or exists without the privity of the Assured.**

1.2 The indemnity payable hereunder shall be:

1.2.1 the amount of the Assured's net loss and any amounts recoverable under Clause 6 herein, collectively not exceeding the sum insured on the Mortgaged Vessel, or

1.2.2 the amount of the unrecoverable claim or part thereof under any of the Owners' Policies and Club Entries

whichever is the lesser amount.

1.3 All the above is subject to the Definitions, Exclusions, Warranties and Conditions below.

2. DEFINITIONS

2.1 Insured Perils

2.1.1 Avoidance of the Owners' Policies and Club Entries or any of them by the Underwriters thereof on the grounds of a misrepresentation or non-disclosure of any material circumstance, whether such misrepresentation or non-disclosure arises from the Assured thereunder or any of such Assured's insurance agents or insurance brokers,

2.1.2 breach of any statutory provision or any express or implied promissory warranty or condition, including without limitation

2.1.2.1 breach of any implied warranty of seaworthiness or legality,

2.1.2.2 breach of Section 39(5) of the Marine Insurance Act of 1906,

2.1.2.3 breach of trading warranties contained in any of the Owners' Policies and Club Entries,

2.1.2.4 breach of any warranty or condition in any of the Owners' Policies and Club Entries in respect of the classification of the Mortgaged Vessel by a Classification Society or any failure to comply with the recommendations of such Society to the extent required by such warranty or condition,

2.1.2.5 breach of any warranty or condition in any of the Owners' Policies and Club Entries which requires compliance with any condition survey, structural survey or P&I Club survey requirements and pursuant to which clauses Underwriters deny a claim,

2.1.2.6 breach of any warranty or condition in any of the Owners' Policies and Club Entries in respect of the ownership, flag, management or charter on a bareboat basis of the Mortgaged Vessel,

2.1.3 failure of the Assured under the hull and machinery policy or the owners, managers or superintendents of the vessel or any of their onshore management to exercise due diligence in respect of any loss or damage to the Mortgaged Vessel where such failure to exercise due diligence entitles the Underwriters of the owner's hull and machinery policy to deny a claim otherwise recoverable thereunder,

2.1.4 any deliberate or fraudulent casting away of or damage to the Mortgaged Vessel,

2.1.5 avoidance of a claim under any of the Owners' Policies and Club Entries or of any of the Owners' Policies and Club Entries by reason of breach of the duty of good faith in respect of such claim, or

2.1.6 the operation of any applicable provision in any of the Owners' Policies and Club Entries which provides for a time limitation on the presentation of claims,

2.1.7 in the event of the total loss of the Mortgaged Vessel, the final judgement or award of the courts or arbitration tribunal agreed to have jurisdiction under the express terms of Owners' Policies and Club Entries (or in the event of their being no such express terms a competent court) following a contested hearing whereby the Owners' claim is not recoverable under either Owners' hull and machinery or war risks policies on the grounds that the loss has not been proved to have been proximately caused by a peril insured under those policies and is not otherwise excluded from payment by any exclusion or other provision therein.

2.2 **Owners' Policies and Club Entries** - means hull and machinery policies on terms equivalent to or wider than the current Institute Time Clauses Hulls or American Institute Hull Clauses, (if taken, increased value policies on terms equivalent to Institute Time Clauses - Hull Disbursements and Increased Value (Total Loss only and Excess Liabilities) or American Institute Increased Value and Excess Liabilities Clauses), war risks on terms equivalent to current Institute War and Strikes Clauses Hulls `- Time and full protection and indemnity risks on conditions equivalent to the rules of a P&I Club that is a member of the International Group of P&I Associations.

2.3 **Net Loss** - means the Assured's loss under the loan agreement to the extent secured by mortgage on the Mortgaged Vessel net of any amounts recovered or recoverable under all security arrangements contained in or collateral to the loan including but not limited to all mortgages (whether on vessels insured hereunder or on other vessels), liens, any floating and fixed charges, security interests, guarantees, insurance policies and pledges.

2.4 **Mortgaged Vessel** - means the vessel mortgaged to the Assured which is listed on the schedule attached hereto.

3. EXCLUSIONS

In no case shall this policy cover:

3.1 any loss or expense arising from or as a result of

3.1.1 the relevant Owners' Policies and Club Entries having been terminated or cancelled or cover suspended or non-payment of claims by the Underwriters or insurance brokers thereof due to non-payment of premium or call,

3.1.2 insolvency or financial default of any of the Underwriters of the Owners' Policies and Club Entries,

3.1.3 inability of any party to transmit funds,

3.1.4 any fluctuation in exchange rates,

3.1.5 the operation of any franchise deductible or provision for self-insurance.

3.2 loss or damage directly or indirectly caused by or contributed to by or arising from:

3.2.1 ionising radiation from or contamination by radioactivity from any nuclear fuel or from any nuclear waste or from combustion of nuclear fuel,

3.2.2 the radioactive, toxic, explosive or other hazardous or contaminating properties of any nuclear installation, reactor or other nuclear assembly or nuclear component thereof,

3.2.3 any weapon of war employing atomic or nuclear fission and/or fusion or other like reaction or radioactive force or matter.

4. WARRANTIES

It is warranted in respect of the Mortgaged Vessel that:

4.1 Owners' Policies and Club Entries have been taken out and, except as a result of the occurrence or existence of an insured peril without the privity of the Assured, shall be maintained throughout the currency of this insurance for an insured value and limit of liability not less than the amount insured hereunder or the amount of the outstanding loan to the extent secured by the Mortgaged Vessel,

4.2 each of the Owners' Policies and Club Entries is endorsed to the extent of the Assured's interest, and

4.3 the Assured has procured and registered a valid first mortgage.

5. CHANGE OF OWNERSHIP OR CONTROL

This insurance will terminate automatically at the time the Assured becomes or is aware of or privy to any change of:

5.1 Classification Society or of change, suspension, discontinuance, withdrawal or expiry of the Mortgaged Vessel's class within that Society,

5.2 ownership, flag, management or control of the Mortgaged Vessel,

unless the Assured gives prompt notice of such change in writing, and agrees to pay an additional premium, if required, and Owners' Policies and Club Entries are maintained.

6. DUTY OF ASSURED (SUE AND LABOUR)

6.1 The Assured shall report in writing to the Underwriters any circumstances which may give rise to a claim under this insurance within 30 days of the Assured's knowledge of such circumstances and shall thereafter keep the Underwriters fully informed of all developments.

6.2 It is the duty of the Assured and their servants and agents to take such measures as may be reasonable for the purpose of averting or minimising a loss which would be recoverable under this insurance.

6.3 The Underwriters will reimburse charges properly and reasonably incurred by the Assured their servants or agents for such measures except for legal costs and expenses incurred by the Assured in relation to any claim under Owners' Policies and Club Entries which shall only be reimbursed in accordance with clause 6.4 herein.

6.4 Subject to the condition precedent that the Mortgaged Vessel is entered in a Freight, Demurrage and Defence Club covering the cost of the owner proceeding against the Owners' Policies and Club Entries, the Underwriters will reimburse those legal costs and expenses incurred by the Assured in pursuing the non-paying Owners' Policies and Club Entries not otherwise recoverable as part of the net loss but only where the Assured can demonstrate to the satisfaction of the Underwriters that it has made every reasonable effort to compel the owner to pursue the non-paying Owners' Policies and Club Entries. This policy shall not pay for legal costs and expenses incurred by the Assured in monitoring the claim against the Owners' Policies and Club Entries.

6.5 Any amounts payable under this clause shall be included within and shall not be additional to the sum insured.

6.6 Measures taken by the Assured or the Underwriters with the object of averting or minimising a loss which would be recoverable under this insurance shall not be considered as a waiver or acceptance of a claim or otherwise prejudice the rights of either party.

7. CLAIMS

7.1 The Assured shall prove a claim under this insurance by:

7.1.1 demonstrating to the satisfaction of the Underwriters, that by reason of the perils insured under clause 2.1 there is no reasonable prospect of the Owners and/or Assured succeeding in the claim against the Underwriters of the Owners' Policies and Club Entries, or

7.1.2 in the event of disagreement between the Underwriters and the Assured by either referring the issue to a sole arbitrator in London to be agreed upon between the Underwriters and the Assured or on final court judgement or arbitration award delivered in favour of the Underwriters of Owners' Policies and Club Entries.

7.2 The Underwriters shall pay any claim hereunder within 3 months of the date that both the claim is proved in accordance with Clause 7.1 and net loss is established.

8. SUBROGATION

8.1 Upon payment to the Assured of a claim hereunder, the Underwriters shall be subrogated to all the rights and remedies of the Assured in respect of such payment.

8.2 It is a condition of this insurance that any payments by the Underwriters shall not be applied by the Assured in or towards discharge or satisfaction of the amount of the outstanding indebtedness.

9. EFFECT OF UNDER INSURANCE

If the loan amounts are not fully insured hereunder at the time of loss then the indemnity payable hereunder including any sue and labour amounts shall be reduced in proportion to the under insurance.

10. AUTOMATIC TERMINATION

10.1 Cover hereunder for loss of or damage to or liability of a Mortgaged Vessel shall terminate in respect of those risks covered by the War and Strikes Clauses of the Mortgaged Vessel

10.1.1 AUTOMATICALLY upon the outbreak of war (whether there be a declaration of war or not) between any of the following

United Kingdom, United States of America, France, the Russian Federation, the People's Republic of China

10.1.2 AUTOMATICALLY in respect of a Mortgaged Vessel in the event of that Mortgaged Vessel being requisitioned either for title or use

10.1.3 7 days after the Underwriters of the Owner's War Risks Insurances or any of them have given notice of cancellation unless such War Risks Insurances have been reinstated prior to or from the expiry of such notice and any new rate of premium and amendments to terms, conditions or warranties, if any, have been agreed, or

10.1.4 7 days after the Underwriters hereon have given notice of cancellation in respect of such risks.

10.2 Cancellation in accordance with Clauses 10.1.3 and 10.1.4 shall become effective on the expiry of 7 days from midnight on the day on which the notice of cancellation is given. The Underwriters agree however to reinstate this insurance subject to agreement between the Underwriters and the Assured prior to the expiry of such notice of cancellation as to new rate of premium and/or conditions and/or warranties.

1/3/97

INDEX